Lecture Notes in Computer Science 11354

Commenced Publication in 1973
Founding and Former Series Editors:
Gerhard Goos, Juris Hartmanis, and Jan van Leeuwen

More information about this series at http://www.springer.com/series/7409

Yong Tang · Qiaohong Zu ·
José G. Rodríguez García (Eds.)

Human Centered Computing

4th International Conference, HCC 2018
Mérida, Mexico, December, 5–7, 2018
Revised Selected Papers

 Springer

Editors
Yong Tang
South China Normal University
Guangzhou, China

Qiaohong Zu
Wuhan University of Technology
Wuhan City, China

José G. Rodríguez García
CINVESTAV
Mexico City, Mexico

ISSN 0302-9743   ISSN 1611-3349 (electronic)
Lecture Notes in Computer Science
ISBN 978-3-030-15126-3   ISBN 978-3-030-15127-0 (eBook)
https://doi.org/10.1007/978-3-030-15127-0

Library of Congress Control Number: 2019933338

LNCS Sublibrary: SL3 – Information Systems and Applications, incl. Internet/Web, and HCI

This Springer imprint is published by the registered company Springer Nature Switzerland AG
The registered company address is: Gewerbestrasse 11, 6330 Cham, Switzerland

Preface

Human-centered computing (*HCC*), affected by the rapid advance of emergent and disruptive technologies, is a term with an ever-changing definition in recent decades. The core concept, however, remains: "understanding human beings and serving human needs." Putting humans at the center, HCC advocates holistic solutions that blur the previously clearly defined boundaries among humans, computing devices, and environments. In particular, along with new developments in artificial intelligence, deep learning, brain–computer interfaces, and human microchip implants, come not only new opportunities but also threats with deeper penetration into every aspect of our everyday work and leisure life.

In light of this, it is our great privilege to present this collection of articles from the 4th International Conference on Human Centered Computing (HCC 2018) which was held during December 5–7, 2018, in Mérida, Yucatán, Mexico. HCC Mérida boasted a two-day program focusing on a "hyper-connected world" that redefines human behaviors as well as human societies. The primary topics of HCC 2018 consisted of "hyper-connectivity," "Data - Making Sense of Data," and "Collaboration."

A unique initiative of the HCC series is the way the series bring the latest research and technological outcomes to the communities where they are most needed – bridging the digital division due to infrastructure, education, and exposure. Working closely with local scholars, HCC enjoyed enthusiastic support that exceeded our expectations from the booming innovative hub of Yucatán Peninsula. We would like to express our gratitude to the conference organization team for their hard work.

The responses to the HCC 2018 call for papers was very positive with submissions from both local and overseas research communities. All submissions went through a very strict reviewing process (at least two peer-reviews and one meta-review by senior Program Committee members) and around 34% were accepted as full papers for oral presentation, with an additional 18 short papers and posters. Unfortunately, many high-quality papers could not be accepted owing to capacity restrictions. The quality of HCC 2018 was ensured by the arduous efforts of the Program Committee members and invited reviewers, to whom we owe our highest gratitude and appreciation.

Finally, we hope the participants enjoyed Mérida, long a destination for people around the world seeking the rich Mayan history and recently for talents looking for innovation. We hope that you benefited from the presentations at the conference and had the chance to foster international collaborations as well as marvel at the natural and man-made wonders near-by.

December 2018

Yong Tang
Qiaohong Zu
José G. Rodríguez García

Organization

Honorary Conference Chair

Mirna Alejandra Manzanilla   Instituto Tecnológico de Mérida, Mexico
   Romero

Conference Chair

Yong Tang   South China Normal University, China

Program Committee Co-chairs

José G. Rodríguez García   CINVESTAV, Mexico
Bin Hu   Lanzhou University, China

Organizing Committee

Mario R. Moreno Sabido   Instituto Tecnológico de Mérida, Mexico
Qiaohong Zu   Wuhan University of Technology, China
Chengzhou Fu   South China Normal University, China

Publication Committee

Bo Hu   Fujitsu Labs of Europe, UK
Philip Moore   Birmingham City University, UK

Secretariat

Jizheng Wan   Coventry University, UK

Program Committee

James Anderson	Luis Carriço	Marco De Sá
Jose Albornoz	Jingjing Cao	Matjaž Debevc
Natasha Aleccina	Qinghua Cao	Luhong Diao
Angeliki Antonio	Guohua Chen	Monica Divitini
Juan Carlos Augusto	Tianzhou Chen	David Dupplow
Roberto Barcino	Yiqiang Chen	Haihong E.
Paolo Bellavista	Lizhen Cui	Xiaozheng E.
Adams Belloum	Aba-Sah Dadzie	Talbi El-Ghazali

James Enright
Henrik Eriksson
Chengzhou Fu
Yan Fu
Shu Gao
José G. Rodríguez García
Mauro Gaspari
Bin Gong
Horacio González-Vélez
Chaozhen Guo
José María Gutiérrez
Chaobo He
Fazhi He
Hong He
Andreas Holzinger
Bin Hu
Cheng Hu
Changqin Huang
Zongpu Jia
Mei Jie
Hai Jin
Lucy Knight
Hiromichi Kobayashi
Roman Laborde
Hanjiang Lai
Thomas Lancaster
Victor Landassuri-Moreno
Liantao Lan
Bo Lang
Agelos Lazaris
Chunying Li

Hua Li
Jianguo Li
Shaozi Li
Wenfeng Li
Xiaowei Li
Zongmin Li
Xiaofei Liao
Hong Liu
Lianru Liu
Lizhen Liu
Yongjin Liu
Alejandro Llaves
Yanling Lu
Hui Ma
Haoyu Ma
Yasir Gilani
Mohamed Menaa
Marek Meyer
Maurice Mulvenna
Mario Muñoz
Aisha Naseer
Tobias Nelkner
Sabri Pllana
Xiaoqi Qin
Klaus Rechert
Uwe Riss
Andreas Schraber
Stefan Schulte
Susan Scott
Beat Signer
Matthew Simpson

Mei Song
Xianfang Sun
Yuqing Sun
Wenan Tan
Menglun Tao
Shaohua Teng
Yinglei Teng
Boris Villazon Terrazas
Coral Walker
Maria Vargas-Vera
Jizheng Wan
Qianping Wang
Yun Wang
Yifei Wei
Ting Wu
Xiaoxue Wu
Zhengyang Wu
Toshihiro Yamauchi
Bo Yang
Yanfang Yang
Linda Yang
Zhimin Yang
Xianchuan Yu
Guanghui Yue
Yong Zhang
Gansen Zhao
Shikun Zhou
Shuhua Zhu
Tingshao Zhu
Gang Zou

Contents

The Design and Development of Central China Power Grid Intelligent Ticket Forming System

Na Hu[1(✉)], Renbin Su[2], Sanbao Chen[1], and Liangxiong Xu[1]

[1] City College, Wuhan University of Science and Technology,
Wuhan 430083, Hubei, China
314407927@qq.com
[2] Central China Electric Power Dispatching and Control Branch of State Grid
Company of China, Wuhan 430077, Hubei, China

Abstract. Based on debugging operation experience of central China power grid, the operation instructions of the debug operation ticket are divided into modules by function, the mechanism of each module ticket is studied one by one. According to the specification of state order in central China power network, the data structure describing the state of various equipment components in power grid is defined, and the state information of equipment is recorded and transferred. According to the characteristics of the debugging charged path, the chain inference rule and error prevention rule based on the network topology are compiled. This paper designs and develops the intelligent ticket-making system of debugging operation in central China power network through the modularization, data-based and practical processing of the ticket-making system. The efficiency and security of debugging operation in central China power grid are improved.

Keywords: Debug ticket · Status order terminology · Status form

1 Introduction

With the rapid development of the central China power grid scale, the new plant equipment is put into operation, the transformation of the old equipment replacement should adjust power supply task is increasing. At the same time, the debugging operation involves many primary equipment, complex secondary protection configuration and complicated switching process. At present, the debugging operation order still needs to be written and checked manually by dispatchers. The work intensity is large, the efficiency of the ticket is low, and there are also security risks, which is shown in articles [1–3]. Therefore, it is urgent to develop an automatic application system with the function of debugging ticket intelligent ticket forming and error checking.

This paper defines a standard data structure that can accurately describe the state of various types of power grid equipment—state form, so that the computer can accurately identify the state of the debugging equipment, and also provide data for the computer reasoning. Meanwhile, the state form also acts as a flow form, which circulates during the generation process of the debugging operation ticket, to record and pass the state

© Springer Nature Switzerland AG 2019
Y. Tang et al. (Eds.): HCC 2018, LNCS 11354, pp. 1–10, 2019.
https://doi.org/10.1007/978-3-030-15127-0_1

information of the debugging equipment. According to the operating habits of switching operation, this paper develops intelligent reasoning rules and error prevention rules based on the topology of the grid. It can intelligently reason the operation sequence of the switch and the isolator during the state transition according to the initial state and the target state, and complete the process of error prevention check. After the modularization, digitization, and process processing of the debugging operation process, combined with the reasoning rules, this paper designs and develops an intelligent debugging operation ticket system for Central China Power Grid. This is different from article [4], which emphasizes the operability of central China.

2 Division of Debugging Operation Ticket Module

According to the experience of debugging operation of Central China Power Grid, the following five steps are generally carried out for both new and replaced equipment: first, commissioning equipment initial status check; second, temporary adjustment of debugging equipment protection; third, test trip operation; fourth, debugging equipment restored to normal protection; fifth, commissioning equipment to resume normal operation mode. Corresponding to different operation steps, the debugging operation ticket can be modularly divided, and the generating mechanism of the operation instructions in each division is analyzed one by one in sequence, which is shown in Fig. 1.

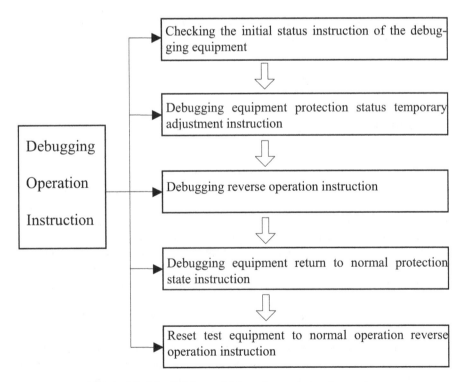

Fig. 1. Module divisions of debugging operation instruction

2.1 Instruction for Checking Initial State

The function of the instruction is to confirm the scope and initial state of all the debugging equipment involved. The scope of debugging tasks and equipment is clear. The initial state of the debugging equipment, according to the operation habit of central China power grid debugging, generally defaults to the cold standby state, and the initial state of the debugging equipment can be directly reasoned.

2.2 Instruction for Temporary Protection State Adjustment

The function of the instruction is to make temporary adjustment to the protection of the debugging equipment to meet the requirement of the protection adaptability under the special connection mode. Generally adjust the charging protection value of the switch on the charged path and exit the reclosing gate. If the debugging path is determined, the switch that needs adjustment protection can be determined. The reasoning process for determining the debug belt path will be explained in detail in the following Sect. 3.

2.3 Instruction for Debugging Reverse Operation

The function of the instruction is to switch the debugging equipment from the initial state to the target state by opening/closing the isolator or switch. Similarly, if the debugging path is determined, according to the principle of side-load charging from power source, first operation knife gate, later operation switch, first operation second, then operation first, etc., can determine the operation sequence and instruction.

2.4 Instruction for Debugging Equipment Return to Normal Protection State

The function of the instruction is to restore the temporary protection of the test equipment to normal state after commissioning. The protection instruction temporarily adjusted in Sect. 2.2 may be restored.

2.5 Instruction for Reversing Operation of Debugging Equipment Return to Normal Operation Mode

The function of the instruction is to convert the debugging equipment from the debugging state to the normal operating state through the operation of the split, closing the switch or switch. In the case that the recovery transmission path is determined, the reverse operation instruction of the operation return to normal operation mode can be determined by using the same reasoning principle as the instruction in Sect. 2.3.

The above analysis of the modularization decomposition of the debug operation ticket and the generating mechanism of each module instruction, modularize and digitize the complex process of operation ticket. The feasibility of computer intelligent generating debugging operation ticket is improved.

3 Establishment of Test Equipment Status Form Based on the State Order Term Specification

In this paper, the data structure describing the state of various power grid equipments is established according to the terms and specifications of the central China power grid dispatching order, hereinafter referred to as the specification of terms, so as to facilitate the computer recognition's equipment state and reasoning operation. The term specification clearly defines the primary state and protective state of power network equipment such as switch, line, bus, transformer in condition of maintenance, cold standby, hot standby, operation, etc.

Take switching equipment as an example below to illustrate the process of building data structures. For switchgear, the term specification defines its cold standby, hot standby and operating state as follows:

Switch cold standby: switch and both sides of the switch are open position; the circuit breaker protection of the switch is in exit state; the reclosing of the switch is in exit state.

Switch hot standby: switch in open position; both sides of the switch are in the closed position; the circuit breaker of the switch is protected in the input state; the reclosing of the switch is in the input state.

Switch operation: switch in open position; both sides of the switch are in the closed position; the circuit breaker of the switch is protected in the input state; the reclosing of the switch is in the input state.

As can be seen, the status of the switch is determined by the following four factors: first, the switch itself is in a split state; second, the split state of the two sides of the switch; third, switching circuit breaker protection; fourth, switch reclosing. So, the data structure of the switch state is set up as shown in following Table 1 (take switch number 5011, two side switch number 50111, 50112 for example): 0 indicates disconnection or exit, 1 indicates a close or input.

Table 1. The data structure of the switch state.

5011 Switching status data	5011 Switch	0/1
	50111 Knife brake	0/1
	50112 Knife brake	0/1
	Circuit breaker protection	0/1
	Reclosing gate	0/1

According to the above data structure, the computer can recognize a specific state of switch, which is crucial to ticket automated generation. For example, the initial state checking instruction in Sect. 2.1 can be generated by the recognition of the state bits in the data structure. Table 2 gives the 5011 switch cold standby status, and Table 3 shows the generation of initial state instruction for debugging equipment according to the 5011 switch cold standby state form.

Table 2. 5011 switch cold standby status form.

5011 switch cold standby status form	5011 Switch	0
	50111 Knife brake	0
	50112 Knife brake	0
	Circuit breaker protection	0
	Reclosing gate	0

Table 3. Generation of initial state instruction for debugging equipment.

Operation task: Check initial status of switch 5011	
Serial number	Operation command
1	Check 5011 switch in open position
2	Check 50111 knife brake in open position
3	Check 50112 knife brake in open position
4	Check 5011 switch circuit breaker protection out
5	Check 5011 turn-off reclosing gate quit

At the same time, the computer can deduce the operation instruction of the equipment state change according to the change of state bit value in the data structure. As shown in following Table 4, by judging the difference between the status bits of the open-turn cold standby form and the hot standby form, the computer can deduce the operation instruction of 5011 switch from cold standby to heat standby state, which is shown in Table 5.

Table 4. 5011 switch hot standby status form.

5011 switch hot standby status form	5011 Switch	0
	50111 Knife brake	1
	50112 Knife brake	1
	Circuit breaker protection	1
	Reclosing gate	1

Table 5. Operation instruction of 5011 switch from cold standby to heat standby state.

Operation task: Switch 5011 from cold standby to hot standby.	
Serial number	Operation command
1	Input 5011 switch circuit breaker full protection
2	Input 5011 switch reclosing
3	Close 50111 switch
4	Close 50112 switch

For other power grid equipment such as line, bus, main transformer and high voltage reactor, low voltage reactance (capacitance), etc., according to the terminology, can also establish corresponding data structure. With the introduction of data structure, data can be used to accurately describe any state of any power grid equipment, which provides a necessary condition for intelligent computer to generate debugging tickets.

4   Intelligent Reasoning of Switching Operation Based on D5000 Network Topology

By changing the status form, we can determine the operation of the switch, switch and protection required by the debugging equipment to switch from the initial state to the target state. However, it is still not possible to determine the sequence of operation of these switches, switches and switches, and of the protection. Based on the topology structure of D5000 system, the inference rules of computer reasoning operation order are established in this paper.

The network topology of D5000 system is a network model generated according to the connection relationship of the electrical components and the electrical state of the grid. The connection mode between the plant and the station is determined through the connection relationship between the devices. This topological relationship provides convenience for inferring the order of switching operations. This is further explained in articles [5–7].

In cases where the operating range of the switch and switch is clear, the order of operation depends on the order in which the switch and the switch are charged. According to the rule that the debugging equipment always starts from the nearest power source (that is, the original electric equipment), this paper adopts the inference rule of chain operation, as follows:

(1) A knife or switch directly connected to a live device on topology will be preferred.
(2) A knife or switch that has determined the order of operation is a live device, while an unoperated knife or switch continues to reason the order of operation in accordance with rule (1);
(3) In accordance with the rules (1) and (2), after reasoning the operation order of the knife brake and switch, reorder the operation order according to knife brake first, switch second; avoid charging other debugging equipment with the switch.
(4) After the test equipment is electrified, the power failure can be restored, according to rule 3, closed switch operation in the opposite order, in order to open the test switch.

Charging 500 kV a AB line and a AC line to check the polarity of road protection, taking two lines as an example: the initial state and target state of the test equipment are shown in Fig. 2.

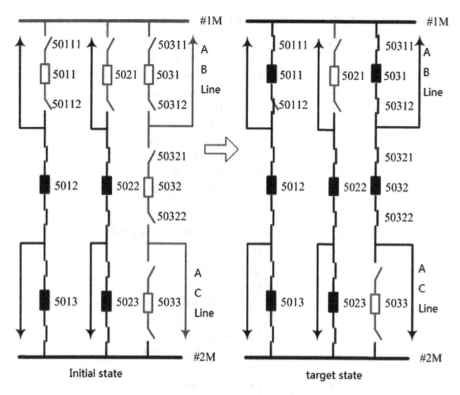

Fig. 2. The diagram of test case.

After comparing the test equipment status form of the initial state and the target state, we can judge the change from the initial state to the target state. It is necessary to close the switch 5032, 5031, 5011and the knife brake 50321, 50322, 50311, 50312, 50111, 50112 these 9 electric components. According to the topology and electric state of electrical components, after searching 9 electrical components according to rule (1), we can judge the priority operation of 50112 knife brake. The 50112-knife gate is then treated as a live device, and the search judgment of the operation order of the remaining 8 electrical components is continued. In this way, according to rules (1) and (2), we can get 9 electrical components operating order as shown in Table 6. Finally, according to the rules (3) and (4), we get the operation sequence shown on the Table 7 below.

Table 6. 9 electrical components operating order.

Serial number	Operation command
1	Close 50112 Knife brake
2	Close 5011 Switch
3	Close 50111 Knife brake
4	Close 50311 Knife brake
5	Close 5031 Switch
6	Close 50312 Knife brake
7	Close 50321 Knife brake
8	Close 5032 Switch
9	Close 50322 Knife brake

Table 7. The reasoning results of operating sequence

Serial number	Operation command
1	Close 50112 Knife brake
2	Close 50111 Knife brake
3	Close 50311 Knife brake
4	Close 550312 Knife brake
5	Close 30321 Switch
6	Close 550322 Knife brake
7	Close 5011 Switch
8	Close 5031 Switch
9	Close 5032 Switch
10	Open 5032 Switch
11	Open 3021 Switch
12	Open 5011 Switch

In the course of debugging the reasoning operation of the operation order, we should also consider the rules of error prevention: the locking function between the primary equipment operation, such as no load broach gate, no ground wire power supply, etc.; the second part of the safety check, test equipment cannot be operated without protection, test switch cannot be put into reclosing function, test switch circuit breaker protection must be put in, test switch temporary charging protection must be put in; overhaul process to prevent errors, which has further explanations in articles [8] and [9], for example, to prevent start-up and maintenance work before reaching the working conditions.

On the basis of modularization decomposition of debug operation ticket, data of test equipment by test equipment status form, and intelligent reasoning of switching operation by network topology, Fig. 3 shows the design block diagram of intelligent ticket system for debugging and operation of central China power grid.

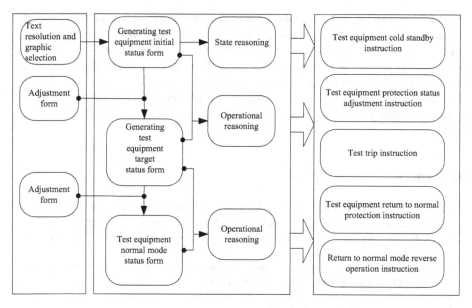

Fig. 3. Design diagram of telligent debugging operation ticket system

By analyzing the debugging text or D5000 system graphical interface selected, the debugging equipment scope input debugging operation ticket system; according to the component type of the debugging equipment, the system automatically matches and establishes the initial state form of the test equipment, which is the default cold standby. According to the information of the initial status form, the system makes state reasoning and automatically generates the initial status instruction of the test equipment. By adjusting the status of the protection, switch and knife gate in the initial status form of the debugging equipment, the drafters generate the target status table which meets the requirements of the test. The system first generates the instruction of the test equipment protection status adjustment according to the change of the protection status mark bit in the status form, and then according to the change of mark position, switch, switch, etc. According to the chain rule, the test operation instruction is generated. Similarly, after the completion of the test, the drafters adjust the status of the protection, switch and knife gate in the equipment status form according to the request of the debugging equipment to return to normal operation. Then the system generates test equipment to resume normal protection instructions and restore the normal way to reverse the operation instructions.

5   Conclusion

It is faster to write the debug operation ticket by using the debug operation ticket intelligent system. It only needs D5000 graphic interface to select or modify the status form sign bit operation, without having to enter the operation instruction manually, improved working efficiency. For improving the safety of the test operation, the process of module conversion and stage operation of the intelligent ticket system requires the participants to check the initial state of the equipment before the test, the state of the test target, the state of the normal way, and the reverse during the test. The operating path should be clearly understood, so that the staff pay more attention to the key nodes of the test and the safety of the switching operation; At the same time, in the reasoning of switching operation instruction generation, the rule of checking the safe operation against debugging operation is added, which can avoid the leakage of the artificial draft.

References

1. Zhuang, W., Chen, G., Zhang, G., et al.: A survey of dispatching operating order system based on expert system in NCC. Electric Power **38**(12), 18–20, 96–100 (2003)
2. Kong, C.: Development and application of the order-sheet expert system. Inf. Commun. **2**, 64–65 (2011)
3. Lin, Y., Wang, J., Liu, Z., et al.: Research of an expert system for generating operation sequences of substation. Relay **33**(24), 59–62 (2005)
4. Zhou, M., Lin, J., Yang, G.: New- type intelligent dispatching operation order system. Autom. Electric Power Syst. **28**(11), 71–74 (2004)
5. Li, Q., Guo, W.: Identification on connection model of intelligent dispatching operation instruction system. Jiangsu Electr. Eng. **25**(1), 54–56 (2006)
6. Dong, Y., Cheng, J., Peng, B., et al.: A method of automatic generation of diagram-database-rules-oder based on bay model. Autom. Electric Power Syst. **39**(3), 84–89 (2015)
7. Lin, X., Ren, J., Zhang, B., et al.: An intelligent dispatching operation-tickets system in electric power system based on network reconfiguration. Relay **407**(7), 143–147 (2012)
8. Zhu, X., Bai, B.: Development of dispatch anti -misoperation system. Electric Power Autom. Equip. **27**(5), 96–100 (2007)
9. Ye, L., Yin, Z.: Intelligent anti-misoperation operation ticket system consider secondary device. Proc. Chin. Soc. Univ. **23**(6), 145–149 (2011)

Research on the Classification and Channel Selection of Emotional EEG

Jie Dang, Sirui Wang, Cancheng Li, Yongzong Wang,
and Hong Peng[✉]

Gansu Provincial Key Laboratory of Wearable Computing,
School of Information Science and Engineering,
Lanzhou University, Lanzhou, China
{dangj16,wangsr16,licch17,wangyz2015,
pengh}@lzu.edu.cn

Abstract. The emotion affect every aspect of our daily lives. Because of the high temporal resolution and low cost, EEG is widely used in the fields of emotion recognition. This paper studies the effects of different EEG feature combinations and channel selections on emotion recognition. Three effective features including differential entropy (DE), 1st difference (1st) and fractal dimension (FD) were extracted from the EEG signals, and their performances in the three situations of four emotion classifications, two emotions classification on valence and two emotions classification on arousal were calculated and compared by SVM. Two channel selection methods, including the mean relief channel selection algorithm and the common channel selection algorithm, were used to select the best channel. The results showed that when selecting the top 10 channels, the accuracy of the four emotional states classification rate was approximately 95%. This is significant for reducing the number of electrodes and reducing the complexity of brain-computer interface applications in the future.

Keywords: EEG · Emotion recognition · Feature combination ·
Channel selection

1 Introduction

It is well established that emotion affects all aspects of people's lives. Emotion recognition [1–3] has become an important issue for researchers in related fields. Because different emotional states lead to different EEG patterns, emotion recognition can be performed based on EEG [4].

In previous studies on the brain-computer interface, many studies have concentrated on EEG signal classification [5]. Due to the practical need for a portable system, researchers have proposed many channel selection algorithms. A large number of researchers are exploring channel selection algorithms based on emotion recognition. In the emotion recognition task, we need to select features that are closely related to emotions. Khalil et al. [6] adopted a mathematical method based on the extracted mean, standard deviation, skewness, kurtosis of the EEG signal, the mean absolute value of the original signal first-difference and the mean absolute value of the first-difference in

© Springer Nature Switzerland AG 2019
Y. Tang et al. (Eds.): HCC 2018, LNCS 11354, pp. 11–22, 2019.
https://doi.org/10.1007/978-3-030-15127-0_2

normalized-signals. They used these six time-domain features and the QDA classifier to classify calm, positive and negative emotions. As a result, three kinds of emotions with a 63.33% accuracy were identified. There are many studies on feature selection in emotional classification. However, few studies focus on channel optimization. Aitzol et al. [7] described a novel algorithm. An estimation of distribution algorithm (EDA) was applied in a reduced range to obtain the optimal channel subset. Experimental results showed that the resulting channel subset was consistent with motor-imagery-related neurophysiological principles. Arvaneh et al. [8] proposed and applied an extended CSP algorithm, called sparse CSP, to select the smallest channel numbers, which formulates the task of channel selection as an optimization problem given the restrictions on classification accuracy. Geng et al. [9] proposed a multi-class common spatial pattern (MCCSP) for EMG channel selection.

There are some studies that explored EEG channel selection in the field of emotion recognition. Ansari et al. [10] proposed a channel selection algorithm based on the synchronization likelihood method to reduce the number of EEG channels for emotion classification from 64 to 5. In recognition of different combinations of positive, middle and negative emotional states, the effect of classification is not obvious. In contrast to these studies, our study focuses on the MRCS algorithm and three kinds of channel selection algorithms that do not rely on subjects. According to MRCS method, we study channel sorting and classification based on these seven feature combinations. Due to individual differences, each subject's optimal channel ordering has greater differences, although the channel selection algorithm can significantly reduce the channel numbers while ensuring the classification effect. We compared three channel selection algorithms that do not rely on subjects and explored the distribution of the common effective channels in the brain.

2 Research Methodology

2.1 Experimental Procedure

Sixteen right-handed male undergraduate and graduate student volunteers (mean age = 22.81, SD = 1.60) from Lanzhou University were recruited for the study. All participates had no history of neurological or psychiatric illness and had normal or corrected to normal vision. To inspire the subjects' emotions, we used pictures from the International Affective Picture System (IAPS) for Chinese students [11]. Before the formal experiment began, we collected the resting EEG data for 60 s. During this 60 s, the subjects were instructed to look at a cross sign presented in the center of the screen and try to relax. The experimental protocol is shown in Fig. 1. The whole task was divided into 4 blocks, each block consisted of 15 trials of the same emotional induction type. The four blocks of emotional induction type were depressed, relaxed, fearful, and joyful. At the end of each block, the subjects were asked to complete a Self-Assessment Manikin (SAM) to score the 15 pictures they had seen before through arousal and valence.

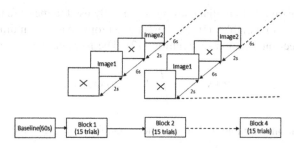

Fig. 1. The experimental protocol.

2.2 Data Preprocessing

The noise in the feeble recording EEG, such as power line interference and ocular artifacts, pose a major problem for EEG interpretation and disposal. Therefore, it is important to eliminate artifacts for extracting pure EEG signals.

The first step was to remove the useless electrodes. Because of our setting, EEG data for the records of three electrodes—HEOG, VEOG and FT10—needed to be abandoned. Therefore, the number of the effective EEG channels was 61. Then, filtering was required. A 1 Hz high-pass filter was used to remove the baseline drift, and a 49–51 Hz bandpass filter was used to remove 50 Hz power frequency interference. A 60 Hz low-pass filter was used to remove EEG irrelevant frequency components. To improve the speed of the operation, raw EEG signals were sampled at 256 Hz. The third step was to extract effective EEG fragments. During each block score and rest time, some collected EEG signals are invalid. It is necessary to remove this part of the invalid EEG data. The fourth step was to remove the eye potentiometer and muscle artifacts using an ICA algorithm. The final step was to remove the base value and extract the emotion-related EEG data. The EEGLAB toolbox was used to preprocess the original EEG signals [12]. With these steps, pure EEG data can be extracted.

2.3 Feature Extraction

Feature extraction is a key step in preprocessing [13, 14]. Many types of features have been used for emotion recognition based on EEG, such as time, frequency, or time-frequency domain features [15]. We chose the following three features:

1st Difference. The mean of the absolute values of the first differences of the raw signal.

$$\delta_x = \frac{1}{T-1}\sum_{t=1}^{T-1}|X(t+1) - X(t)| \tag{1}$$

where X is the raw N-sample EEG signal.

Differential Entropy. Differential entropy is generally used to measure the complexity of continuous random variables [16]. Differential entropy features on different rhythms have a better effect on emotion recognition [17]. The formula is as follows:

$$DE(X) = - \int_X f(x) \log(f(x)) dx \qquad (2)$$

where X is a random variable and f(x) is the probability density function of X.

$$\begin{aligned} DE(X) &= \int_{-\infty}^{+\infty} \frac{1}{\sqrt{2\pi\sigma^2}} e^{-\frac{(x-\mu)^2}{2\sigma^2}} \log\left(\frac{1}{\sqrt{2\pi\sigma^2}} e^{-\frac{(x-\mu)^2}{2\sigma^2}}\right) dx \\ &= \frac{1}{2} \log\left(2\pi e \sigma^2\right) \end{aligned} \qquad (3)$$

It was found that each EEG frequency band approximately satisfied a Gaussian distribution [16]. Therefore, the differential entropy of each EEG frequency band can be calculated according to formula (4). We define the differential entropy of gamma rhythm as DEg.

$$DE_i(X) = \frac{1}{2} \log\left(2\pi e \sigma_i^2\right) \qquad (4)$$

Fractal Dimension. The FD value reflects the nonlinearity of the EEG signal [18]. The fractal dimension calculated by the Higuchi algorithm has proven to have a better classification effect [19].

2.4   Channel Selection Method

Mean-ReliefF-Channel-Selection Algorithm. The main idea of the ReliefF algorithm is similar to the basic idea of the k-nearest neighbor algorithm [20]. According to the correlation between each feature and category, different weights are given to the features, and the features with weights less than a certain threshold are removed. MRCS defines the average weight of all the features belonging to the channel as the weight of the channel [21].

Common Channel-Selection Algorithm. This section gives a brief description of the common channel selection method. We define the names of these three common channel selection algorithms below.

Weighted Addition Method. First, the weight of each channel was calculated, and then this value was normalized to [−1, 1]. The weight of the corresponding channels for all subjects was added, and this result was used as the weight of the common channel, which does not rely on the subjects. Ranking these common channel weights from largest to smallest, we can identify the most important common channels. Common channel weights are calculated as shown (5).

$$W(l) = \sum_{s=1}^{S_n} W(l_s) \tag{5}$$

S_n is the number of the subject, and $W(l_s)$ represents the normalized weight of channel l from the subjects.

Accuracy Weighted Method. The weight of the subjects with higher classification accuracy should take a larger proportion to select the common channel because the purity of data collection and emotional induction intensity from these subjects are higher. First, the accuracy of the emotion classification of a subject based on feature extraction from all EEG channels was calculated, and then the accuracy was used as a weighting term of the corresponding subject channel weight. Common channel weights are calculated as shown in (6).

$$W(l) = \sum_{s=1}^{S_n} W(l_s) * Acc_s \tag{6}$$

Acc_s represents the classification accuracy of subject s in all EEG channels and features. According to formula (6), the accuracy weighting of each channel can be calculated, and then these values are sorted. The optimal channel set can be selected according to the number of common channels.

Rank Weighted Method. The calculation formula of the rank weighted method is shown in (7) [22].

$$W(l) = \frac{1}{S_n} \sum_{s=1}^{S_n} \sum_{p=1}^{|L|} \frac{\varphi(O^s, l, p)}{p} \tag{7}$$

Suppose that p is the order of channel l in channel sort O^s of the subject s, then $\varphi(O^s, l, p) = 1$ where $|L|$ is the number of channels, S_n is the number of subjects, and $W(l)$ is the calculated generic weight of the channel l. According to formula (7), the weight of each channel can be calculated, and then these are sorted. The common channel can be selected according to the requirements of the actual task.

2.5 Data Analysis

The three features were extracted and normalized, and then, the feature vectors composed of the features on 61 channels in the same emotion fragment were calculated. The penalty parameter C and kernel function g were optimized by the grid search method [23]. To calculate the optimal results, the average accuracy rate was calculated after 5 times of 5-fold cross-validation, which was taken as the final result.

3   Results and Discussion

3.1   Emotion Classification Accuracy Result

We divided the classification results into three categories. (1) On the basis of the emotional states, we divided the results into 4 classes: happiness, relaxation, depression, and fear. (2) Emotional valence was classified into two groups, happiness and relaxation in one group, and depression and fear in the other group. (3) Emotional arousal was classified into two groups, depression and relaxation belonged to the low arousal state and happiness and fear belonged to the high arousal state. The specific results are shown in Fig. 2.

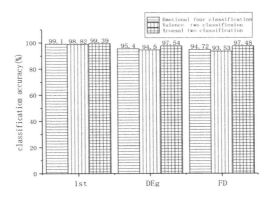

Fig. 2. The average classification accuracy of 16 subjects of three features.

From Fig. 2, we can see that the accuracy rates of the four emotional states classifications of 1st, DEg and FD were 99.1%, 95.4%, and 94.72%, respectively. The accuracy rates of the two valence classifications of 1st, DEg and FD were 98.82%, 94.5%, 93.53%, respectively. The accuracy rates of the two arousal classifications of 1st, DEg and FD were 99.39%, 97.54%, and 97.48%. It is obvious that the method obtained good classification result.

3.2   MRCS Algorithm Channel Selection Result

We randomly divided the four emotional states of each subject into two parts with some data taken in the channel selection set, and the other data taken in the channel validation set. Then, the MRCS algorithm was used to calculate the optimal channel ranking of various feature combinations for every subject in the channel selection set, and the relationship between the accuracy rate of four emotional states classification and the number of channels was calculated on the validation set. The average classification accuracy rates for the 16 subjects are shown in Fig. 3. It can be seen from Fig. 3 that when the number of EEG channels decreased to 10, the classification accuracy of the seven feature combinations reached a relatively high value.

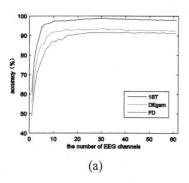 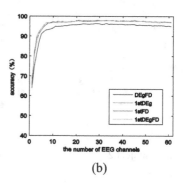

(a)   (b)

Fig. 3. The relationship between the average accuracy of the four emotional states classifications and channel numbers in 16 subjects using a combination of seven features.

By comparing the three feature combinations—1stDEg, 1stFD and 1stDEgFD, it was found that the three curves are difficult to distinguish for the relationship between the average classification accuracy and the number of channels. However, it can be seen that the curve showing the standard deviation of the accuracy rate and the number of channels, in this case the first 20 channels, which were optimal channels, shows that the standard deviation of the 1stDEg combination was smaller, as shown in Fig. 4(b). Therefore, the 1stDEg combination was selected as the optimal combination of the MRCS algorithm for picture induced emotional data. Thus, the channel ranking selected by each subject was used as the basis for channel layout in the design of the portable emotion recognition system. We used the MRCS algorithm and 1stDEg feature combination to calculate the optimal channel ranking for each subject. We randomly selected 6 subjects out of 16 subjects, and the results are shown in Fig. 5.

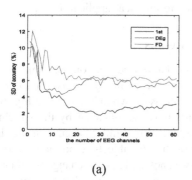 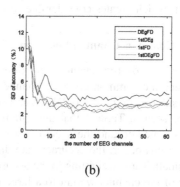

(a)   (b)

Fig. 4. The relationship between the SD of average accuracy of emotion classification and channel numbers in 16 subjects using the combination of the seven features.

The gray level of each circle represents the priority of the corresponding channels in the channel ranking. The deeper the color is, the higher the ranking was. Although there was certain regularity in the channel distribution area with a high ranking, there

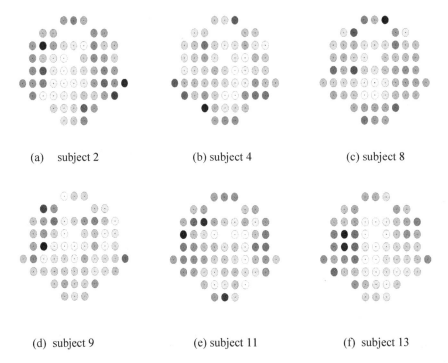

Fig. 5. The distribution map of EEG channels obtained by the MRCS algorithm for 6 subjects as the 1stDEg feature.

was a significant difference in the optimal channel ordering for each subject. Due to individual differences, the optimal channel set among different subjects was not consistent, which creates considerable inconvenience to practical application.

3.3   Common Channel Selection Algorithm Results

On the basis of three channel selection algorithms that did not rely on the subjects, we obtained the common effective channel ranking and the brain region distribution map of 16 subjects under the 1stDEg feature combination, as shown in Table 1 and Fig. 6. It can be seen in Table 1 that the common channel rankings obtained by the weighted addition method and the accuracy weighted method were similar because in the data used in this paper, the accuracy rate difference of the 16 subjects' emotion categories was small. Thus, weighting to the accuracy rate was not sufficient to change the relative order of the channel weight to a large extent. Compared with the other two methods, the channel rankings for the results obtained by the rank weighted method were not consistent. They accounted for a larger proportion in the same channels out of the first ten channels where they were highest. Taking the weighted addition and the rank weighted method as an example, the two had 8 of the same channels in the first ten optimal channels, which were C6, C5, F5, FC5, Oz, TP10, TP9 and C3. The principle of the weighted addition method and the rank weighted method differed greatly, but the

Table 1. The results of the common channel selection methods.

Common channel selection methods	Channel sorting
Weighted addition method	TP9, FC5, C5, Oz, TP10, F5, P7, C3, C6, F3, O1, AF3, PO7, TP8, FT7, O2, T8, T7, CP6, F7, TP7, CP5, FP2, F4, F1, AF7, FC3, FT8, Fp1, P8, PO4, AF4, Fpz, PO8, F6, F2, F8, AF8, FC6, CP4, CP3, C4, P5, Fz, FC4, P6, FC2, PO3, C1, Poz, CP2, Cz, P3, P4, CP1, FC1, C2, P2, CPz, Pz, P1
Rank weighted method	C6, C5, F5, FC5, Oz, TP10, TP9, PO7, C3, O2, FP2, FT7, CP6, F1, O1, P7, F2, AF3, F3, TP7, T8, AF7, CP5, T7, PO4, TP8, F4, F6, P8, FC3, PO8, AF4, F7, Fp1, CP4, FT8, Fpz, CP3, FC2, P5, AF8, C4, F8, Fz, POz, FC4, FC6, PO3, P6, P3, P4, C1, CP2, Cz, FC1, CPz, C2, Pz, CP1, P2, P1
Accuracy weighted method	TP9, FC5, C5, Oz, TP10, P7, F5, C3, C6, F3, O1, AF3, PO7, FT7, TP8, O2, T7, T8, CP6, CP5, TP7, F7, F4, FP2, F1, AF7, FC3, FT8, Fp1, PO4, P8, AF4, Fpz, PO8, F6, F2, F8, AF8, FC6, CP4, CP3, C4, P5, Fz, FC4, P6, FC2, PO3, C1, POz, Cz, CP2, P3, P4, CP1, FC1, C2, P2, CPz, Pz, P1

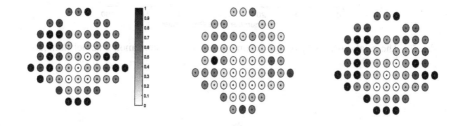

(a)Weighted addition method   (b) Rank weighted method   (c) Accuracy weighted method

Fig. 6. The distribution of the encephalic region beyond the three common channel selection methods.

channels contained in the first ten optimal channels were very similar and reflected that the 1stDEg features above these channels played a very important role in the classification and recognition of the four emotional states. It can be seen in Fig. 6 that the effective channel distribution areas obtained by the three algorithms have very high similarity and that the effective channel distribution areas obtained by the parts with high rankings had high similarity. Additionally, the parts with high ranking were basically located at the left prefrontal and posterior occipital regions, which is consistent with the physiological mechanism of emotion production and confirms the validity of this study to that extent [24].

We adopted the common channel selection algorithm to calculate the relationship between the accuracy of emotion classification and the number of channels on the testing set of 16 subjects. In addition, their accuracy curves and corresponding channel numbers are shown in Fig. 7. In the first 10 common channels, the average accuracy of the four emotional states classification of 16 subjects reached approximately 95%. We calculated the four emotional states classification accuracy of the first 10 common channels obtained by the weighted addition method. In the remaining 51 channels, 10 channels were randomly selected to calculate the accuracy of the four emotional states classification. The relationship between these two accuracy rates is shown in Fig. 8.

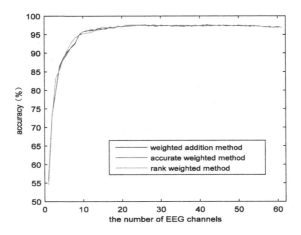

Fig. 7. The relationship of average accuracy rate and the com mon channel numbers.

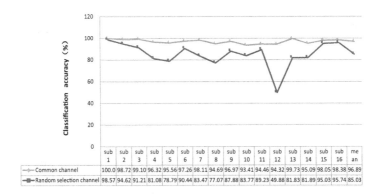

Fig. 8. The accuracy of the four emotional states classification of 10 EEG channels.

4   Conclusions

In this paper, we use an MRCS algorithm to discover the relationship between classification accuracy and channel numbers to select the optimal feature combination 1stDEg. Then, we compared three kinds of common channel selection algorithms including weighted addition, rank weighted and accurate weighted methods. This study has shown that the performance of the three methods was similar using our dataset, and we found that the distribution area was basically located in the left frontal lobe and the posterior occipital lobe. When we chose the best 10 common channels, the accuracy of the four emotional states classification was approximately 95%. This conclusion validates the channel selection algorithm that does not rely on the subject. It has important practice instruction significance for the selection of common channels and is of great significance to the design of a portable emotion recognition system.

Acknowledgment. This work was supported by the National Basic Research Program of China (973 Program) (No. 2014CB744600), the National Natural Science Foundation of China (Grant No. 61632014, No. 61210010, No. 61300231), the National key foundation for developing scientific instruments (No. 61627808), Program of Beijing Municipal Science \& Technology Commission (No. Z171100000117005), Fundamental Research Funds for the Central Universities (lzujbky-2017-kb08, lzujbky-2017-sp25), Talent Innovation Venture Science and Technology Project of Lanzhou (2015-RC-60), Science and technology innovation project.

References

1. Liogiene, T., Tamulevicius, G.: SFS feature selection technique for multistage emotion recognition. In: Proceedings of the 2015 3rd Workshop on Advances in Information, Electronic and Electrical Engineering (AIEEE), Riga, Latvia, pp. 1–4. IEEE (2015)
2. Hu, B., Rao, J., Li, X.: Emotion regulating attentional control abnormalities in major depressive disorder: an event-related potential study. Sci. Rep. **16**(2), 127 (2017)
3. Li, X., Song, D., Zhang, P.: Exploring EEG features in cross-subject emotion recognition. Front. Neurosci. **12**, 162 (2018)
4. Li, X., Hu, B., Sun, S.: EEG-based mild depressive detection using feature selection methods and classifiers. Comput. Methods Programs Biomed. **136**(C), 151–161 (2016)
5. Ameri, R., Pouyan, A., Abolghasemi, V.: EEG signal classification based on sparse representation in brain computer interface applications. In: Proceedings of the 2015 22nd Iranian Conference on Biomedical Engineering (ICBME), Tehran, Iran, pp. 21–24. IEEE (2016)
6. Khlili, Z., Moradi, M.H.: Emotion recognition system using brain and peripheral signals: using correlation dimension to improve the results of EEG. In: Proceedings of the 2009 International Joint Conference on Neural Networks, Atlanta, GA, USA, pp. 1571–1575. IEEE (2009)
7. Astigarraga, A., Arruti, A., Muguerza, J.: User adapted motor-imaginary brain-computer interface by means of EEG channel selection based on estimation of distributed algorithms. Math. Probl. Eng. **2016**(1), 1–12 (2016)
8. Arvaneh, M., Guan, C., Kai, K.A.: Optimizing the channel selection and classification accuracy in EEG-based BCI. IEEE Trans. Biomed. Eng. **58**(6), 1865–1873 (2011)
9. Geng, Y., Zhang, X., Zhang, Y.T.: A novel channel selection method for multiple motion classification using high-density electromyography. Biomed. Eng. Online **13**(1), 102 (2014)
10. Ansari-Asl, K., Chanel, G., Pun, T.: A channel selection method for EEG classification in emotion assessment based on synchronization likelihood. In: Proceedings of the 2007 15th European Signal Processing Conference, Poznan, Poland, pp. 1241–1245. IEEE (2007)
11. Nan, L.X., Xiang, X.A.: Native research of international affective picture system: assessment in university students. Chin. J. Clin. Psychol. **17**(6), 687–689, 692 (2009)

12. Delorme, A., Makeig, S.: EEGLAB: an open source toolbox for analysis of single-trial EEG dynamics including independent component analysis. J. Neurosci. Methods **134**(1), 9–21 (2004)
13. Ackermann, P., Kohlschein, C., Bitsch, J.Á.: EEG-based automatic emotion recognition: feature extraction, selection and classification methods. In: Proceedings of the 2016 IEEE 18th International Conference on e-Health Networking, Applications and Services (Healthcom), Munich, Germany, pp. 1–6. IEEE (2016)
14. Alsukker, A., Al-Ani, A.: Evaluation of feature selection methods for improved EEG classification. In: Proceedings of the 2006 International Conference on Biomedical and Pharmaceutical Engineering, Singapore, Singapore, pp. 146–151. IEEE (2006)
15. Jenke, R., Peer, A., Buss, M.: feature extraction and selection for emotion recognition from EEG. IEEE Trans. Affect. Comput. **5**(3), 327–339 (2017)
16. Shi, L.C., Jiao, Y.Y., Lu, B.L.: Differential entropy feature for EEG-based vigilance estimation. In: Annual International Conference of the IEEE Engineering in Medicine and Biology Society. IEEE Engineering in Medicine and Biology Society, Osaka, Japan, pp. 6627–6630 (2013)
17. Duan, R.N., Zhu, J.Y., Lu, B.L.: Differential entropy feature for EEG-based emotion classification. In: Proceedings of the 2013 6th International IEEE/EMBS Conference on Neural Engineering (NER), San Diego, CA, USA, pp. 81–84. IEEE (2013)
18. Liu, Y., Sourina, O.: Real-time subject-dependent EEG-based emotion recognition algorithm. In: Gavrilova, M.L., Tan, C.J.K., Mao, X., Hong, L. (eds.) Transactions on Computational Science XXIII. LNCS, vol. 8490, pp. 199–223. Springer, Heidelberg (2014). https://doi.org/10.1007/978-3-662-43790-2_11
19. Wang, Q., Sourina, O., Nguyen, M.K.: Fractal dimension based neurofeedback in serious games. Vis. Comput. **27**(4), 299–309 (2011)
20. Peker, M., Arslan, A., Sen, B.: A novel hybrid method for determining the depth of anesthesia level: Combining ReliefF feature selection and random forest algorithm (ReliefF + RF). In: Proceedings of the 2015 International Symposium on Innovations in Intelligent SysTems and Applications (INISTA), Madrid, Spain, pp. 1–8. IEEE (2015)
21. Zhang, J., Chen, M., Zhao, S.: ReliefF-based EEG sensor selection methods for emotion recognition. Sensors **16**(10), 1558 (2016)
22. Mesa, I., Rubio, A., Tubia, I.: Channel and feature selection for a surface electromyographic pattern recognition task. Expert Syst. Appl. **41**(11), 5190–5200 (2014)
23. Chang, C.C., Lin, C.J.: LIBSVM: a library for support vector machines. ACM **2**(3), 1–27 (2011)
24. Ledoux, J.E.: Emotion circuits in the brain. Annu. Rev. Neurosci. **23**(2), 274 (2000)

Research on Real-Time Low Air Image Intelligence Image Acquisition and Processing Methods

Pin Wang[1], Guocheng Zheng[1], Shen Zhang[1], Peng Xu[1],
Zhumei Zhao[1], Haoyang Yu[2], and Yuxiao Yang[2(✉)]

[1] Systems Engineering Research Institute, Beijing 100000, China
wangpin20@163.com, zheng_guocheng@163.com
[2] North China Electric Power University, Baoding 071000, China
yangyuxiao_ncepu@outlook.com

Abstract. In order to solve the problem that all levels of civil aviation meteorological centers are unable to provide all the possible meteorological information services that may reach the low-altitude airspace for general aviation flight, this paper proposes a real-time low-altitude meteorological intelligence image collection and processing method. Based on this method, a real-time low-altitude meteorological intelligence image collection system can be designed and implemented. The real-time low-altitude meteorological intelligence image collection system is an information system that provides real-time low-altitude airspace visibility and weather condition information services for image-based flight photography, transmission, processing, and meteorological information dissemination. It includes the image collection subsystem and image processing subsystem, meteorological information publishing Web Server. Among them: The image collection subsystem is mainly used for automatically taking images of the surrounding environment at regular intervals, and transmitting the photographs and their time and place information to the image processing system installed in the low-altitude navigational information service department through the wireless communication device; the image processing subsystem collects images and related information for further processing; the meteorological information publication Web server mainly provides general airline users with services such as consultation, inquiry, and submission of low-altitude meteorological intelligence information of relevant airspace related to images.

Keywords: General aviation · Meteorological information ·
Information system · Image processing

1 Introduction

Low altitude airspace generally refers to airspace below 3000 m in height. The flight activities conducted in the low-altitude airspace are mainly general aviation flights. Navigation flights are mainly dedicated to the role of industry, agriculture, forestry, tourism, entertainment, sports events, disaster relief, and other fields. Because of the unique nature and special needs of these fields, general aviation flights often do not

© Springer Nature Switzerland AG 2019
Y. Tang et al. (Eds.): HCC 2018, LNCS 11354, pp. 23–34, 2019.
https://doi.org/10.1007/978-3-030-15127-0_3

determine fixed routes in advance, and flight airspace is complex. At the same time, the navigable aircraft flies in the low altitude airspace. The meteorological conditions in this airspace are very complicated and the weather transformation is very frequent. Some sudden extreme weather will have a serious impact on the navigable flight activities. Therefore, the flight demand for meteorological information is very urgent. At present, civil aviation meteorological centers at all levels and airport weather stations mainly provide meteorological information for airports and existing air routes and routes. There are service blind spots, and it is difficult to meet the demand for inconvenient navigable flight routes and complicated airspace conditions for low airborne intelligence. The real-time low-altitude image intelligence image acquisition and processing method proposed in this paper is to provide low-level flight meteorological information to the general aviation flight in real time. The real-time low-altitude and airspace meteorological conditions displayed in images can visually reflect the visibility and weather information of a low-altitude airspace at present, reducing the low-altitude aerial information service blind areas at all levels of meteorological centers will provide strong support for navigable flights. The real-time low-air image intelligence image acquisition system based on this method integrates advanced wireless communication technologies such as 4G, Beidou, satellite communications, and digital radio stations, and has the characteristics of adapting to a complex geographical environment, facilitating layout, and controlling costs.

2 Method

The real-time low-altitude meteorological intelligence image acquisition and processing method proposed in this paper can solve the problem that all levels of civil aviation meteorological centers can not provide all possible meteorological information services that may reach the low altitude airspace for general aviation flight, and provides real-time low-altitude airspace visibility and weather condition information service in the form of images.

According to the above objectives and functions, the real-time low-altitude meteorological intelligence image acquisition and processing method firstly performs meteorological intelligence and image collection, then carries out image processing, and finally releases the meteorological information service based on the Internet. Therefore, a system based on a real-time low-altitude meteorological intelligence image acquisition and processing method includes the image collection system, a image processing system, a weather information publication web server, and related protection facilities.

The image acquisition system includes hardware devices and control software. Hardware devices include cameras, embedded controllers, solar powered devices, 360° pan-tilt heads, and wireless communication devices. The control software consists of a control module and a data communication module.

The image processing system includes a hardware device, a image management software, and a web server. Hardware devices include image management servers, databases, wireless communication devices, high resolution displays. The image management software consists of a image processing module (automatic (default), manual), a central camera control module, a communication module, a database management module, and a display module.

The wireless communication devices of the image collection system and the image processing subsystem are all composed of a 4G communication module, a data transmission radio communication module, a Beidou module, a GPS module, and a 4G communication module.

Protective facilities include camera housings, lightning rods and field protection buildings.

In the method proposed in this paper, the image acquisition system should be set up at the ground commanding height in the airspace that may be reached by navigation flight to establish an unmanned image collection station. The image acquisition system will automatically take the photos of the surrounding environment and photograph the photos and their time and place information. Through the wireless communication equipment transmitted to the image processing system installed in the low altitude navigation information service department, the personnel of the low altitude navigation information service department will further process the images and related information through the image processing system, and save the processing results so that the users can get real-time images and visibility and weather status information by visiting the weather intelligence release web server for its required airspace. The method proposed in this paper is a good complement to the current shortage of meteorological intelligence services for meteorological flight meteorological services, and provides a strong support for the safety of navigation flights.

3 System Design

3.1 Overall Structural Design

The overall structure design of the real-time low-air image intelligence image acquisition system is shown in Fig. 1. The image acquisition system will be deployed in a low-altitude area where no aviation meteorological information service is currently available to establish an unmanned image collection station. The method of deployment is: selecting an elevation point of view in the airspace where there is a demand for meteorological information services, establishing support and protection of the permanent supporting facilities of the image acquisition system, and installing cameras, heads, embedded controllers, and wireless communication devices on the supporting facilities, solar power devices, embedded controllers to install control software.

The camera is mainly based on industrial cameras adapted to various complex meteorological environments and weather conditions. The selection of industrial cameras needs to refer to the overall meteorological conditions of the place where the camera is located, and the flexibility to select an industrial camera that can work stably under such an overall meteorological environment at a reasonable cost depending on the overall change of the local temperature and the severity of the extreme weather that may occur. The selected camera communicates with the embedded controller via the Ethernet interface to implement the embedded controller to control the camera and obtain the captured image.

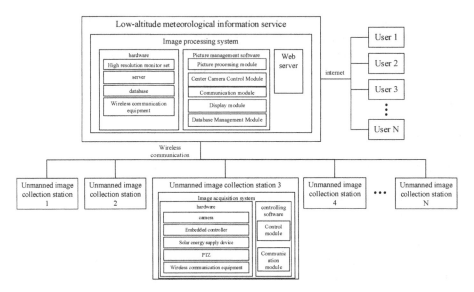

Fig. 1. Overall system structure

The camera is mounted on a stable, adjustable height and 360° pan-tilt head. The PTZ can communicate with the embedded controller through the Ethernet interface to implement the embedded controller's control over the PTZ.

Wireless communication devices include 4G modules, 3G modules, Beidou modules, data transmission radio modules, and satellite communication modules. The actual unmanned image collection station may configure the communication device constituted by one or more of the above-mentioned communication modules according to needs and conditions.

Cameras, PTZs, embedded controllers, and wireless communication devices provide DC power from solar powered devices. The solar power supply device mainly comprises a solar windsurfing board and a battery.

The control software consists of a camera control module and a communication module. The camera control module realizes the control of the camera and the pan-tilt and obtains the basic information of the images and images taken by the camera. The communication module controls the wireless communication device to send the image and the basic information of the image and various device parameters to the designated low-altitude meteorological information service department. Receive various instructions from the Low Air Meteorological Intelligence Service department.

Image processing systems are deployed in the Low Air Meteorological Intelligence Service department. The image processing system includes a hardware device, a image management software, and a web server. Hardware devices include image management servers, databases, wireless communication devices, high resolution displays. The image management software consists of a image processing module, a central camera control module, a communication module, a database management module, and a display module.

The image management server installs image management software, connects wireless communication devices, connects high-resolution display groups, and provides human-machine interaction device interfaces. The database stores images and related information, which is a common database of the image processing system and the web server. The wireless communication device receives images and related information from an unmanned image collection system and sends control instructions.

Image management software image processing module integrates image processing tools, staff information management functions, automatic association of image information and staff information, central camera control module to achieve remote control of unmanned image collection cameras and PTZ, communication module control and receive From the image collection station images, image basic information and various equipment parameters, and send instructions to control the camera, PTZ and other equipment.

The Web server provides consultation, search, submission, and other services for low-altitude airborne intelligence information of related airspaces mainly for images. "Consultation" means that the web server provides a platform for communication and consultation between users and personnel in the low-altitude sector to obtain more definitive information about low-altitude meteorological intelligence services. "Query" refers to the user directly inquires about the image or text information already existing in the image processing system. "Submission" means that users can submit the weather information they have. Through this web service, the image processing system has established a low-altitude air image information exchange and sharing network platform between the low-altitude meteorological information service department and the general aviation users, enabling users to become service providers while obtaining services.

Navigation users can quickly and easily obtain real-time meteorological information in the low-altitude airspace that they may reach via the PC and browsers in various mobile devices.

As can be seen from Fig. 1, the image acquisition system and the image processing system are many-to-one relationship. The image processing system can provide services for multiple navigation users at the same time, real-time low-altitude navigation intelligence collection, processing system integration advanced communication, control, Technologies such as the Internet have constructed a networked and intelligent meteorological information sharing space for low-altitude flights.

3.2 Data Flow Design

Figure 2 shows the data flow diagram of the real-time low air image intelligence image acquisition system. The integrated data in Fig. 2 includes picture data, picture-related information, and device parameter information. The meteorological information integrated data includes picture data, picture related information, and exchange information. The control commands include camera control commands and pan-tilt control commands. The instructions are encrypted during transmission.

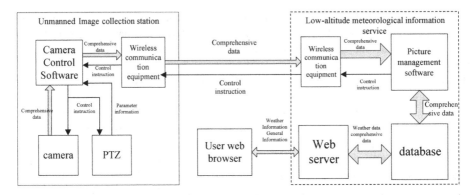

Fig. 2. Data flow diagram

3.3   Acquisition Subsystem Hardware Structure Design

Figure 3 shows the hardware structure of the picture acquisition subsystem. As shown in Fig. 3, 4G and 3G modules are connected to the embedded controller via USB. The GPS module, data radio module, and Beidou module are connected to the embedded controller through the serial port. The camera and PTZ are connected to the embedded controller through the Ethernet interface. The solar power supply device provides DC power for the entire device and passes the parameter information to the embedded controller through the serial port. The maintenance of the picture collection system is accomplished through a communicator.

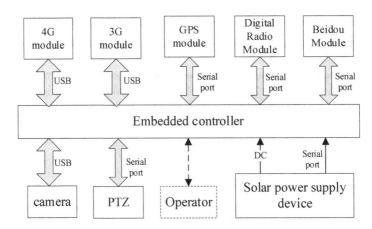

Fig. 3. The hardware structure of the picture acquisition subsystem

3.4   Acquisition System Control Software Design

Figure 4 shows the architecture of the control software for the picture acquisition system. As shown in Fig. 4, the picture acquisition system control software is a three-layer structure and is composed of a display module, a control module, and a communication module.

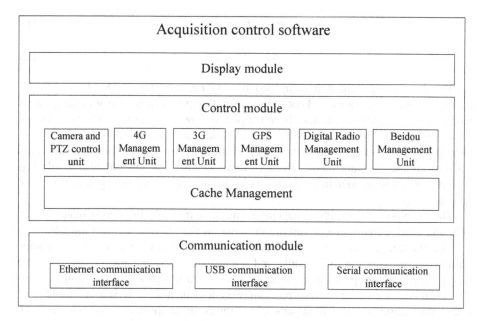

Fig. 4. Control software design architecture

The display module provides a device maintenance interface for the communicator.

The control module consists of a camera and PTZ control unit, a 4G management unit, a 3G management unit, a GPS management unit, a digital radio station management unit, and a Beidou management unit, and is provided with communication buffer management.

The camera and the PTZ control unit complete the acquisition of the device parameter information, pictures and related information, invoke the communication unit to realize the information transmission and reception, and analyze the control commands from the picture processing system to realize the control of the camera and the PTZ. The control of the camera mainly includes the shooting mode (such as full automatic mode, full manual mode, aperture priority mode, shutter priority mode, B mode, etc.), the size of the camera aperture, the speed of the shutter, the focal length of the camera lens, the size of the single image file, and the frequency of shooting. The control of the PTZ mainly includes the selection of the shooting direction and the adjustment of the height of the stand. The frequency of shooting includes two aspects. First, the duration of each $360°$ panorama (T360) and the number of photos taken (S). Second, the time interval (D) for continuous 360-degree panorama shooting at different times. The shooting time (T1) of each single photo is calculated according to formula (1). The horizontal angle (α) of the photographed environment that can be recorded in each photo is calculated by formula (2). Interval D can be flexibly set. For example, from 6 to 18 o'clock, the time interval D = 15 min, 18 o'clock to 6 o'clock the next day, time interval D = 12 h. This setting will not only allow 360-degree panoramas to be shot every 15 min, but will not be shot at night. The control of the pan tilt must meet the camera's shooting requirements.

$$T_1 = T360/S \tag{1}$$

$$\alpha = 360^\circ/S \tag{2}$$

The 3G management unit completes the 3G module startup, shutdown, restart, parameter setting functions, and obtains the operating parameters. The 3G management module parses control instructions from the picture processing system. The instructions from the image processing system include only startup, shutdown, restart, and restoration of default parameter settings. In addition, more complex control instructions come from the communicator. The 3G management unit module also sends the acquired 3G module operating parameters to the picture processing system at regular intervals.

The functions of the GPS management unit, data transmission station management unit, and Beidou management unit are the same as the 3G management module.

The 4G management unit completes the startup, shutdown, restart, and parameter setting functions of the 4G module and obtains the operating parameters. The 4G management module parses control instructions from the picture processing system. The instructions from the image processing system include only startup, shutdown, restart, and restoration of default parameter settings. In addition, more complex control instructions come from the communicator. The 4G management unit module also sends the acquired 4G module operating parameters to the picture processing system at regular intervals.

Communication cache management provides a unified cache management interface for the above wireless management unit. The communication module provides unified Ethernet communication, USB communication, and serial communication interfaces to the upper layer.

3.5   Picture Management System Picture Management Software Design

Figure 5 is a picture management software architecture diagram of a picture processing system. As shown in Fig. 5, the picture management software has a three-layer structure. Display module is the top layer of the independent engineering software structure, and the image processing module and the central camera control module constitute the middle layer. The bottom layer includes the database management module and the communication module.

The display module mainly includes three main interfaces: a monitoring device interface of the collection station, a picture displays interface of the collection station airspace, and a picture processing interface.

The collection station equipment monitoring interface includes a communication equipment monitoring interface, a camera pan tilt monitoring interface, and a solar power supply monitoring interface. The collection station equipment monitoring interface provides site options for switching to different monitoring collection station monitoring interfaces.

The collection station airspace picture display interface displays by default the picture transmitted by the picture collection system of a selected picture collection station and the time information, location information of the picture shooting, and the

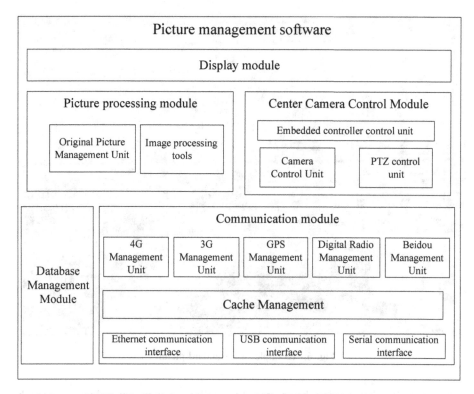

Fig. 5. Picture management software design architecture

direction information of the picture relative to the collection station. The displayed information is the original image and its related information that was shot this time if the last shot and sent to the image processing system has not yet been processed. If the most recently taken picture and information have been processed, the processed picture taken this time and the edited complete navigation information is displayed. The interface can display multiple pictures and related information at the same time. The collection station airspace picture display interface provides pictures and related information query functions using the shooting time period and the name of the collection station as keywords.

The image processing interface provides image processing tools and meteorological information editing interfaces for low air image intelligence service personnel. When a picture processing tool is used to process a picture, a standard picture of the picture will be displayed synchronously in the interface, and the picture is processed with the standard picture as a reference object and the weather information displayed in the picture is manually edited into text information. The text information and the original image-related information, such as time information, location information, and the content of the image with respect to the direction of the collection station, together constitute a complete navigation information. The standard picture is shown in Fig. 6. At the time of the initial completion of the collection station, the standard picture was

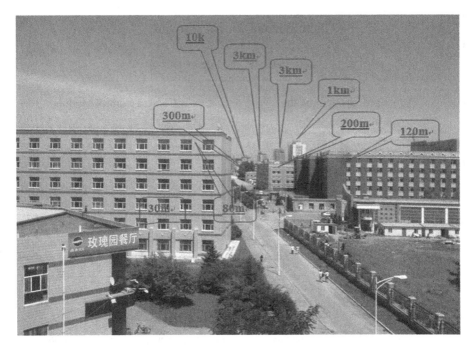

Fig. 6. Example of a standard picture

taken by a staff member using the camera of the collection station and obtained through actual measurement and post-processing. During the shooting of the standard picture, the staff takes pictures of the camera at different focal lengths and at all possible angles. The standard image is processed and its related information is stored in the standard image library.

The picture processing module includes an original picture management unit and a picture processing tool.

The picture management unit provides an interface for acquiring picture and its related information to the airspace display domain of the collection station, receives pictures and related information from the picture collection system, and stores it in the database. The image processing tool integrates picture processing functions such as picture cutting and editing, and provides a function interface for the picture processing interface.

The central camera control module consists of an embedded controller control unit, a camera control unit, and a PTZ control unit.

The embedded controller control unit realizes the function of starting, shutting down and restarting the embedded air unit remotely. When the embedded controller is turned on, off, and restarted, the camera, head, and solar power supply device will also be turned on, off, and restarted under the control of the embedded controller.

The camera control unit realizes the remote control of the camera. The main control of the camera mainly includes the shooting mode, the size of the camera aperture, the speed of the shutter, the focal length of the camera lens, the size of the single image file, and the shooting frequency. The PTZ control unit implements remote control of the PTZ. The control of the PTZ mainly includes the selection of the shooting direction and the adjustment of the stand height. In the foregoing description of the acquisition control software camera and the pan-tilt control unit, the camera and the pan-tilt have been described in detail, and will not be described here.

The database management module provides various interfaces for database operations to its upper layers.

The communication module includes 4G management unit, 3G management unit, GPS management unit, digital radio station management unit, Beidou management unit, communication buffer management, Ethernet communication, USB communication, and serial port communication interface.

The 4G management unit, 3G management unit, GPS management unit, data transmission station management unit, and Beidou management unit provide remote control of the corresponding equipment. The remote control instructions to the device only include startup, shutdown, restart, and restore default parameter setting instructions. Cache management, Ethernet communication, USB communication, serial communication interface with acquisition control software.

The supporting facilities are mainly composed of lightning protection facilities, solar panels, and equipment room. Each supporting facility must be designed to meet the requirements based on the environment in which the acquisition system is located.

4 Deployment Method

The system deployment method is as follows. First, the camera of the unmanned image collection station automatically captures pictures of the airspace where it is located according to the set parameters. The unmanned image collection station sends the pictures and related information to the low-altitude navigation information service department that sets up the picture processing system in real time through the wireless communication device. The low-altitude flight information personnel use the picture processing tools provided by the picture processing system to process the pictures that need to be processed. The processed pictures and edited low-altitude navigation information are stored in the database, and the web server provides services to the navigation users. If the collected images do not require manual processing, they will be automatically saved to the database for access by the navigation user service.

5 Conclusion

In order to solve the problem that all levels of civil aviation meteorological centers are unable to provide all the possible meteorological information services that may reach the low altitude airspace for general aviation flight, this paper proposes a real-time low air image intelligence image acquisition and processing method. Based on this method, a real-time low air image intelligence image acquisition system can be designed and implemented. The real-time low-altitude image intelligence image acquisition system is an information system that provides real-time low-altitude airspace visibility and weather condition information services for the navigation flight and integrates the functions of photo shooting, transmission, processing, and meteorological information release. Including film acquisition subsystem, image processing subsystem, and weather information publishing web server. The real-time low air image intelligence image acquisition system based on the method reduces the blind area of the low air image intelligence service at all levels of meteorological centers. And it has the characteristics of adapting to a complex geographical environment, convenient deployment, and controllable cost. It will certainly provide strong support for navigation flights.

References

1. Li, X.: Research on general aviation weather service. Mod. Naviation **2**(01), 47–54 (2016)
2. Ma, W., Zhou, Y.: The demand and prospect of general aviation meteorological services. J. Civ. Aviat. **2**(01), 47–51 (2018)
3. Pan, M., Chen, Y., Li, Q.: FPGA based on the image acquisition system design. Foreign Electron. Meas. Technol. **31**(03), 58–61 (2012)
4. Liu, L., Shen, M., Sun, Y.: Acquisition system and wireless transmission by 3G for farmland image based on FPGA. Trans. Chin. Soc. Agric. Mach. **42**(12), 187–190 (2011)

Brain Computer Interface Application for People with Movement Disabilities

Sebastián Poveda Zavala[1], José Luis León Bayas[1], Alejandro Ulloa[1], Juan Sulca[1], José Luis Murillo López[1], and Sang Guun Yoo[1,2(✉)] ⓘ

[1] Departamento de Informática y Ciencias de la Computación, Escuela Politécnica Nacional, Quito, Ecuador
{sebastian.poveda, jose.leon01, victor.ulloa02, juan.sulca, jose.murillo01, sang.yoo}@epn.edu.ec
[2] Departamento de Ciencias de la Computación, Universidad de las Fuerzas Armadas ESPE, Sangolquí, Ecuador
yysang@espe.edu.ec

Abstract. The intention of this work is to propose and develop a Brain-Computer Interface solution for people with movement disabilities. The proposed solution will read, monitor and translate brainwaves generated from the Central Nervous System of a person with movement disabilities to replace/rehabilitate his/her natural movements or for allowing him/her to control different devices such as household appliances. The brainwaves used in this work are those produced when a person blinks. We have selected the blinks' related brainwaves since most of the people with movement disabilities (even the most serious) can do this kind of movements. The prototype of the proposed solution created in this work uses a control mechanism based on the raw electroencephalography data extracted from a MUSE headband for manipulating real life household appliances connected to an Arduino system. Experimental results indicated that people with limited training process were capable to use the prototype with manageable level of error.

Keywords: Brain Computer Interface · BCI · Electroencephalography · EEG · Device control · Movement disabilities

1 Introduction

Brain-Computer Interface (BCI) is a system that measures the activities of the Central Nervous System (CNS) of a person and translates such measurements into an artificial output, which is used by the outside world for different purposes [1, 2]. Nowadays, BCI has become a popular research field since it allows the development of different applications in several areas. For example, BCI has been used in the medical area to allow intensive care unit patients to communicate with nurses or doctors through electroencephalograms (EEG) and mobile applications [3]; it also has been used to help people with disabilities or diseases e.g. Parkinson [4] to deliver them the capabilities to perform normal movements or regain autonomy by combining with other technologies such as robotics [5, 6]. Moreover, BCI has been used in psychology to make mental

© Springer Nature Switzerland AG 2019
Y. Tang et al. (Eds.): HCC 2018, LNCS 11354, pp. 35–47, 2019.
https://doi.org/10.1007/978-3-030-15127-0_4

status evaluations [7–9]. The medical applications are not the only ones, BCI also has been used in other areas, such as home automation [9], security [10], neuromarketing [10] and videogames projects (e.g. BrainArena [9] and GATE research project [11]). As mentioned before, BCI has many applications in different areas; however, one of the areas which more help delivers to people is the medical one, since it contributes to improve the life quality of users. According to World Health Organization, between 20 to 50 million of inhabitants around the world suffer to car accident injuries which causes temporal or permanent disabilities [12]. Disabilities Statistics Annual Report of 2016 also indicates that around 10% of USA's population have some sort of disabilities.

In this situation, this work proposes and develops a BCI solution for people with movement disabilities. The proposed solution will monitor and translate brainwaves of a person with movement disabilities to replace/rehabilitate his/her natural movements or for allowing him/her to control different devices such as household appliances.

The rest of the paper is organized as follows. First, Sect. 2 explains some concepts related to BCI and EEG, and details the signals generated when a person blinks. Then, Sect. 3 explains the details of the proposed solution. Finally, Sect. 4 concludes the present paper.

2 Background

BCI is a system that evaluates the Central Nervous System (CNS) activities and translate them into an artificial output that replaces the natural CNS output [13], and there are several technologies for measuring brain activity. Among them, the most common noninvasive technology is the electroencephalogram (EEG) [14]. This method has high temporal resolution and it is capable of measuring changes that occur within milliseconds [15]. The main benefits of EEG are that it is easy to set up, it is portable, and it is relatively cheap. It includes sensors that detects the electrical activities of the brain that are recorded from the scalp [16–18]. As mentioned previously, the activities of the person are reflected in their brainwaves; for example, when a person does a task like blinking, it is possible to see the changes of the raw EEG signals since such movement produces a slight activation of the brainwaves. Eye blinks are normally considered internal artifacts in the EEG, and these artifacts can be good control signals since they can be used in various BCI applications [19].

3 A BCI Solution to Support People with Movement Disabilities

3.1 Definition of the Problem and Development Methodology

As mentioned in the first section, one of the areas where BCI delivers more help to people is the medical one, since it contributes to improve the life quality of users. According to the World Health Organization, between 20 to 50 million of inhabitants around the world suffer injuries due to car accidents [12] producing temporal or permanent disabilities. Disabilities Statistics Annual Report of 2016 also indicates that around 10% of USA's population have some sort of disabilities. With this antecedent, we believe that BCI applications can help people with disabilities by assisting them to execute activities they were not able to. In this aspect, the intention of this work is to create a solution (composed of hardware and software) for people with physical disabilities who cannot move most part of their body but can make facial expressions i.e. eye blinks. The proposed solution will allow the aforementioned people to control devices that they would not normally be able to. For example, a person who cannot turn on and off the lights by him/herself, now by using his/her brainwaves generated when they blink, they will be able to do so. For the development of the present solution, the traditional waterfall model will be used, which defines a series of steps to follow throughout the software project [20, 21] i.e.: (1) requirements, (2) design, (3) implementation, (4) verification, and (5) maintenance. Since the purpose of this work is to provide a functional prototype, the fifth step i.e. maintenance will not be executed.

3.2   Development of the Proposed Solution

3.2.1   Requirements

As mentioned previously, the intention of this work is to create a solution (composed of hardware and software) for people with physical disabilities who cannot move most part of their body but can make facial expressions. The proposed solution will allow the aforementioned people to control devices that they would not normally be able to. In the following, the list of the main requirements of the proposed solution are described:

- The user will use a non-intrusive way of reading brain signals i.e. an EEG device.
- The facial expression used for the present solution will be eye blinks.
- The user will execute a combination of different types of eye blinks to execute several types of commands to control different devices.

3.2.2   Design

The general architecture of the proposed solution is structured with the following components (see Fig. 1).

- User: The person who will use the proposed solution.
- EEG: The electroencephalograms device which will gather the brainwaves of the user. This device will detect the electrical signals produced by the brain and will convert them to analog signals. The analog brainwave signals will be transferred to the feature extraction module.
- Feature extraction: The raw data delivered by EEG needs to be processed. The feature extraction module eliminates noise signals and eliminate those signal not required by the system. In other word, the present systems only take those brain signals that help to identify eye blinks; the module will extract EEG signals produced when a person opens and closes their eyes. The algorithm used to detect different type of eye blinks are explained in details in the next subsection.

- Feature Translation: Once extracted the brainwaves produced when the user blinks, those signals are translated into device control commands understandable by the controller. The encoding system used by this component to translate the signals into commands are detailed in the next subsection.
- Controller: The controller receives the output of the feature translation module (i.e. device control commands) and convert them into signals understandable by the device which the user wants to control e.g. turn on/off of a household appliance.
- Devices: Devices which are controlled by the user using his/her brainwaves.

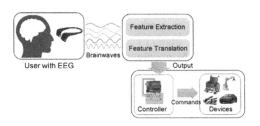

Fig. 1. Architecture of the proposed BCI system

3.2.3   Implementation

The design proposed in the previous sub-section has been implemented using the diagram shown in Fig. 2. The EEG device used for the developed prototype is Muse Headband, a brain sensing headband that measures brain activities via four sensors [23]. The gathered brainwaves is transferred to an application called Muse Monitor [22] installed in a mobile device via Bluetooth protocol. The raw data received by Muse Monitor is streamed in real time via User Datagram Protocol (UDP) connection. The UDP stream is received by a software developed exclusively for the present prototype, which includes the feature extraction and translation functionalities. Once executed the feature extraction and translation processes, the software delivers the output to the Arduino system (controller) which will sent the predetermined signals to control external devices (i.e. household appliances for this prototype). In the following, we explain the details of the implementation.

3.2.3.1 User with EEG and Muse Monitor
As explained before, the EEG device used for the development of the prototype is the Muse Headband. The data gathered by the Muse Headband is transferred to the Developed Software (called as Software in the rest of the paper) via Muse Monitor; in other words, Muse Monitor works as an intermediary between the Muse Headband and the Software. The intermediation of Muse Monitor was necessary since Muse Headband's manufacture company does not deliver the appropriate SDK for the 2016 model.

As shown in Fig. 2, raw data gathered by Muse Headband is sent to Muse Monitor using the Bluetooth protocol, while same data is forwarded to the Software using an UDP connection.

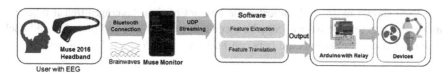

Fig. 2. Diagram of the prototype of the proposed solution

3.2.3.2 Software

Before describing how the Software was implemented, it is important to explain how the raw signals delivered by EEG device have been processed. In the following, we will explain the different algorithms proposed in this work to execute the feature extraction and translation processes of the raw brain signals.

A. Feature Extraction

Signals acquired by EEG need to be understood before anything else is made. Since the proposed system needs to detect the eye blinks of the user, the feature extraction module will extract those signals generated when the user opens and closes their eyes. To understand brainwaves variation when the user opens and closes his/her eyes, different samples of raw EEG signals were recorded and analyzed. The intention of the analysis was to distinguish the different types of blinks i.e. short, medium, and long blinks.

Muse Headband records the brainwave data in microvolts (μV) and the signal range goes from 0 to 1682.815 μV [22]. One of the reason of using the eye blinks signals in the proposed system is because those are one of the clearest ones; this conclusion was reached through several experiments made in this work. In our experiments, we could determinate that, when a user performs a short or medium blink, the signal changes by dropping first to a range below 750 μV and then elevating above 900 μV. This oscillation never enters to a long state of normality (when the oscillation stays between 800 and 900 μV). Figures 3 and 6 show how different type of signals (TP 9 and TP 10) change when a user make short and medium blinks. In contrast to the medium and short blinks, long blinks make the signals to drop below 750 μV, then the signals bounce back to a state of normality, and then they ascend above 900 μV (see Figs. 7 and 8).

Once understood the signal variation in eye blinks, the time duration of short, medium and long blinks in different people was analyzed. For this analysis, 21 people (11 women, 10 men) between the ages of 20 and 60 years old were selected; none of them have participated in other BCI researches. For the mentioned analysis, the experimental population was asked to make a short, medium and long blinks several

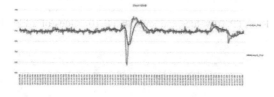

Fig. 3. Raw TP 10 values and the smooth TP 10 values of a short blink

times. Data shown in Table 1 indicates the amount of time (average) that it takes the brainwave signal to drop below 750 µV, ascend above 900 µV, and then descend back to a normal state (oscillations are between 800 µV and 900 µV) (Figs. 4 and 5).

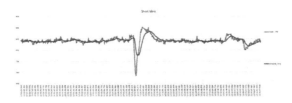

Fig. 4. Raw TP 9 values and the smooth TP 9 values of a short blink

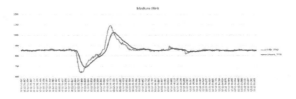

Fig. 5. Raw TP 10 values and smooth TP 10 values of a medium blink

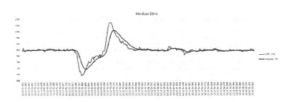

Fig. 6. Raw TP 9 values and smooth TP 9 values of a medium blink

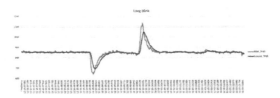

Fig. 7. Raw TP 10 values and smooth TP 10 values of a long blink

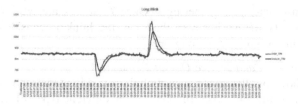

Fig. 8. Raw TP 9 values and smooth TP 9 values of a long blink

Based on Table 1, it is possible to deduce that the short and medium blinks have an average time under 1 s, while long blinks has an average of 2.05 s. Again, based on Table 1, we have noticed that the time differentiation between short and medium blinks, and between medium and long blinks were not significantly large. In this situation, to reduce the level of error between different types of blinks, we have decided only to use short and long blinks as valid signals for the proposed system.

Table 1. The average times of a short blink, medium blink and long blink (seconds)

Participant No.	Average short blink time	Average medium blink time	Average long blink time
1	0.2098	0.7669	1.7263
2	0.2992	0.7569	1.8810
3	0.2536	0.8886	1.4774
4	0.5630	0.7608	1.7421
5	0.3726	0.7508	1.7385
...			
19	0.3026	0.9062	1.9426
20	0.5262	0.9937	2.2891
21	0.3067	0.8315	1.3140
Total average	0.3042	0.9249	2.0543

After executing the experiments to classify different types of blinks, the algorithm for detecting automatically short and long blinks was developed (see Fig. 9). In such algorithm, two time variables i.e. T_1 and T_2 are used. As shown in Fig. 9, the algorithm is executed using the following steps. First, time variables T_1 and T_2 are set to 0. Then, the system starts to read the voltage level of the real time brainwave signals in a variable called as W. If W is lower than 750 μV, T_1 is set with the current time and the system returns to read W; this step allows to detect when the user has closed his/her eyes, and it is possible since the level of voltage generated by brainwaves when a person closes his/her eyes is lower than 750 μV. If W is equal or higher than 750 μV, the system verifies if T_1 has already set; if T_1 is equal to 0, the system returns to read W, since it means that the user has not closed his/her eyes yet. Otherwise, the system verifies if $W > 900$ μV; if $W > 900$ μV, the system sets T_2 with the current time, and it means that the user has open his/her eyes. Once T_1 and T_2 are set, ΔT is calculated where $\Delta T = T_2 - T_1$. If ΔT is lower than 1 s, the blink is considered short one; otherwise, the blink is considered long one.

B. Feature Translation and Device Output Commands

Once analyzed how to differentiate a short blink from a long one, an encoding system for correlating the different types of blinks with different types of commands was developed. For ease of interpretation, we have decided to represent the short blinks with dots (.) and long blinks with dashes (-) in the proposed encoding system.

Before starting explain the encoding system, it is important to explain that, since users must control the external devices with opening and closing their eyes, blink combinations produced in their normal life (when they do not want to activate a command) must be excluded from the system to reduce the possibility of false positives. For this, we have executed several tests with different users to get a combination of blinks that is not common in a daily life blink combination.

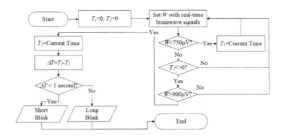

Fig. 9. Detection of short and long blinks

Through many tests in different users, we could notice that three long blinks in a row are not usually presented in a normal blink sequences. Based on this analysis, we have decided to create the following encoding system that included the aforementioned blink combination (see Fig. 10).

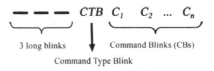

Fig. 10. Proposed encoding system

As shown in Fig. 10, a command always starts with three long blinks in a row. This allows the system to differentiate a command with a normal blink sequence of the user. Once received three long blinks, the system is ready to execute a command. The command is comprised of a Command Type Blink (*CTB*) and n number of Command

Blinks (*CBs*). Since the *CTB* is a blink, it can be a short or a long blink. Depending on the type of *CTB* (i.e. short or long blink), the command can be either a user command or control command. User commands are those commands that allow users to control external devices, while Control Commands are those related to the control of the proposed system that is not included in User commands. Finally, Command Blinks (*CBs*) are the combination of blinks that allows a user to execute a specific activity. Even though that it is quite obvious, it is important to mention that the number of command blinks i.e. value of n depends on the number of external devices controlled by the system, where 2^n = number of devices.

In the developed prototype, we have used the *CTB* = short blink for user commands, *CTB* = long blink for control commands, and two command blinks (i.e. n = 2). Based on this configuration, the combination of commands shown in Table 2 was implemented, which executes the finite state machine illustrated in Fig. 11.

To deliver a better understating of the encoding system, an example is explained. Assuming that a user wants to turn the lights on (i.e. activate device 1) and turn the television on (i.e. activate device 2), he/she must execute the following blink combinations:

1. Activate the system executing the control command ——.. (4 long blinks and 2 short blinks)
2. Store in memory the user command to activate device 1 i.e. ——... (3 long blinks and 3 short blinks)
3. Store in memory the user command to activate device 2 i.e. ——..- (3 long blinks, 2 short blinks and 1 long blink)
4. Execute the stored commands executing the control command ————— (6 long blinks)

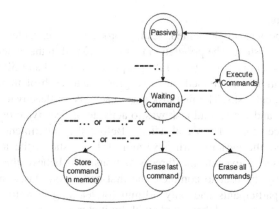

Fig. 11. State diagram of prototype's command pattern

Table 2. Commands used for experiments

Type of command	Description (what the command does)	Blink combination
Control	Activate the system	——..
Control	Erase the last input command	——.-
Control	Erase all command and deactivate the system	———.
Control	Execute the chain of user commands	———
User	Activate device 1	—...
User	Deactivate device 1	—..-
User	Activate device 2	—.-.
User	Deactivate device 2	—.–

Once determined the details of the Feature extraction and translation processes, it is time to explain the development tools used for the creation of the Software. The Feature extraction/translation software was developed using Java JDK 8 and Netbeans IDE 8.2. The developed software makes a UDP connection to Muse Monitor and receives the brainwaves in real time. Once received the raw signals, they are processed following the previously mentioned algorithms. The resulting commands are delivered to a controller developed in an Arduino system.

3.2.3.3 Controller and External Devices
Finally, the commands resulted from the feature translation process of the Software is delivered to the Controller. The controller, which is comprised of an Arduino System with the corresponding relays and actuators, receives the commands and generates different signals to control (e.g. turn on/off) different external devices.

3.2.4 Verification
Once the prototype was implemented, testing process was executed to measure the effectiveness of the system. The population that participated in the experiment included 100 people (80 university students and 20 people between 30 and 50 years old). The experiment population was divided into five groups and each of them was asked to complete two different tasks. The first task was to activate or deactivate one device at a time (e.g. a lamp), and the second task was to activate or deactivate two devices in on command sequence (e.g. a lamp and a fan). Before the experiment, the participant learned how to use the system during 30 min. Table 3 shows the average attempts required by the user to execute the three situations. We believe that 2.53 and 3.47 average attempts by the user to complete the first and second tasks are not bad considering that the participants had only 30 min for learning how to use the system. Experimental results indicated that people with limited training process were capable to use the prototype with acceptable level of error. We believe that the percentage of error will be reduced considerably when the user gets used to the proposed solution.

Table 3. Number of attempts for executing tasks with 30 min training

Group No.	Task 1	Task 2
1	2.33	3.33
2	2.67	3.33
3	2.33	3.67
4	2.67	3.67
5	2.67	3.33
Average	2.53	3.47

Table 4. Number of attempts for executing tasks with 2 days training

Participant No.	Task 1	Task 2
1	2	3
2	2	4
3	2	4
4	1	3
5	2	2
Average	1.8	3.2

After executing the first experiment, a second one was executed to understand if a prolonged training of system usage would lead to a reduced level of error. For this test the population was reduced to 5 users and they were allowed to use the system for two days. This experiment was conducted with the hypothesis that an extended use of the system by the user could reduce the number of attempts to execute the same tasks of experiment 1 since they would become accustomed to interacting with the system. The results of this experiment are shown in Table 4. The second experiment demonstrated that a longer training time can reduce the number of attempts to execute the pre-established commands. The average for executing Task 1 was 1.8 attempts (reduction of 0.73 attempts i.e. improvement of 28.85%) while the average for executing Task 2 was 3.2 attempts (reduction of 0.27 attempts i.e. improvement of 14.44%). Based on the results of the second experiment, it could be possible to assume that the level of error will be gradually reduced while the user got used to the system.

4 Conclusion and Future Works

In the present work, a BCI solution that uses brainwaves generated by eye blinks as a method for controlling different devices was presented. This paper has proposed different algorithms for the feature extraction and translation processes and implemented a functional prototype to show the real benefits of the system. The developed prototype allows the control of real life household appliances using brainwaves extracted from a MUSE headband. Experimental results indicated that people with limited training process were capable to use the prototype with acceptable level of error. In the future, we will present an improved system that will differentiate a voluntary blink combinations from non-voluntary ones using artificial intelligence techniques. This improvement will reduce the time and number of attempts to send a command to a device. We hope the present and improved systems could help users with various types of paralysis to gain a certain degree of independence.

Acknowledgments. The authors would like to thank to Corporación Ecuatoriana para el Desarrollo de la Investigación y Academia - CEDIA for the financial support given to the present research, development, and innovation work through its GT program, especially for the IoT and Smart Cities GT fund.

References

1. Vourvopoulos, A., Niforatos, E., Hlinka, M., Skola, F., Liarokapis, F.: Investigating the effect of user profile during training for BCI-based games. In: Proceedings of the 2017 9th International Conference on Virtual Worlds and Games for Serious Applications (VS-Games), Athens, Greece, pp. 117–124. IEEE (2017)
2. Wolpaw, J.R., Birbaumer, N., McFarland, D.J., Pfurtscheller, G., Vaughan, T.M.: Brain–computer interfaces for communication and control. Clin. Neurophysiol. **113**(6), 767–791 (2002)
3. Dehzangi, O., Farooq, M.: Wearable Brain Computer Interface (BCI) to assist communication in the Intensive Care Unit (ICU). In: Proceedings of 2018 IEEE International Conference on Consumer Electronics (ICCE), Las Vegas, NV, USA, pp. 1–4. IEEE (2018)
4. Pringsheim, T., Jette, N., Frolkis, A., Steeves, T.D.L.: The prevalence of Parkinson's disease: a systematic review and meta-analysis. Mov. Disord. **29**(13), 1583–1590 (2014)
5. Chaudhary, U., Birbaumer, N., Ramos-Murguialday, A.: Brain–computer interfaces in the completely locked-in state and chronic stroke. Prog. Brain Res. **228**, 131–161 (2016)
6. Foresi, G., Freddi, A., Iarlori, S., Monteriu, A., Ortenzi, D., Pagnotta, D.P.: Human-robot cooperation via brain computer interface. In: Proceedings of 2017 IEEE 7th International Conference on Consumer Electronics – Berlin (ICCE-Berlin), Berlin, Germany, pp. 1–2. IEEE (2017)
7. Borisov, V., Syskov, A., Tetervak, V., Kublanov, V.: Mobile brain-computer interface application for mental status evaluation. In: Proceedings of 2017 International Multi-Conference on Engineering, Computer and Information Sciences (SIBIRCON), Novosibirsk, Russia, pp. 550–555. IEEE (2017)
8. Tang, J., Zhou, Z.: A shared-control based BCI system: for a robotic arm control. In: Proceedings of 2017 First International Conference on Electronics Instrumentation & Information Systems (EIIS), Harbin, China, pp. 1–5. IEEE (2017)
9. Abdulkader, S.N., Atia, A., Mostafa, M.S.M.: Brain computer interfacing: applications and challenges. Egypt. Inform. J. **16**(2), 213–230 (2015)
10. van Erp, J., Lotte, F., Tangermann, M.: Brain-computer interfaces: beyond medical applications. Computer **45**(4), 26–34 (2012)
11. Nijholt, A., van Erp, F.J.B., Heylen, D.K.J.: BrainGain: BCI for HCI and games. In: Proceedings of AISB 2008 Symposium Brain Computer Interfaces and Human Computer Interaction, Aberdden, Scotland, United Kingdom, pp. 32–35. The Society for the Study of Artificial Intelligence and Simulation of Behaviour (2008)
12. World Health Organization. http://www.who.int/features/factfiles/roadsafety/en/. Accessed 19 July 2018

13. Wolpaw, J.R.: Brain-computer interfaces: something new under the sun. In: Brain-Computer Interfaces: Principles and Practice, pp. 3–12. Oxford University Press, Oxford (2012)

14. He, B.: Neural Engineering, 2nd edn. Springer, New York (2013). https://doi.org/10.1007/978-1-4614-5227-0

15. Zhang, P., Jamison, K., Engel, S., He, B., He, S.: Binocular rivalry requires visual attention. Neuron 71(2), 362–369 (2011)

16. Salabun, W.: Processing and spectral analysis of the raw EEG signal from the MindWave. Prz. Elektrotechniczny 90(2), 169–174 (2014)

17. Wolpaw, J.R., McFarland, D.J.: Control of a two-dimensional movement signal by a noninvasive brain-computer interface in humans. Proc. Natl. Acad. Sci. U. S. A. 101(51), 17849–17854 (2004)

18. Doud, A.J., Lucas, J.P., Pisansky, M.T., He, B.: Continuous three-dimensional control of a virtual helicopter using a motor imagery based brain-computer interface. PLoS One 6(10), 1–10 (2011)

19. Chambayil, B., Singla, R., Jha R.: Virtual keyboard BCI using Eye blinks in EEG. In: Proceedings of 2010 IEEE 6th International Conference on Wireless and Mobile Computing, Networking and Communications (WiMob), Niagara Falls, ON, Canada, pp. 466–470. IEEE (2010)

20. Royce, W.W.: Managing the development of large software systems: concepts and techniques. In: Proceedings of the 9th International Conference on Software Engineering, Monterey, California, USA, pp. 328–338. IEEE (1987)

21. Ali Munassar, N.M., Govardhan, A.: IJCSI Int. J. Comput. Sci. 7(5), 95–101 (2010)

22. Muse Monitor. http://www.musemonitor.com. Accessed 19 July 2018

23. Muse. http://www.choosemuse.com, last accessed July 19 2018

Research on Configuration Arrangement of Spatial Interface in Mobile Phone Augmented Reality Environment

Ye Dai[1](✉) and Wenjun Hou[2]

[1] School of Digital Media and Design Arts,
Beijing University of Posts and Telecommunications, Beijing 100876, China
daiyer@foxmail.com
[2] Beijing Key Laboratory of Network Systems and Network Culture,
Beijing University of Posts and Telecommunications, Beijing 100876, China

Abstract. Recently, due to the release of ARKit and ARCore, the mobile phone can achieve stable motion track via COM (concurrent odometry and mapping). By this way, mobile phones can display the interface in the real space like the multimodal interface in VR (virtual reality) glass, users can browse and interact with interface not only by touching screen but also by motion modal (moving and rotating mobile phone). This paper studied the optimal arrangement of spatial interface and the influence of various layout parameters on browsing efficiency. Through the test platform program, users can adjust the layout parameters (shape, distance, and size) of the spatial interface, and do some browsing tasks in each layout type. The result of interview feedback data analysis showed that the main factors affected by layout parameter on the browsing experience are screen size, movement rate and mapping deformation. And calculated their alternative relationship with layout parameters. Finally, through the recorded data of browsing speed and satisfaction coming from test platform, it figured out the best value and critical point for each layout parameters.

Keywords: Spatial interface · Augmented reality · Multimodal interface · Mobile phone interaction

1 Introduction

In order to display more information, the screen size of mobile phone had been expanding in recent years. Recently, due to the advent of ARKit and ARCore, user can get very stable SLAM (simultaneous localization and mapping) experience in augmented reality through its mobile phone. Mobile phone can track motion and display interface in real space stably, by using a process called concurrent odometry and mapping to understand where the phone is relative to the world around it.

Recently, some mobile browsing applications tried to display interface in this way. For example, the AR version of Twitter display a large amount of interface into the space, as shown in Fig. 1.

© Springer Nature Switzerland AG 2019
Y. Tang et al. (Eds.): HCC 2018, LNCS 11354, pp. 48–59, 2019.
https://doi.org/10.1007/978-3-030-15127-0_5

Fig. 1. Spatial layout of Twitter information

2 Related Research

2.1 Augmented Reality Interaction Design

With the maturity of augmented reality technology, research on interaction and inter-face design in augmented reality scenarios had gradually increased [1]. It was a problem to be solved that use the latest technology to construct a reasonable interaction mode and bring a better user experience. Previous research in the augmented reality interaction design were focusing on novel interaction modal, such as gesture interaction [2, 3], wearable device collaborative interaction [4, 5], semantic interaction [6, 7], etc.

2.2 Spatial Interface

At present, the research on the layout of spatial interface was mainly based on head-mounted display devices. Lin et al. measured the interactive feedback performance of the AR spatial interface of the head-mounted display device [8], and Costello studied the layout rules of the spatial interface of the VR scenario [9]. However, the interaction between the head-mounted device and the mobile terminal was quite different in some aspects, such as the fatigue of moving equipment and the sense of sedimentation caused by field of view.

2.3 Augmented Reality Interface on Mobile

At present, the research on mobile augmented reality interface was mainly focus on 2D screen interface rather than spatial interface, as the SLAM on mobile phone was not stable enough in previously. Thus to achieve stable localization, most previous research on spatial interface for mobile devices was based on specific visual feature. Liu et al. performed SLAM localization through feature patterns to spatially locate by identifying perspective changes in the specified pattern [10]. However, this method need ensure that the specific visual feature is always in the field of view of the camera, making it difficult for the user to move the rotating mobile phone freely. Therefore, the spatial interface needs to be attached to the visual feature, which has a large constraint on the spatial interface layout.

3   User Test

3.1   Goal and Process

Nowadays, due to the advent of ARKit and ARCore, it is possible to achieve stable SLAM on normal mobile phone device without constraint of special visual features. Consequently, it was convenient to display spatial multimodal interface. However, the uniform specification and research results for the layout rule of spatial interface on mobile phones were scarce, it's still uncertain that which layout way (distance, shape and physical size) could create a better browsing experience. This paper aimed to analyze the how spatial interface browsing experience is affected by their shapes, distances and sizes, and figure out the reasonable values of each parameter.

The content of this paper was mainly divided into three parts: Firstly, it introduced the test process, including test environment, test tasks and test platform program. Next it showed the influence of various arrangement parameters on browsing experience, through the feedback data obtained from participants' browsing experience interview. Finally, it shows the spatial interface layout form conclusions and the appropriate range of various parameters value, which is obtained through the analysis of speed data record by test platform.

3.2   Participants

Thirty participants (14 males and 16 females) took part in the testing, with ages of 23.1 years old on average (SD = 3.4). All participants had experience of using augmented reality mobile applications. They were with normal vision or corrected vision in this test. And they all took some practice to be familiar with the operation on spatial multimodal interface.

3.3   Task

Participants need browse spatial interface to find out correct icon which was specified in task. The spatial interface layout mode has four basic shapes (flat, vertical cylinder, horizontal cylinder, sphere) in tasks, as shown in Fig. 2. Participants can change interface shape type and adjust others parameters (distance and physical size). Program recorded participants' speed in each task. And after each task, we interviewed about the experience of that layout type. Then we used the speed data and interview results to analysis the rule of each interface layout parameter (shape, distance and physical size) effected by browsing satisfaction and performance, and figured out suitable range of the layout parameters.

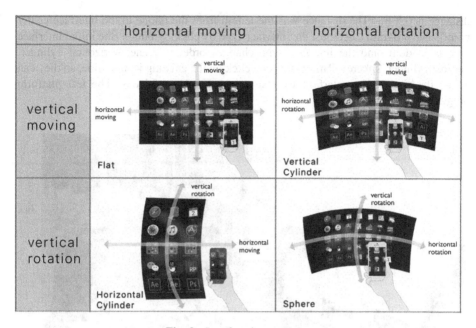

Fig. 2. Interface layout shape

Traditional interface contents in browsing tasks used to be text, list, picture or icon. In our test tasks, the performance in vertical and horizontal direction needed to be compared, some contents like text and list that have visual direction guidance when browsing (from left to right, from top to bottom) aren't suitable for this task. Icon content hasn't visual direction guidance, it's selected as the interface content in the task. Each icon content is uppercase letters, and the screen projection size is 70 pixels (it's projection of physical size rather than fixed screen size, and it will be smaller as the distance increases).

3.4 Environment

Participants tested in sitting posture, the height of the table and chair was 0.75 m and 0.44 m, and the distance between the table and chair was 0.35 m. The height of the spatial interface center point from the ground was 1 m, which was close to the height of handheld mobile phone in sitting posture for most people. The device used in the test was iPhone7 (system version is iOS11.2.1).

3.5 Platform Program

The test platform program was developed by the Unity engine. The augmented reality rendering and interface logic were processed in conjunction with Unity-ARKit-Plugin and UGUI.

During the test, the measured distance can be adjusted by clicking plus and minus buttons, interface distance will change 0.05 m with each click. The initial default distance is 0.5 m. (The distance was the length from the phone to the interface center point). The program can adjust the interface type by clicking button (the initial interface type is random, and the interface type changes order is plane, horizontal cylinder, vertical cylinder, sphere). Participants can click on the existing icon to hide, so they can adjust the size of the interface by change the number of icons. The test platform program was shown in Fig. 3.

Fig. 3. Program for the experimental platform

4   User Experience Factors

Through the test interview, its result showed that the main browsing experience factors affected by spatial interface layout parameters (distance, physical size and shape) were screen size, movement rate and mapping deformation.

4.1   Screen Size

The icon pixel size on mobile phone screen depended on both the distance and physical size (proportional to physical and inversely proportional to distance). Therefore, some interfaces were the same in screen pixel size but different in distance and physical size.

While further size was twice from closer interface's and further distance was twice from the closer interface, their pixel sizes reflected on mobile phone screen were the same in both the visible icon size and icons number. In this test, icon physical size was computed dynamically, to make sure that every distance interface's reflected screen sizes were 70px, as shown in Fig. 4.

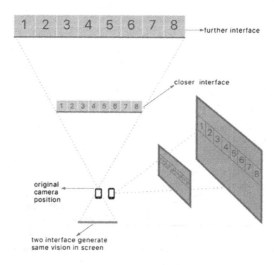

Fig. 4. Fix icons to same screen pixel size in different distance

4.2 Movement Rate

Although in each distance we can adjust icon physical size to fix screen pixel size, it didn't mean distance hasn't effect on user experience. Interface in different distance groups had difference moving change rate. It represented reflected size changing degree while distance was changing.

For example, when mobile phone got close to interface, the visible icons size would be larger and icons numbers would be less. But the change degree was different between the further interface and closer interface, as shown in Fig. 5.

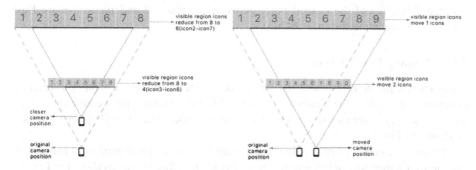

Fig. 5. The effect of different distances on the movement rate

According to users' feedback in the test, with distance increasing, user experience on movement rate can be divided into three parts, as shown in Fig. 6.

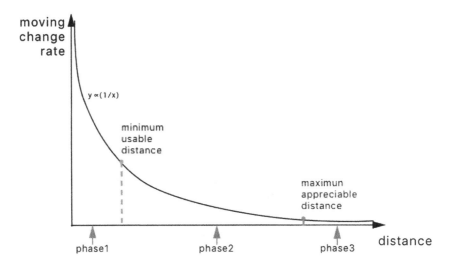

Fig. 6. The relationship between the rate of change of movement and the distance

Phase 1: in this distance range, the moving change rate was too large to use, user felt the interface was enlarged too much while their hands just moved a little space.

Phase 2: in this distance range, interface distance is appreciable, users felt interface's zooming and moving while their hands moved, and can estimate the distance by moving change rate.

Phase 3: in this distance range, interface distance was unappreciable, as the moving change rate was too small that user even couldn't perceive the change of interface while their hands moved.

The critical point between the phase 1 and phase 2 named as minimum usable distance (MUD), and the critical point between the phase 2 and phase 3 named as maximum appreciable distance (MAD).

4.3 Mapping Deformation

While mobile phone rotates and the normal of flat interface wasn't parallel to camera direction, flat interface the vision in screen would be deformed in perspective rules. The arc interface (cylinder and sphere) wouldn't be deformed when mobile phone rotated, as shown in Fig. 7.

According to users' feedback in the test, with the perspective degree increasing, user experience on mapping deformation can be divided into three parts, as shown in Fig. 8.

Phase 1: in this angle range, the change of interface deformed size was too small that user hard to recognize content had perspective deformation.

Phase 2: in this angle range, user can recognize the perspective deformation but can still discern the content clearly.

Phase 3: in this angle range, the perspective deformation was too serious to discern the content.

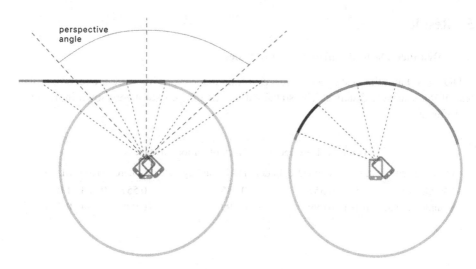

Fig. 7. Deformation in different shape types

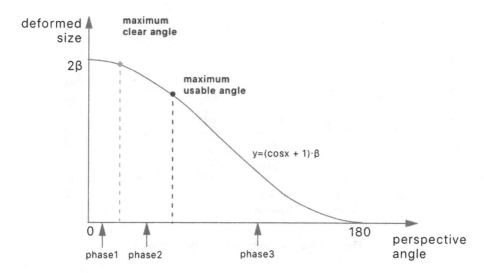

Fig. 8. The relationship between deformation size and perspective angle

The critical point between the phase 1 and phase 2 named as maximum clear angle (MCA), and the critical point between the phase 2 and phase 3 named as maximum usable angel (MUA). In arc direction (one direction in cylinder and two directions in sphere), the degree of deformation didn't change with perspective angle, the MCA and MUA in arc direction were the optimum angle and maximum angle of arm rotation when holding a mobile phone.

5 Results

5.1 Distance Data of Various Shape Types

MUD, MAD and the best speed distance of four shape types interface are record, in the test of the interface distance. The statistical results of the data were shown in Table 1 and Fig. 9.

Table 1. Descriptive statistics of various distances

Distance type	Flat	Vertical cylinder	Horizontal cylinder	Sphere	MUD	MAD
Sample mean	0.556	0.557	0.565	0.557	0.323	1.012
Sample variance	0.101	0.099	0.099	0.098	0.094	0.200

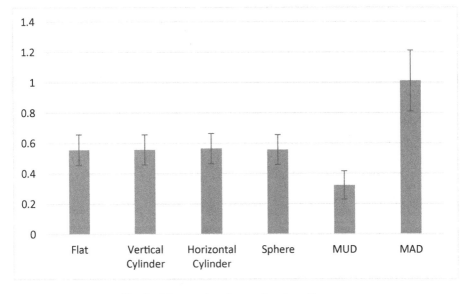

Fig. 9. Mean and variance of various distances

It can be seen from the distance statistics that the optimal distance results of each shape type of interface were similar. To figure out whether they had correlation, this paper analyzed the variance of the best speed distance between every shape type groups. Experimental null hypothesis: H0 – there is no difference, H1 – there is a significant difference (Table 3).

In the single factor homogeneity test of variance, P value was $0.997 > 0.05$ (Table 2), which indicate the variance was homogeneous, and analysis of variance can be performed. The test construction method of variance analysis was performed by selecting the LSD method, and the significance P value is $0.746 > 0.05$, which meant the null hypothesis (H0).

Table 2. Test of homogeneity of variance

Levene statistic	df1	df1	Sig.
0.15	3	116	0.997

Table 3. ANOVA

	Sum of squares	df	Mean square	F	Sig.
Between groups	0.014	3	0.005	0.411	0.
Within groups	1.271	116	0.011		
Total	1.285	119			

Table 4. Descriptive statistics at various angles

Interface type	Flat	Vertical cylinder	Horizontal cylinder	Sphere
Horizontal MUA mean	0.556	0.557	0.565	0.557
Horizontal MUA variance	0.101	0.099	0.099	0.098
Horizontal MCA mean	73.15	99.53	77.29	97.98
Horizontal MCA variance	19.74	37.53	29.79	31.24
Vertical MUA mean	76.16	66.92	102.99	99.29
Vertical MUA variance	12.24	13.53	14.73	13.74
Vertical MCA mean	57.98	53.26	75.51	74.86
Vertical MCA variance	15.01	11.84	19.69	17.26

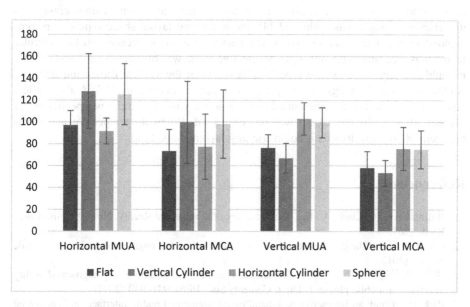

Fig. 10. Mean and variance of various perspective angles

5.2   Angle Data of Various Interfaces

In the test of the interface perspective angle, the MUA and MCA in both height and width of four shape types interfaces were recorded. The statistical results were shown in Table 4 and Fig. 10.

6   Conclusion

This article aimed to explore the appropriate spatial interface layout parameters (distance, size and shape) in the mobile phone augmented reality. Through user test and data analysis, it found three main user experience factors affected by layout parameters, which were screen size, movement rate and mapping deformation. And figured out the best value and critical point for this parameters.

Since this layout parameters values had individual influence for different people, to ensure that this values were suitable for most people, it was recommended to use 95% value ($\mu \pm 2\sigma$).

The sample mean of the most suitable distance was 0.56 m, and the MUD and MAD above 95% were 0.511 m and 0.612 m. Therefore, it was recommended that the distance of spatial interface keep within the range of 0.51–0.61 m.

The horizontal direction (width) was a linear interface (flat, horizontal cylinder), the MUA and MCA above 95% were 32.34° and 70.13°. The vertical direction (height) was a linear interface (flat, vertical cylinder), the MUA and MCA above 95% were 28.76° and 46.54°.

When choosing of interface shape type, it depended on the interface size. It was recommended that when the horizontal dimension interface content is small enough, as its perspective angle was within 32.34°, the horizontal layout shape type was recommended to use linear layout. When the perspective angle exceeds 70.13°, circular layout was recommended. When its perspective angle was between 32.24° and 70.13°, it could use any layout shape type. Similarly, when the vertical dimension interface content was small enough, as its perspective angle was within 28.76°, the vertical layout shape type was recommended to use linear layout. When the perspective angle exceeds 46.45°, circular layout was recommended. When its perspective angle was between 28.76° and 46.45°, it could use any layout shape type.

References

1. Billinghurst, M., Clark, A., Lee, G.: A Survey of Augmented Reality. Now Publishers Inc., Breda (2015)
2. Chun, W.H., Höllerer, T.: Real-time hand interaction for augmented reality on mobile phones (2013)
3. 준영, 박한훈, 박정식, et al.: Implementation of hand-gesture-based augmented reality interface on mobile phone. J. Digit. Contents Soc. 16(6), 941–950 (2011)
4. Park, H.: iHand: an interactive bare-hand-based augmented reality interface on commercial mobile phones. Opt. Eng. 52(2), 7206 (2013)

5. Choi, J., Park, H., Park, J., et al: Bare-hand-based augmented reality interface on mobile phone. In: IEEE International Symposium on Mixed and Augmented Reality, pp. 275–276. IEEE (2012)
6. Bartie, P.J., Mackaness, W.A.: Development of a speech-based augmented reality system to support exploration of cityscape. Trans. GIS **10**(1), 63–86 (2006)
7. Goose, S., Sudarsky, S., Zhang, X., et al.: Speech-enabled augmented reality supporting mobile industrial maintenance. Pervasive Comput. IEEE **2**(1), 65–70 (2003)
8. Lin, S., Cheng, H.F., Li, W., et al.: Ubii: physical world interaction through augmented reality. IEEE Trans. Mob. Comput. **PP**(99), 1 (2017)
9. Costello, A., Tang, A.: An egocentric augmented reality interface for spatial information management in crisis response situations. In: Shumaker, R. (ed.) ICVR 2007. LNCS, vol. 4563, pp. 451–457. Springer, Heidelberg (2007). https://doi.org/10.1007/978-3-540-73335-5_49
10. Liu, J., Yang, D.: A new type of interactive interface design based on ARToolKit. Energy Procedia **13**, 9642–9647 (2011)

Char-Level Neural Network for Network Anomaly Behavior Detection

Sheng Wang[1], Jiaming Song[2(✉)], and Ruixu Guo[3]

[1] Northen Institute of Electronic Equipment of China, Beijing 100191, China
sheng.w2000@sina.com
[2] Beijing University of Posts and Telecommunications (BUPT),
West Tucheng Road on the 10th, Beijing, China
songjiaming2015@163.com
[3] Chengdu Zhongke Hexun Science and Technology Co., Ltd., Chengdu, China
53303160@qq.com

Abstract. With the rapid development of the Internet, various attacks against network servers have been increasing. At present, most of the network protection measures are mainly aimed at attacks on the network layer and the transport layer. There is almost no protection against attacks at the application layer, but more and more attacks against the web are completed through the application layer. Traditional intrusion detection methods rely too much on rule matching, and there is a problem of high false positive rate. In view of the shortcomings of traditional network intrusion detection, this paper introduces char-level neural network method into the field of Network anomaly behavior detection, and the experimental data is the http requests parsed from the collected web logs. The experiment results show that, compared to traditional machine learning models, the char-level neural network performs better in term of detecting anomaly intrusion.

Keywords: Anomaly behavior detection · Char-level · Neural network ·
HTTP request

1 Introduction

At present, more and more attacks against the application layer, most network protection technologies, such as firewall [1, 2], access control [3], secure routing [4, 5], intrusion detection [6, 7], etc. In generally, it is for the protection of the transport layer and the network layer. The data of the application layer is not included in the detection range, which leads to the traditional network security devices are unable to protect against application layer attacks such as SQL injection attacks and application layer DDOS attacks [8]. Due to the important position of Web services in network services, Web services become the main target of application layer attacks, so it is urgent to introduce new detection methods to effectively identify attacks against Web services [9].

Currently in web attack detection, there are two main methods of malicious behavior detection: the misuse detection and the anomaly detection [10]. Snort [11] a network intrusion detection system based on a rule-based language combining

Y. Tang et al. (Eds.): HCC 2018, LNCS 11354, pp. 60–68, 2019.
https://doi.org/10.1007/978-3-030-15127-0_6

signature methods. However, this misuse detection method requires researchers to have expertise and just supports the detection of known attacks. What's worse, as the attack signatures increase, it costs too much time. For anomaly detection, Joshi and Geetha used a supervised machine learning method based on the Naive Bayes to detect the SQL injection [12]. Yu proposed a hybrid web log intrusion detection model that merged with misuse detection and anomaly detection [13]. Kruegel extracted statistics features and character features from http request parameters, excluding the request path, to construct the normal behavior model [14]. The disadvantage of this machine learning based abnormal behavior detection method is the difficulty in constructing features.

In view of the limitations of these detection methods and the abnormal behavior often appear as the insertion of specific characters in the http request, this paper proposes a char-level neural network method to automatically identifying network anomaly http request. The technique does not need to construct the features and it just needs to use the http request data of the web attack behavior to train the neural network. After the training is completed, the web request behavior can be identified and classified.

2 Data Model and Framework Overview

2.1 Data Model

In this paper, we mainly focus on HTTP requests logged by most common web servers (such as Apache). Besides, our anomaly detection model analyzes GET requests which use parameters to pass values to the server. A complete log includes the IP, the time stamp, the method, the referrer, the response code, the URI and so on. The URI includes most information of a log and is the main concern for anomaly detection.

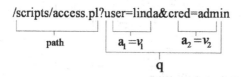

Fig. 1. The structure of an HTTP request

The complete structure of an URI is showed in the Fig. 1. An URI can be expressed as the composition of the request path (*path*) and request parameter (*q*). The request path is the path to the desired source. The request parameter is used to pass parameters to the server which is leaded by "?" character. A request parameter consists of an ordered list of n pairs of attributes and their corresponding values. That is $q = (a1, v1);$ $(a2, v2); \cdots (an, vn)$. This paper concentrates on the request path and request parameter.

2.2   Framework Overview

Figure 2 shows the overall framework for anomaly behavior detection based on HTTP requests. The framework consists of four parts: log collection, log parsing, char2vec, anomaly behavior classification.

Log Collection. Web servers generate a large amount of logs in order to record runtime information and system states. Each log contains the IP, the time stamp, the method, the referrer, the response code, the URI and so on. Since web logs certainly record the user behaviors, analyzing logs are increasingly important for anomaly behavior detection. Figure 2 shows a few web logs extracted from the Apache server.

Log Parsing. In most cases, attackers often insert malicious strings or codes into HTTP requests, such as SQL injections, Cross-site scripting and so on. Therefore in this paper, we need to parse the HTTP requests from web logs.

Char2vec. In this part, we use char2vec method to transform the HTTP requests to vectors. Because hackers often use a combination of specific characters and numbers to attack the web servers, this paper converts each HTTP request character into a vector, which is called char2vec.

Anomaly Behavior Classification. This process puts the vectors into the char-level neural network method to automatically identifying network anomaly HTTP requests. The method does not need to construct the features and it just needs to use the http request data of the web attack behavior to train the neural network. After the training is completed, the model can be used to predict the next web request behavior.

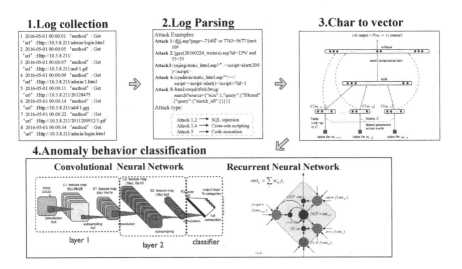

Fig. 2. Framework for anomaly behavior detection based on web logs.

3 Network Anomaly Behavior Detection Based on Char-Level Neural Network

At present, the commonly used analysis methods in network behavior analysis include probability-based methods, machine learning-based methods, data mining-based methods, and neural network-based methods. After considering the advantages and disadvantages of various methods, this paper adopts a neural network-based method, which uses Char-level Neural Network as the HTTP behavior classification module to classify the extracted HTTP behavior.

3.1 Char to Vector Method

Generating feature vectors is the basic process in natural language process. Word2vec model is often used to map word to a semantic space to get vectors. Word2vec is implemented by means of neural network. Considering the context of text, there are two models, CBOW and Skip-gram, which are similar in the training process.

In terms of anomaly behavior detection, attackers are interested in inserting specific characters to attack, this paper decide to use char2vec model, which is similar to word2vec. The skip-gram model will be explained in the following.

The skip-gram model uses a character as input to predict the characters around it. The training process are as follows.

Building a Vocabulary. This paper treats HTTP requests as textual content. Firstly, we need to count all occurrences of characters to generate a vocabulary. Then transferring each character into one-hot code.

Generating Training Samples. This part need to decide the window-size in order to generate 2*window-size training samples for each character, such as $(i, i - window)$, $(i, i - window + 1), \ldots, (i, i + window - 1), (i, i + window)$.

Training Model. The neural network iteratively trains a certain number of times, and obtains the parameter matrix of the input layer to the hidden layer. The transposition of each row in the matrix is the character vector of the corresponding character (Fig. 3).

3.2 Network Anomaly Behavior Detection Based on Char-Level Neural Network

For a training sample set (x_i, y_i), the neural network algorithm can provide a complex and nonlinear hypothesis model $h_{w,b}(x)$ with parameters W, b, which can be used to fit our data. The neural network consists of a large number of neurons. Figure 4 shows the structure of each neuron.

This process can be expressed as $h_{w,b}(x) = f(W^T x) = f(\sum_{i=1}^{3} W_i x_i + b)$, where $f(x)$ is called the activation function. In generally, the activation function is sigmoid function or tanh function. A neural network is the joining of many single "neurons" so that the output of one "neuron" can be the input of another "neuron". The following Fig. 5 is a basic neural network structure.

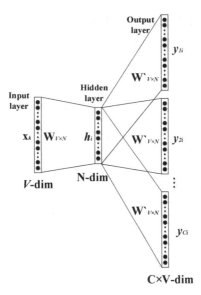

Fig. 3. Char2vec model based on Skip-gram

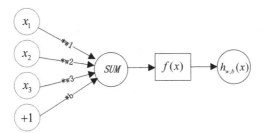

Fig. 4. Single neuron structure

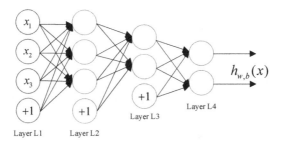

Fig. 5. The structure of neural network

The meaning of the parts in Fig. 5 are as follows. The leftmost layer L1 of neural network is called the input layer, and the rightmost layer L4 is called the output layer. The input vector x_i is obtained using the char2vec method above, and the output vector $h_{w,b}(x)$ is the result of the predicted anomaly behavior. What's more, the middle layers L2 and L3 are called the hidden layers, since we can't observe their values in the training sample set. Besides, the circle labeled "+1" is called the offset node.

Specifically, this paper uses two kinds of char-level neural network structures, namely convolutional neural networks [15] and recurrent neural networks [16] to classify network anomaly behavior.

4   Experiment and Result Analysis

4.1   Experiments Datasets and Environment

In this experiment, the web log data is collected in the Chinese government websites, which contains nine types of attacks, for example, SQL injections, Cross-site scripting, Code execution, SSRF vulnerability and so on. Besides, the data is provided by a web security company, which cooperates with us. Moreover, the labels are obtained based on known signatures.

We run all experiments on a 64-bit Windows 7 operating system. The experimental tool is a Python-based TensorFlow package.

4.2   Experiments Results

The experiment results show that, using the char-level convolutional neural network method in this paper, the recognition accuracy of network anomaly behavior reaches 100% on training set, 99.74% on testing set. What's more, we visualizes the accuracy and loss values of convolutional neural network in Fig. 6a, b using the Tensorboard module in TensorFlow. In the figure, the horizontal axis represents the number of iterations of the char-level neural work, and the vertical axis is the accuracy and loss values respectively. It is clear that the char-level neural network can finally reach convergence. Similarly, the accuracy and loss values of using Long Short Term Memory (LSTM) recurrent neural network show in Fig. 7a, b. The recognition accuracy of network anomaly behavior reaches 95.85% on training set, 94.53% on testing set.

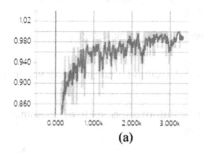 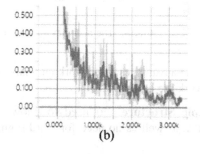

(a)   (b)

Fig. 6. The accuracy (a) and loss (b) of convolutional neural network on the testing sets

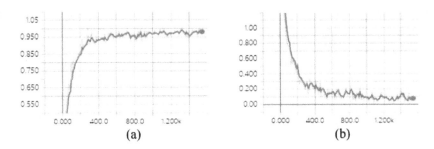

Fig. 7. The accuracy (a) and loss (b) of LSTM neural network on the testing sets

Next, we compare the classification effects of machine learning algorithms for anomaly behavior detection and char-level neural network. We extract the length of request path, the number of request parameter, the maximum length of request parameter, frequency of uppercase/lowercase letters, frequency of digitals, frequency of special characters and character entropy as input vectors of machine learning algorithms [13].

Compared to traditional machine learning models, such as Linear Regression (LR), Naïve Bayes (NB), Support Vector Machines (SVM) and K nearest neighbor (KNN), Fig. 8 show the char-level neural network model has a higher accuracy rate in term of detecting anomaly intrusion.

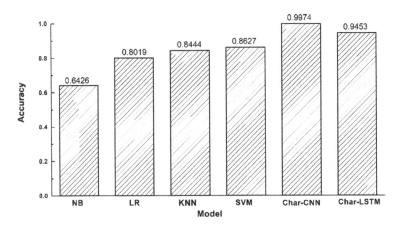

Fig. 8. The accuracy results of different models on the testing set.

The advantages of network anomaly behavior detection based on char-level neural network are as follows.

Simple Structure. No need to know structural information such as grammatical syntax, as well as no need for word segmentation.

Low Dimension. The vocabulary dimension of the character level is lower than the word, phrase, and sentence level.

Wide Application. Basically applicable to all languages.

5 Conclusion

Char-level neural network for anomaly behavior detection monitors the system from the perspective of application layer, so that some hidden attacks can be found, and the detection system can also change the analysis of each packet into attackers' behavior, thereby reducing the load on the system. A series of the experiment results show the char-level neural network has good anomaly behavior detection performance. Furthermore, the char-level neural network can be applied to other anomaly detection field.

References

1. Jiang, L.: Protection of computer network security based on firewall technology. China Comput. Commun. (2017)
2. Hao, W.: Network safety protection based on firewall technology. Commun. Technol. **7**, 010 (2007)
3. Sun, D., Liu, W., Ren, P., et al.: Reputation and attribute based dynamic access control framework in cloud computing environment for privacy protection. In: International Conference on Natural Computation, Fuzzy Systems and Knowledge Discovery, pp. 1239–1245. IEEE (2016)
4. Yao, L., Man, Y., Huang, Z., et al.: Secure routing based on social similarity in opportunistic networks. IEEE Trans. Wirel. Commun. **15**(1), 594–605 (2016)
5. Wu, Y., Xiang, Z., Feng, R., et al.: Secure routing based on node trust value in wireless sensor networks. Chin. J. Sci. Instrum. **33**(1), 221–228 (2012)
6. Xiang, L.I.: Local network intrusion detection algorithm based on empirical mode decomposition. J. Southwest China Norm. Univ. (2016)
7. Zhao, Y.: Network intrusion detection system model based on data mining. In: IEEE/ACIS International Conference on Software Engineering, Artificial Intelligence, Networking and Parallel/Distributed Computing, pp. 155–160. IEEE (2016)
8. Ranjan, S., Swaminathan, R., et al.: DDoS-resilient scheduling to counter application layer attacks under imperfect detection. In: Proceedings of the 25th IEEE International Conference on Computer Communications, pp. 1–13. IEEE, Piscataway (2006)
9. Lai, J.Y., Wu, J.S., Chen, S.J., et al.: Designing a taxonomy of web attacks. In: Proceedings of the 3rd International Conference on Convergence and Hybrid Information Technology, pp. 278–282. IEEE, Piscataway (2008)
10. Gollmann, D.: Computer Security, 2nd edn.
11. Caswell, B., Beale, J., Baker, A.: Snort Intrusion Detection and Prevention Toolkit. Syngress, Rockland (2006)
12. Joshi, A., Geetha, V.: SQL injection detection using machine learning. In: International Conference on Control, Instrumentation, Communication and Computational Technologies, pp. 1111–1115. IEEE (2014)
13. Yu, J., Tao, D., Lin, Z.: A hybrid web log based intrusion detection model. In: International Conference on Cloud Computing and Intelligence Systems, pp. 356–360. IEEE (2016)

14. Kruegel, C., Vigna, G., Robertson, W.: A multi-model approach to the detection of web-based attacks. Comput. Netw. **48**(5), 717–738 (2005)
15. Li, Z., Qin, Z., Huang, K., Yang, X., Ye, S.: Intrusion detection using convolutional neural networks for representation learning. In: Liu, D., Xie, S., Li, Y., Zhao, D., El-Alfy, E.-S. (eds.) ICONIP 2017. LNCS, vol. 10638, pp. 858–866. Springer, Cham (2017). https://doi.org/10.1007/978-3-319-70139-4_87
16. Kim, J., Kim, J., Thu, H.L.T., et al.: Long short term memory recurrent neural network classifier for intrusion detection. In: International Conference on Platform Technology and Service, pp. 1–5. IEEE (2016)

A Flexible Doubly Clamped Beam Energy Harvester with a Standard Rectifier Electric Circuit

Jie Mei[1(✉)], Huafeng Shi[1], Dingfang Chen[1], Lijie Li[1,2], Wenfeng Li[1], and Qiong Fan[1]

[1] Wuhan University of Technology, Wuhan 430063, China
meijieben@foxmail.com, 1070271795@qq.com,
cadcs@126.com, liwf_cn@126.com, 903518345@qq.com
[2] Swansea University, Swansea SA1 8EN, UK
l.li@swansea.ac.uk

Abstract. While wearable electronics are rapidly developing nowadays, it is greatly limited by the power solutions. Flexible piezoelectric energy harvester presents advantages of high energy density, compact architecture, and easy integration with MEMS, which provides an attractive prospect to power these next generation electronics. Since the flexible devices are usually devised with wavy, island-bridge, and precisely controlled buckling structures, the doubly clamped beam structure for energy harvesting application is analytically studied in this paper. Combine with Euler-Bernoulli beam theory and separation variable method, the analytical expression for output voltage is derived. By conducting the analytical simulation, it is found that the output power is related with the geometry dimensions, external excitation and load resistances. For further validation, experiment is systematically studied. By connecting the standard rectifier electric circuit with the energy harvesting device, it is found that a 0.1uF capacitor can be fully charged in 0.15 s, and the charged output voltage is about 2.5 V, which are successfully used for powering LEDs.

Keywords: Flexible energy harvester · Doubly clamped beam ·
Rectifier electric circuit

1 Introduction

Since flexible energy harvester provides a prospective solution for powering wearable electronics, remote and mobile environment sensors, and implantable biomedical devices, it has gained more and more attentions during the last decade for advantages of light weight, high flexibility and facile fabrication process. There are several sources that can be converted to electric power, which include mechanical vibration energy, thermal energy and solar energy. Regarding with the mechanical energy conversion process, piezoelectric, triboelectric, electromagnetic and electrostatic energy transducing mechanisms are intensively discussed. Among them, the flexible piezoelectric energy harvester is usually the potential choice for researchers because of its high

© Springer Nature Switzerland AG 2019
Y. Tang et al. (Eds.): HCC 2018, LNCS 11354, pp. 69–78, 2019.
https://doi.org/10.1007/978-3-030-15127-0_7

energy density, compact architecture, straightforward micromachining process and easy integration with MEMS.

In current researches, the cantilever beam with a tip mass is the typical structure for energy harvesting since it can bear large deformation with a lower excited frequency comparing with other geometry structures. However, for flexible energy harvesting device, they are usually fabricated with wavy, island-bridge, and precisely controlled buckling structures to improve the flexibility and stretchability. Therefore, the doubly clamped beam model rather than the cantilever beam structure is selected for analyzing the flexible piezoelectric energy harvester. In 2013, Pillatsch [1] presented a passively self-tuning mechanism of a clamped-clamped beam for wideband energy harvesting utilizing the sliding proof mass. In 2014, Liang [2] connected the doubly clamped beam with rectangular frame to further reduce the resonant frequency using asymmetrical proof mass and supporting mass. In their works, the reduced resonant frequency is 150 Hz and 165 Hz with output power of 0.992 mW and 0.844 mW respectively. Zheng conducted an experimental analysis of the PVDF doubly clamped beam energy harvester and found that its output power is almost twice as large as that of the sandwich beam with vibration frequency being ranged from 1 to 30 Hz at an acceleration of 0.1 g [3]. Kashyap [4] derived an analytical model for the doubly clamped unimorph segmented piezoelectric energy harvester with Euler-Bernoulli beam assumptions. In 2016, Emad [5] utilized the stretch strain of the doubly clamped beam structures to harness the ambient vibration energy, in which the geometry structure only composed of piezoelectric material without the elastic substrate. In their simulation results, it showed highly nonlinear phenomena with an output of 4 μW from vibration of 0.5 g at 70 Hz. In 2018, Damya [6] fabricated a miniature piezoelectric doubly clamped energy harvester with proof mass loading at beams center using MEMS technology, which can output voltage of 4 V and power of 80 μW for wireless sensor nodes. Zhou [7] characterized the length effects of piezoelectric layer on the output performance of doubly clamped beam energy harvester under random excitation.

In above studies, the reason for researchers focusing on the doubly clamped energy harvester is because more uniform stress distribution is expected in the clamped-clamped structure comparing with cantilever beam. In order to lower the working frequency, proof masses are attached in all above works. However, the alternative way to reduce the resonant frequency is to utilize the plastic substrate. In this paper, a doubly clamped beam energy harvester with flexible plastic substrate without proof mass is devised. In order to further validate the efficiency, a standard rectifier circuit is connected to power light emitting diodes (LED).

In the paper, the structure is outlined as follows: the analytical model of doubly clamped energy harvester is put forward in Sect. 2. In Sect. 3, simulation and experimental results are provided to prove the validity of the analytical method. The conclusion remarks are finally made in the Sect. 4.

2 Doubly Clamped Energy Harvester Model

The schematic figure of the doubly clamped piezoelectric energy harvester is shown as Fig. 1(a), where one end of the beam is fixed to the frame, and the other end is clamped and excited with applied external force $P(x, t)$ along the x-coordinate direction. The harvester is composed of PVDF piezoelectric layer and PVC flexible substrate. The initial length of the energy harvester is L. When it works, it will be compressed periodically. The corresponding deflection of arbitrary point in the piezoelectric layer is denoted as $w(x, t)$. Electrodes are set at the top and bottom surfaces of PVDF material.

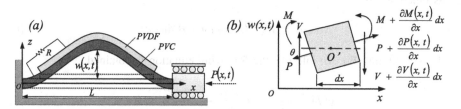

Fig. 1. (a) Schematic figure to show the working mechanism of the flexible doubly clamped beam energy harvester (b) Bending deformation map of finite element in piezoelectric layer

If the doubly clamped beam deformation conforms to the Euler-Bernoulli beam theory, the finite element deformation map of the piezoelectric layer is shown as Fig. 1 (b). According to the equilibrium equation in transverse direction, formula (1) can be deduced.

$$\rho A(x) dx \frac{\partial^2 w(x, t)}{\partial t^2} = -(V + dV) + V + (P + dP)\sin(\theta + d\theta) - P\sin\theta \quad (1)$$

In Eq. (1), ρ is the mass density per unit length, $A(x)$ is the cross-section area, V denotes the shear force, and θ is the rotation angle.

Similarly, according to the moment equilibrium status, Eq. (2) is deduced as follows,

$$(M + dM) - (V + dV)dx - M = 0 \quad (2)$$

From the relationship between rotation angle and transverse deflection, Eq. (3) is formulated as,

$$\sin(\theta + d\theta) \approx \theta + d\theta = \theta + \frac{\partial \theta}{\partial x} dx = \frac{\partial w(x, t)}{\partial x} + \frac{\partial^2 w(x, t)}{\partial x^2} dx \quad (3)$$

Combining Eqs. (1)–(3) and ignoring the higher order items, it is then derived as Eq. (4),

$$(EI)_{comp} \frac{\partial^4 w}{\partial x^4} + \rho A \frac{\partial^2 w}{\partial t^2} - P \frac{\partial^2 w}{\partial x^2} = 0 \quad (4)$$

Where $(EI)_{comp}$ denotes the equivalent bending stiffness of the composite beam. It is expressed as (5),

$$(EI)_{comp} = \frac{E_c w_c t_c^3 + E_p w_p t_p^3}{12} + \frac{E_c E_p w_c w_p t_c t_p \left(t_c + t_p\right)^2}{4\left(E_c w_c t_c + E_p w_p t_p\right)} \tag{5}$$

Where E_c, t_c, w_c are Young's modulus, thickness and width of substrate respectively.

If utilizing the separation variable method, the transverse displacement $w(x,t)$ can be assumed as (6),

$$w(x,t) = W(x)(A \cos \omega t + B \sin \omega t) \tag{6}$$

By substituting Eq. (6) into (4), the following equation is obtained,

$$(EI)_{comp} \frac{d^4 W}{dx^4} - P \frac{d^2 W}{dx^2} - \rho A \omega^2 W = 0 \tag{7}$$

Then the general form of solution to Eq. (7) is derived as (8),

$$W(x) = A\cosh sx + B\sinh sx + C\cos sx + D\sin sx \tag{8}$$

In Eq. (8), A, B, C and D are constants. Combining with boundary conditions of the doubly clamped beam energy harvester, $\sin sl \sin sl = 0$, the mode shape function is derived as (9),

$$W(x) = \sin\left(\frac{n\pi}{l}x\right) \quad n = 1, 2, 3, \cdots \tag{9}$$

And the nature frequency is calculated as Eq. (10),

$$\omega_n = \frac{\pi^2}{l^2} \sqrt{\frac{(EI)_{comp}}{\rho A}} \sqrt{n^4 - \frac{n^2 F_0 l}{\pi^2 (EI)_{comp}}} \tag{10}$$

In order to solve the forced vibration of doubly clamped beams, separation variable method is utilized as Eq. (11),

$$w(x,t) = \sum_{n=1}^{\infty} W_n(x) q_n(t) \tag{11}$$

Where $W_n(x)$ denotes the mode shape at nth natural frequency, and $q_n(t)$ represents the corresponding generalized coordinate.

If (11) is substituted into Eq. (4), then (4) can be rewritten as (12),

$$(EI)_{comp} \left(\sum_{n=1}^{\infty} \frac{d^4 W_n(x)}{dx^4} q_n(t) \right) + \rho A \sum_{n=1}^{\infty} W_n(x) \frac{d^2 q_n(t)}{dt^2} = P(x,t) \frac{d^2 W_n(x)}{dx^2} q_n(t)$$

(12)

Through integrating above equation from 0 to 1, it can be simplified into the following form with orthogonality conditions.

$$\frac{d^2 q_n(t)}{dt^2} + \omega_n^2 q_n(t) = \frac{1}{\rho A b} Q_n(t)$$

(13)

Where $Q_n(t) = \int_0^l P(x,t) W_n(x) dx$, and $b = \int_0^l W_n^2(x) dx$.
Then the solution to Eq. (13) is obtained as (14).

$$q_n(t) = A_n \cos \omega_n t + B_n \sin \omega_n t + \frac{1}{\rho A b \omega_n} \int_0^t Q_n(\tau) \sin \omega_n(t - \tau) d\tau$$

(14)

In (14), the first two terms represent the vibration state under transient conditions, and the third term represents the vibration state under steady state. Therefore, when solving the dynamic model, Eq. (14) can be simplified to (15).

$$q_n(t) = \frac{1}{\rho A b \omega_n} \int_0^t Q_n(\tau) \sin \omega_n(t - \tau) d\tau$$

(15)

Combining Eqs. (9) and (15), the deflection can be derived as (16),

$$w(x,t) = \frac{2F_0}{\rho A l} \sum_{n=1}^{\infty} \frac{1}{\omega_n^2 - \omega_0^2} \sin \frac{n\pi}{2} \sin \frac{n\pi x}{l} \sin \omega_0 t$$

(16)

Then the stress of an arbitrary point on the beam can be expressed as (17),

$$\sigma(x,t) = E_p y' \frac{\partial^2 w(x,t)}{\partial x^2}$$

(17)

Where E_p is the elastic modulus of piezoelectric material.
Substituting the formula (16) into (17), the deflection can be rewritten as (18),

$$\sigma(x,t) = -\frac{2F_0 E_p y'}{\rho A l} \sum_{n=1}^{\infty} \frac{1}{\omega_n^2 - \omega_0^2} \left(\frac{n\pi}{l} \right)^2 \sin \frac{n\pi}{2} \sin \frac{n\pi x}{l} \sin \omega_0 t$$

(18)

Because the piezoelectric vibrator is operating at low frequency, it suggests that $\omega_n \gg \omega_0$. Then (18) can be rearranged as (19),

$$\sigma(x,t) = -\frac{2F_0 E_p y'}{\rho A l} \sum_{n=1}^{\infty} \frac{1}{\omega_n^2} \left(\frac{n\pi}{l}\right)^2 \sin\frac{n\pi}{2} \sin\frac{n\pi x}{l} \sin\omega_0 t \qquad (19)$$

Substituting the above formula into the piezoelectric constitutive equation $D_3 = d_{31}T_1 + \varepsilon_{33}^T E_3$, the electric displacement is obtained as (20),

$$D_3(x,t) = -\frac{2d_{31}F_0 E_p y'}{\rho A l} \sum_{n=1}^{\infty} \frac{1}{\omega_n^2} \left(\frac{n\pi}{l}\right)^2 \sin\frac{n\pi}{2} \sin\frac{n\pi x}{l} \sin\omega_0 t + \varepsilon_{33}^T E_3 \qquad (20)$$

By integrating (20) across the electrode surface area, the generated charge can be deduced as (21),

$$Q_3 = \frac{2\pi w_p d_{31} F_0 E_p y'}{\rho A l^2} \sum_{n=1}^{\infty} \frac{n}{\omega_n^2} \left[\cos\frac{n\pi a}{l} - \cos\frac{n\pi(a+l_p)}{l}\right] \sin\frac{n\pi}{2} \sin\omega_0 t + \frac{w_p l_p \varepsilon_{33}^T}{t_p} V_3(t) \qquad (21)$$

Where y' represents the distance from an arbitrary point to the neutral plane, l and l_p are lengths of the substrate and piezoelectric material respectively. a is the distance from the piezoelectric material to the length boundary. w_p denotes the width of piezoelectric layer, and t_p is the thickness of piezoelectric layer.

Then current formula can be derived by integration of (21) with respect to time.

$$\begin{aligned}
I &= \frac{dQ(t)}{dt} \\
&= \frac{2\omega_0 \pi w_p d_{31} F_0 E_p y'}{\rho A l^2} \sum_{n=1}^{\infty} \frac{n}{\omega_n^2} \left[\cos\frac{n\pi a}{l} - \cos\frac{n\pi(a+l_p)}{l}\right] \sin\frac{n\pi}{2} \sin\omega_0 t + C_p \frac{dV(t)}{dt}
\end{aligned} \qquad (22)$$

Where $C_p = \frac{w_p l_p \varepsilon_{33}^T}{t_p}$.

Combining Eqs. (21) and (22), the induced potential can be obtained as (23),

$$V(t) = \frac{2\pi w_p d_{31} F_0 \omega_0 E_p y'}{\rho A l^2 \sqrt{C_p^2 + \frac{1}{R_L^2}}} \sum_{n=1}^{\infty} \frac{n}{\omega_n^2} \left[\cos\frac{n\pi a}{l} - \cos\frac{n\pi(a+l_p)}{l}\right] \sin\frac{n\pi}{2} \sin(\omega_0 t + \varphi) \qquad (23)$$

Since high-order vibration modes hardly impact the induced voltage, for the convenience of calculation, only the fundamental mode term is considered in the output equation. Then (23) is rewritten as (24),

$$V_0(t) = \frac{2w_p l^2 d_{31} F_0 \omega_0 E_p y'}{\pi^3 (EI)_{comp} \sqrt{C_p^2 + \frac{1}{R_L^2}} \left(1 - \frac{F_0 l}{\pi^2 (EI)_{comp}}\right)} \left[\cos\frac{\pi a}{l} - \cos\frac{\pi(a+l_p)}{l}\right] \sin\omega_0 t \qquad (24)$$

From Eq. (24), it can be deduced that the output potential is determined by the geometry dimensions, external excited force and load resistance.

3 Simulation and Experiment Results

In order to prove the feasibility of the model, both analytical analysis and experimental characterization are conducted. Table 1 shows the parameters of the doubly clamped energy harvester. Assuming that the externally applied load F0 is 3N, and the excited frequency is 30 Hz. From Eq. (24), the varying trend of deduced output voltage and output power versus time are shown as Fig. 2(a) and (b) respectively, where the connected load resistance is 5 MΩ. In the simulation results, the amplitude of output voltage is around 3.67 V, and the instantaneous power output is 2.7 μW.

Table 1. Parameters for the doubly clamped energy harvester

l	l_p	a	w_c	w_p	t_c
30 mm	20 mm	5 mm	10 mm	10 mm	3 mm
t_p	E_c	E_p	d_{31}	ε_{33}	
2 mm	2900 MPa	3900 MPa	21 pc/N	12 F/m	

(a) 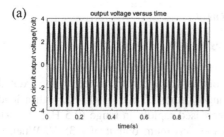 (b)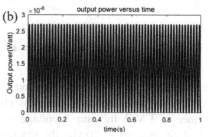

Fig. 2. Simulation results of output voltage and power by utilizing the analytical model

For further study of doubly clamped energy harvester model, the prototype has been fabricated. Figure 3(a) shows the experiment setup which include devices of the computer interface, signal generator (AFG3021C), power amplifier, vibrating shaker (Modal 50A Vibration Exciter), standard AC-DC rectifier circuit and the data acquisition card. Figure 3(b) is the enlarged view of the prototype.

From Fig. 3(b), it is observed that the flexible energy harvester is doubly clamped. When the shaker is excited by the signal generator, it will move one end of the device sinusoidally along the axial direction, while the other end is kept fixed. Since both the piezoelectric patch and the substrate are both polymers, the device is easily deformed into curve structure. As the piezoelectric layer is strained, the mechanical energy will be converted into electric energy through the direct piezoelectric effect. Figure 4 shows the result of the output voltage that have been captured by the data acquisition card. From the results, it shows that the amplitude of the output voltage is around 3 V, and the resonant frequency is 30 Hz, which agrees well with the analytical results.

Fig. 3. (a) Experimental setup, in which ① is the computer, ② is the signal generator, ③ is power amplifier, ④ is vibrating shaker, ⑤ is the doubly clamped energy harver, ⑥ is the AC-DC rectifier circuit and ⑦ is data acquisition card; (b) Enlarged view of the energy harvesting device.

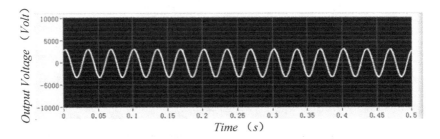

Fig. 4. Waveform captured by the data acquisition card

Figure 5(a) shows the varying trend of output voltage with respect to the excited frequency. As the excited frequency is changed from 0.5 Hz to 100 Hz with a load resistance of 3 MΩ, the output voltage is increased rapidly in range (0.5 Hz, 30 Hz). When it climbed up to 30 Hz, the reached the maximum output of 2.35 V. When it continued increasing the excited frequency, the induced voltage will be decreased. The same varying trend is observed when external load resistance is changed to be 4 MΩ and 5 MΩ. Figure 5(b) depicts the relationship between output power and external load resistance. From the trendline, a peak value of 4.8 μW is found, where the optimum load resistance of 3 MΩ is determined.

Although the output voltage across the electrodes are alternative signals, it can be rectified into DC signal through the standard bridge circuit. Figure 5(c) represents charging process of capacitor by the flexible energy harvester under different excitation frequency, where the capacitance is 0.1 uF. From the results, it is observed that the capacitor can be fully charged in 0.15 s, and the charging speed is independent with the excited frequency. Since the rectified output voltage in the standard AC-DC circuit is about 2.5 V, it can be utilized to power electronics. Figure 5(d) shows the possibility of the flexible doubly clamped energy harvester to power LEDs, which can make the flexible doubly clamped energy harvester present more attractive prospects in the future.

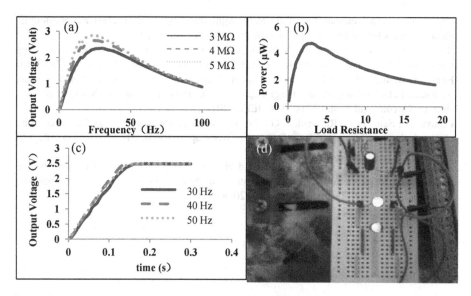

Fig. 5. (a) DC Output voltage versus frequency; (b) output power versus load resistance; and (c) output voltage in the charging process through standard rectifier electric circuit; and (d) applications of the energy harvester to power LEDs.

4 Conclusion

In the paper, the analytical model of the flexible energy harvester is provided based on the Euler-Beam theory and separation variable method. In order to prove the feasibility of the model, both simulation and experimental analysis are provided. Through comparison, the simulation results of 3.67 V and instantaneous power 2.7 µW coincides well with experiments. For further study of doubly clamped energy harvester model, the authors have also fabricated the prototype. By connecting the device with a standard AC-DC rectifier electric circuit, the 0.1 uF capacitor can be fully charged up to 2.5 V in 0.15 s, through which LEDs is lit to show its attractive prospects.

Acknowledgements. This paper was supported by the National Natural Science Foundation of China (Grant No. 51805395) and Natural Science Foundation of Hubei Province (20181j001: Interfacial Defects Initiation Mechanism of Flexible Laminated Thin Film Energy Harvester and its Fabrication Process).

References

1. Pillatsch, P., Miller, L.M., Halvorsen, E., et al.: Self-tuning behavior of a clamped-clamped beam with sliding proof mass for broadband energy harvesting. In: Journal of Physics: Conference Series, PowerMEMS 2013 Conference, vol. 476, p. 2068 (2013)
2. Liang, Z., Chundong, X., Bo, R., et al.: A low frequency and broadband piezoelectric energy harvester using asymmetrically serials connected double clamped–clamped beams. Jpn. J. Appl. Phys. **53**, 087101 (2014)

3. Zheng, Y., Wu, X., Parmar, M., et al.: Note: high-efficiency energy harvester using double-clamped piezoelectric beams. Rev. Sci. Instrum. **85**(2), 102 (2014)
4. Kashyap, R., Lenka, T.R., Baishya, S.: A model for doubly clamped piezoelectric energy harvesters with segmented electrodes. IEEE Electron Device Lett. **36**(12), 1369–1372 (2015)
5. Emad, A., Mahmoud, M.A.E, Ghoneima, M., et al.: Modeling and analysis of stretching strain in clamped-clamped beams for energy harvesting. In: IEEE International Midwest Symposium on Circuits and Systems, pp. 1–4. IEEE (2017)
6. Damya, A., Sani, E.A., Rezazadeh, G.: An innovative piezoelectric energy harvester using clamped–clamped beam with proof mass for WSN applications. Microsyst. Technol., 1–9 (2018). 2017
7. Zhou, X., Gao, S., Jin, L., et al.: Effects of changing PZT length on the performance of doubly-clamped piezoelectric energy harvester with different beam shapes under stochastic excitation. Microsyst. Technol. (11), 1–15 (2018)

Optimization of Node Deployment in Wireless Sensor Networks Based on Learning Automata

Yu Zhang, Shufen Liu, and Lu Han[(✉)]

Jilin University, Changchun, China
zyu16@mails.jlu.edu.cn, {liusf,hanlu}@jlu.edu.cn

Abstract. In order to solve the problem of node deployment in wireless sensor networks, this paper proposes a method based on learning automaton to optimize node deployment. Determine node deployment location by the experience gained from the continuous communication between learning automata and the random environment. Select the optimal position sets among the selectable positions. Learning automata can quickly reach full coverage and significantly reduce deployment costs. The simulation results reveal that our developed approach can not only reduce the sensor node power consumption effectively, but also prolong the network lifetime in Wireless Sensor Networks (WSNs).

Keywords: Learning automata · Node deployment · Wireless sensor network · Optimization

1 Introduction

Wireless Sensor Networks (WSNs) are distributed sensor networks. The rapid growth of WSNs in recent years has benefited from the rapid development of Micro-Electro-Mechanism System (MEMS), System-on-Chip (SOC), infinity communications and low-power embedded technologies. The advantages of WSNs such as robustness, scalability and distribution, and self-organization are widely used in many fields such as military, medical and environmental monitoring.

The design of WSNs is a very complicated process that requires stability, timeliness and continuity. Node deployment is a core issue in WSNs, including deployment cost, detection capability and life cycle. The sensor node has a small battery which can't be replaced. The key part of extending the network's life cycle is energy conservation. The primary goal of node deployment is to cover the monitored targets and maintain connectivity of the network. Each node of the sensor has to transmit and receive data, transmit information to the sink through a multistage jump, and then transmit it to the final user via a network or satellite. The problem is that sensors closer to sink will be used more often as a relay station for data communications, and will use the battery more quickly.

Many types of algorithms have been applied to resolving node deployment issues. For example, using the multi-objective evolutionary algorithm, the approximate problem of the Pareto front (PF) is transformed into a number of single-target optimization problems using the decomposition method [1]. Ant colony algorithm significantly improves load balance by sacrificing a small coverage cost [2]. By the virtual

© Springer Nature Switzerland AG 2019
Y. Tang et al. (Eds.): HCC 2018, LNCS 11354, pp. 79–85, 2019.
https://doi.org/10.1007/978-3-030-15127-0_8

force's accurate self-deployment algorithm, virtual gravity is constructed between the attraction and repulsive force to automatically generate the path between sink and target [3]. The learning automata used in this paper is a model of the interaction between stochastic automata and stochastic environment with variable structure. It splits the coverage problem into many rounds, and the result of each round is a solution. As the algorithm continues, the learning automaton can choose the best solution to find the location suitable for the node deployment in the current network environment.

2 Basic Algorithm

In this paper, the low cost and connectivity of the network should be ensured when the coverage problem based on the grid. By designing an algorithm to solve the problem, the deployed nodes need to cover the specified area and ensure low cost. The deployed node is connected to at least one node. The problem of node coverage is solved by learning automata to avoid energy-hole, reduce deployment costs and speed up deployment.

(1) Stochastic Automaton
A stochastic automaton is a six tuple $\{x, \Phi, \alpha, p, A, G\}$, where x is the input set, $\Phi = \{\Phi_1, \Phi_2, \ldots, \Phi_s\}$ is the set of internal, $\alpha = \{\alpha_1, \alpha_2, \ldots, \alpha_r\}(r < s)$ is output or action set [4]. p is an action probability vector that manages the selection of each state. A is an algorithm which is also called the update strategy or the optimization scheme which generates $p(n + 1)$ from $p(n)$. $G : \Phi \to \alpha$ is an output function. G is a stochastic function and deterministic.

(2) Environment
Environment has random response characteristics. Input the action into the environment and receive the response [5]. The response is binary $\{0, 1\}$, 0 is non-penalty mechanism and 1 is penalty mechanism. $C_i(i = 1, \ldots, n)$ is penalty probabilities. Each element in C is related to α. If C_i does not depend on n, then the environment is considered stable, otherwise it is not stable. It is assumed here that C_i is unknown.

(3) Learning Automata
Learning automaton is stochastic automaton that runs in a random environment. It updates its behavior according to the input of the environment. These responses from environment are all random, so the probability vector $p(n)$ is also random [6]. The input from the environment to learning automata can be divided into P model, S model and Q model. If the input set is binary, such as $\{0, 1\}$, the model is P model. The input set of the Q model is a finite set of different symbols. The input set of the S model is the random variable in the [0, 1] interval (Fig. 1).

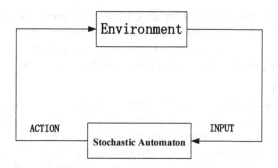

Fig. 1. The principle of learning automata

(4) Optimization Scheme

The optimization scheme of updating the action probability is the key factor to determine the performance of the learning automaton. Generally an optimization scheme can be represented as

$$p(n+1) = T(p(n), \alpha(n), x(n)) \tag{1}$$

T is an operator. $\alpha(n)$ and $x(n)$ represent the action and input of automata at time n. Learning automata can be divided into optimal or suboptimal according to the property, or it can be divided into linear, non-linear and mixed types according to the properties of functions.

The basic idea of optimization scheme is very easy to understand. If the learning automata chooses an action α_i at n time, it happens non-penalty input, then the action probability $p_i(n)$ will be improved, and all the other parts in $p(n)$ will be reduced, and vice versa.

In general, in the case of n, the action probability is α_i.

$$p_j(n+1) = p_j(n) - f_j(p(n)), \ for \ x(n) = 0 \tag{2}$$

$$p_j(n+1) = p_j(n) + g_j(p(n)), \ for \ x(n) = 1 \ (j \neq i) \tag{3}$$

The algorithm for $p_i(n+1)$ is to be fixed so that $p_k(n+1)$ $(k = 1, **, r)$ add to unity. Thus

$$p_i(n+1) = p_i(n) + \sum_{j \neq i} f_j(p(n)), \ for \ x(n) = 0 \tag{4}$$

$$p_i(n+1) = p_i(n) - \sum_{j \neq i} g_j(p(n)), \ for \ x(n) = 1 \tag{5}$$

Non-negative continuous function $f_j()$ and $g_j()$ have $p_k(n+1) \in (0,1)$ for all $k = 1, \ldots, r$ when $p_k(n) \in (0, 1)$.

By setting $f_i(p) = \alpha p_j, g_j(p) \equiv 0$ for all j, we get the linear reward-inaction (L_{R-I}) scheme. Ignore the environmental penalty input, the probability of action of these inputs remains unchanged. However, that replacing the penalty by inaction in the L_{R-P} scheme totally changes the performance from expediency to ε-optimality.

(5) System Model

First, make some assumptions about the current wireless sensor network. A sink is randomly deployed in the network, and sensor nodes and receivers will not change after deployment. The nodes of all sensors have the same initial energy. Sink has unlimited energy. The links in the wireless sensor network are symmetrical, and the nodes can calculate the distance between them according to the intensity of the signal. Figure 2 shows an example of a simple node deployment problem.

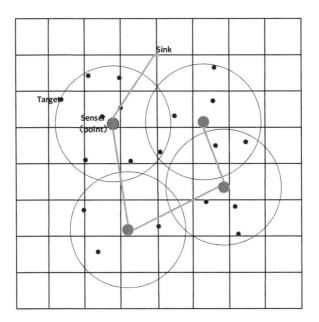

Fig. 2. Node deployment issues.

In the binary WSNs, the sensor field is composed of discrete grid points, the deployed sensor can detect the interest points in the sensing radius. You need to select as few nodes as possible from the candidate grid points so that all targets can be covered by the deployed nodes. Each member of the collection can be connected to sink by one or more jumps.

3   Algorithm Implementation

Through understanding learning automata, the following solutions are proposed for node deployment in wireless sensor networks. Assign an automata for each node to monitor all targets which need to be covered. Choose a position as a starting point. From the starting point, the automata deploys the next node after selecting an action from the action set. Calculate the number of coverage targets within the node. The proposed algorithm iteratively finds different possible coverage targets of the network, and depending on the response received from the environment, the coverage targets are penalized or rewarded. If the starting point is not the end point, choose a node with the maximum coverage targets as a new starting point. The automata updates the action set and chooses a new action. Update the set of uncovered targets. Record the route which is selected by the automata. Loop through the above operations until the target set is empty. The optimal solution is obtained by comparing the solutions obtained each turn.

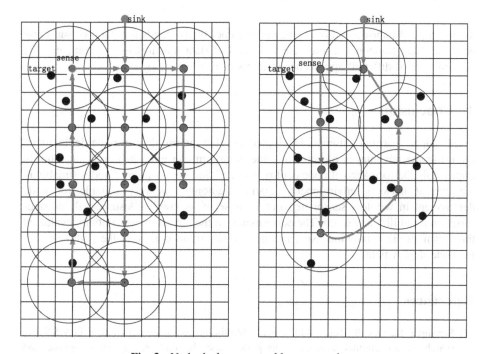

Fig. 3. Node deployment problem comparison

As shown in Fig. 3, the left picture shows how to solve the problem of node deployment in general situation. The right picture shows the learning automata to solve the same node deployment problem. It is obvious that learning automata is more stable than other methods, which can save the deployment cost and prolong the life cycle of

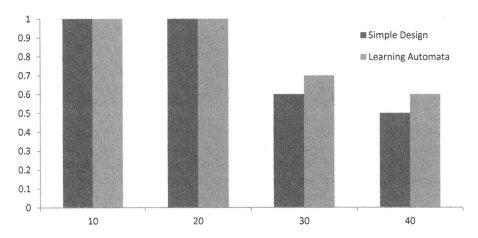

Fig. 4. Comparison of average network lifetime of different approaches

the network. As shown in Fig. 4, the vertical coordinate stands for Ratio of surviving nodes and the horizontal coordinate is Round of data transmission. It compares the performance of different approaches in terms of residual energy.

4   Conclusion

In this paper, learning automata is adopted to solve the problem of node deployment. By using the advantages of automata, the algorithm can efficiently find the route to improve the performance of node deployment and find the position of nodes suitably in the current network environment. Learning automata is very suitable for node deployment in wireless sensor network because of its good robustness, global optimization capability and so on. The performance of the algorithm was initially evaluated through multiple simulation experiments. The algorithm optimizes the path selection for node deployment.

References

1. Sengupta, S., Das, S., Nasir, M.D., et al.: Multi-objective node deployment in WSNs: in search of an optimal trade-off among coverage, lifetime, energy consumption, and connectivity. Eng. Appl. Artif. Intell. **26**(1), 405–416 (2013)
2. Liu, X., He, D.: Ant colony optimization with greedy migration mechanism for node deployment in wireless sensor networks. J. Netw. Comput. Appl. **39**(1), 310–318 (2014)
3. Yang, M.H., Cao, Y.D., Tan, L., et al.: A precision deployment algorithm in mobile sensor network. Trans. Beijing Inst. Technol. **29**, 27–31 (2009)
4. Cai, Y.L., Lou, W., Li, M.L.: Cover set problem in directional sensor networks. In: Proceedings of the IEEE International Conference on Future Generation Communication and Networking, Washington, DC, USA, pp. 274–278 (2007)

5. Chakrabarty, K., Iyengar, S.S., Qi, H., Cho, E.: Grid coverage for surveillance and target location in distributed sensor networks. IEEE Trans. Comput. **51**(12), 1448–1453 (2002)
6. Guo, X.M., Zhao, C.J., Yang, X.T., et al.: A deterministic sensor node deployment method with target coverage based on grid scan. Chin. J. Sens. Actuators. **25**(1), 104–109 (2012)
7. Misra, S., Kumar, M.P., Obaidat, M.S.: Connectivity preserving localized coverage algorithm for area monitoring using wireless sensor networks. Comput. Commun. **34**(12), 1484–1496 (2011)
8. Li, D., Liu, W., Cui, L.: EasiDesign: an improved ant colony algorithm for sensor deployment in real sensor network system. In: Proceedings of IEEE Globecom 2010, December 2010
9. Lin, F.Y.S., Chiu, P.L.: A near-optimal sensor placement algorithm to achieve complete coverage/discrimination in sensor networks. IEEE Commun. Lett. **9**(1), 43–45 (2005)
10. Zhang, Y.Z., Wu, C.D., Cheng, L., et al.: Research of node deployment strategy for wireless sensor in deterministic space. Control Decis. **25**(11), 1625–1629 (2010)

Research on Embedded and Monitoring Test Technology of C⁴ISR System

Wenyuan Xu[1], Wenjin Yang[3], Chunyu Li[2], Shuo Shi[4],
Shengxiao Zhang[1(✉)], Hao Li[1], and Haoyang Yu[5]

[1] Systems Engineering Research Institute, Beijing 100094, China
asheng2003@qq.com
[2] Naval Radar and Sonar Repair Plant, Qingdao 266001, China
[3] Military Commission of Combat System in Beijing, Beijing 100094, China
[4] Beijing Information Technology Co., Ltd., Beijing 100094, China
[5] North China Electric Power University, Baoding 071003, China

Abstract. After the C⁴ISR system is delivered to the user, it is inspected during the equipment support maintenance process that the equipment has low generality, weak automation, and difficult maintenance. This paper proposes the C⁴ISR system embedded and monitoring test technology, designed componentization, standardization, Integrated embedded and monitoring test environment system architecture, including system usage patterns and logical structures; detailed description of key technologies such as XML (Extensible Markup Language) interface model adaptive technology, component-based system efficient and flexible integration technology and memory-based database data interaction technology; and an embedded and monitoring environment case and system application flow of a C⁴ISR system, including embedded test process and monitoring test process. The test environment has been successfully applied in many projects, which can realize the recurrence, positioning, confirmation and verification of interface and process problems encountered in the C⁴ISR system maintenance guarantee process.

Keywords: Embedded test · Monitoring test · C⁴ISR · Componentization

1 Introduction

The C⁴ISR system is the center of networked and informationized operations, and is the "force multiplier" [1]. Its equipment quality will directly affect the outcome of a war. The C⁴ISR system involves the command and control, intelligence reconnaissance, early warning test, communication and navigation, electronic countermeasures, comprehensive support, and distribution of multiple military resources such as combat personnel, and heterogeneous complex military information systems [2], presenting features such as multiple fields and complex structures [3], versatility, frequent information interaction, and high real-time requirements. After the C⁴ISR system is delivered, on the one hand, the system involves a wide variety of external systems and platforms, complex interactive links, and a variety of interface information. There are many types, many ways of link interaction, and the test equipment is not universal,

© Springer Nature Switzerland AG 2019
Y. Tang et al. (Eds.): HCC 2018, LNCS 11354, pp. 86–92, 2019.
https://doi.org/10.1007/978-3-030-15127-0_9

which makes the system maintenance work difficult. On the other hand, with the increase of the training frequency of the C^4ISR system, the equipment usage rate is high, but the test equipment is inconvenient to cause the system maintenance. In the end, the demand for maintenance tasks is urgent, some factories have heavy repair tasks, system maintenance personnel lack experience, lack of maintenance equipment or backward technology, and the degree of automation is not strong enough to effectively guarantee the normal use of equipment.

Therefore, this article designed a componentized, cross-platform C^4ISR system simulation test environment system framework from how to quickly build C^4ISR system test environment, shorten the development cycle [4]; how to improve the generalization and automation of simulation test, improve the test level of C^4ISR system, realize the recurrence, positioning, validation, verification of interface and process problems encountered in the C^4ISR system maintenance guarantee process, and developed a typical C^4ISR system test environment suitable for a unit, the environment has been successfully applied in many projects.

2 C^4ISR System Test Environment Architecture Design

2.1   Using Pattern Design

According to the maintenance needs of the C^4ISR system, two modes of embedded and monitoring maintenance modes are provided, including.

Monitoring Test Mode. Accessing the test device in the C^4ISR system network, recording the information exchanged by the system, and performing structured analysis on the recorded information based on the standard system message protocol, and assisting the system maintenance personnel analyze the reason of C^4ISR system interface failure.

Embedded Test Mode. By simulating the interface and process of the system associated with the C^4ISR system, assisting the system ensures maintenance personnel analyze the C^4ISR system interface and process failure causes.

2.2   System Logic Structure Design

The architecture of the C^4ISR system test environment [5] should follow the following principles:

a. The test environment should adopt a componentized development method to assemble and independently maintain each module, so as to quickly build a test environment according to the test requirements, facilitate the reuse of modules, improve the sharing and inspection environment, research and development, maintenance efficiency, and reduce development and maintenance costs;

b. The interface test data supports one-time editing and multiple use, providing one-button interface test to improve maintenance efficiency;

c. The test environment should record, store, manage and query the test problems in a unified manner to facilitate knowledge precipitation and sharing;

d. The test environment should be portable, support Windows and domestic operating systems (such as the Kirin) and other operating systems to adapt to various test requirements and improve R&D efficiency;

e. The test hardware environment must be portable and adapt to the use of different spaces;

f. The architecture design should not only make full use of the existing achievements, but also take into account the development needs of future test technologies.

Combined with the C^{4}ISR system test requirements, based on the component architecture design method, a tailorable interface test environment architecture for multi-professional fields is established, as shown in Fig. 1.

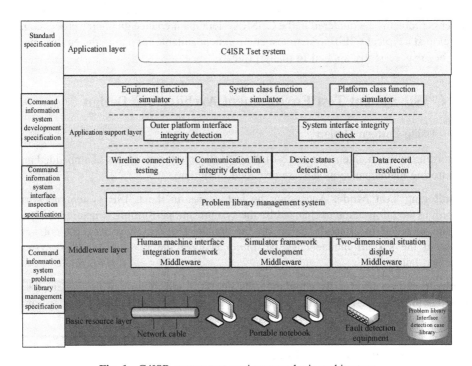

Fig. 1. C4ISR system test environment logic architecture

The architecture consists of four levels, from bottom to top, the base layer, the middleware layer, the application support layer, and the application layer among them:

- The base layer provides an operating environment, including a network cable, a portable notebook, a fault test device, a problem library/interface test case library etc.
- The middleware layer includes a human-machine interface integration framework middleware, a simulator framework development middleware, and a two-dimensional situation display middleware.
- Application support layer consists of equipment function simulator, system function simulator, platform function simulator, external platform interface integrity test,

system interface integrity test, wired connectivity test, communication link integrity test, equipment state inspection, data collection analysis, and problem library management systems. The equipment function simulator provides functions such as navigation and radar [2, 6]; the system function simulator provides the associated command system functions; the platform function simulator provides the external platform system functions; and the wired connectivity test provides wired network connectivity test; The problem library management system provides problem libraries, case additions, queries, modifications, and deletions.

- The application layer is mainly based on the test target, and the human-machine interface integration framework is used to assemble to form the C^4ISR test system.

3 Key Technologies Implemented by the System

3.1 XML-Based (Extensible Markup Language) Interface Model Adaptive Technology

An important task of C^4ISR system simulation test is the match of the message interface. However, due to different equipment and different development time of the same equipment, the format of the operational protocol is different. If the message is docked through hard code, different equipment and interface docking code are different, resulting in code duplication of development, multiple versions of maintenance difficult and inefficient. To this end, an interface method for combat protocol interface based on XML [7] is proposed, which shields the relationship between code and specific message format, and realizes the change requirements of the protocol through the configuration file. Among them, an example of the combat message abstract model is as follows:

```
<Protocol Name="TimeMessage>
<Element name="NUMBER", Type=" unsigned int", Length="1"/>
<Element name="ID", Type=" unsigned int ", ValidValue="128", Length="1"/>
<Element name="LENGTH", Type=" unsigned int ",
ValidValue="88"Length="2"/>
<Element name="TIME", Type=" unsigned int ", Length="4">
<BasicInfo LSB="0.1" Unit="MS"/>
</Element>
<Element name="Control1",Type="CONTROL" Length="4">
<subelement name="spair", Begin="0", End="0" Type="spair"/>
<subelement name="OperationType", Begin ="1", End="2" Type="Code">
<EncodeItem Type ="COMMON", Value="0" Content="one"/>
<EncodeItem Type ="COMMON ", Value="1"Content ="two"/>
<EncodeItem Type ="COMMON ", Value="4" Content="custom"/>
</subElement>
</Element>
</Procotol>
```

3.2 Component-Based System Efficient and Flexible Integration Technology

The test environment function module includes monitoring test and embedded simulation functions. The functions are complex. The system software needs to be designed and integrated by layered and componentized architecture technology. The mode of flexible organization forms different system functions. Firstly, according to the test requirements, the test function is divided into network class and business class interface plug-ins. The plug-in function is independently developed according to the plug-in interface development specification provided by the human-machine interface. The provided interfaces include plug-in loading, plug-in uninstallation, plug-in running/stopping, plug-in initialization, plug-in reverse-initialize, etc. then, according to the test object, the function to be tested is configured; finally, according to the configured test function, each plug-in is started to provide a test system for the system maintenance personnel.

3.3 Data Interaction Technology Based on In-Memory Database

The application layer communication protocol refers to a format for transmitting messages between application processes on different end systems defined by a program developer. The data in the system is transmitted on the basis of the application protocol. The data interaction technology based on the in-memory database can automatically resolve the application layer communication protocol and provide a data interaction cache based on the in-memory database. With the dynamic parsing function of the message format, the user dynamically defined message format can be automatically parsed into an application. The data parsing and interaction technology based on in-memory database provides a set of efficient and reliable management mechanism for experimental verification system data analysis and real-time data storage.

4 C^4ISR System Verification Environment Application Case

A C^4ISR system provides embedded and monitoring analog fault test, while providing network line test, functional and physical connections to the system as shown in Figs. 2 and 3.

(1) In the case of monitoring test, the system is connected to the C4ISR system network, the integrated test device records the interface information of the C4ISR system interaction, and based on the standard C4ISR system message protocol, the recorded information is structured and analyzed;

(2) In the embedded simulation, the integrated test equipment simulates the equipment system, the operational command system and the platform type interface, and performs interface check with the C^4ISR system.

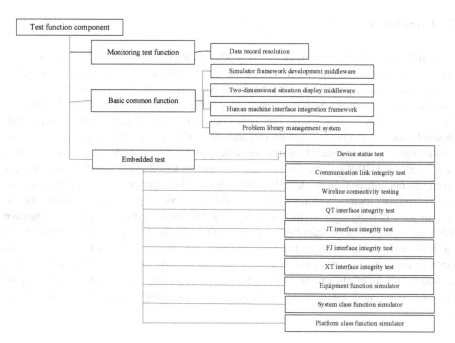

Fig. 2. C4ISR system function chart

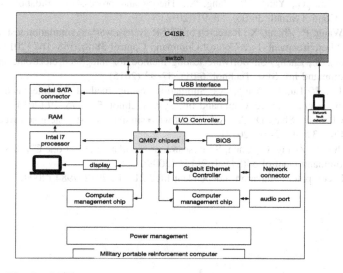

Fig. 3. C4ISR system test environment physical connection diagram

5 Conclusion

In accordance with the requirements of modularization, serialization and standardization of test equipment, this paper studies the test equipment. According to the principle of "uniform design, parallel development and comprehensive integration", based on the modular and component design concept, adopts the unified design idea and follows the stable application framework design, relies on standard information protocol and link channel mode to realize the design research of C^4ISR system test equipment, achieve standardization and general design of test equipment, and develop a set of test equipment for C^4ISR system for fast and convenient to solve the interface and process problems encountered in the C^4ISR system during the maintenance guarantee process, and reproduce, locate, confirm and verify the problem. It plays an important role in system joint debugging, interface docking, and function verification, and can improve the maintenance support capability of the factory for the C^4ISR system. However, as the maintenance capability is further improved, the interface and process test are difficult to meet the requirements. It is necessary to add the one-button performance test and other test functions on the existing basis, so as to realize the rapid improvement of the inspection and maintenance capability and effectively ensure the use of the equipment.

References

1. Wang, W., Liu, G., Yang, H., Yang, X.: Theory and process of validation for C4ISR. Command Control Simul. **36**(06), 88–91 (2014)
2. Peng, H., Wang, P., Zhang, X.: Research of C^4ISR system warfare simulation test technology in battlefield environment. Fire Control Command Control **38**(8), 158–168 (2013)
3. Fu, J., Wang, J., Zhang, S.: Software testing method for integrated electronic information system. Command Inf. Syst. Technol. **6**(01), 87–92 (2015)
4. Sun, X., Fu, J., Lu, L., Yang, X.: Simulation testing method for integrated electronic information system software. Command Inf. Syst. Technol. **5**(4), 75–79 (2014)
5. Fang, L., Chen, H., Shao, D.: Analysis of software simulation test system construction. Ship Electron. Eng. **32**(6), 12–16 (2012)
6. Peng, J., Shu, C., Zhang, G., Chen, S.: Research on simulative test of one C^4ISR system. Fire Control Command Control **34**(S1), 169–171 (2009)
7. Fan, W.: Is computer simulation independent life? **21**(24), 7982–7984 (2009)

Embedded Data Processing Platform Resource Scheduling Research

Qiang Cui[1], Shufen Liu[2], Qifeng Xu[2], and Tie Bao[2(✉)]

[1] System Engineering Research Institute of China State, Beijing, China
[2] College of Computer Science and Technology,
Jilin University, Changchun, China
baotie@jlu.edu.cn

Abstract. When the embedded data processing platform is dispatching resource, it should first estimate the weight of each processing nodes. This paper uses the grey prediction model and the weights of rotation algorithm to schedule resource for embedded data processing platform. And resource scheduling is verified through the system test of the embedded data processing platform.

Keywords: Embedded system · Resources scheduling ·
Grey forecasting model · Weights of polling

1 Introduction

With the development of technology, embedded technology has been integrated into our life. Medical electronics, intelligent furniture, logistics management, electric power control, even electronic watches, cars, aircraft, satellites and all the devices with digital interfaces and program control are marked by the embedded system [1]. The resource scheduling of embedded platform aims to regulate, measure, analyze and use various resources reasonably and effectively. Effective resource scheduling algorithm can improve operation speed and operation efficiency better. Based on task completion time and resource load balance, literature [2] finds out the optimal resource scheduling scheme through particle swarm optimization algorithm. Literature [3] uses gray-scale prediction model to realize resource scheduling of virtual machine system through network benchmark test and polynomial modeling. Literature [4] constructs a problem based on logic model and applies the algorithm's satisfiability problem to solve the resource scheduling problem. Literature [5] proposes a Delay scheduling algorithm (RFD) based on resource prediction, which is based on the prediction method of resource availability to schedule jobs reasonably. In this paper, we use grey prediction model to predict the load of CPU, calculate the weight according to the prediction results, rearrange the CPU queue and use the weight polling algorithm to realize resource scheduling.

Y. Tang et al. (Eds.): HCC 2018, LNCS 11354, pp. 93–99, 2019.
https://doi.org/10.1007/978-3-030-15127-0_10

2 Research and Design for Resource Scheduling

2.1 Gray-Scale Prediction Model

The gray-scale prediction model is a prediction method for the long term description of the development law of things through a small amount of incomplete information and establishing the grey differential prediction model. The data of grey-scale prediction is the inverse result of the prediction value obtained by the GM (1, 1) model of data generation.

The GM (1, 1) model assumes that the utilization rate of a node's CPU processing node at N continuous time can be denoted as,

$$W^0 = \left(\omega_1^0, \omega_2^0, \omega_3^0 \cdots, \omega_N^0 \right) \tag{1}$$

The GM (1, 1) model assumes that the utilization rate of a node's CPU processing node at N continuous time can be denoted as,

$$W^0 = \left(\omega_1^0, \omega_2^0, \omega_3^0 \cdots, \omega_N^0 \right) \tag{2}$$

Then the data are added to the data sequentially as follows,

$$\omega_1^1 = \omega_1^0 \tag{3}$$

$$\omega_2^1 = \omega_1^0 + \omega_2^0 \tag{4}$$

$$\vdots$$

$$\omega_N^1 = \omega_1^0 + \omega_2^0 + \cdots \omega_N^0 \tag{5}$$

A new series can be generated by a cumulative addition, and the new series is denoted as follows,

$$W^1 = \left(\omega_1^1, \omega_2^1, \omega_3^1 \cdots, \omega_N^1 \right) \tag{6}$$

It also can be denoted simply as following:

$$\omega^{(1)}(i) = \left\{ \sum_{j=1}^{i} \omega^0(j) | i = 1, 2 \cdots N \right\} \tag{7}$$

By accumulating, the vibration and fluctuation of raw data can be weakened. Then the latter item is subtracted from the previous item,

$$\Delta \omega^1(i) = \omega^1(i) - \omega^1(i-1) = \omega^0(i) \tag{8}$$

where $i = 1, 2, \cdots N, \omega^0(0) = 0$. ω^1 satisfies the first order ordinary differential equation,

$$\frac{d\omega^1}{dt} + a\omega^1 = \mu \tag{9}$$

Where μ is the value of development of gray, which is the constant input to system. The initial condition of the differential equation is, when $t = t_0$, $\omega^1 = \omega^1(t_0)$, its solution is,

$$\omega^1(t) = \left[\omega^1(t_0) - \frac{\mu}{a}\right]e^{-a(t-t_0)} + \frac{\mu}{a} \tag{10}$$

The discrete value of samples that are sampled equidistantly

$$\omega^1(k+1) = \left[\omega^1(1) - \frac{\mu}{a}\right]e^{-ak} + \frac{\mu}{a} \tag{11}$$

For a cumulative sequence, W^1, this paper estimates the constant μ and a by the least square method. Because $\omega^1(1)$ is the initial value, $\omega^1(2), \omega^1(3) \cdots \omega^1(N)$ are substituted for the equation respectively, and the differential is substituted for differential, and the samples are sample equidistantly, $\Delta t = (t+1) - t = 1$. Then it has,

$$\frac{\Delta\omega^1(2)}{\Delta t} = \Delta\omega^1(2) = \omega^1(2) - \omega^1(1) = \omega^0(2) \tag{12}$$

For the same reason,

$$\frac{\Delta\omega^1(N)}{\Delta t} = \omega^0(N) \tag{13}$$

It can be deduced at the same time,

$$\omega^0(2) + a\omega^1(2) = \mu \tag{14}$$

$$\omega^0(3) + a\omega^1(3) = \mu \tag{15}$$

$$\cdots\cdots\cdots\cdots$$

$$\omega^0(N) + a\omega^1(N) = \mu \tag{16}$$

The $ax^1(i)$ can be moved to the right, and its vector product form is denoted as following,

$$
\begin{cases}
\omega^0(2) = [-\omega^1(2), 1]\begin{bmatrix} a \\ \mu \end{bmatrix} \\
\omega^0(3) = [-\omega^1(3), 1]\begin{bmatrix} a \\ \mu \end{bmatrix} \\
\cdots\cdots \\
\omega^0(N) = [-\omega^1(N), 1]\begin{bmatrix} a \\ \mu \end{bmatrix}
\end{cases}
\tag{17}
$$

Then the matrix expression is

$$
\begin{bmatrix} \omega^0(2) \\ \omega^0(3) \\ \vdots \\ \omega^0(N) \end{bmatrix}
=
\begin{bmatrix}
-\frac{1}{2}[\omega^1(2)+\omega^1(1)] & 1 \\
-\frac{1}{2}[\omega^1(3)+\omega^1(2)] & 1 \\
\vdots \\
-\frac{1}{2}[\omega^1(N)+\omega^1(N-1)] & 1
\end{bmatrix}
\begin{bmatrix} a \\ \mu \end{bmatrix}
\tag{18}
$$

The three matrices of the upper equation are expressed as W, B, U respectively.

$$
W = BU \tag{19}
$$

The least square estimation is

$$
\hat{U} = \begin{bmatrix} \hat{a} \\ \hat{\mu} \end{bmatrix} = (B^T B)^{-1} B^T W \tag{20}
$$

The corresponding equation of time can be obtained by substituting the estimated value â and μ̂ into the equation.

$$
\hat{x}^1(k+1) = \left[x^1(1) - \frac{\hat{\mu}}{\hat{a}} \right] e^{-\hat{a}k} + \frac{\hat{u}}{\hat{a}} \tag{21}
$$

When k = 1, 2, ... , N − 1, the upper equation is a fitting value. When K >= N, the upper equation is the predicted value. It is equivalent to a cumulative fitting value, which is reduced by subtraction, and then the predicted value is obtained.

2.2 Weighted Rounded Robin Algorithm

Owing to the embedded platform of data processing is a number of isomorphic servers in function, and the data processing process, the processing capacity of components is not very different from each other. However, it needs to consider some real-time situations, such as CPU utilization rate, memory utilization rate, network node congestion and so on. Therefore, this paper uses the Weighted Rounded Robin algorithm. Because the processing capability of the data processing platform is the same, and the Weighted Rounded Robin is preferred. But with each dynamic assignment of the task, the real-time state of the processors will change.

First, the received task queue enters the centralized scheduling server, the resource management component in this paper, second, assigns weights to each server in advance according to the real-time status of the processor, and last assigns the task to the task by the corresponding weight value. Therefore, a reasonable determination of the weight is the key to the allocation of resources. By processing the historical data of the nodes, the CPU load rate of the corresponding nodes can be estimated, and each CPU weight queue is determined according to the CPU load rate.

The Weighted Rounded Robin algorithm uses the node weight to represent the processing performance of the node. The algorithm assigns each processing node by the order of weight value and polling scheduling, and the processing nodes with high weight value are able to handle more task requests with lower weight than the weight value. The algorithm process can be described as follows:

This paper sets the processing node in the cluster as $N = \{N0, N1, \ldots, Nn - 1\}$, where the weight value of node Ni can be denoted by $W(Ni)$, and i represents the ID of the last selected sever. $T(Ni)$ represents the amount of Ni currently allocated.

$\sum T(N_i)$ represents the amount of tasks that are required to be assigned currently. $\sum W(N_i)$ represents the sum of the weights of the nodes. Then, it has:

$$W(N_i) \Big/ \sum W(N_i) = T(N_i) \Big/ \sum T(N_i) \tag{22}$$

Computation of weights is an important factor in achieving load balancing. When initializing the system, the operator needs to set the initial weight DW (Ni) for each node according to the conditions of each node.

The dynamic weights of the processing nodes are determined by various parameters in the running state, which mainly include CPU resources, memory resources, response time and the number of modules to run. For data processing, CPU utilization and memory utilization are relatively important. In order to express the weight of each factor, a constant coefficient is introduced, ω_i. This paper uses this constant to represent the importance of the influencing factors, where $\sum \omega_i = 1$. Therefore, the weight value formula of each node Ni can be described as:

$$\text{LOAD}(N_i) = \omega_1 \cdot \text{Lcpu}(N_i) + \omega_2 \cdot \text{Lmemory}(N_i) + \omega_3 \cdot \text{Lprocess}(N_i) \tag{23}$$

Where Lf(Ni) represents the amount of load of a node Ni. The upper equation can represent the CPU utilization ratio, memory utilization ratio and process number. The dynamic weight of the resource scheduling component is run periodically through the state of the program, and the final weight of the system can be calculated through the node weight.

$$W_i = A * DW(N_i) + B * (\text{LOAD}(N_i) - DW(N_i)) * 1/3 \tag{24}$$

In this formula, if the dynamic weight is just equal to the initial weight value and the final weight is unchanged, the load state of the system is just in the ideal state, equal to the initial state DW (Ni). In the case of the constant number of nodes, the task is not dispatched, or the task with low priority is unloaded, and then the task is sent.

3   Validation of Resource Scheduling System

After completing the design and implementation of ship resource scheduling components, we need to carry out a comprehensive system testing of resource scheduling
components. In the process of development, each module is unit tested. In this section,
the black box testing method is mainly used to verify the functional requirements and
non-functional requirements proposed in the requirements.

3.1   Resource Allocation Test

Resource generation is divided into two steps, load forecasting and resource generation.
Load prediction verification: First of all, the normal CPU queue is used for load
forecasting, weight calculation, and CPU queue is rearranged by weight size. Resource
generation: The CPU queue is configured in turn according to the task, and the resource
scheduling plan is displayed in the table below (Table 1). However, as a ship display
control terminal component, resource scheduling components retain the ability of
manual configuration by operators simultaneously. The configuration information is
stored in the message queue at this time.

Table 1.   The error of fitting of each time.

Time	1	2	3	4	5
Error (%)	0	1.37310	−1.85130	−0.45378	0.87294

3.2   Resource Distribution Test

When the device clicks on the right key, it can be successfully sent to the resource
scheme. The embedded data processing platform framework gives the resource
scheduling component feedback message according to the load condition after receiving
the message of the load message. At this point, the resource scheduling component will
prompt the operator, and the operator needs to choose whether to continue executing the
scheduling. If necessary, the resource scheduling component will continue to load. If it
is not needed, the queue will be emptied and the components loaded will be unloaded.

3.3   Test of Gray-Scale Prediction Module

Embedded internal data platform collects real-time CPU state information in real time,
and predict the data through AR model. The prediction curve is shown as follows,
Fitting error value of CPU is shown as following:

The parameters, a $= -0.00917$, $\mu = 42.76613$, where the predictive values of time
6, 7, 8 are shown in Table 2.

By predicting the predicted value of CPU load balance, the lower value is the actual
observed CPU value. It can be seen from the table that the actual value is not much
different from the predicted value, so the application of the grey prediction model to the
resource scheduling system has a good prediction result.

Table 2. Grey model predicted value and actual value.

Time	6	7	8
Predictive value	45.01805	45.43289	45.85162
Actual value	45	44	46

4 Conclusion

In this paper, the grey prediction model and the Weighted Rounded Robin algorithm are used to schedule the embedded data processing platform. In the system, the gray prediction model is used to predict the load of the normal CPU queue, and the weight value is calculated. The weight value is used to rearrange the CPU queue, and the weights are polled according to the load message condition. Scheduling algorithm is used for resource scheduling. The constant coefficient is introduced in the Weighted Rounded Robin algorithm, which indicates the importance of the influence factors, thus making the system priority scheduling to important resources and improving the efficiency of the system. Through the resource scheduling of the component testing system and the collection of CPU state information, the prediction value is compared. The results show that the grey prediction model is applied to the resource scheduling system with better prediction results.

References

1. Furuichi, T., Yamada, K.: Next generation of embedded system on cloud computing. Procedia Comput. Sci. **35**, 1605–1614 (2014)
2. He, P., Wu, L., Song, K., Cao, Y.: Resource scheduling in embedded cloud computing based on particle swarm optimization algorithm. Electron. Des. Eng. **22**(10), 88–90 (2014)
3. Deng, X.: The resources of the virtual system dynamic allocation policy. J. Hangzhou Dianzi Univ. **4**, 39–42 (2013)
4. Gorbenko, A., Popov, V.: Task-resource scheduling problem. Int. J. Autom. Comput. **9**(4), 429–441 (2012)
5. Zhang, J.: Embedded Software Designs. Xian University of Electronic Science and Technology Press, Xian (2008)
6. Zolfaghar, K., Aghaie, A.: A syntactical approach for interpersonal trust prediction in social web applications: combining contextual and structural data. Knowl. Based Syst. **26**, 93–102 (2012)
7. Caarls, W.: Skeletons and asynchronous RPC for embedded data and task parallel image processing. IEICE Trans. Inf. Syst. **89**(7), 2036–2043 (2006)
8. Li, H., Luo, L., Zhou, Y.: Design of radar data processing software based on Ruihua embedded real-time operating system. Fire Control Radar Technol. **2018**(1), 18–26 (2018)
9. Park, H., Choi, K.: Adaptively weighted round-robin arbitration for equality of service in a many-core network-on-chip. Comput. Digit. Tech. IET **10**(1), 37–44 (2016)
10. Ravi, S., Kocher, P.: Security as a new dimension in embedded system design. In: Design Automation Conference, pp. 753–760 (2004)

Wi-Fi Attention Network for Indoor Fingerprint Positioning

Ting Zhang[1]($\boxtimes$), Yi Man[1], Xiaoshi Fang[2], Ligang Ren[3], and Yue Ma[1]

[1] Beijing University of Posts and Telecommunications, Beijing 100876, China
zhangtingzt@bupt.edu.cn
[2] Instrumentation Technology and Economy Institute, Beijing 100055, China
[3] China United Network Communications Corporation,
Beijing Branch, Beijing 100038, China

Abstract. With the rapid development of human society, the scope and complexity of urban living space are increasing, and the demand for space position information is constantly improved by human activities. Especially in recent years, with the advent of the mobile Internet and the rapid popularization of smart phone terminals, the lifestyle and behavior habits of people are undergoing a huge change. Therefore, the positioning technology is also increasingly received by people. Among them, Wi-Fi fingerprint positioning is the most popular way. In this paper, we proposed Wi-Fi Attention Networks (WAN) for Wi-Fi fingerprint indoor positioning. In the WAN, a bidirectional LSTM is used to get a representation vector of Wi-Fi by summarizing the contextual information. Then, an attention mechanism is used to extract the Wi-Fi words which are important to the representation of Wi-Fi sequences and get a high-level representation vector. Finally, a fully connected network is used for classification. Experimental results demonstrate that WAN performs better than traditional machine learning methods on the publicly available dataset.

Keywords: Wi-Fi fingerprint positioning · Bidirectional LSTM · Attention networks

1 Introduction

In recent years, with the advent of the mobile Internet and the rapid popularization of smart phone terminals, the lifestyle and behavior habits of people are undergoing a huge change. Nowadays, people in daily life habitually through LBS (Location Based Service) to find restaurants, banks, takeout and so on. Therefore, positioning technology has become an increasingly significant problem.

At present, GPS (Global Positioning System) is the most widely used positioning system. However, in complex indoor environments, due to factors such as weak signal, environmental noise and multipath interference, GPS's positioning accuracy deteriorates rapidly and even cannot be located [1]. In order to solve the positioning problem in complex indoor environments, most of the solutions are based on local wireless sensor network [2]. In recent years, Wi-Fi wireless cards are installed on most

Y. Tang et al. (Eds.): HCC 2018, LNCS 11354, pp. 100–106, 2019.
https://doi.org/10.1007/978-3-030-15127-0_11

smartphones, tablets and laptops, providing a ready-made hardware platform for indoor positioning. Therefore, Wi-Fi fingerprint positioning has become the most widely used indoor fingerprint positioning [3].

1.1 Problem Definition

The Wi-Fi fingerprint positioning, as shown in Fig. 1, is roughly divided into two phases: the offline phase and the online phase. In the offline phase, by recording the RSS of Access Points (APs) at the Reference Points (RPs), a fingerprint database is built in the positioning area and the positioning model is trained. In the online phase, feature extraction is performed on the real-time Received Signal Strength (RSS) of Aps through the positioning model to estimate the device position of current moment. The positioning model generally adopt machine learning methods, which can divided into classification method which discrete result can be obtained and regression method which the continuous position coordinates can be obtained. In order to simplify the problem, in this paper, we adopt classification positioning model, and we propose a positioning model based on deep learning in offline phase, which has good performance on the publicly available dataset.

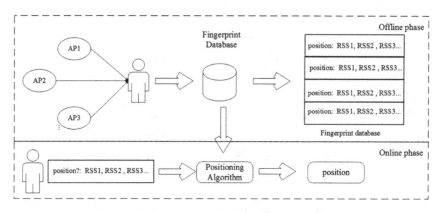

Fig. 1. Technology of Wi-Fi fingerprint positioning

2 Related Work

In recent year, some scholars have introduced machine learning into Wi-Fi finger-print positioning, such as KNN [4], WKNN [5] and SVM [6]. In this method, it mainly training a classification model based on machine learning to extract the mapping relationship between Wi-Fi RSS and device location. Wang et al. [7] used a four-layer FCN (fully connected network) for fingerprint positioning. The network structure is pre-trained by stacking and self-encoders, and the global fine-tuning is performed by backpropagation. The feature is automatically extracted from the wavy wireless signal and linearly transformed to calculate the position of target. FCN model can achieve higher positioning accuracy and enhances system robustness, but the multi-layer FCN has too many parameters, which need a large number of data to train the parameters. More importantly, FCN does not take advantage of positional information between RSS.

If we sort each AP and RSS pairs according to the RSS value, the connection between each AP and RSS pairs and its surrounding pairs is relatively close, just like the adjacent word in a sentence. Although in other fields, deep learning model such as CNN, LSTM and Attention model perform well in sequence modeling, it is still not well applied in Wi-Fi fingerprint positioning. Therefore, in the paper, we proposed a fingerprint positioning method based on Bi-LSTM and Attention model.

3 Wi-Fi Attention Network

The overall architecture of the Wi-Fi Attention Network is show in Fig. 2 which refers to Yang's Hierarchical Attention architecture [8]. It consists of two part: a Wi-Fi sequence encoder and a Wi-Fi attention layer. We introduce the details in the following sections.

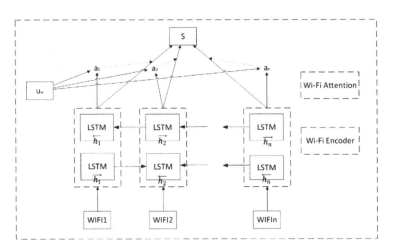

Fig. 2. The architecture of Wi-Fi Attention Network

3.1 Wi-Fi Encoder

Each Wi-Fi sequence contains several APs and its RSS, drawing on the idea of text classification, we treat an AP and its RSS pairs as a Wi-Fi word, and treat each Wi-Fi sequence as a Wi-Fi sentence. Assume that a Wi-Fi sentence contains n Wi-Fi words w_i, with $i \in [0, T]$ represents the i_{th} Wi-Fi word in the Wi-Fi sentence. We first embed the Wi-Fi words to vectors through an embedding matrix W_e, $x_i = W_e * w_i$. Then, we use a bidirectional LSTM [9] to get representation vector of Wi-Fi words by summarizing information from both directions, in this way, we can incorporate the contextual information of Wi-Fi sentence. The bidirectional LSTM contains the forward LSTM which get the Wi-Fi sentence s from x_1 to x_n and a backward LSTM which reads from x_n to x_1.

$$x_i = W_e * w_i \tag{1}$$

$$\overrightarrow{h_i} = \overrightarrow{LSTM}(x_i) \tag{2}$$

$$\overleftarrow{h_i} = \overleftarrow{LSTM}(x_i) \tag{3}$$

We obtain a complete representation vector for as given word w_i by concatenating the forward hidden state hit and backward hidden state h_i. In this way, h_i can summarizes the information of the whole Wi-Fi sentence centered around w_i.

$$h_i = \left[\overrightarrow{h_i}, \overleftarrow{h_i}\right] \tag{4}$$

3.2 Wi-Fi Attention

Not all Wi-Fi words have equivalent effect on the representation of the Wi-Fi sentence. Therefore, we introduce attention mechanisms to extract Wi-Fi words which are important to the representation of the Wi-Fi sentence, and summarize the representation of those Wi-Fi words according to its significance to form a Wi-Fi sentence vector.

$$u_i = \tanh(W_w h_i + b_w) \tag{5}$$

$$\alpha_i = \frac{\exp(u_i^T u_w)}{\sum_i \exp(u_i^T u_w)} \tag{6}$$

$$s = \sum_i \alpha_i h_i \tag{7}$$

At first, we use a single-layer MLP to get the u_i as a hidden representation of h_i. Then, we calculate the importance of this Wi-Fi words as the similarity between u_i and u_w which is a context vector and get a normalized weight vector α_i through a softmax function. Finally, we calculate the Wi-Fi sentence representation vector s as a weighted sum of Wi-Fi words. The context vector u_w is the high-level representation of the Wi-Fi sentence, which is randomly initialized and learned together during training process.

3.3 Wi-Fi Classification

The Wi-Fi sentence representation vector s is a high-level representation of this Wi-Fi sequence and can be used as features for the Wi-Fi classification:

$$p = softmax(W_s s + b_s) \tag{8}$$

Then, we choose then max p as the classification result of this Wi-Fi sequence.

4 Experiments

4.1 Dataset

We evaluate the effectiveness of our Wi-Fi Attention Network model on the public dataset of BDCI 2017 Task 1, to accurately locate the user's store in the mall [10]. This dataset has 1.1 million training sample and 48.4 thousand test sample, the target of this competition is to locate the user's current store accurately. The dataset including three kind of information. The first information is the user information, including user id and its historical consumption record. The second information is the shop information, such as average consumption price of this shop. The last information is the consumption information, including Wi-Fi sequence information, longitude and latitude information. In order to evaluate the effectiveness of our Wi-Fi Attention Network model, we only use the Wi-Fi sequence information. This Wi-Fi sequence information is Wi-Fi list consist of several AP information, for each AP, it has three items: id, signal and flag. Among them, id is the unique identifier of APs, it can distinguish between different APs. Signal is the RSS of APs, which vary from −104 to 0, 0 represent the max value of AP's RSS. Flag indicates whether this AP is connected by the device. We use 80% of the dataset for training, 10% for validation and the remaining 10% for test.

4.2 Preprocessing

We split Wi-Fi sequence into Wi-Fi words as shown in Fig. 3. The original Wi-Fi sequence has ten AP information, each AP information has three parts: id, signal and flag. In order to achieve good effect, we sort each AP information in descending order according to the RSS value. Then we can get the preprocessed Wi-Fi sequence.

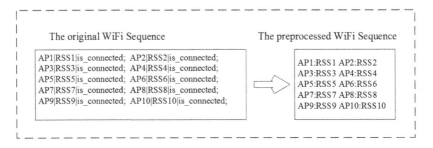

Fig. 3. The preprocessing of Wi-Fi sequence

4.3 Evaluation

We train our Wi-Fi Attention Network on the preprocessed dataset and compare it with other models. The hyper parameters as shown in Table 1, which are tuned on the validation dataset. In our model, we set the LSTM dimension to be 100, in this situation, the bidirectional LSTM produce a 200 dimensions vector. And we set the Wi-Fi word embedding dimension to be 100. For training, we use RMSProp optimizer, batch size of 128, training epochs of 20 and learning rate of 0.001.

Table 1. Table parameters of WAN.

Parameters	Values	Explanation
Embedding dim	100	Wi-Fi word embedding dimension
Size of LSTM	100	Bidirectional LSTM dimension
Epochs	20	Model training epochs
Learning rate	0.001	Adjustment speed of loss function
Batch size	128	Sample size in one training
Optimizer	RMSProp	Optimizer function

In order to prove the effect of our model, we compare it with several methods, including traditional approaches such as KNN (Bahdanau et al. 2014) and SVM (Zhu et al. 2013) and other neural network model, LSTM, GRU (Bahdanau et al. 2014) without attention structure. The experimental results are shown in Table 2, we refer our models as WAN (Wi-Fi Attention Network), and the result show that WAN gives the best performance across this dataset on precision, recall and F1 score. From Table 2 we can see that WAN outperforms the traditional approaches KNN and SVM model by 5.0% and 4.2% respectively on precision. More importantly, WAN can significantly improve over LSTM 1.5% and GRU 1.4% on precision.

This result proves our hypothesis that better performance can be obtained by incorporating knowledge of Wi-Fi fingerprint sequence structure in the WAN architecture. The intuition underlying WAN is that not all parts of a Wi-Fi fingerprint sequence are equally relevant for indoor positioning and determining the significant parts involves modeling the interactions of the Wi-Fi parts, not just their presence in isolation. That is to say, if we sort each AP according to its RSS value, the connection between AP and its surrounding AP is relatively close. This relationship can help us improve the performance of Wi-Fi fingerprint positioning.

Table 2. The target of different architecture.

	Methods	Precision	Recall	F1
Bahl et al. 2012	KNN	86.34	83.95	85.13
Zhu et al. 2013	SVM	87.12	84.78	85.93
Bahdanau et al. 2014	LSTM	89.72	85.93	87.78
Bahdanau et al. 2014	GRU	89.83	86.44	88.10
This paper	WAN	**91.29**	**89.84**	**90.56**

5 Conclusion

In this paper, we proposed Wi-Fi Attention Networks (WAN) for Wi-Fi fingerprint indoor positioning. In the WAN, a bidirectional LSTM is used to get a representation vector of Wi-Fi by summarizing the contextual information. Then, an attention mechanism is used to extract the Wi-Fi words which is important to the representation of Wi-Fi sequences and get a high-level representation vector. Finally, a fully connected network is used for classification. Experimental results demonstrate that WAN performs than other models on the publicly available dataset.

Acknowledgement. This work is supported by the National Key R&D Program of China (No. 2018YFB1201500), the National Natural Science Foundation of China under (Grant No.61471055), and the Beijing Natural Science Foundation under Grant No. L171011, Beijing Major Science and Technology Special Projects under Grant No. Z181100003118012 and China Railway with project No. BX37.

References

1. Sithole, G., Zlatanova, S.: Position, location, place and area: an indoor perspective. ISPRS Ann. Photogramm. Remote Sens. Spat. Inf. **III-4**, 89–96 (2016)
2. Xia, S., Liu, Y., Yuan, G., et al.: Indoor fingerprint positioning based on Wi-Fi: an overview. Int. J. Geo Inf. **6**(5), 135 (2017)
3. Gentile, C., Alsindi, N., Raulefs, R., et al.: Geolocation Techniques: Principles and Applications. Springer, New York (2012). https://doi.org/10.1007/978-1-4614-1836-8
4. Bahl, P., Padmanabhan, V.N.: RADAR: an in-building RF-based user location and tracking system. Proc. IEEE INFOCOM **2000**(2), 775–784 (2000)
5. Ni, L.M., Liu, Y., Lau, Y.C., et al.: LANDMARC: indoor location sensing using active RFID. Wirel. Netw. **10**(6), 701–710 (2004)
6. Zhu, Y.J., Deng, Z.L., Liu, W.L., Xu, L.M., Fang, L.: Multi-classification algorithm for indoor positioning based on support vector machine. Comput. Sci. **39**(4), 32–35 (2012)
7. Wang, X., Gao, L., Mao, S., et al.: DeepFi: deep learning for indoor fingerprinting using channel state information. In: Wireless Communications and NETWORKING Conference, pp. 1666–1671. IEEE (2015)
8. Yang, Z., Yang, D., Dyer, C., et al.: Hierarchical attention networks for document classification. In: Conference of the North American Chapter of the Association for Computational Linguistics: Human Language Technologies, pp. 1480–1489 (2017)
9. Bahdanau, D., Cho, K., Bengio, Y.: Neural machine translation by jointly learning to align and translate. Computer Science (2014)
10. Information on http://www.datafountain.cn/?spm=5176.100067.444.1.62dd1226aO75xl#/ competitions/279/intro

Reinforcement Learning Based Cooperation Transmission Policy for HetNets with CoMP Technology

Nan Lin[1,2,3], Yifei Wei[1,2,3]($\boxtimes$), Mei Song[1,2,3], Chunping Hou[1,2,3], and Ligang Ren[1,2,3]

[1] School of Electronic Engineering, Beijing University of Posts and Telecommunications, Beijing 100876, People's Republic of China
{mjjeje,weiyifei,songm}@bupt.edu.cn
[2] Department of Communication Engineering, School of Electrical and Information Engineering, Tianjin University, Tianjin 300072, China
hcp@tju.edu.cn
[3] China United Network Communications Corporation, Beijing Branch, Beijing 100876, People's Republic of China
renlg5@chinaunicom.cn

Abstract. In order to ensure the performance of cell edge users and meet the explosive growth of people's data services, Coordinated Multiple Points technology (CoMP) has become a key technology in the evolution of wireless networks. Cooperative cell selection is a major support point of CoMP multi-point collaboration technology. However, multi-differentiated clusters and radio resource optimization at this stage still cannot meet the requirements of cell edge users for signal reception quality and rate, and improvement of system performance is also facing serious challenges. This paper focuses on the cooperative transmission technology of base stations in CoMP system. This paper focuses on the cooperative transmission technology of green micro base stations in CoMP system. Using the method of reinforcement learning (RL), a semi-dynamic collaboration scheme that can be learned autonomously is proposed. Simulations show that this solution has a significant effect on improving system throughput and improving edge user rates.

Keywords: CoMP · Cooperative cell · SARSA · Reinforcement learning

1 Introduction

With the proliferation of mobile devices and the quality of service required by users, communication systems require high capacity. One of the effective methods is through collaborative communication technology [1, 2]. Each base station cooperates with each other to provide services for users, which greatly solves the problems of base station coverage blind area and user services, including the user edge cell user transmission rate. However, interference from other base stations severely degrades the performance of the communication system [3]. Therefore, the base station can be grouped to enable cooperative transmission. The base stations that participate in the service together

© Springer Nature Switzerland AG 2019
Y. Tang et al. (Eds.): HCC 2018, LNCS 11354, pp. 107–118, 2019.
https://doi.org/10.1007/978-3-030-15127-0_12

constitute a collaborative cell. The use of a coordinated beam forming scheme [4] can effectively eliminate inter-cell interference. In order to provide better services for users, we focus on the composition of collaborative communities in this article.

In a multi-cell system, cooperation processing between base stations provides a relatively high system gain by reducing inter-cell interference [5] compared to single-cell processing, and when all the base stations in the system are collaborating the system will achieve an optimal performance [6]. First, an important issue is non-ideal channel state information (CSI) between base stations. Because it is subject to frequency bands, feedback channels in frequency division multiplex (FDD) systems and time division multiplex (TDD) system noise can utilize these frequency bands. Second, the delay in the loopback and the frequency band limit the base station. The sharing of data and CSI. Therefore, in order to deal with such issues, the collaborative community will use a center. A Central Control Unit (CCU) controls the base stations in a coordinated cell and randomly collaborates at any time. Medium scheduling user collection. Although the users at the edge of the collaborative cell are still subject to strong inter-cluster interference, they do so in clusters. The cooperative transmission technology can effectively eliminate the interference in the collaborative cell.

Currently, there are mainly three ways to select collaborative cells: statically selecting cooperating cells [7–9], dynamically selecting cooperating cells [10–12] and semi-dynamically selecting cooperating cells. Static selection of cooperative cells is based on geographical location, and according to certain requirements, a fixed number of neighboring cells are selected to participate in cooperation. Usually several selected cells have relatively serious internal interference, which helps to reduce the number of cells Interference. Dynamically selecting cooperating cells depends on the user's channel feedback. According to the feedback interference, the control unit dynamically selects a suitable cooperating cell for the user. In this case, each user's cooperating cell may be different, but this is for each user. For the most part, it also minimizes inter-cell interference. Semi-dynamically selecting a coordinated cell refers to firstly selecting a relatively large pre-cooperative set (candidate set) for the user in advance, and then the user dynamically selects and participates in collaboratively processing the cells in the pre-cooperative set to jointly process, and the number of cells that eventually collaborate is equal to or less than the number of pre-coordinated cells in the cell.

There are several studies related to collaborative cells- selecting systems. The authors in [13] considered a static subdivision method that includes block diagonalization, and the precoder of each cluster is used to zero the interference of other clusters near the cluster boundary to the user. The authors in [14] proposes a more flexible solution that adapts to changes in network conditions by changing the cluster size in real time. In order to serve the set of users for scheduling, the authors in [15, 16] respectively optimized the clustering by maximizing the rate of reaching the users and minimizing the interference power. The author in [17] proposed a base station negotiation algorithm and considered a fixed cluster size to design the clustered form. However, existing researches are only for purely static clustering or purely dynamic clustering. In [18], the problem of base station clustering and beam former design was solved by the optimization problem. However, the computational complexity of dynamically selecting cooperative cells is too high and the practical application is not extensive. They have not taken into account that purely static clustering cannot satisfy the dynamic changes caused by the movement of UEs and the complexity of purely dynamic clustering. That is, they cannot improve system performance while meeting the low complexity of the system.

However, the management of semi-dynamically selecting cooperating cells has not been studied further. Therefore, the research in this paper proposes an adaptive semi-dynamic collaborative cell selection method based on reinforcement learning in CoMP scenarios. Reinforcement learning not only reduces the computational complexity of the proposed algorithm greatly, but also the system can adaptively change the cooperative cell mode through the reinforcement learning algorithm. Moreover, the concept of exploration is widely used. This means that learners sometimes choose a non-optimal action to update existing knowledge and adapt to different environments. Therefore, the proposed algorithm can effectively serve the exploration of edge users [19].

The rest of this paper is organized as follows. The system models and problem formulation are described in Sects. 2 and 3. The stochastic optimization problem with the SARSA reinforcement learning algorithm is solved in Sect. 4. Some numerical results are presented in Sect. 5. Finally, we conclude this study in Sect. 6.

2 System Model

In order to improve the system throughput and improve the service quality of the edge users, we consider the following scenarios. As illustrated in Fig. 1, in a multi-cell system, many small base stations (SBSs) and users (UEs) are included, and SBS and UE are associated. Each SBS and UE are equipped with a single antenna. In order to avoid the interference caused by other SBSs, the SBSs forms a cooperating cell. It is assumed that all SBSs are connected to the macro base station (MBS) through a low-latency backhaul network. MBS decides the cooperation and deployment of the cooperating cell.

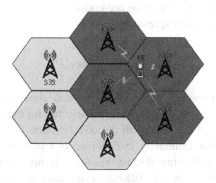

Fig. 1. Pre-collaboration set under green macro base station management.

For the selection of cooperating cell, the system is first divided into several pre-cooperative communities L_i according to the priori information of the system geographical location, and then in the pre-cooperative community, in each time slot t, different collaborative cells and UEs are generated in each cell. It is assumed that, at each time slot t, the pre-cooperative set L_i produces $U = \{1, 2, \ldots, u\}$ of cooperative cells, each of which contains $N = \{1, 2, \ldots, n\}$ of base stations and $K = \{1, 2, \ldots, k\}$ of users.

In order to eliminate interference between cooperating communities, we use zero-forcing precoding. Within each collaborative cell U_j, the signal received by the UE k from the SBS n is

$$y(t) = \sum_{n \in U_j} \sqrt{p^{k,n}} \varpi^{k,n} h^{k,n} + \sum_{n \notin U_j} \sqrt{p^{k,n}} \varpi^{k,n} h^{k,n} + z \tag{1}$$

Among them, $\sqrt{p^{k,n}}$ is the power received by the user k from the base station n, which $\omega^{k,n} h^{k,n}$ represents the precoding vector and the channel vector from the base station n to the user k, respectively. z is additive Gaussian white noise with a mean of 0 and a variance of σ^2.

Then for a CoMP user, the Signal to Interference Ratio (SINR) can be expressed as

$$\gamma_n(t) = \frac{|\sum_{n \in U_j} \sqrt{p^{k,n}} \varpi^{k,n} h^{k,n}|^2}{|\sum_{n \notin U_j} \sqrt{p^{k,n}} \varpi^{k,n} h^{k,n}|^2 + \sigma^2} \tag{2}$$

We think that the total network bandwidth is divided into sub-channels with bandwidth B, and each user is assigned a sub-channel. Because each of the pre-cooperative sets is similar, we only consider the total system throughput at time t in a pre-cooperative set. Therefore, the overall throughput of collaborative users in a pre-cooperative set is

$$C_{sum}(t) = \sum_{k=1}^{K} \sum_{n=1}^{N} B log_2(1 + \gamma_n(t)) \tag{3}$$

Assume l is the number of UEs in the pre-cooperative set, the average throughput of the pre-cooperative set can be express as

$$C_{average}(t) = \frac{1}{l} C_{sum}(t) \tag{4}$$

The cooperative community grouping mentioned above means that a relatively large pre-collaboration set (candidate set) is pre-selected for the user, and then the user dynamically selects the cells involved in the collaboration in the pre-cooperative set. The number of cells that ultimately collaborate is equal to or less than the pre-cooperative set. In the following discussion, there is no intersection between each cooperating cell, each SBS can only belong to one cooperating cell.

3 Problem Formulation

Most of the researches on RL are based on the theoretical framework of discrete and finite state Markov decision process (MDP). MDP simply mean that an agent takes an action to change his state to obtain a cycle of interaction between reward and environment. MDP's strategy depends entirely on the current state, which is also a manifestation of its Markov nature. The Markov decision process is described by the five-tuple (S, A, Pr, R, β), where: S is a finite set of states, A is a finite set of actions, P is the state transition probability, and R is the reward function, β, is a discount factor used to calculate cumulative returns. Note that unlike the Markov process, the state transition probability of the Markov decision process is that $Pr_{ss'}^a = Pr[S_{t+1} = s' | S_t = s, A_t = a]$ and $R_{ss'}^a = E[r_{t+1} | S_t = s, A_t = a, S_{t+1} = s']$ are the transition probability and the expectation of an immediate reward from the state S_t to the state S_{t+1} after selecting action A_t respectively. Specifically, our optimization problems are as follows

a. State space: The composition of the collaborative community is taken place at the beginning of every time slot. We define s_t be the network state and evolve across time slots $t = \{1, 2, \cdots\}$ with the Markovian property and S denote the state space. Our system state is related to SINR of the users in the pre-cooperative cell, and thus $s_t \in S$ can be expressed as

$$st = \{\gamma_1(t), \gamma_2(t), \cdots \gamma_n(t)\} \tag{5}$$

b. Action space: Let us consider that (u, n) represents cooperative cell pair where u and n are the number of cooperative cells in pre-collaboration set for CoMP and the number of SBSs per cooperative cell respectively. Hence, the action can be defined as

$$at = (u, n) \tag{6}$$

c. Reward: Based on the state definition, we use the average throughput represent the reward in (4). The reward function of the learning algorithm is defined as

$$Rt = Caverage(t) \tag{7}$$

According to the timely reward R, we can get the state action value function $Q(s, a)$, which refers to the discount award obtained by the strategy constructed by the sequence of actions executed in the current state.

4 Proposed Algorithm

4.1 Reinforcement Learning

Reinforcement learning is a branch of machine learning. It refers to the process in which an agent makes actions according to the environment in the current state and gets timely returns. This process diagram is shown in Fig. 2. With the above definition of state and reward function, we can apply SARSA algorithm to adaptively select

Fig. 2. Reinforcement learning process.

collaborative cells. The proposed algorithm consists of two phases, the optimal decision process and the SARSA algorithm.

A. Optimal Decision Process

The key to solving the problem is to make the decision π to choose the optimal set of collaborative cells. All executable decision sets are $\Pi = \{\pi_1, \pi_2, \cdots\}$. To determine the quality of each decision, evaluate the given policy π by defining a value function or an action value function. Under the strategy, we define $V^\pi(s)$ as a value function in the state s and satisfy the Bellman equation:

$$
\begin{aligned}
V^\pi(s) = E_\pi[R_t|s_t = s] &= E_\pi[\sum_{k=0}^{\infty} \beta^k R_{t+k+1}|s_t = s] \\
&= E_\pi[R_{t+1} + \beta V^\pi(s')|s_t = s],
\end{aligned}
\tag{8}
$$

Where $E_\pi[.]$ represent the reward expectations for a given strategy π, s' is the next state, and t is any time slot. Similarly, in state s, we make action a according to policy π and get an action value function:

$$
\begin{aligned}
Q^\pi(s, a) = E_\pi[R_t|s_t = s, a_t = a] &= E_\pi[\sum_{k=0}^{\infty} \beta^k R_{t+k+1}|s_t = s, a_t = a] \\
&= E_\pi[R_{t+1} + \beta V^\pi(s')|s_t = s, a_t = a]
\end{aligned}
\tag{9}
$$

In the MDP problem, There is a partial order relationship in the policy of convergence: if $\pi' \geq \pi$, then $Q'_\pi(s) \geq Q_\pi(s), \forall s \in S$.

And there is a theorem, For any MDP: There is always an optimal strategy π^* that is better than any other strategy, or at least as good, all optimal decisions reach the optimal value function, $V^{\pi^*}(s) = V^*(s)$ All optimal decisions reach the optimal action value function, $Q^{\pi^*}(s, a) = Q^*(s, a)$. From this we can get the optimal Bellman equation:

$$
Q * (s, a) = E[R_{t+1} + \beta max Q^*(s', a')|s_t = s, a_t = a]
\tag{10}
$$

Where $Q^*(s, a)$ describes the long-term optimal value that comes with being in a state and performing an action, that is, after performing a particular action in this state, consider all possible subsequent states and always select the most in these states. The long-term value of excellent actions to implement. Then the optimal policy $\pi^*(s, a) = argmax Q^\pi(s, a)$. Therefor, in order to get the optimal policy π is equivalent to find the optimal action value function $Q^*(s, a)$.

B. SARSA Algorithm

We will solve the mentioned MDP issues through SARSA to find an online optimal strategy. SARSA is based on the idea of value iteration, and its optimal strategy π^* can be found by estimating $Q^\pi(s_t, a_t)$. In addition, SARSA is a table-based learning algorithm in which the user stores a $Q(s_t, a_t)$ table for each state action pair before selecting a cooperating cell. We can obtain this form through the following process. We can define the alternating sequence of state actions as the plot shown in Fig. 3. From the initial state s_1, the user selects the action a_1 for the current state, and then obtains the immediate reward R_2 and moves to the next state s_2. Repeat the above process until the end of the state.

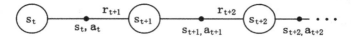

Fig. 3. Sarsa algorithm.

In SARSA algorithm, the user updates the action-state value (Q value) according to current state, reward, and learning rate

$$Q(s_t, a_t) \leftarrow Q(s_t, a_t) + \alpha_t [R_{t+1} + \beta Q(s_{t+1}, a_{t+1}) - Q(s_t, a_t)] \tag{11}$$

Where α_t represents the learning rate factor and β is the discount factor. When the episode is executed many times, the $Q(s, a)$ of any state-action pair in the table tends to a stable value, which means that the number of explorations is large enough. In other words, the user is already experienced enough to find the optimal strategy.

4.2   Exploitation and Exploration

A standard reinforcement learning algorithm must include exploration and exploration. Exploring helps the agent to fully understand its state space, and to use it to help the agent find the optimal sequence of actions. Its exploration and exploration are reflected in random strategies. In this article, we used the $\varepsilon - greedy$ strategy:

$$\pi(a|s) = \begin{cases} 1 - \varepsilon + \frac{\varepsilon}{\|A(s)\|}, & \text{if } a = argmax_a Q(s, a) \\ \frac{\varepsilon}{\|A(s)\|}, & \text{if } a \neq argmax_a Q(s, a). \end{cases} \tag{12}$$

The exploitation strategy is a greedy strategy:

$$\pi_{exploitation} = 1 - \varepsilon + \frac{\varepsilon}{\|A(s)\|}, \text{if } a = argmax_a Q(s, a) \tag{13}$$

Exploration is a uniform random strategy, i.e.:

$$\pi_{exploration} = \frac{\varepsilon}{\| A(s) \|}, if \ a \neq argmax_a Q(s,a) \tag{14}$$

Where $\|A(s)\|$ denotes the number of elements in $A(s)$, When episode is large enough, the $Q(s,a)$ can be well learned, so we don't need to explore further. Then, as the episode frequency increases, ε can gradually decrease.

Our proposed algorithm is described as follows.

SARSA Algorithm

Initialize $\beta, \alpha, \varepsilon$ and all $Q(s,a)$

Repeat

Initialize s_1

Choose $a_1 = (u_1, n_1)$ from s_1 using $\varepsilon - greedy$ policy

Repeat

Take action a_1 ,obverse next state s_{t+1} and

immediate rewards R_{t+1} based on (7)

Choose next action a_{t+1} from s_{t+1}

Update $Q(s_t, a_t)$ as (11)

Update current state $s_{t+1} \leftarrow s_t$

Update current action $a_{t+1} \leftarrow a_t$

Until terminal state s_N

Until terminal episode

---

5   Numerical Results

This section uses the Monte Carlo method to simulate all of the above algorithms to verify their performance. Consider here a pre-cooperative set base station group downlink communication system with cell number $N = 61$, and each base station is separately distributed in the center of each cell, set cell radius $R_a = 500$ m, define cell center to cell edge The hexagonal ring region having a distance of 0.8 R to R is the cell edge region. Here, only the influence of large-scale fading on the clustering process is considered, that is, the channel parameter $h_{nk} = PL_{nk} * \emptyset_{nk}$ is defined. Which PL_{nk} represents the path fades, $\emptyset_{nk}$ represents the shadow effect, it obeys the lognormal distribution service, and its variance is 8 dB, $PL_{nk} = PL_0 * d_{nk}^{-\lambda}$, where d_{nk} represents the user's distance to the base station, and λ is the path loss factor of 3.76, PL_0 is a constant whose value is proportional to path fading. The system bandwidth is 1.5 MHz, Monte Carlo times are 500, and the learning parameters $\alpha = 0.05, \beta = 0.7$.

In the case of learning parameters, we evaluate and compare the following solutions through trial and error. The algorithms involved in the comparison are:

(1) Non-CoMP
    Each small base station serves users in its cell only and does not serve users in other cells.
(2) Static CoMP
    Static CoMP cooperation is based on the geographical location of three adjacent cells as a cooperative cluster. Edge users in the cluster are served by three adjacent cell base stations, and RR resource block scheduling algorithms are used. Its state is fixed and does not change with the change of the network.
(3) Dynamic CoMP
    With rich search capabilities, actions change dynamically over time, maximizing the overall speed of the SBS.
(4) RL based Semi-Dynamic CoMP
    The above-mentioned semi-dynamic cooperation scheme based on reinforcement learning.

Figure 4 shows the CDF curves of system throughput under clustering conditions. It can be seen from the figure that the CDF curves of semi-dynamic CoMP and Dynamic CoMP are located in Non CoMP and Static CoMP, indicating that semi-dynamic CoMP good system throughput performance.

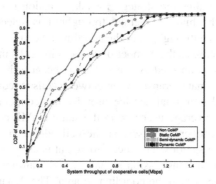

Fig. 4. The CDF curves of system throughput

The above results are discussed. First, in a non-CoMP scenario, nearly one third of the users in a multi-cell network environment are in the cell edge area, and users at the edge of the cell experience strong inter-cell interference. Therefore, its system throughput performance is very poor, then, due to the limitations of the location of receiving static CoMP collaboration, although the system throughput performance is improved, it is not significant. It is worth discussing the dynamic cooperation scheme and the proposed algorithm, the multi-cell system environment is transformed into a system environment between cells and cells. The inter-cell interference within a cell can be well resolved. Finally, the adaptive semi-dynamic cooperating algorithm based on greedy algorithm is more adaptive than the static CoMP family scenario. The matching between multiple cells and scheduling can better meet the needs of users. Therefore, the system throughput is the large and the system performance is better.

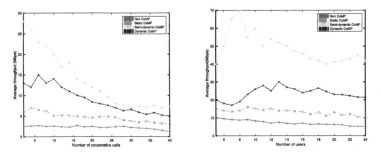

Fig. 5. Average user rate for each algorithm.

Figure 5 shows statistics of the average user throughput of cooperating cells when the system reaches a steady state. It can be seen from the figure that the average user throughput of the collaborative cells in the algorithm proposed in this paper is the best, because the algorithm proposed in this paper is to choose the (u, n) adaptively after statically selecting the pre-cooperative set. According to the average throughput of the reward value and the Q value, the strategy for taking action is continuously updated, and the strategy with the highest average throughput is selected. As a result, the probability that the average throughput of the coordinated cell is large. And the average throughput does not increase as the number of cooperating cells increases. Static CoMP is only better than direct-connected systems. This is because in the scenario of static CoMP cooperation, coordinated transmission between cells is required. In this case, the resources allocated to the original service user by the cooperative base station are occupied. Therefore, the average user rate in the static CoMP cooperation scenario is relatively low, but the cooperative users at the cell edge can receive services from multiple cooperative base stations. The received signal is a superposition of a plurality of coordinated base station transmission signals, which can greatly improve the performance of the cell edge user in the coordinated area. The dynamic collaboration and semi-dynamic collaboration are based on the system status and the coordination scheme is continuously adjusted, and the performance of cell-edge users is more significant. However, since the semi-dynamic CoMP solution based on the reinforcement learning algorithm is more inclined to select a solution with a large average throughput, and the pre-cooperative set is selected according to a geographical location in advance, the computational complexity is greatly reduced. Therefore, because semi-dynamic CoMP continuously learns and continuously adjusts the coordination scheme to select the behavior of maximizing system throughput, it can better adapt and perfectly match the channel state, thereby meeting the requirements for improving system performance.

6   Conclusion

This paper mainly proposes a reinforcement learning collaboration scheme under the CoMP scenario. We propose a new collaboration solution that uses enhanced learning algorithms to improve the performance of the edge users, and our algorithm also greatly reduces the computational complexity and simplifies the problem. In practical applications, in scenarios where the communication range is large and the users are dispersed, in order to reduce the performance degradation of a few edge users and increase the average user service quality, the solution proposed in this paper can be effectively solved. In addition, the simulation results show that the algorithm proposed in this paper has advantages in improving the system throughput, and over time, this advantage will become more and more obvious.

Acknowledgment. This work is supported by the National Natural Science Foundation of China (61571059).

References

1. Marsch, P., Fettweis, G.: On base station cooperation schemes for downlink network MIMO under a constrained backhaul. IEEE (2008)
2. Kusashima, N., Garcia, I.D., Sakaguchi, K., et al.: Fractional base station cooperation cellular network. In: International Conference on Information, Communications and Signal Processing. IEEE Press, pp. 99–103 (2009)
3. Simsek, M., Bennis, M., Güvenç, İ.: Learning based frequency- and time-domain inter-cell interference coordination in HetNets. IEEE Trans. Veh. Technol. **64**(10), 4589–4602 (2015)
4. Pennanen, H., Tölli, A., Latva-Aho, M.: Decentralized robust beamforming for coordinated multi-cell MISO networks. IEEE Signal Process. Lett. **21**(3), 334–338 (2014)
5. Karakayali, M.K., Foschini, G.J., Valenzuela, R.A.: Network coordination for spectrally efficient communications in cellular systems. IEEE Wirel. Commun. **13**(4), 56–61 (2006)
6. Gesbert, D., Hanly, S., Huang, H., et al.: Multi-cell MIMO cooperative networks: a new look at interference. IEEE J. Sel. Areas Commun. **28**(9), 1380–1408 (2010)
7. Boccardi, F., Huang, H.: Limited downlink network coordination in cellular networks. In: IEEE, International Symposium on Personal, Indoor and Mobile Radio Communications, pp. 1–5. IEEE (2007)
8. Venkatesan, S.: Coordinating base stations for greater uplink spectral efficiency in a cellular network. In: IEEE, International Symposium on Personal, Indoor and Mobile Radio Communications, pp. 1–5. IEEE (2007)
9. Marsch, P., Fettweis, G.: Static clustering for cooperative multi-point (CoMP) in mobile communications. IEEE International Conference Communications, pp. 1–6. IEEE (2011)
10. Feng, M., She, X., Chen, L., et al.: Enhanced dynamic cell selection with muting scheme for DL CoMP in LTE-A. In: Vehicular Technology Conference, pp. 1–5. IEEE (2010)
11. Wang, L.C., Yeh, C.J.: Cell grouping and autonomous channel assignment for cooperative multi-cell MIMO systems. In: IEEE, International Symposium on Personal, Indoor and Mobile Radio Communications, pp. 1432–1436. IEEE (2009)
12. Seki, Y., Takyu, O., Umeda, Y.: Performance evaluation of user selection based on average SNR in base station cooperation multi-user MIMO. In: Radio and Wireless Symposium, pp. 133–138. IEEE (2010)

13. Zhang, J., Chen, R., Ghosh, A., et al.: Networked MIMO with clustered linear precoding. IEEE Trans. Wirel. Commun. **8**(4), 1910–1921 (2009)
14. Papadogiannis, A., Gesbert, D., Hardouin, E.: A dynamic clustering approach in wireless networks with multi-cell cooperative processing, pp. 4033–4037 (2008)
15. Moon, J.M., Cho, D.H.: Inter-cluster interference management based on cell-clustering in network MIMO systems. In: Vehicular Technology Conference, pp. 1–6. IEEE (2011)
16. Liu, J., Wang, D.: An improved dynamic clustering algorithm for multi-user distributed antenna system. In: International Conference on Wireless Communications & Signal Processing, pp. 1–5. IEEE (2009)
17. Zhou, S., Gong, J., Niu, Z., et al.: A decentralized framework for dynamic downlink base station cooperation. In: IEEE Conference on Global Telecommunications, pp. 3640–3645. IEEE Press (2009)
18. Hong, M., Sun, R., Baligh, H., et al.: Joint base station clustering and beamformer design for partial coordinated transmission in heterogeneous networks. IEEE J. Sel. Areas Commun. **31**(2), 226–240 (2013)
19. Chung, B.C., Cho, D.H.: Semidynamic cell-clustering algorithm based on reinforcement learning in cooperative transmission system. IEEE Syst. J. **PP**(99), 1–4 (2017)

Capacity Estimation of Time-Triggered Ethernet Network Based on Complex Network Theory

Qing Wang[1] , Liping Teng[1] , Wenxing Hong[2(✉)] , Huimin Wu[1] ,
Beichen Li[1] , and Guihua Liu[3]

[1] School of Electrical and Information Engineering, Tianjin University,
Tianjin, China
{Wang,tlp512}@tju.edu.cn, whmtju@outlook.com, relidin@126.com
[2] Automation Department, Xiamen University, Xiamen, China
hwx@xmu.edu.cn
[3] Tianjin Jinhang Computing Technology Research Institute, Tianjin, China
chuyuxi@tju.edu.cn

Abstract. Time-Triggered Ethernet (TTE) is a new Ethernet network communication technique based on mixed time service and event service. Transmission flow is classified by time critical characteristic as Time-Triggered (TT) flow, Rate-Constrained (RC) flow and Best-Effort (BE) flow. And compound traffic partition scheduling method is used for reasonable time planning of real-time and non-real-time traffic. This research further estimates the capacity of TTE network on the basis of the existed complex network modal and relative capacity theory, with partition scheduling characteristic of TTE. Furthermore, according to the characteristic that the edge with maximum betweenness is most likely to be congested, the relationship between capacity of TTE network and the edge betweenness can be established and utilized to estimate the capacity. Then, reasonability of the capability estimation method can be verified by relative experiment. With this research, it is more efficient to design and plan TTE network flow.

Keywords: Time-triggered ethernet network · Complex network ·
Network capacity

1 Introduction

Time-Triggered Ethernet (TTE) network, combining high real-time service and traditional best-effort service, is a new Ethernet network communicating technic with characteristic of high speed, real-time and error tolerance, which has high value in the application of transmission technique in the aerospace field. TTE network utilizes three types of basic facilities including Synchronization

Supported by Civil Areaspace Fundation No.[2016]1299.

Master (SM), Synchronization Client (SC) and Compression Master (CM) to guarantee the time synchronization of network [1–3]. As to assure the real-time performance and safety of network, TTE network classifies traffic flow by time-critical characteristics as Time-Triggered (TT) traffic, Rate-Constrained (RC) traffic and Best-Effort (BE) traffic [4,5].

Using compound traffic partition scheduling mode with reasonable time planning and schedule design, TTE network enables three traffic flows with diverse transmission rules and priorities transmitted in an efficient way avoiding delay and collision loss of important information in time windows under the circumstance of link confliction. In [6], off-line and dynamic generation method for periodic schedule is introduced, with compound traffic partition scheduling mode. This research further estimates network on the basis of the partition scheduling method.

Complex network represents various real network modal abstractly. Graph theory is used to show network structure while mathematical model is for constructing network model. In [7] and [8], Erdos and Renyi introduce random probability to complex network and raise Erdos-Renyi (ER) random network modal. Also, in [9], Watts and Strogatz raise the WS modal. Barabasi and Albert raise a standard modal called BA scale-free modal in [10] and [11], whose degree distribution of the generated network is the power law form of power law exponent 3. Furthermore, they raise another scale-free modal called AB which can generate arbitrary power law exponents. In this research, BA modal is used as analysis foundation.

The initial definition of network capacity is: if such a queuing system with multiple service nodes and buffering queues is equivalent to a queuing system with only one service node (including multiple service desks) and one buffer queue, the network topology capacity is defined as the critical threshold of the node data transmission rate when the buffer queue length of the queuing system is changed from limited to infinite. Under this threshold, the queue length is stable, and above this threshold, the system buffer queue length is unstable, where the stable state is defined as the system buffer queue length does not increase infinitely with the simulation time. In [12], network capacity is analysed and the optimization strategy of network capacity is raised. And in [13], network capacity is estimated based on complex network theory while the estimation formula for maximum edge betweenness of network with the random fault is given.

In this paper, the capacity estimation formula based on complex network derived in reference [13] is extended and applied to the capacity estimation of TTE network. Based on the theory of complex networks, the BA scale-free network model is constructed. The edge betweenness in complex networks are selected as the key parameters to measure the capacity of TTE networks. Since the maximum edge betweenness is the ratio of the number of the shortest paths in the shortest path routing strategy, it represents the centrality and importance of the edges in the network. The larger the betweenness of edges is, the more traffic needs to be assumed in the network. According to the scheduling characteristics of TTE network, the network capacity of TTE network is discussed and estimated by partition in this paper, and the relationship between TTE network capacity and related parameters is constructed based on complex network model. The results are simulated and verified.

2   Complex Network Model

The complex network abstracts the actual network nodes and connections into the nodes and edges in graph theory, and utilizes $G = (V, E)$ to represent a network, V to represent the number of nodes in the network, and E to represent the number of edges. In a network, the degree of a node represents the number of edges connected by the node to all other nodes in the network. The degree of the node, represented as k can reflect the centrality of the node, that is, the greater the degree, the more nodes are connected to the node. The betweenness is classified as edge betweenness and node betweenness. The edge betweenness is defined as the proportion of all shortest paths in the network that pass through that node to the total number of shortest paths. And, the definition of node betweenness is the proportion of the total number of shortest paths in the network that pass through that side of all the shortest paths in the network. Both betweenness and degree descript the centrality of a node or edge in a network.

From [14] and [15], the node betweenness of network node v is defined as:

$$B(v) = \sum_{s \neq t, s \neq v} \frac{\sigma_{st}(v)}{\sigma_{st}} \qquad (1)$$

where the σ_{st} represents the number of shortest paths from s to t in the network while the $\sigma_{st}(v)$ is the number of paths passing node v among the shortest paths.

The edge betweenness is defined as:

$$B(l_{st}) = \sum_{s \neq t, s \neq v} \frac{\delta_{st}(l_{st})}{\delta_{st}} \qquad (2)$$

where the δ_{st} represents the number of shortest paths from s to t in the network while the $\delta_{st}(l_{st})$ is the number of paths passing the edgeamong the shortest paths.

Based on BA scale-free network as a network analysis model, BA scale-free network is to make complex network start from a few nodes and increase the number of nodes through the process of priority connection until it increases to a very large scale. The concrete process is as follows:

First, starting with the number of m_0 nodes, these nodes consist a fully connected network, and each time step adds a new node to the network. At the same time, the newly added nodes are connected to the nodes in the original network.

Connection probability between newly added nodes and existing nodes:

$$P_i = k_i / \sum_j k_j \tag{3}$$

where k_j is the degree of all the existed nodes [12].

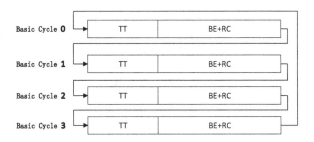

Fig. 1. Periodic scheduling diagram

3 Traffic Scheduling Model of TTE Network

TTE network information is classified by the time importance of traffic flow as three types: TT information is required to be qualified with high real-time ability. And, the waiting time, delay and jitter of TT information are fixed. Furthermore, all the TT information is transmitted on time with the pre-designed schedule, which enjoys the higher priority than the other two; RC information requires the guarantee of largest bandwidth interval, which is normally transmitted in the spare space after TT transmission; BE information is transmitted in the left space of network.

In Fig. 1, the TTE periodic scheduling diagram of mixed traffic partition mode shows that several Basic Cycle (BC) compose a Matrix Cycle (MC) of TT. And, BC is the greatest common divisor while MC is the least common multiple of all the TT flow period. From the vertical perspective, the periodic scheduling diagram reveals that n BC are connected end to end in parallel so that the total time of n BC is equal to the time of MC. From the lateral perspective, each BC is made up of the first half TT frame and last half RC plus BE frame.

4 Estimation of TTE Network Capacity

Due to the characteristics of the partitioned scheduling model, the traffic in TTE network reflects different bandwidth occupation in different basic periods. Therefore, the traffic estimation of TTE network should be corresponding to the scheduling and distribution of traffic, and it is also piecewise. Here, we estimate the capacity of different traffic types in TTE networks.

There are N nodes in the network. The line between the two nodes indicates that there is a link between the two nodes, the initial bandwidth of the link is C, the communication between any two nodes in the network is established, and there is the shortest path between the two nodes. If the bandwidth in the communication process is M, the initial bandwidth C of all communication links between two nodes is occupied by bandwidth M. When the bandwidth in a link becomes zero, that is, the side is congested, which is the number of communications calls in the network, defined as the network capacity [13].

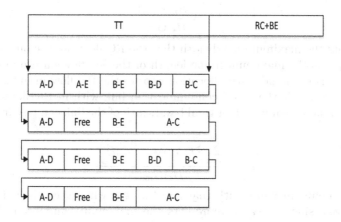

Fig. 2. Conflict-free composite table

4.1   TT Traffic

As shown in Fig. 2, according to the principle of contraction at the left end, the smaller the TT frame is in the left time period, the larger the space for transmitting other information will be. Different TT tasks with the same period $2^m * BC$ ms are arranged at longitudinal intervals with $2^{m-1} - 1$ on the periodic schedule.

The frame length range of TT traffic which has the highest priority is 64bytes to 1518bytes. In the TT time period, the time resources are divided into individual time slots, which are equal in length, but the TT frame length varies within the frame length range, TT frames of different lengths can take up one or more time slots.

TT information is transmitted in physical channels because of its high real-time performance and partition scheduling. So the bandwidth of TT information communication depends on the frame length of TT information and the length of time slice.

Since TT information is transmitted within a specified time period and without multiple calls simultaneously occupying the communication channel, the network capacity of the TT part is fixed and no congestion occurs.

4.2   RC Traffic

The transmission of RC flow in TTE network must follow the AFDX protocol. According to the AFDX protocol, there are many virtual links (VL) in the network, many VLs occupy the same physical link, and different RC flows occupy different VL channels. The transmission needs to satisfy the requirement of BAG maximum bandwidth interval. RC traffic is sent over virtual link in the free area where TT traffic is sent. For a single RC flow, the bandwidth consumed by the transmission is the ratio of the maximum frame length to the maximum bandwidth interval of the RC information [16]:

$$M_i = \frac{S_{RC_i,\max}}{BAG_{RC_i}} \tag{4}$$

M_i represents the maximum bandwidth that the RC flow can use on the virtual link. $S_{RC_i,\max}$ is the maximum frame length of the RC flow on the virtual link and BAG_{RC_i} is the bandwidth allocation gaps of the RC flow in milliseconds.

In TTE networks, there is a communication link with m RC flows and the maximum available bandwidth or total bandwidth of the link can be expressed as

$$M_m = \frac{\sum\limits_{i=1}^{m} S_{RC_i \cdot \max} \bullet \frac{BAG_{\max}}{BAG_i}}{BAG_{\max}} \tag{5}$$

M_m is the maximum bandwidth that can be used when there is m RC information on one side. $BAG_{\max}$ represents the maximum bandwidth allocation interval in M RC flows.

In TTE networks, the bandwidth occupied by each call communication of RC flow is equal to the probability from occurrence of M_i to M_m, and the size of the two values depends on the maximum frame length of the RC flow and the bandwidth allocation interval in the network. The expectation for bandwidth per call is

$$E(M) = \sum\limits_{j=1}^{n} \frac{\sum\limits_{i=1}^{m} S_{RC_i,\max} \bullet \frac{BAG_{\max}}{BAG_i}}{BAG_{\max}} \bigg/ m \tag{6}$$

$$m, n = 1, 2, 3 \ldots\ldots m$$

In a network with N nodes, the number of shortest paths is N (N−1), and the maximum number of edge betweenness in the network is B^*, so the probability that the edge with maximum betweenness is selected by the communication is

$$B^*/N(N-1) \tag{7}$$

In this network, the network capacity which represents the maximum network rate from unblocked to congested network R is:

$$R = \frac{CN(N-1)}{(\sum\limits_{j=1}^{n} \frac{\sum\limits_{i=1}^{m} S_{RC_i,\max} \bullet \frac{BAG_{\max}}{BAG_i}}{BAG_{\max}} \big/ m) \bullet B^*} \tag{8}$$

$$m, n = 1, 2, 3 \ldots\ldots m$$

R is the maximum number of times that can be called in the time period of sending RC flow in TTE network, and it is also the number of calls when the initial bandwidth of the network is exhausted. It is defined as the capacity of the network and represents an extreme value of the network from smooth communication to congestion.

4.3   BE Traffic

In TTE network, BE traffic has the lowest priority among the mixed traffic, and it is also the common unsecured information. In the process of capacity estimation, it can be regarded as a traffic distributed uniformly in each communication bandwidth service. The bandwidth M of each call communication can be regarded as the bandwidth value of equal probability in a bandwidth range. The expectation in this communication process is $E(M)$. The initial bandwidth of all links is C, and the maximum number of times that the network can be called is [13]:

$$R = \frac{CN(N-1)}{E(M) \bullet B^*} \tag{9}$$

5   Result of Simulation and Analysis

According to the BA scale-free network model of complex network, two groups of initial parameters are set up: $m_0 = 3, m = 1$ and $m_0 = 5, m = 4$. Because the initial parameters of the model are different from the number of increasing node points, the network size can be extended to any size N. The same network size may has different network structure and maximum edge betweenness. In this chapter, the network capacity of RC traffic and BE traffic in TTE network under different initial model parameters and different network size are calculated and compared.

The bandwidth allocation gaps for RC flows in TTE networks is shown in Table 1, with a frame length range of 64 bytes-1518 bytes. The network capacity is calculated at different network sizes compared with the initial parameters $m_0 = 3, m = 1$ and $m_0 = 5, m = 4$. The calculated results are shown in Fig. 3. With different initial conditions, it can be found that the network capacity shows a similar overall trend with little differences. The solid line represents the network capacity of RC traffic at different sizes under initial conditions $m_0 = 3, m = 1$. When N = 50, the network capacity is larger than that of N = 500, and when N = 100, the speed of network size change is lower than that of maximum edge betweenness, so the capacity decreases. The dashed line represents the capacity of RC Traffic Networks with different size. After a brief decline at N = 100, network capacity has been on the rise, and eventually surpassed the one at N = 50. The network capacity is closely related to the network size N and the maximum edge betweenness. Also, the maximum edge betweenness and the network model are related to the initial parameters. So, a variety of factors act on the network capacity.

Table 1. Bandwidth allocation gaps of RC flows

RC	BAG(ms)	RC	BAG(ms)	RC	BAG(ms)	RC	BAG(ms)
RC1	2	RC11	2	RC21	8	RC31	2
RC2	16	RC12	2	RC22	8	RC32	32
RC3	8	RC13	4	RC23	2	RC33	8
RC4	16	RC14	8	RC24	4	RC34	16
RC5	2	RC15	2	RC25	4	RC35	4
RC6	8	RC16	2	RC26	2	RC36	16
RC7	8	RC17	8	RC27	16	RC37	8
RC8	4	RC18	32	RC28	4	RC38	4
RC9	4	RC19	4	RC29	8		
RC10	8	RC20	16	RC30	2		

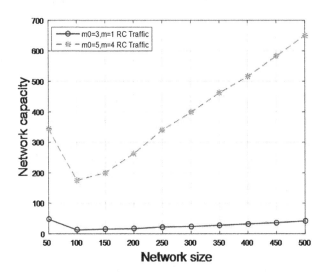

Fig. 3. Network capacity of RC traffic

The trend of BE Traffic is shown in Fig. 4, and the overall trend is the same as that of RC flow. The solid line represents the network capacity of BE traffic at different scales under initial condition $m_0 = 3, m = 1$ and the dashed line represents the network capacity under initial condition $m_0 = 5, m = 4$.

The contrast between the two flows under different initial conditions is shown in Fig. 5. Due to the constraints of the maximum frame length and the maximum bandwidth interval, the network capacity of the RC traffic is always lower than that of the RC traffic under networks of two different initial parameters. When $m_0 = 3, m = 1$, the difference of network capacity of RC traffic and BE traffic is small, which indicates that under this type network structure, the advantage and the network capacity of RC traffic transmission can not be improved. When

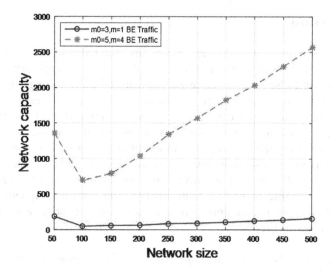

Fig. 4. Network capacity of BE traffic

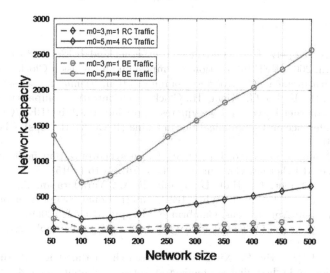

Fig. 5. Network capacity contrast of RC and BE traffic

$m_0 = 5, m = 4$, the network capacity of RC traffic and BE traffic with any network size are both larger than that under $m_0 = 3, m = 1$. Furthermore, the advantages of both flows are strengthened while the drawbacks are weaken.

Because of the characteristics of mixed traffic in TTE network, different network structure and scale have different influence on network capacity. According to the analysis results, the reasonable network can be further constructed to improve the mixed traffic capacity and achieve the purpose of network design and traffic planning, then improve the network performance.

6   Conclusion

According to the theory of complex network and the calculation method of network capacity, this paper extends the method of network capacity analysis of TTE network, and discusses the network capacity of TTE network according to the partition scheduling mode. The relationship between TTE network capacity and network size, maximum edge betweenness is analyzed, and the theoretical simulation and verification of the calculation results are carried out. With the results of simulation analysis, we can select different network size and network structure according to different network requirements, at the same time, It is also possible to optimize and improve the network capacity of TTE network mixed traffic according to the relationship between network capacity and various complex network parameters, which is also the next step of improvement and research work in future.

Acknowledgment. This work was supported by the Civil Areaspace Fundation under Grant No.[2016]1299.

References

1. Ademaj, A., Kopetz, H.: Time-triggered ethernet and IEEE 1588 clock synchronization. In: 2007 IEEE International Symposium on Precision Clock Synchronization for Measurement, Control and Communication, Vienna, pp. 41–43 (2007)
2. Zhang, Y., He, F., Lu, G., Xiong, H.: Clock synchronization compensation of time-triggered ethernet based on least squares algorithm. In: 2016 IEEE/CIC International Conference on Communications in China (ICCC Workshops), Chengdu, pp. 1–5 (2016)
3. Lan, J., Xiong, H.G., Qiao, L.I., et al.: Clock synchronization fault-tolerance in time-triggered Ethernet. Comput. Eng. Des. **26**, 11–16 (2015)
4. Steiner, W., Bauer, G., Hall, B., Paulitsch, M., Varadarajan, S.: TTEthernet dataflow concept. In: 2009 Eighth IEEE International Symposium on Network Computing and Applications, Cambridge, MA, pp. 319–322 (2009)
5. Zurawski, R.: Industrial Communication Technology Handbook, 2nd edn. CRC Press, Florida (2014)
6. Liu, W.C., Li, Q., He, F., Xiong, H.G.: Research on time-triggerd-ethernet synchronization and scheduling mechanism. Aeronaut. Comput. Tech. **41**(4), 122–127 (2011)
7. Erdos, P., Renyi, A.: On random graphs. Publ. Math. **6**, 290–297 (1959). Debrecen
8. Albert, R., Barabási, A.L.: Statistical mechanics of complex networks. Rev. Mod. Phys. **74**(1), 47 (2002)
9. Watts, D.J., Strogatz, S.H.: Collective dynamics of "small-world" networks. Nature **393**(6684), 440–442 (1998)
10. Barabási, A.L., Albert, R., Vespignani, A.: Emergence of scaling in random networks. Science **286**(5439), 509–512 (1999)
11. Barabási, A.L., Albert, R., Jeong, H.: Mean-field theory for scale-free random networks. Physica A **272**(1), 173–187 (1999)
12. Man, L.: Research on Optimization Strategy of Complex Network Transmission Capacity. Beijing Jiaotong University, Beijing (2016)

13. Guo, D.C., Liang, M.G., Wang, G., Wang, L.: Network capacity based on complex network theory. J. Beijing Jiaotong Univ. **35**(3), 123–127 (2011)
14. Zhou, T., Liu, J.G., Wang, B.H.: Notes on the algorithm for calculating betweenness. Chin. Phys. Lett. **23**(8), 2327–2329 (2006)
15. Zhao, L., Lai, Y.C., Park, K.: Onset of traffic congestion in complex networks. Phys. Rev. E **71**(2), 26125–126122 (2005)
16. Dai, Z., He, F., Zhang, Y.J., Xiong, H.G.: Real-time path optimization algorithm of AFDX virtual link. Acta Aeronautica Et Astronautica Sinica **36**(6), 1924–1932 (2015)

A New Communication P System Model Based on Hypergraph

Wenjuan Li and Xiyu Liu[✉]

Business School, Shandong Normal University, Jinan, Shandong, China
2814916406@qq.com, xyliu@sdnu.edu.cn

Abstract. The purpose of this paper is to propose a new kind of P system based on hypergraph. This paper proposes the concept of hypergraph in mathematical space and combines hypergraph with cell membrane to construct a new P system. Specifically, this new membrane system incorporates the interrelationship of hyperedges in hypergraphs and introduces the concept of weight and directionality of hypergraphs. At the same time, we have written new, more convenient rules for this new membrane system. This will be the first combination of hypergraph and membrane calculation, which brings a new theoretical method for membrane calculation. Through the simulation of the register machine, the computational completeness of the hypergraph P system is proved.

Keywords: Hypergraph · Membrane computing ·
Computational completeness

1 Introduction

Membrane calculation is the thought put forward by Professor Păun in his visit to Finland in 1998. It is a branch of natural calculation. Its purpose is to establish a distributed parallel computing model with good computational performance by referring to and simulating the way chemical substances are handled by cells, tissues, organs or other biological structures. Membrane computing has been shown to have the computational power of the equivalent Turing machine. By using the separation characteristics of the cell membrane to design the computational framework, the membrane calculation solves many computational problems effectively. Membrane computing has many successful applications in the fields of biology, biomedicine, linguistics, computer graphics, economics, approximate optimization, and cryptography. Therefore, many scholars pay close attention to it.

In machine learning problem settings, we usually assume that there is a paired relationship between the objects we are interested in. The edge of a hypergraph contains more than one node, so it contains more information than an ordinary graph. Using hypergraphs to represent the complex relationships between objects of interest to us can accurately describe the relationships between objects while ensuring the accuracy of the machine learning algorithms.

© Springer Nature Switzerland AG 2019
Y. Tang et al. (Eds.): HCC 2018, LNCS 11354, pp. 130–142, 2019.
https://doi.org/10.1007/978-3-030-15127-0_14

Inspired by the above research, this paper focuses on the combination of hypergraph and membrane computing. This article incorporates the hypergraph into the membrane structure to create a new P system. Specifically, this new membrane system incorporates the interrelationship of hyperedges in hypergraphs and introduces the concept of weight and directionality of hypergraphs. At the same time, we have written new, more convenient rules for this new membrane system. The computational completeness is proved by simulating the ADD, SUB and HALT instruction module. In this new membrane system, the membrane itself has added many new features and functions, so it is more convenient for us to calculate.

2 Hypergraph

In this section, we present some general concepts of Hypergraph which can be applied to the form of the membrane structure.

2.1 Hypergraph Without Orientation

As we all know, in normal graphs, one edge can only be linked to two vertices. But in hypergraphs, its edges can be connected by vertices of any number, and we usually call it hyperedges or links.

Definition 1. In mathematical definition of hypergraphs, a hypergraph is a generalization of a graph, whose edges contain an arbitrary number of vertices. A hypergraph H can be represented by a set of vertices and a set of hyperedge: $H = (V, E)$. V is a set of elements, called vertices or nodes. E is a set of hyperedges, and a hyperedge can be formally represented by a non-empty subset of V. Therefore E is a set of subsets of V.

Figure 1 shows the contrast between the normal graph and the hypergraph.

For example, in Fig. 1, the hypergraph H can be represent by a set of vertices and a set of hyperedge as follows:

$H = (V, E)$

$V = \{v_1, v_2, v_3, v_4, v_5, v_6, v_7\}$

$E = \{e_1, e_2, e_3, e_4\} = \{\{v_1, v_2, v_3\}, \{v_2, v_3\}, \{v_3, v_4, v_5\}, \{v_6\}\}$

$d = | e_j |$ means the number of vertices in the hyperedge, and we call it the degree of the hyperedge. $r(H) = \max | e_j |$ is the rank of H and $s(H) = \min | e_j |$ is the subjacent rank of H.

Matrix representation of hypergraphs

$$H(V, E) = \begin{cases} 1, \text{ for } v \in e \\ 0, \text{ for } otherwise \end{cases} \qquad (1)$$

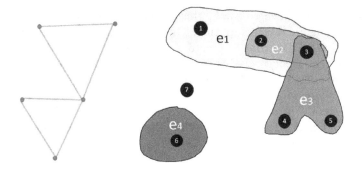

Fig. 1. Normal graph and hypergraph

2.2 Hypergraphs with Orientation

If the direction is defined on the edge of the hypergraph, it is called a directed hypergraph. Directed hypergraph is an extension to the concept of hypergraph.

Definition 2. Let hypergraph $H = (V, E)$, if $E = \{e_1, e_2, ..., e_m\}$ is the set of directed edges, we call H directed hypergraph.

If we define weight on the vertices or the hyperedges, we call the hypergraph H **weight to vertices hypergraph** or **weight to edges hypergraph**. The **primal graph** of hypergraph H is a graph whose vertices are the same as the vertices of H, and it also contains the edges between all pairs of vertices. Sometimes we also call it the **Gaifman graph** of the hypergraph.

Below is an example of hypergraphs with orientation (Fig. 2).

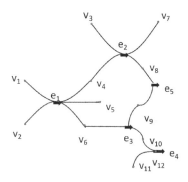

Fig. 2. An example of hypergraph with orientation

We define a matrix with $\{-1, 0, 1\}$ to represent the directed hypergraph.

$$[a(i,j)]_{m \times n} = \begin{cases} -1, & \text{for } v_i \in T(e_j) \\ 1, & \text{for } v_i \in H(e_j) \\ 0, & \text{for } otherwise \end{cases} \tag{2}$$

3   A New Membrane Structure Based on Hypergraph

3.1   Traditional P Systems

A membrance system is composed of membrane structure, object multisets and evolution rules. We will use the membrane to protect the reactor. Let m, m', m'' be membranes of a P system, the followings are some concepts of the basic operations of membranes. **vicinal membrance**: m, m' are vicinal only if $m' \subset m$ and there is no m'' like $m' \subset m'' \subset m$. **skin membrane**: there is an unique skin membrane in each membrane system, and the skin membrane has no upper vicinal membranes. **elementary membrane**: the elementary membrane has no lower vicinal membranes. **district in membrane**: This concept is pretty obvious to the elementary membrane, but for other kind of membrane, the district refers in particular to the space between the membrane and its nearest vicinal membrance. **degree**: number of membranes; **sibling membranes** m, m': if m'' is a common upper vicinal for both m and m'.

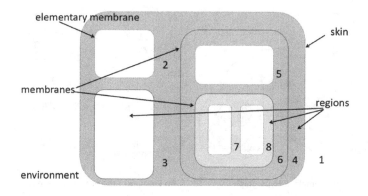

Fig. 3. A cell-like membrane

A Parentheses Expression
Membrane structures can be represented by matching parentheses. We can express Fig. 3 by the following parentheses:

$$[[]_2[]_3[[]_5[[]_7[]_8]_6]_4]_1$$

The second component of the membrane system is the multiset of objects. The evolution rules of objects is provided by the rewriting rules of the multiset. The construct of a P system with symport/antiport rules for M is:

$$\Pi = (O, T, C, \mu, \omega_1, \dots, \omega_m, (R_1, \dots, R_m), i_0)$$

For a set U, a multiset over U is a mapping $M : U \to N$, where N is the set of nonnegative integers. For $a \in U$, $M(a)$ is the multiplicity of a in M.

(i) O is the alphabet, the elements of it is called objects; (ii) $T \subset O$ is the alphabet of terminal objects; (iii) $C \subset O - T$ is the catalyst, the elements of it do not change during evolution and do not produce new characters, but some evolutionary rules must have its participation; (iv) μ is a membrane structure that contains a degree of m; (v) $\omega_1, \ldots, \omega_m$ are the multisets of objects contained by the region i of membrane structure μ; (vi) $R_1, \ldots, R_m$ are finite sets of symport and antiport rules, $R_i (i = 1, ..., m)$ is associated with the m membranes of μ; (vii) i_0 is the input/output mark of membrance.

Rule: $u \to v$, u is a string on O, v is over O_{tar}, $O_{tar} = OTAR$, $TAR = \{here, out, in\}$.

3.2   Membrane Structures Based on Hypergraph

In order to facilitate us to better combine hypergraph and membrane, we will stress a containment relationship to the hyperedge.

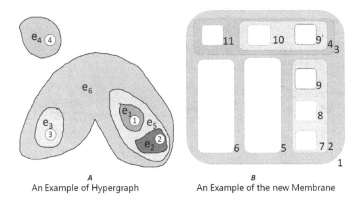

A
An Example of Hypergraph

B
An Example of the new Membrane

Fig. 4. Membrane structures based on hypergraph

As we can see in Fig. 4,
$e_6 = \{v_1, v_2, v_3\}$, $e_5 = \{v_1, v_2\}$
$e_6 \cap e_5 = \{v_1, v_2\} = e_5$, $e_6 \cup e_5 = \{v_1, v_2, v_3\} = e_6$
so that, $e_5 \subseteq e_6$
In this case, we will express e_6 in other way: $e_6 = \{v_3, e_5\} = \{e_3, e_5\}$.
so, in Figure A of Fig. 4
$H(V, E)$
$V = \{v_1, v_2, v_3, v_4\}$
$E = \{e_1, e_2, e_3, e_4, e_5, e_6\}$
$\quad = \{\{v_1\}, \{v_2\}, \{v_3\}, \{v_4\}, \{v_1, v_2\}\{v_1, v_2, v_3\}\}$
$\quad = \{\{e_1\}, \{e_2\}, \{e_3\}, \{e_4\}, \{e_1, e_2\}\{e_3, e_5\}\}$

For e_i, e_j, if there has no edge e_m like $e_i \subset e_m \subset e_j$, we define e_j as the upper edge to e_i, and e_i as the lower edge to e_j.

Now we describe membrane structures based on Hypergraph. First of all, in order to convint us to better combine hypergraph and membrane, we just consider H as a subhypergraph of a primal graph. Let H be a weight to edges directed hypergraph, and the hypergraph H is a subhypergraph of a primal graph. As described above, a hypergraph H is denoted by a set of vertices and a set of edges. $H = (V, E)$, $E = \{e_1, e_2, \ldots, e_m\}$, we define the hypergraph H as the skin membrane of a membrane structure, and the hyperedges as the membranes. Obviously, a hyperedge with only one vertex is the elementary membrane. Now we consider a hypergraph H with each vertex has its own edge, and we will extend the concept of accessible from the vertices to the hyperedges. Then we can say if the edges e_i and e_j are accessible, it will be sure that the edges on the chain has one common upper edge.

3.3   Hypergraph P System

We combined the traditional P system with the hypergraph to create a new P system and named it **Hypergraph P System** (H-P System). In this new P system, we also define a cross-like relationship between the membranes in addition to the included relationships. The membrane utilizes the inclusion relationship between the super-edges, which is more conducive to calculation than the conventional membrane.

In Figure B of Fig. 4, we present an example of a new membrane system. We can also use the following method to represent it.

$$[[[]_7[]_8[]_9]_2[[[]_9[]_{10}]_4[]_{11}]_3[]_5[]_6]_1$$

As we can see, in this kind of membrane structure, one membrane may have two or more different upper vicinal membrances.

Definition 3.1. A P system based on a hypergraph H, called a hypergraph P system, with antiport and symport rules is a construct

$$\Pi = (m, O, T, C, H, Ew_i, \omega_1, \ldots, \omega_m, A(i,j)((R_1\rho_1), \ldots, (R_m\rho_m)), i_0)$$

m is the degree of membrane,

O is the alphabet, the elements of it is called objects,

$T \subset O$ is the alphabet of terminal objects,

$C \subset O - T$ is the catalyst, the elements of it do not change during evolution and do not produce new characters, but some evolutionary rules must have its participation,

$H = (V, E)$, V is the set of the vertices and E is a set of hyperedges. H represents the relationships between membranes of the hypergraph P system,

Ew_i is the weight of membrance i, the default value is 1,

$\omega_1, \ldots, \omega_m$ are the multisets of objects contained by the region i of membrane structure,

$A(i,j)$ is the accessible matrix to the membrances (edges),

$((R_1\rho_1), \ldots, (R_m\rho_m))$ are finite sets of symport and antiport rules, $R_i(i = 1, \ldots, m)$ is associated with the m membranes, ρ is the partial ordering relationship of the rules R, which is called the priority relation. ρ indicates the priority order of the execution of rules R,

i_0 is the input/output mark of membrance.

If rules are totally ordered as $r_1 > r_2 \ldots > r_n > r_{n+1} > \ldots$. Only the system has reached a stable configuration with respect to the rule r_n can the rule r_{n+1} start application.

3.4   Rules in Hypergraph P Systems

Now we describe the communication rules in hypergraph P systems. For our purpose in this paper, we have written several major rules in a hypergraph P system. Each type of rule may have operators such as out, in, here. The superscript of an element means it will go out to a upper membrane (for superscript u) or to a lower membrane (for superscript l).

For the rule r_i: $((x_u, y_l), out; (\alpha_u, \beta_l), in)$, means for the membrane m, x will go out to the upper membranes, y will go out to the lower membranes; α, β will come in to the membrane m from the upper membranes and the lower membranes.

For the rule r_j: $((x_p, y_q), out; (\alpha_p, \beta_q), in)$, means for the membrane m, x will go out to the membrane p, y will go out to the membrane q; α, β will come in to the membrane m from the membrane p and the membrane q.

For the rule r_i: $((x^i, y^j), out; (\alpha^h, \beta^k), in)$, means for the membrane m, the number of elements x and y piercing the membrane were i, j and the number of elements α and β going into the membrane were h, k.

For the rule r_i: $((x, y), out; (\alpha, \beta), in)$, means for the membrane m, if m has the upper membranes and the lower membranes, x, y will go out to the upper membranes (for the skin membrane, it will go out to the environment) and the lower membranes at the same time; α, β will come in to the membrane m from the one who has the elements.

For the membrane m, if the orientation on it is (-1), the membrane will dissolve when the objects in it have been ran out or it have reached a stable state. And if the orientation is $(+1)$, the membrane will retain even if there is no object in it. The default value for our definition is positive.

For a membrane m, if the weight on it equal to one, then any number of elements can go through the membrane, the number of the elements depend on the rules. If the weight is a non-zero integer, then the elements can pass through the membrane m only when the number of elements is equal to the weight of the membrane m.

3.5   Configuration and Computation

In this section, we describe the configuration and computation of hypergraph P systems. In this paper, we will never change the structure of the membrane.

A simple P-system configuration refers to the state of the system by specifying the objects and rules associated with each membrane. We usually call the initial state *initial configuration*. So, the initial configuration of the system is composed of the multisets which is represented by ω_i in Π.

In this paper, the membrane calculation follows the principle of maximal parallelism. The principle of maximum parallelism means that rules should be used in parallel with maximum extent. That is, all the rules that can be used must be used. An object can only be used by a rule that selects by priority (non-deterministic selection rules if priority ρ_i is empty); any object that can be used by the rule must choose a rule that evolves according to the rule.

This evolution, called computation, is done by applying rules in the membrane. Rules have the form of rewrite rules, or other processes, such as passing objects through the membrane. Rules are used in a cell in each time unit. If there is no rules or promoters in a membrane m, then the objects in it will never change.

We say that the computation halts if the system has reached a configuration that has no rules available. If the application in a system no longer changes the membrane structure and the strings/object within the membrane structure, then we can say that the system is stable, even if some rules are still in use. The computation is successful if and only if it halts, or it is stable. The result of a halting/stable computation is the number described by the multiplicity of objects present in the cell i_0 in the halting/stable configuration.

4   Computational Completeness

What we want to prove is that the hypergraph-based P-system can generate all recursive set of countable numbers (their family is represented by NRE). In order to achieve this goal, we simulated the register machines, which is known to be as same as Turing machine.

A register machine is a construct $M = (m, H, l_0, l_h, I)$, where m is the number of registers, H is the set of instruction labels, l_0 is the start label (assigned to an ADD instruction), l_h is the halt label (labeling a instruction HALT), and I is the set of instructions; each label from H labels only one instruction from I, thus precisely identifying it. The instructions are of the following forms:

- $l_i : (ADD(r), l_j, l_k)$ (add 1 to register r and then go non-deterministically to one of the instructions with labels l_j, l_k),
- $l_i : (SUB(r), l_j, l_k)$ (if register r is non-zero, then subtract 1 from it, and go to the instruction with label l_j; otherwise, go to the instruction with label l_k),
- $l_h : HALT(the halt instruction)$.

The registration machine M calculates the number n in the following manner: All registers are set to null initially (i.e., storing the number zero), and then we will start our calculate. We apply the instruction with label l_0 and then we continue to apply the instructions in the manner shown on the label (made possible by the contents of the registers); If we reach the halt instruction, then the number stored in the first register at that moment will be calculated by M. $N(M)$ represents the set of all numbers computed by M. It is well known that a register machine is a feature of a home NRE and can be obtained even if we force the first register to not be canceled during the calculation.

While preserving the generality, we can assume that in a halt configuration, all the registers are different from the first one, they are all empty, and in the calculation we only increase the contents of the output register, not cancel it.

A register machine can also work in the accepting mode: we introduce a number n in the first register while other registers are empty, we start with the label l_0 to calculate. If the computation finally halts, then the number n is accepted.

In order to show that the hypergraph-based P-system can correctly simulate machine M, we only need to show $NRE \subseteq N(\prod)$. To do this, we describe the NRE in a register-machine fashion. Let $M = (m, H, l_0, l_h, I)$ be a register machine, we describe a hypergraph-based communication P system with antiport and symport rules which simulates the register machine M, we consider register 0 as the one where the result is obtained, and it will never be decremented during the computation.

We construct a hypergraph P system with membrances and rules. Each membrance has its own weight and orientation, the weight on the membrane can be used to control the number of the elements entering and out of the membrane at the same time, the membrane with the orientation of (-1) will dissolve when the objects in it have been ran out or it have reached a stable state.

In turn, simulating M means to simulate the ADD instructions and the SUB instructions. Thus, we will have a type of modules associated with ADD instructions, one associated with SUB instructions, and one dealing with the spiking of the output neuron (a FIN module). We can see the modules of the types in Fig. 4 respectively. We label the membranes with $1, 2, \ldots, m$ for those who associated with the registers of M, label l_i, l_j, l_k for instructions. The membranes labeled c_1, c_2, c_3 is associated with the l_i recognition ADD instruction and SUB instruction. Let's assume that only if the element in the register increases by two, the register's value is incremented by one. If the element in the register is only increased by 1, the value of the register will not increase. At each point, if the register r has n values, then the membrane r will contain 2n elements.

Simulating an ADD instruction $l_i : (ADD(r), l_j, l_k)$ − module ADD (Part A in Fig. 5).

for $l_i, r_1 : (a_l, out)$
for $c_1, r_2 : (a_l, out)$
for $c_2, r_3 : (a_l, out)$
for $c_3, r_4 : (a_{l_j}, out)$ $r_5 : (a_{l_k}, out)$
for $l_j, r_6 : (a^2 \rightarrow a)$
for $l_k, r_7 : (a^2 \rightarrow a)$

The first instruction marked as l_0 is an ADD instruction. Suppose we're on a step, when we must emulate instructions $l_i: (ADD(r), l_j, l_k)$ there is an

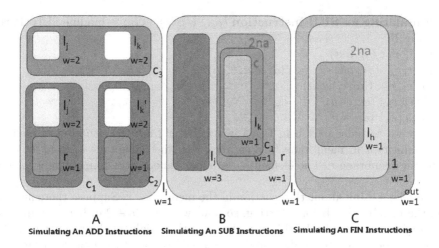

A
Simulating An ADD Instructions

B
Simulating An SUB Instructions

C
Simulating An FIN Instructions

Fig. 5. Simulating instructions

element a in membrane l_i (similar to the initial configuration), in addition to those related to the register of membranes, any other membranes won't appear element. Having the element a in it, membrane l_i starts work. Element a will enter the membrane c_1, c_2, and c_3 according to the rule r_1.

In the next step, both membrane c_1 and c_2 will send a element a to the membrane r by the rules r_2, r_3 respectively. And then we will have two more elements in the membrane r, so that the value of the register r increases 1. Because there is no rules in the membrane r, so it will never continue.

Now our problem is to make sure that the system can non-deterministically select the instruction l_j or l_k. We can find that the element a in c_1, c_2 will also go into the membranes l_j, l_k by the rules r_2, r_3 respectively, as the same time, the rule r_4 or r_5 worked in c_3 is going to send one element a at random to membrane l_j or l_k. Both weights of the membranes l_j and l_k are equal to 2, means that only two elements go into the membranes at the same time, so they can pass it. If we use the rule $r_4 : (a_{l_j}, out)$, means that the system will selects the instruction l_j, then the membrane l_j will receive two element a, and the rule r_6 will work one step later. Because of the lack of element a, rule r_5 will not be applied. Similarly, if we use the rule $r_5 : (a_{l_k}, out)$, means that the system will selects the instruction l_k, then the membrane l_k will receive two element a, and the rule r_7 will work one step later. In this way, we complete the simulation of non-deterministical selection of instruction r_j or r_k.

Therefore, from the rule r_1 start work in membrane l_i, we pass to firing non-deterministically one of membrane l_j or l_k, while also increasing by 2 the number of element a from membrane r. So that we have simulating an ADD instruction successfully.

Simulating an SUB instruction l_1: $(SUB(r), l_j, l_k)$ − module SUB (Part B in Fig. 5).

for l_i, r_1 : (a_l^2, out)

for r, r_2 : (a_l, out) r_3 : $(a_l^3, out; c, in)$

for c_1, r_4 : $((a_l, c_l), out)\mid_c$

for l_j, r_5 : $(a^3 \rightarrow a)$

for l_k, r_6 : $(a \rightarrow a^2)\mid_c$

Now we consider part b in Fig. 4, we will star from the situation of having two elements a in membrane l_i, a catalyst c in the membrane c_1, and $2n$ elements in membrane r. The elements a of membrane l_i goes immediately to the membrane r.

Now let's think about the reaction in the membrane r. Suppose there are $2n$ a in the membrane r before starting, so now, we have $2n + 2$ a. If $n = 0$, means that register r is empty. In this case, because the weight of the membrane l_j is 3, the rule r_3 cannot be applied, and rule r_4 will put one the element a from the membrane l_i into the membrane c_1. And then, the element a and catalyst c will go into the membrane l_k by applying rule r_4. So the rule r_6 will be applied in the next step.

If $n \neq o$, we know that the register r is non-zero. After the first step, we now have $2n + 2$ a in the membrane r. According to the principle of maximum parallelism, rule r_2, r_3 will proceed simultaneously. We're going to send four element a into the membrane c_1, send three element a into the membrane l_j, and get the catalyst c from the membrane c_1. At the same time, we will subtract 1 from the membrane r. In the next step, there will be three element a in the membrane l_j, the reaction $(a^3 \rightarrow a)$ will begin immediately. However, the reaction in the membrane c_1 is terminated because the catalyst c in membrane c_1 is taken into the membrane r.

In this way, we can simulate SUB instructions.

Ending a computation module FIN (Part C in Fig. 5).

for l_h, r_1 : (c, out)

for $1, r_2$: $(a_u^2, out)\mid_c$

for out, r_3 : $a^2 \rightarrow a$

Upon reaching the halting instruction of register machine M, the membrane labeled l_h sends a catalyst c to register 1, membrane 1 has $2n$ elements a and begin to send element a to the membrane out under the action of the catalyst c. In the membrane out, the number of element a will be reduced by half by the rule r_3. By this way, the number of elements a in the membrane out is now exactly equal to the value of register 1 at the end of the computation of M.

The result of the calculation can also be defined as the number of steps that pass between the first and second element of the output film sent to the environment; from the previous proof it is easy to get the necessary changes to stop the module.

5   Conclusions

In this paper, we introduce and investigate a new kind of P system based on Hypergraph. We combined the concept of a hypergraph with a membrane system to create a new membrane system and write the rules for communication. The computational completeness is proved by simulating the ADD, SUB and HALT instruction module. In particular, this new membrane system incorporates the interrelationship of hyperedges in hypergraphs and introduces the concept of weight and directionality of hypergraphs. In this new membrane system, the membrane itself has added many new features and functions, so it is more convenient for us to calculate.

Acknowledgements. Project is supported by National Natural Science Foundation of China (61472231, 61502283, 61876101, 61802234, 61806114), Ministry of Eduction of Humanities and Social Science Research Project, China (12YJA630152), Social Science Fund Project of Shandong Province, China (16$BGLJ$06, 11$CGLJ$22), China Postdoctoral Project (40411583).

References

1. Păun, G., Rozenberg, G., Salomaa, A.: Membrane Computing. Oxford University Press, New York (2010)
2. Adleman, L.M.: Molecular computation of solutions to combinatorial problems. Science **266**(5187), 1021–1024 (1994)
3. Lipton, R.J.: DNA solution of hard computational problems. Science **268**(5210), 542–545 (1995)
4. Păun, G., Rozenberg, G., Salomaa, A.: DNA Computing, New Computing Paradigms. Springer, Heidelberg (2010)
5. Ishdorj, T.-O., Leporati, A., Pan, L., Zeng, X., Zhang, X.: Deterministic solutions to QSAT and Q3SAT by spiking neural P systems with pre-computed resources. Theoret. Comput. Sci. **411**(25), 2345–2358 (2010)
6. Linqiang, P., Alhazov, A.: Solving HPP and SAT by P systems with active membranes and separation rules. Acta Informatica **43**(2), 131–145 (2006)
7. Linqiang, P., Martín-Vide, C.: Solving multidimensional 0-1 knapsack problem by P systems with input and active membranes. J. Parallel Distrib. Comput. **65**(12), 1578–1584 (2005)
8. Ishdorj, T.-O., Leporati, A., Pan, L., Wang, J.: Solving NP-complete problems by spiking neural P systems with budding rules. In: Păun, G., Pérez-Jiménez, M.J., Riscos-Núñez, A., Rozenberg, G., Salomaa, A. (eds.) WMC 2009. LNCS, vol. 5957, pp. 335–353. Springer, Heidelberg (2010). https://doi.org/10.1007/978-3-642-11467-0_24
9. Linqiang, P., Zeng, X., Zhang, X., Jiang, Y.: Spiking neural P systems with weighted synapses. Neural Process. Lett. **35**(1), 13–27 (2012)
10. Păun, G.: A quick introduction to membrane computing. J. Logic Algebraic Program. **79**(6), 291–294 (2010)
11. Robin, F.: Users guide to discrete Morse theory. Séminaire Lotharingien de Combinatoire **48**, 1–35 (2002). Article B48c

12. Han, J., Kamber, M.: Data Mining. Concepts and Techniques. Morgan Kaufmann Publishers, Higher Education Press, Beijing (2002)
13. Ceterchi, R., Martin-Vide, C.: P systems with communication for static sorting. GRLMC Report 26, Rovira I Virgili University (2003). Cavaliere, M., Martín-Vide, C., Paun, Gh. (eds.)
14. Minsky, M.: Computation Finite and Infinite Machines. Prentice Hall, Englewood Cliffs (1967)
15. The Hypergraph Web Page. https://en.wikipedia.org/wiki/Hypergraph

Non-Orthogonal Multiple Access (NOMA) in Providing Services for High-Speed Railway and Local Users in DownLink MIMO System

Guannan Ma$^{(\boxtimes)}$, Yang Liu, Yifan Liu, and Yue Ma

School of Electronic Engineering Technology,
Beijing University of Posts and Telecommunications, Beijing 100876, China
maguannan@bupt.edu.cn

Abstract. In this paper, we contribute to investigate the problem that providing service for High Speed Railway (HSR) in multiple-input multiple-output non-orthogonal multiple access (MIMO-NOMA) system to increase HSR rate. A novel system model is proposed to serve both HSR and local users. Coordinated Regularized Zero Forcing (CRZF) precoding is applied to deal with inter-cluster and inter-cell interference with considering the effect of Doppler Shift of HSR and the feedback delay of the Channel State Information (CSI). The proof of the performance of proposed model can outperform conventional MIMO-NOMA and MIMO-OMA is provided. At last, simulation is performed to demonstrate the correctness of the theory.

Keywords: Non-orthogonal multiple access (NOMA) ·
Multiple-input multipleoutput (MIMO) ·
Coordinated Regularized Zero Forcing (CRZF) ·
Signal-to-interference-plus-noise (SINR)

1 Introduction

Recently, NOMA has been proposed as a promising technique to enhance spectrum efficiency in next generation networks [1]. The basic idea of NOMA is to serve multiple users at the same time, frequency and code, but with different power levels. These users with worse channel conditions (i.e., weak users) can decode their higher-power-level signals directly by treating others signals as noise. In contrast, those users with better channel conditions (i.e., strong users) adopt the successive interference cancellation (SIC) technique for signal detection.

On the other hand, current cellular systems are only powerful in providing high-data-rate wireless access services for low-mobility (<10 km/h) users. When users move with a high velocity (>120 km/h), the achievable data rates of those systems drop significantly [2]. All of these increase the demand to consider

© Springer Nature Switzerland AG 2019
Y. Tang et al. (Eds.): HCC 2018, LNCS 11354, pp. 143–155, 2019.
https://doi.org/10.1007/978-3-030-15127-0_15

high-mobility users during developing communication systems. However, to the best of our knowledge, there are no previous works which studied NOMA scheme considering HSR. Therefore, we investigate the NOMA solution that provides services both for HSR and local users.

The application of MIMO to NOMA systems was presented in [3] by Ding et al., where a new design of precoding and detection matrices for MIMO-NOMA is proposed, the closed-form expressions of rate gap between MIMO-NOMA and MIMO-OMA was also derived and the outage probability of the users of a MIMO-NOMA cluster was evaluated. The concept of MIMO-NOMA has been validated by using systematic implementation in [4–6], which demonstrates that the use of MIMO can outperform conventional MIMO-OMA. A non-orthogonal multiple access based zero-forcing beamforming (NOMA-ZFBF) system has been designed to enhance the sum capacity in [7], where a clustering and power allocation algorithm was also been proposed to reduce interference from other beams as well as from the other user sharing the BF vector.

There are several tough problems to provide reliable communication for HSR including that severe Doppler frequency shift, high penetration loss, outdated channel feedback information, frequent handover processes, different channel statistics. Various ways was proposed to overcome some of these problems. The high-penetration loss and large simultaneous group handover processes are significantly reduced with the two-hop network architecture, in which an access point (AP) installed in the train cabin serves as a mobile relay (MR) [2].

In this paper, to solve the problems mentioned above building on these advantages inherent in NOMA and multiuser BF systems, we propose a coordinated MIMO-NOMA system model to enhance the HSR rate and provide services for HSR and local users at the same time.

Compared to these existing works, the contributions of this paper are as follows:

- We consider the problem of providing services for HSR in MIMO-NOMA systems for the first time. Propose a new system model that can serve both HSR and local users and increase the rate of HSR.
- In the proposed system model, we consider both inter-cell interference and inter-cluster interference and the coordinated RZF precoding method is applied to eliminate them.
- We analyze the performance of proposed system, the proof of proposed model can improve the rate of HSR compared with MIMO-NOMA and MIMO-OMA is given.
- Simulation is performed to verify the theory. And simulation of the proposed program using RZF and ZF precoding is compared to demonstrate that RZF can compensate for noise inflation in the low signal-to-noise-ratio (SNR) regime.

The rest of the paper is organized as follow. In Sect. 2, we propose the system model of this paper. In Sect. 3, we describe the CRZF precoding scheme used in the paper. In Sect. 4, we provide the proof of performance comparison between MIMO-NOMA and MIMO-OMA. In Sect. 5, we perform simulation to verify theory. In Sect. 6, we conclude the report and discuss about the future work.

2   System Model

In this section, we introduce the system model, which includes the network model, the channel model and the transmission model.

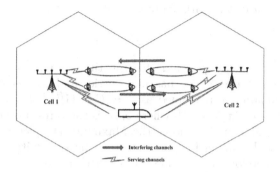

Fig. 1. Network model

2.1   Network Model

The system contains multiple cells, each cell has single base station (BS). Assuming that L cells cooperate to serve HSR, the BS of each cell serves U local users (moving speed <10 km/h, denoted by u = 1, 2, ..., U) and a passing HSR (moving speed >120 km/h denoted by user V). We consider a new downlink multi-BS cooperative MIMO-NOMA solution for serving both local users and HSR users. Each BS is equipped with N_t antennas. Each user is equipped with a single antenna. All the cells are interconnected via backhaul links assumed to be error free without delay. For HSR, we apply a two-hop architecture [8,9] to avoid frequent handovers and severe vehicle penetrative loss. The MR needs to reallocate the acquired resources to many users in the cabin. The MR referred to below all refer to the HSR. We consider the intra-cluster interference, inter-cluster interference and inter-cell interference at the same time, and analyze the performance of the proposed cooperative MIMO-NOMA solution. This paper uses two adjacent cells as an example for analysis, that is L = 2 as shown in the Fig. 1.

2.2   Channel Model

The local users and the MR experience the same kind of large-scale fading as they are in the same environment. Suppose user is located at a distance d from the BS, then the path loss of user is

$$L(d) = 10\alpha \log_{10}(d) \tag{1}$$

where α is the path loss exponent.

Small-scale fadings of local users and the MR are different. Suppose for local user u, u = 1, ..., U, the Doppler spread is zero. We consider a flat fading MIMO channel with N_t transmit antennas and single receive antennas. The timevarying channel at time t is represented by $1 \times N_t$ matrix $\mathbf{h_t}$. The entries of $\mathbf{h_t}$ are assumed to be independent identically distributed (i.i.d) and $h_{i,j} \sim \mathcal{CN}(0,1)(i = 1, 1 \leq j \leq N_t)$.

For the MR, to characterize the outdated CSI, the channel time variation is described by the first-order Markov process [10]

$$\mathbf{h}_t = \rho\mathbf{h}_{t-\tau} + \mathcal{K}\Xi_t \tag{2}$$

where τ denotes the delay, ρ stands for the time correlation coefficient and $K = \sqrt{1 - \rho^2}$. The term Ξ also has i.i.d entries and $\Xi_{i,j} \sim CN(0,1)$. And the temporal correlation ρ is defined as $\rho = E[\mathbf{H}_t\mathbf{H}_{t-\tau}^H]$. In Jake's model for simplicity $\rho(f_d\tau) = J_0(2\pi f_d\tau)$ [11], where $J_0(\cdot)$ is the zero-th order Bessel function of the first kind and f_d denotes the Maximum Doppler Frequency shift. Due to the outdated channel $\mathbf{h}_{t-\tau}$ is known at the transmitter instead of the true channel $\mathbf{h}_t$, therefore, we assume the feedback is given to the transmitter after the HSR receiver predicts the true channel state $\mathbf{h}_t$ according to the above formula, and assume that BS can obtain perfect CSI under this condition, so that the precoding of the transmitter is effective.

2.3 Transmission Model

To further improve the rate of HSR, each BS allocates more antennas and power resources to MR based on the cooperation of multiple base stations to provide services for MR. Based on the above analysis and combined with MIMO-NOMA system, we propose the following transmission model.

Local users of each cell are divided into $M_u = U/2$ clusters, M_u antennas of the base station are allocated to transmit the signals of local users in the same cluster, and N_v antennas are allocated to provide services for MR. In this case, the N_vL antennas of the L BSs provide services for the MR to ensure the reliability of the HSR and increase the rate. In addition, for ease of analysis, we assume that there are two local users in each cluster, and actually it can be more than two. Let $N = (M_u + N_v)$, then the signal $\mathbf{x}_l \in N \times 1$ at the base station before precoding can be expressed as Eq. 3.

$$\mathbf{x}_l = \begin{bmatrix} x_{l,1} \\ \vdots \\ x_{l,M_u} \\ x_{l,M_u+1} \\ \vdots \\ x_{l,N} \end{bmatrix} = \begin{bmatrix} \sqrt{p_{l,1,1}}s_{l,1,1} + \sqrt{p_{l,1,2}}s_{l,1,2} \\ \vdots \\ \sqrt{p_{l,M_u,1}}s_{l,M_u,1} + \sqrt{p_{l,M_u,2}}s_{l,M_u,2} \\ \sqrt{p_{l,M_u+1,v}}s_{l,M_u+1,v} \\ \vdots \\ \sqrt{p_{l,N,v}}s_{l,N,v} \end{bmatrix} \tag{3}$$

Where $x_{l,n}$ represents the signal for the n-th cluster in the l-th cell, and $s_{l,n,k}$ represents the information bearer signal for the k-th local user in the n-th cluster in the l-th cell, $p_{l,n,k}$ represents the power of the k-th local user in the n-th cluster in the l-th cell, where $s_{l,v}$ represents the information bearer signal sent by the l-th base station to the HSR, and $p_{l,n,v}$ represents the power allocated by the n-th cluster to the HSR from the l-th BS. From Eq. 3, we can find that the signal sent by the BS to the local users is the superposition of two user signals in the cluster, and the signal sent to the MR is only the signal of HSR, which means that HSR occupy a whole cluster, so that they will not be influenced by the intra-cluster interference. In addition, the base station allocates N_v antennas exclusively for MR to increase diversity gain. Since the transmission signal vector of the base station is processed by the precoding matrix and then transmitted to each user through the wireless channel, the transmission signal processed by the precoding matrix can be further expressed as

$$\widetilde{\mathbf{x}}_l = \mathbf{G}_l \mathbf{x}_l \qquad (4)$$

Without loss of generality, we focus on the local user and the MR served by the l-th BS and adjacent BS. The $1 \times N_t$ channel vector between the k-th local user n-th cluster in the l-th cell and the serving BS is given by $\mathbf{h}_{l,l,n,k}$. The interfering channel vector between the k-th user n-th cluster in the l-th cell and the z-th interfering BS is denoted by $\mathbf{h}_{z,l,n,k}$, where $z \neq l$. The downlink received signal at the k-th user n-th cluster in the l-th cell is given by

$$y_{l,n,k} = \mathbf{h}_{l,l,n,k} \mathbf{G}_l \mathbf{x}_l + \sum_{\substack{z=1 \\ z \neq l}}^{L} \mathbf{h}_{z,l,n,k} \mathbf{G}_z \mathbf{x}_z + n_{l,n,k} \qquad (5)$$

Where the $\mathbf{h}_{z,l,n,k}$ is the result of the combined effect of path loss and small-scale fading. $n_{l,n,k}$ is an additive Gaussian white noise with a mean value of 0 variance σ_n^2.

According to the cooperative MIMO-NOMA system model, the signal received by MR serviced by adjacent BSs can be expressed as

$$\mathbf{y}_v = \sum_{l=1}^{L} \mathbf{h}_{l,v} \mathbf{G}_l \mathbf{x}_l + n_{l,v} \qquad (6)$$

The n-th column of $\mathbf{G}_l$ is shown as $\mathbf{g}_{l,n} \in N_t \times 1$, represents the precoding vector serving the n-th cluster generated by the l-th base station, and $\mathbf{g}_{l,v} \in N_t \times 1$ represents the precoding vector serving the MR generated by the l-th base station. It is worth noting that the superimposed signals sent by the base station to local users in different clusters are different, therefore, the precoding vectors are different. And the signals sent to HSR users are the same, therefore, the precoding vectors are the same. Then $\mathbf{G}_l$ can be expressed as

$$\mathbf{G}_l = [\mathbf{g}_{l,1}, \cdots, \mathbf{g}_{l,N}] = [\mathbf{g}_{l,1}, \cdots, \mathbf{g}_{l,M_u}, \mathbf{g}_{l,v}, \cdots, \mathbf{g}_{l,v}] \qquad (7)$$

The signal models of the local users and MR above mentioned can be represented as:

$$y_{l,n,k} = \mathbf{h}_{l,l,n,k}\mathbf{g}_{l,n}x_{l,n} + \sum_{\substack{m=1 \\ m \neq n}}^{N} \mathbf{h}_{l,l,n,k}\mathbf{g}_{l,m}x_{l,m} + \sum_{\substack{z=1 \\ z \neq l}}^{L}\sum_{n=1}^{N} \mathbf{h}_{z,l,n,k}\mathbf{g}_{z,n}x_{z,n} + n_{l,n,k}$$

(8)

$$y_v = \sum_{l=1}^{L}\sum_{m=M_u+1}^{N} \mathbf{h}_{l,v}\mathbf{g}_{l,v}\sqrt{p_{l,m,v}}s_{l,v} + \sum_{l=1}^{L}\sum_{m=1}^{M_u} \mathbf{h}_{l,v}\mathbf{g}_{l,m}x_{l,m} + n_v \qquad (9)$$

Each user in the cluster will be interfered by other users in the cluster. According to the NOMA principle, the strong user in the cluster can eliminate the interference of the weak user by performing SIC, the weak user directly demodulates the received signal by treating the strong user signal as interference. For the HSR represented by Eq. 9, since there is no intra-cluster interference, the HSR only needs to eliminate the interference of local users in other clusters through precoding. The specific cooperative precoding processing will be introduced in Sect. 3.

The channel state is very important for implementing NOMA. According to our hypothesis, the first user in each cluster is a strong user, and the second user is a weak user, so the channel gain is sorted as follows

$$\mathbf{h}_{l,l,n,1} \geq \mathbf{h}_{l,l,n,2} \qquad (10)$$

According to the NOMA principle, the power distribution of users in a cluster are as follows

$$p_{l,n,1} \leq p_{l,n,2} \qquad (11)$$

Based on the above signal model, the signal-to-interference-plus-noise-ratio (SINR) for the weak user in the n-th cluster is given by Eq. 12.

$$SINR_{l,n,2} = \frac{|\mathbf{h}_{l,l,n,2}\mathbf{g}_{l,n}|^2 p_{l,n,2}}{|\mathbf{h}_{l,l,n,2}\mathbf{g}_{l,n}|^2 p_{l,n,1} + \sum_{\substack{m=1 \\ m \neq n}}^{N}|\mathbf{h}_{l,l,n,2}\mathbf{g}_{l,m}|^2 p_{l,m} + \sum_{\substack{z=1 \\ z \neq l}}^{L}\sum_{m=1}^{N}|\mathbf{h}_{z,l,n,2}\mathbf{g}_{z,m}|^2 p_{z,m} + \sigma^2}$$

(12)

where $p_{z,m}$ represents the total power allocated to m-th cluster in z-th cell. The strong user in the n-th cluster needs to decode the weak user signal with poor channel status before decoding his own signal. The weak user signal will be demodulated at the strong user with the SINR given in Eq. 13.

If the signal of the weak user in the n-th cluster is successfully demodulated, that is $log(1 + SINR_{l,n,1}^{l,n,2}) > R_{l,n,2}$, where $R_{l,n,2}$ is given by

$$SINR_{l,n,1}^{l,n,2} = \frac{|\mathbf{h}_{l,l,n,1}\mathbf{g}_{l,n}|^2 p_{l,n,2}}{|\mathbf{h}_{l,l,n,1}\mathbf{g}_{l,n}|^2 p_{l,n,1} + \sum_{\substack{m=1 \\ m \neq n}}^{N}|\mathbf{h}_{l,l,n,1}\mathbf{g}_{l,m}|^2 p_{l,m} + \sum_{\substack{z=1 \\ z \neq l}}^{L}\sum_{m=1}^{N}|\mathbf{h}_{z,l,n,1}\mathbf{g}_{z,m}|^2 p_{z,m} + \sigma^2}$$

(13)

$$R_{l,n,2} = log(1 + \frac{|\mathbf{h}_{l,l,n,2}\mathbf{g}_{l,n}|^2 p_{l,n,2}}{|\mathbf{h}_{l,l,n,2}\mathbf{g}_{l,n}|^2 p_{l,n,1} + \sum\limits_{\substack{m=1 \\ m \neq n}}^{N} |\mathbf{h}_{l,l,n,2}\mathbf{g}_{l,m}|^2 p_{l,m} + \sum\limits_{\substack{z=1 \\ z \neq l}}^{L} \sum\limits_{m=1}^{N} |\mathbf{h}_{z,l,n,2}\mathbf{g}_{z,m}|^2 p_{z,m} + \sigma^2})$$

(14)

Then the strong user in the n-th cluster will subtract the weak user signal from the received signal. At this time, the strong user's received signal in the l-th cluster can be expressed as

$$y_{l,n,1} = \mathbf{h}_{l,l,n,k}\mathbf{g}_{l,n}\sqrt{p_l, n, 1}s_{l,n,1} + \sum\limits_{\substack{m=1 \\ m \neq n}}^{M_u} \mathbf{h}_{l,l,n,k}\mathbf{g}_{l,m}x_{l,m}$$

$$+ \sum\limits_{m=M_u+1}^{N} \mathbf{h}_{l,l,n,k}\mathbf{g}_{l,m}x_{l,v} + \sum\limits_{\substack{z=1 \\ z \neq l}}^{L}\sum\limits_{m=1}^{N} \mathbf{h}_{z,l,m,k}\mathbf{g}_{z,m}x_{z,m} + n_{l,n,1} \quad (15)$$

Thus it's received SINR can be expressed as

$$SINR_{l,n,1} = \frac{|\mathbf{h}_{l,l,n,1}\mathbf{g}_{l,n}|^2 p_{l,n,1}}{\sum\limits_{\substack{m=1 \\ m \neq n}}^{N} |\mathbf{h}_{l,l,n,1}\mathbf{g}_{l,m}|^2 p_{l,m} + \sum\limits_{z=1,z \neq l}^{L}\sum\limits_{m=1}^{N} |\mathbf{h}_{z,l,n,1}g_{z,m}|^2 p_{z,m} + \sigma^2}$$

(16)

For HSR, the SINR expression is as follows

$$SINR_v = \frac{\sum\limits_{z=1}^{L}\sum\limits_{m=M_u+1}^{N} |\mathbf{h}_{z,v}\mathbf{g}_{z,v}|^2 p_{z,m,v}}{\sum\limits_{z=1}^{L}\sum\limits_{m=1}^{M_u} |\mathbf{h}_{z,v}\mathbf{g}_{z,m}|^2 p_{z,m} + \sigma^2}$$

(17)

3   Coordinated Regularized Zero Forcing Precoding

In this paper we use Coordinated Regularized Zero Forcing precoding (CRZF) to design the downlink precoding vectors. The CRZF means that the BS in each cell not only applies RZF to the channels of the users in own cell but also considers its interfering channels to users in the adjacent cell, thus mitigating or suppressing the interference it caused to those users.

It is worth noting that in the MIMO-NOMA model we proposed, the channels processed by each BS includes the channels of local users that is divided into M_u clusters, the interfering channels from the BS to other cell users, and the channels from the BS to the MR. That is to say, each BS processes the interference of $M_u L + 1$ users using the precoding matrix.

The transmitter needs to decide whether the channel of the strong user or the channel of the weak user in the same cluster is used to generate the precoding vector, the selected user can completely eliminate the interference from other clusters and the other users whose channel is not used to generate the precoding vector cannot. In order to correctly implement SIC, we select strong users with better channel state to generates the precoding vector.

Let $\hat{\mathbf{h}}_{l,z,n}$ denote the estimation of the CSI of the strong user located at n-th cluster in z-th cell acquired by the l-th BS and used to generate precoding vector. Consider the Doppler shift of HSR, the MR receiver uses Eq. 2 to predict the channel state after delay, which includes precoding matrix generation and information transmission delay, and returns it to BS, such that the BS can precode for the real MR channel. By the same way, the CSI of MR acquired by the l-th BS is $\hat{\mathbf{h}}_{l,v}$. Note that the channel estimation here refers to small-scale fading, and the channel estimation available for the l-th BS can be expressed as follows:

$$\hat{\mathbf{H}}_l = [\mathbf{T}_1, \cdots, \mathbf{T}_z, \cdots, \mathbf{T}_L, \cdots, \mathbf{T}_v]^H \in \mathbb{C}^{(LM_u+1) \times N_t} \qquad (18)$$

Where $\mathbf{T}_z = \left[\hat{\mathbf{h}}_{l,z,1}^H, \cdots, \hat{\mathbf{h}}_{l,z,M_u}^H \right]$ represents the estimate of CSI for all clusters in the z-th cell acquired by the l-th BS. Considering the CRZF precoding scheme, the precoding matrix is:

$$\hat{\mathbf{G}}_l = \hat{\mathbf{H}}_l^H \hat{\mathbf{W}}_l / \sqrt{\xi_l} \qquad (19)$$

where, $\hat{\mathbf{W}}_l = \left(\hat{\mathbf{H}}_l \hat{\mathbf{H}}_l^H + \alpha_l \mathbf{I}_{(LM_u+1)} \right)^{-1} \in \mathbb{C}^{(LM_u+1) \times (LM_u+1)}$, the regularization parameter for the l-th BS is denoted by α_l, ξ_l is a normalization scalar to fulfill the power constraint which given below:

$$\xi_l = \parallel \hat{\mathbf{W}}_l \parallel_F^2 / N_t. \qquad (20)$$

The precoding vector generated by the l-th BS and transmitted to the n-th cluster of l-th cell is the n-th column of $\mathbf{G}_l$:

$$\mathbf{g}_{l,n} = \hat{\mathbf{g}_{l,n}} / \sqrt{\xi_l} \qquad (21)$$

where $\hat{\mathbf{g}}_{l,n}$ is the n-th column of $\hat{\mathbf{H}}_l^H \hat{\mathbf{W}}_l$. The precoding vector generated by the l-th BS and sent to HSR is the $M_u + 1$ column of $\mathbf{G}_l$:

$$\mathbf{g}_{l,v} = \hat{\mathbf{g}_{l,v}} / \sqrt{\xi_i} \qquad (22)$$

where $\hat{\mathbf{g}}_{l,v}$ is the $LM_u + 1$ column of $\hat{\mathbf{H}}_l^H \hat{\mathbf{W}}_l$.

4   System Performance Analysis

This section first presents the capacity performance analysis of the proposed cooperative MIMO-NOMA system serving both high-speed railway and local users, including the rates of local strong user, weak user, and HSR, and total system capacity performance analysis. Then, we give a comparison of the HSR rate in the proposed system model and the HSR rate that can be provided in the traditional MIMO-NOMA and MIMO-OMA schemes. We prove that the rate of HSR of the proposed scheme is strictly better than the rate that the traditional MIMO-NOMA scheme can provide for HSR, that is to say, the proposed system model for HSR and local users can effectively improve the rate of HSR at almost no loss of local user rate.

4.1 Performance of the Proposed Solution

According to the Eqs. 12 and 16 the strong user rate and the weak user rate in the n-th cluster in the l-th cell of the proposed system model are given by:

$$R_{l,n,1} = \log_2(1 + SINR_{l,n,1}) \tag{23}$$

$$R_{l,n,2} = \log_2(1 + SINR_{l,n,2}) \tag{24}$$

The rate of HSR R_v:

$$\log(1 + SINR_v) = \log(1 + \frac{\sum\limits_{z=1}^{L}\sum\limits_{m=M_u+1}^{N} |h_{z,v}g_{z,v}|^2 p_{z,m,v}}{\sum\limits_{z=1}^{L}\sum\limits_{m=1}^{M_u} |h_{z,v}g_{z,m}|^2 P_{z,m} + \sigma^2}) \tag{25}$$

The total rate of local users in cell l is:

$$R_l = \sum_{n=1}^{M_u}(R_{l,n,1}+R_{l,n,2}) \tag{26}$$

The total rate of local users and high-speed railway in cell l is: $R_l + R_v$. The total rate of local users for the entire system is:

$$\sum_{l=1}^{L} R_l = \sum_{l=1}^{L}\sum_{n=1}^{N}(R_{l,n,1}+R_{l,n,2}) \tag{27}$$

4.2 Comparison with MIMO-NOMA

In this part, we give a comparison of the HSR rate in the proposed system scheme under the same resource allocation and the HSR rate that can be provided in the traditional MIMO-NOMA, which proves that the rate of proposed scheme is strict better than the rate of the traditional MIMO-NOMA solution for HSR. According to the principle of proposed model above, it is assumed that the power allocated to MR from each antenna serving for MR is $p_{l,n,v}$. Then the total power allocated by a BS to the HSR is $N_v p_{l,n,v}$. Then it is also assumed that the same power allocated to the MR in the traditional MIMO-NOMA scheme.

In the traditional MIMO-NOMA system, the rate that can be provided to HSR is given by Eq. 28.

$$R_{con\_v}^{NOMA} = \log(1 + \frac{|h_{l,v}g_{l,n}|^2 p_{l,n,v}}{\sum\limits_{\substack{m=1\\m\neq n}}^{N} |h_{l,v}g_{l,m}|^2 p_{l,m} + |h_{l,v}g_{l,n}|^2 p_{l,n,1} + \sum\limits_{\substack{z=1\\z\neq l}}^{L}\sum\limits_{m=1}^{N} |h_{z,v}g_{z,m}|^2 p_{z,m} + \sigma^2}) \tag{28}$$

Since it is difficult to completely eliminate inter-cluster interference. Then $\sum\limits_{m=1,m\neq n}^{N}|\mathbf{h}_{l,v}\mathbf{g}_{l,m}|^2 p_{l,m} > 0$ and $\sum\limits_{\substack{z=1\\z\neq l}}^{L}\sum\limits_{m=1}^{N}|\mathbf{h}_{z,v}\mathbf{g}_{z,m}|^2 p_{z,m} > 0$. Then there are HSR rate:

$$R_{con\_v}^{NOMA} < \log(1 + \frac{|\mathbf{h}_{l,v}\mathbf{g}_{l,n}|^2 N_v p_{l,n,v}}{|\mathbf{h}_{l,v}\mathbf{g}_{l,n}|^2 p_{l,n,1} + \sigma^2}) \qquad (29)$$

For the rate of HSR in our proposed scheme Eq. 25, since the precoding matrix is generated by directly utilizing the channel of the HSR, and the Doppler shift of the HSR is considered, the generated precoding vector can completely eliminate the inter-cluster interference. Then $\sum\limits_{l=1}^{L}\sum\limits_{m=1}^{M_u}|\mathbf{h}_{l,v}\mathbf{g}_{l,m}|^2 p_{l,m} = 0$. The high-speed railway rate can be written as below

$$\log(1 + \frac{\sum\limits_{l=1}^{L}\sum\limits_{m=M_u+1}^{N}|\mathbf{h}_{l,v}\mathbf{g}_{l,v}|^2 p_{l,m,v}}{\sigma^2}). \qquad (30)$$

For equal power comparison, assuming $L = 1$, each antenna is assigned the same power, then

$$\log(1 + \frac{\sum\limits_{m=M_u+1}^{N}|\mathbf{h}_{l,v}\mathbf{g}_{l,v}|^2 p_{l,m,v}}{\sigma^2})$$

$$= \log(1 + \frac{|\mathbf{h}_{l,v}\mathbf{g}_{l,v}|^2 N_v p_{l,m,v}}{\sigma^2}) > \log(1 + \frac{|\mathbf{h}_{l,v}\mathbf{g}_{l,n}|^2 N_v p_{l,n,v}}{|\mathbf{h}_{l,v}\mathbf{g}_{l,n}|^2 p_{l,n,1} + \sigma^2}) \qquad (31)$$

$$R_v > R_{con\_v}^{NOMA} \qquad (32)$$

4.3   Compared with MIMO-OMA Scheme

A scheme based on conventional MIMO-OMA can be described as follows. The MIMO-OMA transmission consists of K time slots. During each time slots, M users one from each cluster, are served simultaneously based on the same manner as described for MIMO-NOMA, as a result, during the first time slot, the 1-st user in the n-th cluster will served, In this way, the rate available to high-speed railway is given by

$$R_v^{OMA} = \frac{1}{2}\log(1 + \frac{|\mathbf{h}_{l,v}\mathbf{g}_{l,v}|^2 N_v p_{l,v}}{\sum\limits_{\substack{m=1\\m\neq n}}^{N}|\mathbf{h}_{l,v}\mathbf{g}_{l,m}|^2 p_{l,m,1} + \sum\limits_{\substack{z=1\\z\neq l}}^{L}\sum\limits_{m=1}^{N}|\mathbf{h}_{z,v}\mathbf{g}_{z,m}|^2 p_{z,m} + \sigma^2}) \qquad (33)$$

It is worth noting that the $1/2$ in front of (33) is because a conventional MIMO-OMA system with N transmit antennas requires two time slots to support 2M users, while the proposed cooperated MIMO-NOMA system with N transmit antennas can support 2M users during a single time slot. Similarly, inter-cluster interference can be assumed $\sum_{\substack{m=1 \\ m \neq n}}^{N} |\mathbf{h}_{l,n,1}\mathbf{g}_{l,m}|^2 p_{l,m,1} = 0$ and inter-cell interference

are $\sum_{\substack{z=1 \\ z \neq l}}^{L} \sum_{m=1}^{N} |\mathbf{h}_{z,v}\mathbf{g}_{z,m}|^2 p_{z,m} = 0$, then:

$$R_v^{OMA} = \frac{1}{2}\log(1 + \frac{|\mathbf{h}_{l,v}\mathbf{g}_{l,v}|^2 N_v P_{l,v}}{\sigma^2}) \tag{34}$$

$$R_v = \log(1 + \frac{\sum_{m=M_u+1}^{N} |\mathbf{h}_{l,v}\mathbf{g}_{l,v}|^2 p_{l,m,v}}{\sigma^2}) = \log(1 + \frac{|\mathbf{h}_{l,v}\mathbf{g}_{l,v}|^2 N_v p_{l,m,v}}{\sigma^2}) > R_v^{OMA} \tag{35}$$

In summary, we prove that in the case of assigning the same system resources to HSR, the rate of HSR in the proposed system model is strictly greater than that of traditional MIMO-NOMA and MIMO-OMA. So we can say that the proposed system model for high-speed railway and local users can effectively improve the rate of high-speed rail users.

5   Numerical Results

In this section, we present the simulation result of the proposed system and compare it with traditional MIMO-OMA and MIMO-OMA schemes. Figure 2 compares the HSR's maximum achievable rate achieved by proposed MIMO-NOMA scheme to that achieved by MIMO-NOMA and MIMO-OMA scheme for varying transmit power. For proposed scheme Fig. 2 considers the case in which there are four local users grouped into two clusters, with two local users in each cluster, and a MR. We can see that the HSR rate of proposed scheme is always higher than the other two and the gap become lager as the transmit power increase. Figure 2 confirms the accuracy of the analytical results developed in Sect. 5 that proposed MIMO-NOMA can outperform MIMO-NOMA and MIMO-OMA, particularly at high SNR.

Figure 3 demonstrates the rate of the HSR, strong user and weak user in the proposed MIMO-NOMA system. As can be seen from the figure, the rate of HSR is higher than that of the strong user at the same distance to base station, because the number of antennas serving high-speed rail users is more than that of strong users, and the high-speed rail users are served by two BSs simultaneously, which proves that our proposed MIMO-NOMA scheme is effective for increasing the speed of high-speed rail users.

Figure 4 Shows the total cell capacity achieved by proposed MIMO-NOMA, traditional MIMO-NOMA and MIMO-OMA scheme when the number of users and transmit antennas are different. From Fig. 4, it can be found that MIMO-NOMA scheme also absolutely outperforms MIMO-OMA in system capacity. In addition, one can also find that the total cell capacity of proposed scheme is slightly lower than that of the MIMO-NOMA solution with the growth of

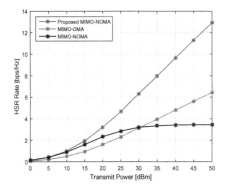

Fig. 2. HSR rate of proposed MIMO-NOMA, MIMO-NOMA and MIMO-OMA

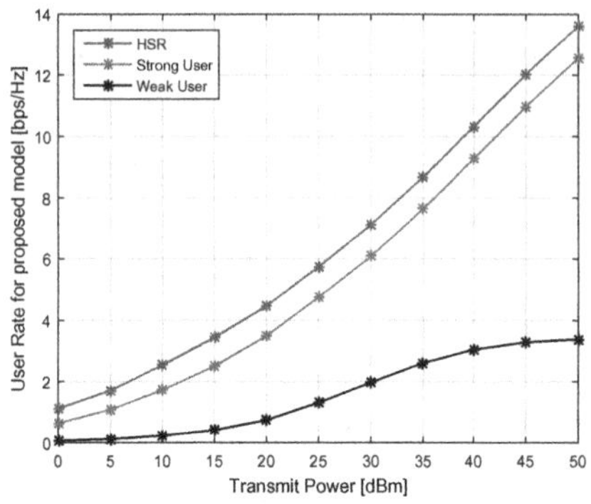

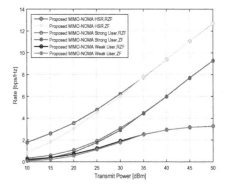

Fig. 3. Rate of strong user, weak user and HSR in proposed MIMO-NOMA scheme

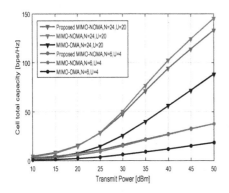

Fig. 4. Cell total capacity in our proposed system.

Fig. 5. HSR rate with RZF and ZF is applied in proposed system.

the number of transmit antennas, it is because that the BS can allocate more antennas to MR to enhance the rate of HSR when antennas increase, which result of the total cell capacity a little decrease. Obviously it is worthwhile to sacrifice the total capacity for a huge increase in HSR rate.

Figure 5 shows the rate of HSR when RZF and ZF is applied in proposed system which well prove the compensation effect of RZF in the low SNR regime.

6 Conclusion

In this paper, we propose a novel system model to provide service for HSR and local users at the same time in downlink MIMO-NOMA system. The CRZF precoding is applied to eliminate inter-cluster and inter-cell interference with considering Doppler shift of HSR and delay of transmission CSI when generate precoding matrix. We theoretically prove that the high-speed rate in the proposed scheme is strictly superior to the MIMO-NOMA and MIMO-OMA schemes. At last, simulation is performed to prove the theoretical derivation in the paper.

Acknowledgment. This work is supported by the National Natural Science Foundation of China under (Grant No. 61427801).

References

1. Yu, Y., Chen, H., Li, Y., Ding, Z., Vucetic, B.: Antenna selection for MIMO-NOMA networks, pp. 1–6 (2016)
2. Zhang, C., Fan, P.: Providing services for the high-speed train and local users in the same OFDMA system: resource allocation in the downlink. IEEE Trans. Wireless Commun. **15**(2), 1018–1030 (2016)
3. Ding, Z., Adachi, F., Poor, H.V.: The application of MIMO to non-orthogonal multiple access. IEEE Trans. Wireless Commun. **15**(1), 537–552 (2015)
4. Chen, X., Benjebbour, A., Lan, Y., Li, A.: Impact of rank optimization on downlink non-orthogonal multiple access (NOMA) with SU-MIMO. In: IEEE International Conference on Communication Systems, pp. 233–237 (2014)
5. Lan, Y., Benjebboiu, A., Chen, X., Li, A.: Considerations on downlink non-orthogonal multiple access (NOMA) combined with closed-loop SU-MIMO. In: International Conference on Signal Processing and Communication Systems, pp. 1–5 (2014)
6. Higuchi, K., Kishiyama, Y.: Non-orthogonal access with random beamforming and intra-beam SIC for cellular MIMO downlink. In: Vehicular Technology Conference, pp. 1–5 (2014)
7. Kim, B., et al.: Non-orthogonal multiple access in a downlink multiuser beamforming system. In: Military Communications Conference, MILCOM 2013, pp. 1278–1283. IEEE (2013)
8. Barbu, G.: E-train-broadband communication with moving trains technical report-technology state of the art. Int. Tech. Rep., Union Railways, Paris, France (2010)
9. Wang, J., Zhu, H., Gomes, N.J.: Distributed antenna systems for mobile communications in high speed trains. IEEE J. Sel. Areas Commun. **30**(4), 675–683 (2012)
10. Narula, A., Lopez, M.J., Trott, M.D., Wornell, G.W.: Efficient use of side information in multiple-antenna data transmission over fading channels. IEEE J. Sel. Areas Commun. **16**(8), 1423–1436 (1998)
11. Jakes, W.C., Cox, D.C.: Microwave Mobile Communications. IEEE Press, New York (1993)

Research on Multi-agent Distributed Supply Chain Information Collaboration Based on Cloud Environment

Qiaohong Zu$^{(\boxtimes)}$ and Ya Liu$^{(\boxtimes)}$

School of Logistics Engineering, Wuhan University of Technology,
Wuhan 430063, People's Republic of China
zuqiaohong@foxmail.com, 1409715698@qq.com

Abstract. The coordination and cooperation of supply chain can not only maximize the benefits of the whole supply chain, but also monitor the production management status of members in real time. It can find and solve problems timely and effectively, so as to maintain stable development of the supply chain. In order to realize the coordination of the supply chain, the first key problem to be solved is the information synergy between the enterprises in the supply chain. Only by achieving information synergy and ensuring the transmission of information in the supply chain timely and effectually, the members can make a better and faster decision in the current network era. The information collaboration of supply chain was studied in depth, and the common distributed supply chain was chosen as the object.

Keywords: Supply chain · Information collaboration · Multi-agent system

1 Introduction

With the intensification of competition in the global market and the rapid development of information technology, enterprises must maintain their competitiveness by accelerating their own informationization process and cooperating with other enterprises in the supply chain. Enterprises play their respective advantages and form complementary advantages, so that the benefits of the entire supply chain can be maximized [1]. To achieve synergy of supply chain, the most important thing is to achieve information coordination.

Research on the coordination of supply chain information has been analyzed by many scholars at home and abroad. However, as the informationization of the supply chain continues to deepen, the information resources also show some new features, such as more widespread distribution, larger data volume, more diverse data categories, etc. The consequence of the huge amount of information is that it is difficult to store and centralize data. Data storage and centralized processing. Therefore, enterprises need to improve their ability to cope with market risks, enhance their core competitiveness, and handle this information better and faster.

© Springer Nature Switzerland AG 2019
Y. Tang et al. (Eds.): HCC 2018, LNCS 11354, pp. 156–168, 2019.
https://doi.org/10.1007/978-3-030-15127-0_16

2 Research of Multi-agent Based Supply Chain Information Collaboration

2.1 Multi-agent Information Collaborative Framework

After analyzing the multi-agent modeling ideas and organizational structure, this paper uses the intermediary structure to build a multi-Agent system. Each subsystem represents a business entity, with a mediation-centric federal structure model, which is divided into functional Agents and information coordination Agents. Each subsystem is uniformly determined by the Information coordination center Agent. After analyzing the information flow of each Agent, the service function of the structural Agent and the functional Agent are designed, and the supply chain information coordination framework is established.

2.2 Internal Structure Design of Multi-agent

Structural division is based on multi-agent information collaborative framework diagram, the rules of division are: independence, hierarchy, and order. After the structure is divided, the functional modules of each Agent need to be designed. This paper selects the structural Agent and some Agents in the functional Agent to design the business process, as follows:

(1) Manufacturer Agent
   The business process of the manufacturer Agent is:
   (a) According to the raw material inventory situation and the downstream distributor's demand order, the production plan is generated. After the production plan is completed, it is submitted to the system for review. The system evaluates the order according to the equipment and other conditions, and then generates a production plan, otherwise it is fed back to the distributor.
   (b) According to the production plan, the system produces and checks the production situation at any time, then it evaluates the completion status, and predicts whether the task can be completed on time. If the forecast cannot be completed on time, it will feedback to the manufacturer and re-establish the production plan.
   (c) After the production of the product is completed, it is fed back to the distributor Agent via the information coordination center Agent.
   (2) Inventory Agent
   The inventory Agent mainly evaluates the order quantity, balances the best inventory, completes the product in and out of the product according to the order, and updates the product inventory information.

2.3 Supply Chain Demand and Inventory Analysis Based on Multi-agent System

In order to facilitate simulation analysis, this paper only considers a simple three-level supply chain, that is, each level of enterprise only considers one situation, while ignoring the role of customers and so on. Based on the previous literature, this paper adds price parameters, and the price-sensitive function predicts the change of demand. The introduced price fluctuation factors can better simulate the operation process of the supply chain. It is assumed that the price-sensitive demand function conforms to the AR(1) process, while the market demand is linearly related to price [2–5]. In the price-sensitive demand function, the demand function is established by considering the first-order linear correlation between the product price and the retailer, and the demand function as follows:

$$f_t = F(R_t, \varepsilon_t) = a - pR_t + \varepsilon_t \tag{1}$$

Where a is the product demand of the market in t period, p is price sensitivity coefficient and p > 0, ε_t is demand disturbance term, it is assumed to be an independent random variable and is normally distributed with time, that is $\varepsilon_t \sim N(0, \sigma^2)$, $\text{Cov}(R_t, \varepsilon_t') = 0 (\forall t \text{ and } t')$.

This paper assumes that the market price is the AR(1) process:

$$R_t = \mu + \rho R_{t-1} + \eta_t \tag{2}$$

Where μ is non-negative constant; ρ is price autoregressive coefficient and $-1 < \rho < 1$, η_t is price perturbation term, an independent random variable with the same normal distribution as ε_t, that is $\eta_t \sim (0, \delta^2)$. In this model, assuming that p_t, η_t are uncorrelated, then from Eqs. (1) and (2), the demand perturbation term ε_t and the market price η_t are each independent and independent of time.

In this paper, the retailer's inventory strategy uses order-up-to, and demand forecasting uses MMSE technology. Let's take a retailer as an example to analyze the retailer's ordering process:

The inventory level of the retailer at the beginning of t period is as follows:

$$S_t = S_{t-1} + d_{t-1} - f_{t-1} \tag{3}$$

$$d_t = S_{t+1} - S_{t+L_1-1} + f_{t+L_1-1} \tag{4}$$

$$d_t = \hat{F}_t^{L_1} - \sum_{i=1}^{L-1} d_{t-i} - S_t \tag{5}$$

Where $\hat{F}_t^{L_1}$ is the forecast value of the lead time L_1, $\sum_{i=1}^{L-1} d_{t-i}$ is the actual order quantity of the retailer. It can be obtained from Eqs. (4) and (5):

$$S_t = \hat{F}_{t-L_1}^{L_1} - F_{t-L_1}^{L_1} \tag{6}$$

It can be seen from the Eq. (6) that the error $\hat{F}_{t-L_1}^{L_1} - F_{t-L_1}^{L_1}$ of the retailer's demand forecast in the $t - L_1$ period is the stock level of the t period. According to the Order-up-to inventory strategy, the actual order quantity d_t of the retailer at the beginning of t period is equal to the difference between the target inventory level at the beginning of t period and the actual inventory level at the end of $t-1$ period:

$$d_t = y_t - (y_{t-1} - f_{t-1}) \tag{7}$$

Similarly, according to the Order-up-to inventory strategy, the target inventory y_t can be expressed as:

$$y_t = \hat{F}_t^{L_1} + \omega \hat{\sigma}_t^{L_1} \tag{8}$$

Where ω is safety factor, $\hat{\sigma}_t^{L_1}$ is retailer demand forecasting error.

From the Eqs. (7) and (8), the actual order quantity of the retailer in the t period is:

$$d_t = \hat{F}_t^{L_1} - \hat{F}_{t-L_1}^{L_1} + f_{t-1} + \omega(\hat{\sigma}_t^{L_1} - \hat{\sigma}_{t-1}^{L_1}) \tag{9}$$

In the MMSE forecast, the expected value of historical data is used to represent the predicted values for future periods:

$$\hat{f}_{t+i} = E(f_{t+i}|f_{t-1}, f_{t-2}, \ldots) \tag{10}$$

When the model conforms to AR(1),

$$\hat{f}_{t+i} = E(f_{t+i}|f_{t-1}) \tag{11}$$

Since the price function conforms to the AR(1) process,

$$\hat{R}_{t+i} = E(R_{t+i}|R_{t-1}) = \frac{1 - \rho^{i+1}}{1 - \rho} \mu + \rho^{i+1} R_{t-1} \tag{12}$$

And because demand is a linear function of price, thus:

$$\hat{f}_{t+i} = a - p\hat{R}_{t+i} = a - b\left(\frac{1 - \rho^{i+1}}{1 - \rho} \mu + \rho^{i+1} R_{t-1}\right) \tag{13}$$

The retailer demand forecast $\hat{F}_t^{L_1}$ is:

$$\hat{F}_t^{L_1} = \sum_{i=0}^{L_1 - 1} \hat{f}_{t+i} = L_1 \mu_f + \frac{p\rho}{1 - \rho} \alpha_{L_1} \mu - b\rho \alpha_{L_1} R_{t-1} \tag{14}$$

Where $\alpha_{L_1} = \frac{1 - \rho^{L_1}}{1 - \rho}$, obtained from Eqs. (14) and (9):

$$d_t^{MMSE} = -p\rho \alpha_{L_1}(R_{t-1} - R_{t-2}) + f_{t-1} + \omega(\hat{\sigma}_t^{L_1} - \hat{\sigma}_{t-1}^{L_1}) \tag{15}$$

Since $\hat{\sigma}_t^{L_1}$ does not change with time and is constant, thus:

$$d_t^{MMSE} = -p\rho\alpha_{L_1}(R_{t-1} - R_{t-2}) + f_{t-1} \tag{16}$$

3 Collaborative Simulation Analysis of Supply Chain Information Based on Multi-agent

3.1 Supply Chain Collaborative Simulation Model Hypothesis

The realization of the supply chain information collaborative process is affected by many aspects. In order to better realize the simulation effect and highlight the main problems, this paper makes a simple treatment of the complex situation, ignoring the factors that have little influence on the simulation situation, and makes the necessary assumptions as follows [6]:

(1) Assuming that the supply chain starts from the seller, the role of the customer is ignored, and the demand is predicted by the seller.
(2) The multi-level and multi-members in the complex supply chain are ignored, the supply chain memberships are simplified, and only the three-level supply chain composed of individual manufacturers, individual distributors and individual sellers in a single supply chain are studied.
(3) Using abstract service, manufacturing, procurement, and inventory management Agents to represent functional Agents for research.

3.2 Modeling Analysis of Agents in Supply Chain Under Information Sharing

From the second section, this paper introduces the price factor, and assumes that the demand model relies on a price-sensitive function, and the price is assumed to be a stable process.

The demand is:

$$f_t = F(R_t, \varepsilon_t) = a - pR_t + \varepsilon_t \tag{17}$$

The function of the price is:

$$R_t = \mu + \rho R_{t-1} + \eta_t \tag{18}$$

The target inventory function is:

$$k_t^i = \hat{F}_t^{L_i} + \omega\hat{\sigma}_t^{L_i} \tag{19}$$

Where $\hat{F}_t^{L_i}$ is the demand forecast value under the lead time L, ω is the safety factor, $\hat{\sigma}_t^{L_i}$ is the lead time demand error, i take 1 or 2, representing the retailer and the distributor respectively.

The value of the lead time demand forecast error is independent of time and its value is a constant. By order-up-to inventory strategy of second quarter and $\hat{\sigma}_t^{L_i}$ is also independent of t:

$$d_t^1 = \hat{F}_t^{L_1} - \hat{F}_{t-1}^{L_1} + f_{t-1} \tag{20}$$

$$d_t^2 = \hat{F}_t^{L_2} - \hat{F}_{t-1}^{L_2} + d_t^1 \tag{21}$$

(1) Retailer Agent. The retailer Agent issues an order request to the distributor agent at the beginning of t period.

As the price conforms to the stationary process, the MMSE can predict the price of the t + i period in this paper:

$$\hat{R}_{t+i} = E(R_{t+i}|R_{t-1}) = \frac{1 - \rho^{i+1}}{1 - \rho}\mu + \rho^{i+1}R_{t-1} \tag{22}$$

The retailer demand forecast $\hat{F}_t^{L_1}$ is:

$$\hat{F}_t^{L_1} = \sum_{j=0}^{L_1-1} \hat{f}_{t+j} = \sum_{j=0}^{L_1-1} \left(a - p\hat{R}_{+j}\right) = L_1\mu_f + \frac{p\rho}{1-\rho}\alpha_{L_1}\mu - p\rho\alpha_{L_1}R_{t-1} \tag{23}$$

Where $\alpha_{L_1} = \frac{1-\rho^{L_1}}{1-\rho}$, obtained from Eqs. (20) and (23):

$$d_t^1 = -p\rho\alpha_{L_1}(R_{t-1} - R_{t-2}) + f_{t-1} \tag{24}$$

$$d_{t+1}^1 = (1 - \rho)\mu_f + \rho d_t^1 + \varepsilon_t - \rho\varepsilon_{t-1} - p\alpha_{L_1+1}\eta_t + p\rho\alpha_{L_1}\eta_{t-1} \tag{25}$$

Therefore, the order quantity for the next t + i period is:

$$\begin{aligned} d_{t+1}^1 = &\left(1 - \rho^i\right)\mu_f + \rho^i d_t^1 + \varepsilon_{t+i-1} - \rho^i\varepsilon_{t-1} - p\alpha_{L_1+1}\eta_{t+i-1} \\ &- p\rho(\alpha_{L_1+1} - \alpha_{L_1})\eta_{t+i-2} - p\rho^2(\alpha_{L_1+1} - \alpha_{L_1})\eta_{t+i-3} - \cdots - p\rho^{i-1}(\alpha_{L_1+1} - \alpha_{L_1})\eta_t \\ &+ p\rho^i\alpha_{L_1}\eta_{t-1}, i = 1, 2, \ldots \end{aligned} \tag{26}$$

The retailer's target inventory is:

$$\begin{aligned} k_t^1 &= \hat{F}_t^{L_1} + \omega\hat{\sigma}_t^{L_1} \\ &= L_1\mu_f + \frac{p\rho}{1-\rho}\alpha_{L_1}\mu - p\rho\alpha_{L_1}R_{t-1} + \omega\sqrt{L_1\sigma^2 + \frac{p^2}{(1-\rho)^2}\left(L_1 + \frac{\rho(1-\rho^{L_1})(\rho^{L_1+1} - \rho - 2)}{1-\rho^2}\right)\delta^2} \end{aligned} \tag{27}$$

(2) Distributor Agent. The distributor Agent receives the order quantity from the seller and orders the goods from the upstream company according to its own inventory level, assuming the order period is L_2. At this point, the distributor's demand forecast from the retailer is:

$$\hat{F}_t^{L_2} = \sum_{i=1}^{L_2} \widehat{d_{t+i}^1} \tag{28}$$

Considering that in the case of information sharing and non-sharing, the information obtained by the distributors from the downstream is different, and the demand forecast will also produce different results. The following are the information sharing and non-sharing.

(A) Information is not shared (Customer demands)

When the information is not shared, the information obtained by the distributor is only the order quantity of the retailer. According to the order quantity of the retailer's t period, the distributor uses MMSE to make predictions, and the order quantity of the retailer in the $t+i(i = 1, 2, \ldots)$ period is:

$$\hat{d}_{t+1}^{1,N} = E(d_{t+i}^1 | d_t^1) = (1 - \rho^i)\mu_f + \rho^i d_t^1 \tag{29}$$

According to Eqs. (28), (29), the predicted value of the distributor's demand is:

$$\hat{F}_t^{L_2,N} = \sum_{i=1}^{L_2} \hat{d}_{t+i}^{1,N} = (L_2 - \rho\alpha_{L_2})\mu_f + \rho\alpha_{L_2}d_t^1 \tag{30}$$

Where $\alpha_{L_2} = \frac{1-\rho^{L_2}}{1-\rho}$, the order quantity of the distributor obtained by Eq. (22) is:

$$d_t^{2,N} = \hat{F}_t^{L_2,N} - \hat{F}_{t-1}^{L_2,N} + d_t^1 = \alpha_{L_2+1}d_t^1 - \rho\alpha_{L_2}d_{t-1}^1 \tag{31}$$

When the information is not shared, the demand forecast error $\hat{\sigma}_t^{L_2,N}$ of lead time L_2 is:

$$\hat{\sigma}_t^{L_2,N} = \sqrt{\left[L_2 + \rho^2(\alpha_{L_2})^2\right]\sigma^2 + p^2\left[\sum_{i=1}^{L_2}(\alpha_{L_1} + L_2 - i + 1)^2 + \rho^2(\alpha_{L_1})^2(\alpha_{L_2})^2\right]\delta^2} \tag{32}$$

Therefore, under non-information sharing, the distributor's target inventory is:

$$k_t^{L_2,N} = (L_2 - \rho\alpha_{L_2})\mu_f + \rho\alpha_{L_2}d_t^1$$

$$+ \omega\sqrt{\left[L_2 + \rho^2(\alpha_{L_2})^2\right]\sigma^2 + p^2\left[\sum_{i=1}^{L_2}(\alpha_{L_1} + L_2 - i + 1)^2 + \rho^2(\alpha_{L_1})^2(\alpha_{L_2})^2\right]\delta^2}$$

$$\tag{33}$$

(B) Information sharing (Customer needs)

When the information is shared, $\varepsilon_{t-1} = d_{t-1} - (a - pR_{t-1})$ and $\eta_{t-1} = R_{t-1} - (\mu + \rho R_{t-2})$ can be known through the price model and the demand model. So the forecast value of the retailer's order quantity during the $t+i(i = 1, 2, \ldots)$ period is:

$$\hat{d}_{t+i}^{1,Y} = E\left(d_{t+i}^1 | d_t^1\right) = \left(1 - \rho^i\right)\mu_f + \rho^i d_t^1 - \rho^i \varepsilon_{t-1} + p\rho^i \alpha_{L_1} \eta_{t-1} \tag{34}$$

Therefore, the distributor's demand forecast value is:

$$\hat{F}_t^{L_2,Y} = \sum_{i=1}^{L_2} \hat{d}_{t+i}^{1,Y} = \left(L_2 - \rho\alpha_{L_2}\right)\mu_f + \rho\alpha_{L_2} d_t^1 - \rho\alpha_{L_2}\varepsilon_{t-1} + p\rho\alpha_{L_1}\alpha_{L_2}\eta_{t-1} \tag{35}$$

According to Eqs. (35), (22), the order quantity of the distributor under information sharing is:

$$\begin{aligned}
d_t^{2,Y} &= \hat{F}_t^{L_2,Y} - \hat{F}_{t-1}^{L_2,Y} + d_t^1 \\
&= \alpha_{L_2+1} d_t^1 - \rho\alpha_{L_2} d_{t-1}^1 - \rho\alpha_{L_2}(\varepsilon_{t-1} - \varepsilon_{t-2}) + p\rho\alpha_{L_1}\alpha_{L_2}(\eta_{t-1} - \eta_{t-2})
\end{aligned} \tag{36}$$

Under information sharing, the demand forecast error $\hat{\sigma}_t^{L_2,Y}$ of the lead time L_2 is:

$$\hat{\sigma}_t^{L_2,Y} = \sqrt{L_2\sigma^2 + p^2\left(\sum_{i=1}^{L_2}(\alpha_{L_1+L_2-i+1})^2\right)\delta^2} \tag{37}$$

Therefore, under information sharing, the distributor's target inventory is:

$$\begin{aligned}
k_t^{L_2,Y} &= \left(L_2 - \rho\alpha_{L_2}\right)\mu_f + \rho\alpha_{L_2} d_t^1 \\
&\quad - \rho\alpha_{L_2}\varepsilon_{t-1} + p\rho\alpha_{L_1}\alpha_{L_2}\eta_{t-1} + \sqrt{L_2\delta^2 + p^2\left(\sum_{i=1}^{L_2}(\alpha_{L_1+L_2-i+1})^2\right)\delta^2}
\end{aligned} \tag{38}$$

(3) Manufacturer Agent. The manufacturer receives the order from the distributor and determines the total production during the t period based on its own inventory level. Modeling and analysis of Agents in supply chain enterprises:

(A) Service Agent receives goods information from enterprises downstream of the supply chain and transmits the information to the production agent. Let M(t) be the order quantity of the downstream enterprise obtained at time t, and the expression is as follows:

$$D(t) = \sum_{c=1}^{t_c} d(c, t) \tag{39}$$

Where c is customer code, t_c is total number of customers, t is the time for orders arrive at service Agent, $d(c, t)$ is customer order quantity at time t.

(B) Manufacturing Agent is responsible for meeting the market demand. Assuming the manufacturing lead time is T_i, then the number of products in the $t + T_i$ period is expressed as:

$$P(t+T_i) = S^* - \left[S(t+1) + \sum_{\eta=1}^{T_i-1} (t+\eta) - V(t+1) - T_i * \hat{D}_t(t) \right] \quad (40)$$

Where $P(t+T_i)$ is production of t + T time, S^* is target inventory.

(c) Procurement Agent is responsible for determining the procurement task. Its implementation process is expressed as follows:

$$\hat{Z}(t) = Z(t) \quad (41)$$

$$\hat{Z}(t+\eta) = Z(t+\eta-1) + \hat{G}(t+\eta-1) \quad (42)$$

$$R(t+\eta) = \max\left[(Z^* - \hat{Z}(t+\eta)), 0 \right] \quad (43)$$

Where $Z(t)$ is stock of raw materials at time t, $\hat{Z}(t)$ is forecasted stock quantity of raw material inventory at time t, $\hat{G}(t)$ is the predicted quantity of total raw materials at time t, $R(t)$ is the quantity of raw materials purchased at the time t, Z^* is target stock of raw materials.

The procurement agent is mainly responsible for the procurement of raw materials, and the amount of orders depends on the purchase amount of raw materials. The stock of raw materials is calculated as follows:

$$Z(t+1) = Z(t) - \sum_{c=1}^{t_c} y(c,t) + P(t) \quad (44)$$

Where $Z(t)$ is stock of raw materials at time t, t_c is total number of customers, c is customer code, $y(c,t)$ is the number of items that are given to customers at t time, $P(t)$ is production at time t.

(4) Inventory management agent. Assuming that S(t) is the inventory quantity of the enterprise at time t, R(t) represents the order quantity of an upstream enterprise at time t, and Q(t) represents the replenishment quantity of goods at time t. Then the goods inventory at time t is expressed as:

$$S(t+1) = S(t) - \sum_{i=1}^{c} R_i(t) + Q(t) \quad (45)$$

3.3 Simulation Analysis of Supply Chain Information Collaborative

The process of establishing a model in the supply chain is analyzes. When the company shares information and does not share information, its inventory and order quantity will reflect the changes in the supply chain.

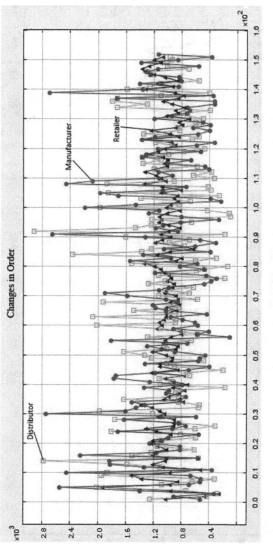

Fig. 1. Supply chain order changes under non-information sharing

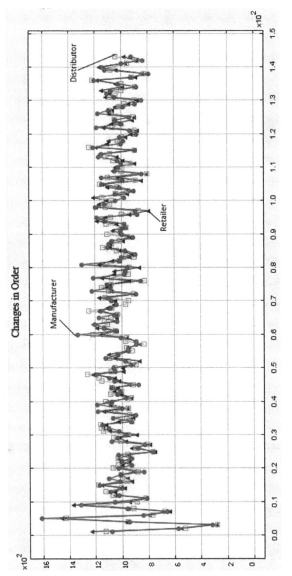

Fig. 2. Change of order in supply chain information cooperation

Analysis of simulation results: (1) Analysis of simulation results of multi-agent system in supply chain enterprises without information coordination. This paper simulates the changes of the supply chain three-level enterprises with the market. In the following figures, the abscissa represents time, unit time/week, and the ordinate represents the order quantity or inventory quantity. Order volume change Fig. 1 shows that the fluctuation of order volume is step by step, the retailer's order fluctuations are the smallest, followed by the distributors, the manufacturers' fluctuations are the biggest, and the subordinate enterprises' small fluctuations will lead to sharp fluctuations in their superior supply chain. The violent fluctuations of the chain. The change in the inventory of the company is the same as the change in the order quantity.

(2) Analysis of simulation results of multi-agent system under the cooperation of supply chain information. Through the simulation analysis of the changes of orders and inventory of supply chain enterprises under non-information collaboration, this paper studies the influence of supply chain enterprises that implement information coordination on the orders and inventory of member companies. In the simulation, the model parameter VS is changed from 0 to 1, so that the system adopts the information collaboration strategy when performing order and inventory simulation. The simulation results are shown in Fig. 2.

From the above figures, it can be found that after the supply chain enterprises choose to implement the information collaboration strategy, the retailer's order and inventory status are not affected, but the order volume and inventory of distributors and manufacturers have undergone tremendous changes. In the information collaboration mode, regardless of how the retailer's order volume or inventory volume fluctuates, the trend of distributors and manufacturers is always closely aligned with the retailer, as shown in Fig. 2.

4   Conclusion

In the era of rapid economic development, supply chain management is also inseparable from the cooperation of all member companies. The key of this is the information collaboration of supply chain. Under this background, this paper made an in-depth study on the information collaboration of distributed supply chain, established a three-level supply chain simulation model and used Repast simulation software to carry out simulation analysis.

References

1. Yang, J.: The construction of the E-commerce platform system in collaboration with national supply chain. In: Proceedings of the 2nd International Conference on Politics, Economics and Law (ICPEL 2017), p. 5. Shandong University (2017)
2. Hosoda, T., Disney, S.M.: A delayed demand supply chain: incentives for upstream players. Omega **40**(4), 478–487 (2011)
3. Yu, F., Xue, L., Sun, C., Zhang, C.: Product transportation distance based supplier selection in sustainable supply chain network. J. Clean. Prod. **137**, 29–39 (2016)

4. Hussain, M., Shome, A., Lee, D.M.: Impact of forecasting methods on variance ratio in order-up-to level policy. Int. J. Adv. Manuf. Technol. **59**(1–4), 413–420 (2012)
5. Klug, F.: The internal bullwhip effect in car manufacturing. Int. J. Prod. Res. **1**(51), 303–322 (2013)
6. Cigolini, R., Pero, M., Rossi, T., Sianesi, A.: Linking supply chain configuration to supply chain performance: a discrete event simulation model. Simul. Model. Pract. Theory **40**, 1–11 (2014)

Mobile Internet Mobile Agent System Dynamic Trust Model for Cloud Computing

Weijin Jiang[1,2,3(✉)], Yirong Jiang[4], Yang Wang[1], Jiahui Chen[1],
Yuhui Xu[1], and Li'na Tan[1]

[1] Institute of Big Data and Internet Innovation, Mobile E-business Collaborative
Innovation Center of Hunan Province, Hunan University of Commerce,
Changsha 410205, China
jlwxjh@163.com, 18508488203@163.com, 810663304@qq.com,
363168449@qq.com, 18785024@qq.com
[2] Key Laboratory of Hunan Province for New Retail Virtual Reality
Technology, Hunan University of Commerce, Changsha 410205, China
[3] School of Computer Science and Technology,
Wuhan University of Technology, Wuhan 430073, China
[4] Tonghua Normal University, Tonghua 134002, China
307553803@qq.com

Abstract. Aiming at the security and trust management of mobile agent system in cloud computing environment, based on the simple public key infrastructure (SPKI) trust mechanism, a mobile agent system objective trust peer management model (MAOTM) is proposed to solve the problem of identity authentication, operation authorization and access control in the mobile agent system; On this basis, the Human Trust Mechanism (HTM) is used to study the subjective trust formation, trust propagation and trust evolution law, and the subjective trust dynamic management algorithm (MASTM) is proposed. Based on the interaction experience between the mobile agent and the execution host and the third-party recommendation information to collect the basic trust data, the public trust host selection algorithm is given. The isolated malicious host algorithm and the integrated trust degree calculation algorithm realize the function of selecting the trusted cluster and isolating the malicious host, so as to enhance the security interaction between the mobile agent and the host. The simulations of the proposed algorithms are verified and proved to be feasible and effective.

Keywords: Cloud computing · Mobile agent system · Subjective trust · Objective trust · Dynamic trust management · Mobile internet

1 Introduction

Mobile Agent technology is an emerging technology that is multi-disciplinary and at the forefront of international research. It is also a product of the combination of Agent and Internet in the field of artificial intelligence [1, 2]. In the cloud computing environment, the problem of resource allocation of trust security domain in mobile agent system is studied [3, 4]. It can track international cutting-edge technology, enrich the theoretical system of cloud computing and mobile agent model, and enhance the

© Springer Nature Switzerland AG 2019
Y. Tang et al. (Eds.): HCC 2018, LNCS 11354, pp. 169–180, 2019.
https://doi.org/10.1007/978-3-030-15127-0_17

security performance of agent application system. It can also promote the further development of China's distributed application technology and information security technology, and promote the application of mobile agent technology in various fields in the cloud computing environment. Therefore, it is of great theoretical and practical significance to study the security trust problem in the design of mobile agent system in cloud computing environment.

2 Related Work

Currently, the trust management model is generally divided into two categories. The first category is the objective rational model. It uses a rational and accurate method to express and deal with complicated trust relationships, and has objective and static management characteristics. The second category is the subjective empirical model, which considers trust to be a subjective judgment of a particular level of specific characteristics or behavior of the object. The method of fuzzy sets is used to study direct trust management [5]. Because trust is somewhat uncertain, fuzzy logic is used to deal with the subjectivity and uncertainty of subject characteristics and behavior cognition, and make decisions based on the determined trust strategy to provide effective support for judging trust status. Trust and the issue of non-trust communication. Guba-oriented e-commerce system presents a complete set of belief propagation algorithm, part of the expression of trust between individuals for predicting the level of trust between any two objects. The Simple Public Key Infrastructure (SPKI) method was proposed by Carl Ellison and Bill Frantz in 1996 [6], which was standardized by the IETF in 1999. The SPKI standard now in use is a mixed version of SPKII/SDSI.

The complexity of trust verification is reduced from the design principle. Compared with the existing PKI-based trust management model, the trust certificate is designed based on SPKI+RBAC, which reduces the complexity of trust verification. It can not only meet the requirements of trust transfer and verification of mobile agent system, but also control the attenuation of trust caused by excessive trust chain [7–9].

3 Dynamic Management Method of Subjective Trust in Mobile Agent System

3.1 Mobile Agent System Subjective Trust Dynamic Management Model

Subjective Trust Dynamic Management Model
The composition of the mobile agent system supervisor trust model is shown in Fig. 1 [9, 10]. Bottom-up: The host is in an open and dynamic network environment. The mobile agent platform (MAP) is located on the host, providing an Execution Environment (EE) for the mobile agent. The MAP is not necessarily unique on a host, but more than one. The subjective trust management model is located in the MAP and consists of three trust components: They are the trust formation component, the trust propagation component, and the trust evolution component. These management

mechanisms provide a trust interaction context for the interaction entity, and monitor the interaction process and behavior of the entity. The trust formation component mainly implements the collection and calculation of trust data, the trust propagation component mainly implements the protocol exchange of trust data, and the trust evolution component mainly realizes the update of the trust data.

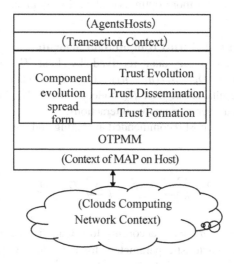

Fig. 1. Subjective trust dynamic management model

Basic Trust Data Collection

1. Recommended trust basic data collection

If the host H_a does not have sufficient understanding of the host H_x, envisaging the host H_a want to interact with the host H_x in order to evaluate the degree of trusted host H_x, and the host H_a queries a group of hosts $\{H_1, H_2, \ldots, H_K\}$ for trust information about host H_x. The recommendation information between the entities defined here is a set of "basic data" about the interaction behavior rather than the comprehensive calculation result of the degree of trust. The purpose of this is to eliminate the accumulation of errors generated during the comprehensive calculation of the degree of trust, and the impact of the trust preferences of the host $\{H_1, H_2, \ldots, H_K\}$. Under normal circumstances, the number of interactions between the host H_i and entity H_x is n, the number of successes n_1, the number of failures n_2, $D_{ki} = n_1/(n_1 + n_2)$, $(n = n_1 + n_2)$.

2. Direct trust data collection

Assume that the host H_a has to interact n times with the host H_x directly, successes n_1, failures n_2, $(n = n_1 + n_2)$. $D_{ax} = n_1/(n_1 + n_2)$ to reflect the host H_a and host H_x interactive behavior is good or bad, which is defined as a direct trust underlying data.

3.2   Recommended Trust Computation Algorithm

We divide the continuous system runtime into equally spaced inspection periods, each of which is called a "time frame" and is represented by $\tau(\tau = 1, 2, \ldots, n)$. Then the interactive behavior of the interactive host is transformed into the quantitative calculation of trust degree. The Gaussian probability distribution theory is used to improve the average algorithm, and a more optimized algorithm is given. The algorithm is as follows.

Algorithm 1. Recommended trust computation algorithm.
   Initialization: let the basic data received by host H_x about host H_a be: $\{D_1, D_2, \ldots, D_k\}$, where: $D_i = n_1/(n_1 + n_2)$, $(0 \leq D \leq 1)$. n_1 is the number of positive interaction results about host H_x collected from M_i during the inspection period, and n_2 is the number of negative interactions.
   Step 1: The data on the host recommended averaging and variance are calculated as follows:

$$\bar{D} = \frac{1}{k}\sum_{i=1}^{k} D_i, S^2 = \frac{1}{k}\sum_{i=1}^{k}(D_i - \bar{D})^2 \tag{1}$$

   Step 2: Order: $\mu = \bar{D}$, $\sigma^2 = S^2$, according to a Gaussian distribution theory, that $K(\mu, \sigma^2)$ use as the characteristic parameters for a random variable T, T obtained probability density functions $p(x)$, which (μ, σ^2) are called expectation and variance of the Gaussian distribution. When $\mu = 0$, $\sigma^2 = 1$, the time, T is called the standard normal distribution.

$$p(x) = \frac{1}{\sigma\sqrt{2\pi}}e^{-\frac{(x-\mu)^2}{2\sigma^2}}(\sigma > 0), (-\infty < x < +\infty) \tag{2}$$

   Step 3: the possibility of random variable T in $(-\infty, v)$, $(v, +\infty)$ appearing in the range can be obtained. Wherein, $P(\leq V)$ T indicates the possibility of appearing in the range of v or less, $P(> v)$ T indicates the possibility of occurring within the range of greater than v.

$$P(\leq v) = \frac{1}{\sigma\sqrt{2\pi}}\int_{-\infty}^{\frac{v-\mu}{\sigma}}e^{-\frac{x^2}{2}}dx, \ \ P(> v) = \frac{1}{\sigma\sqrt{2\pi}}\int_{\frac{v-\mu}{\sigma}}^{\infty}e^{-\frac{x^2}{2}}dx$$

For a given interval value (v_1, v_2), T appears in the specified range possibilities:

$$P(v_1, v_2) = \frac{1}{\sigma\sqrt{2\pi}}\int_{\frac{v-\mu}{\sigma}}^{\frac{1-\mu}{\sigma}}e^{-\frac{x^2}{2}}dx, \ \ (v_1 < v_2) \tag{3}$$

Step 4: Then the variable T in the specified range (v, 1), the possibility [0, 1] appear respectively:

$$P(v,1) = \frac{1}{\sigma\sqrt{2\pi}} \int_{\frac{v-\mu}{\sigma}}^{\frac{1-\mu}{\sigma}} e^{-\frac{x^2}{2}} dx, P(0,1) = \frac{1}{\sigma\sqrt{2\pi}} \int_{\frac{0-\mu}{\sigma}}^{\frac{1-\mu}{\sigma}} e^{-\frac{x^2}{2}} dx \qquad (4)$$

Step 5: Then calculated variables (v, 1) within the scope of the likelihood ratio in the range [0, 1] and T:

$$P_{ax}(v) = \frac{P(v,1)}{P(0,1)} = \int_{\frac{v-\mu}{\sigma}}^{\frac{1-\mu}{\sigma}} e^{-\frac{x^2}{2}} dx \Big/ \int_{\frac{v-\mu}{\sigma}}^{\frac{1-\mu}{\sigma}} e^{-\frac{x^2}{2}} dx \qquad (5)$$

Step 6: The host H_a recommended level of trust on the host H_x ($0 < v < 1$) defined as the ratio (v) is $P_{ax}(v)$ ($0 < v < 1$). Referred to as $T_{x-rec}(v)$, $T_{x-rec}(v) = P_{ax}(v)$ ($0 < v < 1$), where v is the calculated threshold value.

3.3 Comprehensive Calculation of Trust

Considering the direct trust level T_{x-dir} of H_a to H_x, and the recommended trust level T_{x-rec} of H_x collected by H_a, define a variable that comprehensively measures trust, called "Trust Degree". "Trust Degree" is used to quantitatively represent the trustworthiness of interactive hosts in a mobile agent system, which means the possibility that host H_a interacts with another host H_x in a given environment to obtain a positive result (trust: $T_x = 1$). T_x is a comprehensive measure of trust that is used to quantitatively evaluate the current level of trustworthiness of an entity in a mobile agent system and to predict the degree of trust in the next interaction with that entity. The following is a "trust" comprehensive calculation process.

During the same study period, the direct trust level T_{x-dir} and the recommended trust strength T_{x-rec} are weighted and summed, where ρ is the confidence coefficient.

$$T_x = \rho\, T_{x-dir} + (1-\rho)T_{x-rec}(0<\rho\leq 1) \qquad (6)$$

The algorithm implements two functions: first, if entity H_a repeatedly interacts with entity H_x and finds that H_x continues to maintain good behavior (affirmative event), H_x's trust degree Tx will continue to grow, tending to a maximum of 1, and if entity H_x has malicious behavior, its trust will drop rapidly; second, if there is a large change in the degree of trust between time frames n-1 and n, this large amount of change H_x^n will have a large effect, and vice versa. $\sigma(\tau)$ known as the coefficient update control the update speed of trust.

$$T_x^{n+1} = T_x^{n-1} + \sigma(\tau)(T_x^n - T_x^{n-1}) \qquad (7)$$

4   Simulation Experiment Analysis

The key properties and parameter selections of Eqs. (1) to (7) given in the above dynamic trust metrics and evaluation methods are verified by a series of simulation experiments [12–15].

4.1   Experimental Conditions

Experiment 1 examines the algorithm of Eq. (1), and the result is shown in Fig. 2. Take host H_a to investigate multiple interactions with H_x as an example to illustrate [11]: The more bad interactions of host H_x, the higher the value converted to the underlying trust data D_{ax}. If H_x always maintains good behavior, when the interaction time reaches $\tau = 50$ frames, the value of D_{ax} is almost equal to 1; conversely, the more bad interactions of host H_x, the lower the value converted to the base trust data D_{ax}. If the bad behavior persists, the value of D_{ax} is almost equal to 0 when the interaction time reaches $\tau = 50$ frames. If the host's H_x interaction behavior is mixed, the value of D_{ax} fluctuates around 0.5. Therefore, D_{ax} in Eq. (1) can correctly reflect the degree of behavior of H_a in the series of interactions with host H_x. When K does not understand H_x, H_a considers $D_{ax} = 0.5$.

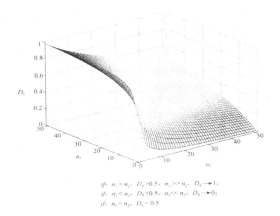

Fig. 2.   Verify Dij(n1, n2), (t = 1, 2, …,50)

Experiment 2 examines the algorithm of Eq. (5), and the result is shown in Fig. 3. Let the trust demand threshold be T_o, and let $v = T_o$, you can see the change of the direct (or recommended) trust degree T_x of the host H_x when calculating the base value v (depending on the collected data is direct or recommend trust data). It can be seen that the higher the T_o value, the higher the expectation requirement for the host satisfying $T_x > T_o$, and the fewer the number of hosts satisfying the condition. For example, when the calculation base value $v =$ trust threshold $T_o = 0.8$, it can be seen from Fig. 3 that only the candidate host whose mathematical expectation value of the host interaction behavior $\mu > 0.6$ can be included in the trusted interaction object.

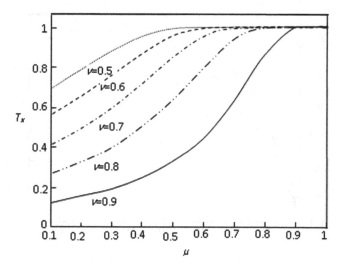

Fig. 3. The relationship between the degree of Hx trust and the mathematical expectation of its behavior at different v values

Experiment 3 verifies the result is shown in Fig. 4. The main consideration here is that if the host H_a is a new member, it wants to interact with H_x. Since there is no direct empirical data, the initial value of the confidence factor is 0, and the trust information must first be collected from the third-party entity. As the interaction experience increases, the confidence factor gradually increases. (a, b) is a pair of constants, a is the maximum confidence factor, and b is used to control the growth rate of the confidence factor according to different situations. The experimental results show that when (a, b) takes different values, when the number of interactions between H_a and H_x increases gradually, the confidence coefficient ρ gradually increases smoothly, and the confidence coefficient approaches the maximum value. In the experiment, three sets of data were selected for comparison. When the time frame $\tau = 40$, the confidence coefficient almost reached their respective maximum values.

Experiment 4 verifies the results are shown in Fig. 5. The amount of change between the two time frames is $0 < \Delta T_{ax} < 1$, and the parameter w controls the evolution speed of T_{ax}. The experimental results show that when the coefficient w = 1, $\sigma(\tau)$ min = 0, the evolution rate is the slowest, and $\sigma(\tau)$ max = 0.46 has the fastest evolution speed. At this time, the maximum influence of ΔT_{ij} on the T_{ax} value in the next time frame is 0.46. As can be seen from Fig. 9, when the parameter is selected w = 2, the trust will get faster the speed of evolution, but the stable region of the algorithm ΔT_{max} is around 0.77. When choosing w = 1.5, ΔT_{max} is around 1. If compared with w = 2, Tax has a slower rate of evolution; if compared with w = 1, T_{ax} has a faster rate of evolution.

The experimental results obtained according to the trust degree synthesis algorithm (14) are shown in Fig. 6. It can be seen that the algorithm has a slow rising fast falling feature. If host H_x continues to maintain good interaction behavior for multiple time frames, host H_x can gradually gain high trust; after obtaining high trust, H_x suddenly

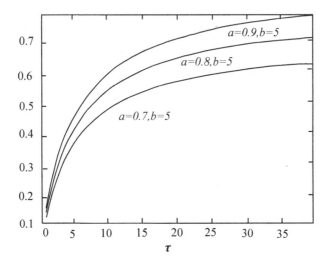

Fig. 4. ρ (τ) changes in the trend

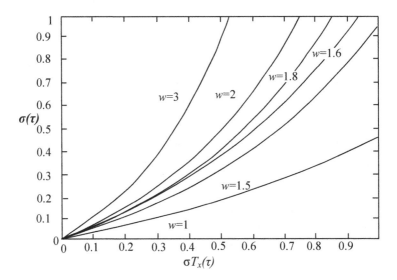

Fig. 5. Change trend of σ (τ) at different w

implements malicious behavior in the interaction to obtain illegal benefits, his trust will decline rapidly, be seen by H_a and spread through trust, so that the trusted group is isolated. If H_x wants to restore the higher trust he once had, it needs to make long-term efforts to maintain good interactions in order to restore his "trust". This main feature of the trust degree comprehensive calculation algorithm effectively suppresses the malicious behavior of the host.

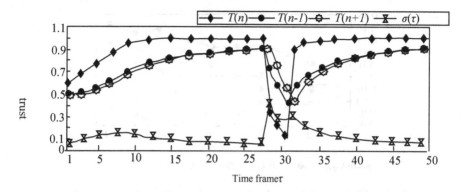

Fig. 6. Changes in trust trends ($w = 1.5$)

Given an algorithm for host H_a to isolate the host M_k (k = 1, 2, ...) with bad or malicious recommendation behavior based on direct empirical data on host H_x. The simulation experiment results are illustrated as follows in Figs. 7, 8 and 9: it can be seen from Fig. 7 that the recommendation data of the recommender M_3 to the host H_x is the best with the direct empirical data of the host H_a to H_x. Depending on the algorithm used, H_a can conclude that: M_3 is more believable, and M_1 and M_2 may be suspected of malicious recommendations.

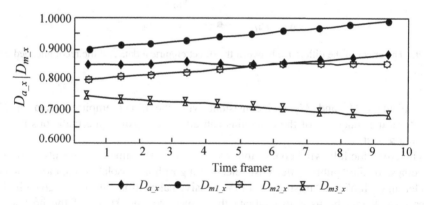

Fig. 7. Comparison of direct empirical data with recommended data

The effect of the difference between the direct empirical data and the recommended data on T_{m_k} is shown in Fig. 9.

It can be seen from Fig. 9 that according to equations in the course of the investigation, the recommendation data of the recommender 3 to the host H_x and the direct empirical data of the host H_a to H_x are the best. Host H_a's direct trust to him gradually increases, and the degree of improvement depends on the degree of consistency. The higher the consistency, the faster the increase; the recommendation data of

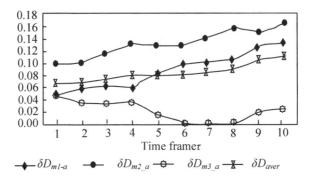

Fig. 8. Differences between direct empirical data and recommended data

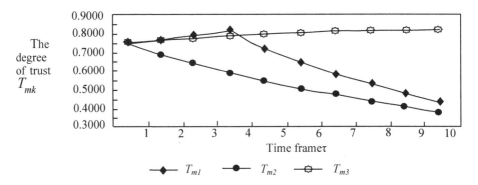

Fig. 9. The effect of the difference between the direct empirical data and the recommended data on T_{m_k}

recommenders M_1 and M_2 is inconsistent with the direct empirical data of H_a. Therefore, the functions of the equations can effectively isolate the malicious recommendation host.

Therefore, the following conclusions can be drawn: the simulation results verify the correctness of the "public letter host selection algorithm", "isolated malicious recommender algorithm" and "trust level comprehensive calculation algorithm" given in this paper [16]. It can be used to evaluate the subjective trust status of the host to be interacted in the mobile agent system, and predict the trustworthiness of the host to be interacted in the next time frame. The series of algorithms given can stimulate the trusted host and isolate the malicious host, which has the function of "punishing evil and promoting good". It can effectively manage the subjective trust dynamic management of the mobile agent system.

5   Conclusion

Divide the trust problem in the mobile agent system into objective trust and subjective trust to divide and conquer. Based on the analysis of the use of SPKI related certificates, this paper studies the problem of dynamic management of subjective trust in mobile agent system under the objective trust management framework of SPKI-based mobile agent system. The trust requirement of the entity (host or mobile agent) in the mobile agent system is analyzed, and a subjective trust dynamic management model consisting mainly of three trust components is proposed. Among them, the trust forms a component and completes the collection of trust data; the trust propagation component completes the communication of trust and the exchange of trust data; trusts the evolution component and completes the update of the trust data. The quantitative representation method of trust in the mobile agent system is given. Based on the basic ideas of description and metric trust proposed in the Josang network trust management model, two basic concepts of Evidence Space and Opinion Space are introduced in the mobile agent system. In the fact space of the mobile agent system, the "good or bad result of entity interaction behavior" is transformed into the "level of credibility of the entity" in the concept space. Using Gaussian Probability Distribution Theory, the method of changing the degree of trust of the host in the mobile agent system is given. "Trust" is used to indicate the degree of trust that Host H_a considers Host H_x within a specified time frame in the Mobile Agent system to evaluate and predict the next secure interaction with Host H_x. A subjective trust dynamic management algorithm is proposed. Finally, through a set of simulation experiments, the feasibility of the proposed algorithm to measure the degree of host trust in the mobile agent system is verified, and the effectiveness of the trust group to improve the security of the interaction in the mobile agent system is verified.

Acknowledgments. This work was supported by the National Natural Science Foundation of China (61472136; 61772196), the Hunan Provincial Focus Social Science Fund (2016ZDB006), Hunan Provincial Social Science Achievement Review Committee results appraisal identification project (Xiang social assessment 2016JD05). The authors gratefully acknowledge the financial support provided by the Key Laboratory of Hunan Province for New Retail Virtual Reality Technology (2017TP1026).

References

1. Boss, G., Malladi, P., Quan, D., et al.: Cloud computing [EB/OL]. http://download.boulder. ibm.com/ibmdl/pub/software/dw/wes/hipods/Cloud_computing_wp_final_8Oct.pdf
2. Cloud Computing Forum & Workshop [EB/OL]. http://www.nist.gov/itl/cloud.cfm. Accessed 20 Aug 2010
3. Gray, R., Cybenko, G., Kotz, D., et al.: Mobile agents and state of the art. In: Bradshaw, J. (ed.) Handbook of Agent Technology. AAAI/MIT Press, Cambridge (2012)
4. Busi, N., Padovani, L.: A distributed implementation of mobile nets as mobile agents. In: Steffen, M., Zavattaro, G. (eds.) FMOODS 2005. LNCS, vol. 3535, pp. 259–274. Springer, Heidelberg (2005). https://doi.org/10.1007/11494881_17

5. Tang, W., Chen, C.: Subjective trust management model based on fuzzy set. Ruan Jian Xue Bao/ J. Software **14**(8), 1401–1408 (2003)
6. Jiang, S., Li, J.: For P2P e-commerce system based on trust mechanism reputation. Ruan Jian Xue Bao/J. Software **18**(10), 2551–2563 (2007)
7. Chen, H., Sun, J., Liu, C., Li, H.: A lightweight secure and trusted virtual execution environment. SCIENCE CHINA Information Sciences **42**(5), 617–633 (2012)
8. Lange, D.B., Oshima, M.: Seven good reasons for mobile agents. Commun. ACM **42**(3), 88–89 (2015)
9. Claessens, J., Preneel, B., Vandewalle, J.: How can mobile agents do secure electronic transactions on untrusted hosts? a survey of the security issues and the current solutions. ACM Trans. Inter. Tech. **3**(11), 28–48 (2016)
10. Ma, H.D., Yuan, P.Y., Zhao, D.: Research progress on routing problem in mobile opportunistic networks. Ruan Jian Xue Bao/J. Software **26**(3), 600–616 (2015). (in Chinese). http://www.jos.org.cn/1000-9825/4741.htm
11. Jiang, W., Xu, Y., Zhang, L.: Research on knowledge reuse dynamic evolvement model based on multi-agent system component. Syst. Eng. Theory Pract. **33**(10), 2663–2673 (2013)
12. Jiang, W., Xu, Y., Wang, X.: Active learning of pair-wise constraints in semi-supervised clustering. J. Syst. Sci. Math. Sci. **33**(6), 708–723 (2013)
13. Jiang, W.: Research on transaction security mechanism of mobile commerce in mobile internet based on MAS. Int. J. Secur. Appl. **9**(12), 289–302 (2015)
14. Liu, X., Li, J.-B., Yang, Z.: Atask collaborative execution policy in mobile cloud computing. Chin. J. Comput. **40**(2), 364–377 (2017)
15. Hu, H., Liu, R., Hu, H.: Multi-objective optimization for task scheduling in mobile cloud computing. J. Comput. Res. Dev. **54**(9), 1909–1919 (2017)
16. Jiang, W.: Multi agent system-based dynamic trust calculation model and credit management mechanism of online trading. Intell. Autom. Soft Comput. **22**(4), 639–649 (2016)

Single Image Super-Resolution by Parallel CNN with Skip Connections and ResNet

Qimei Wang and Feng Qi[(⌧)]

Business School, Shandong Normal University, Jinan, China
474458464@qq.com, cliff@sdnu.edu.cn

Abstract. Image super-resolution has always been a research hotspot in the field of computer vision. Recently, image super-resolution algorithms using convolutional neural networks (CNN) have achieved good performance. But the existing methods based on CNN usually has to many (20–30) convolution layers, which has a large amount of calculations. In response to this problem, this paper proposes a lightweight network model based on parallel convolution, skip connections and ResNet. Parallel convolution means that different sizes of convolution kernels are set in the same convolutional layer to extract image features of different scales. In addition, in order to reduce the loss of image details, we combine the input and output of the previous layer as the input of the next layer, which is the skip connection. We also borrowed the idea of Residual Net. The network learns the residuals between high-resolution images and low-resolution images. Therefore, the algorithm proposed in this paper not only achieves the most advanced performance, but also achieves faster calculations.

Keywords: Image super-resolution · Parallel convolution · Skip connection · Residual net

1 Introduction

Due to the limitation of the equipment in real life, sometimes the image we obtain is low resolution and cannot meet our needs. In order to solve this problem, we can use Image Super-Resolution technology.

Image Super-Resolution refers to the recovery of high resolution images from a lowresolution image or image sequence. Image Super-Resolution is divided into Single Image Super-Resolution (SISR) and Multiple Image Super-Resolution (MISR). In this paper, we mainly focus on Single Image Super-Resolution (SISR). At present, SISR mainly includes three methods: interpolation-based method [1–4], reconstruction-based method [5–8], and learning-based [9–13] method. In recent years, with the wide application of artificial neural networks and the development of deep learning, the idea of deep learning has been introduced into the field of image super-resolution and many classical methods have been proposed. The current classic algorithm contains SRCNN [9], FSRCNN [10], VDSR [11], DRCN [12], SRDenseNet [13], RED [14].

SRCNN is proposed by Dong [9] et al. SRCNN consists of three convolutional layer, which are image feature extraction layer, nonlinear mapping layer and reconstruction layer. Then, Dong [10] and others proposed the FSRCNN algorithm.

Y. Tang et al. (Eds.): HCC 2018, LNCS 11354, pp. 181–187, 2019.
https://doi.org/10.1007/978-3-030-15127-0_18

FSRCNN is an improved algorithm for SRCNN. It eliminates the steps of bicubic interpolation and adds a deconvolution layer to the final layer to amplify the image. Kim [11] and others proposed the VDSR algorithm, drawing on the idea of the ResNet to speed up the network convergence. In addition, Kim [12] and others also proposed the DRCN algorithm. The DRCN algorithm applies the existing recurrent neural network to the image super-resolution field for the first time. Tong [13] et al. proposed the SRDenseNet algorithm, and SRDenseNet links SISR and dense connections for the first time.

In this paper, we propose a new Image Super-Resolution model called CPCSCR in which ideas of skip connection [13], parallel convolution [15] and Residual Net [16] were employed. This network mainly includes four advantages. First, this algorithm directly processes the original image, eliminating the pre-processing steps and retaining the details of the image. Second, parallel convolution is used to extract image features of different scales. Third, skip connection enables the network to more fully learn the image features. Fourth, the network learns the residuals between high-resolution images and low-resolution images, achieves more efficient calculations. Therefore, this model is lighter, less computational, and better performing than the models we mentioned above.

2   Method

As shown in Fig. 1, CPCSCR is a fully convolutional neural network with feature extraction network and reconstruction network. We use the original image as input, extract features of different scales of the image through the parallel convolution module, and then import the extracted features into the reconstruction layer to reconstruct the image details. In addition, our model learns the residuals between low-resolution images and high-resolution images.

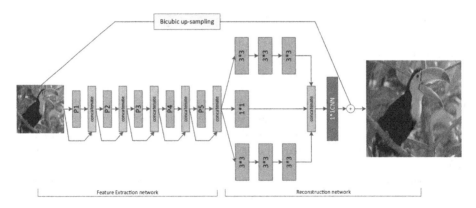

Fig. 1. CPCSCR network overall architecture

2.1   Feature Extraction Network

As shown as Fig. 2, feature extraction layers consists of five parallel convolutional modules, each modules contains 1*1 [17] and 3*3 kernels, bias and Parametric ReLU. Five modules have the same structure. The input and output of each module are spliced and used as input for the next module.

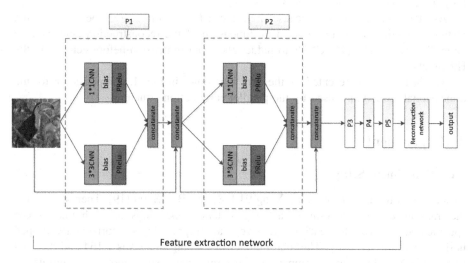

Fig. 2. Construction of parallel convolution modules

In the case of feature extraction, the parallel convolutional proposed by Google [15] is typically used. The parallel convolutional means process an image with multiple different convolution kernels simultaneously and splice different feature maps together. In CPCSCR, we use 1*1 and 3*3 convolutional kernels. The role of the 1*1 convolution kernel is to control the number of feature maps so that it can be easily concatenated with the feature map generated by the 3*3 convolution kernel. Because the original image is the input of the network, the size of the original image is smaller than the image after preprocessing, so a large-sized convolution kernel is not required [10]. A 3*3 convolution kernel is sufficient to cover the entire image information.

In addition, we have found that the network has a good performance when the network contains 5 modules. Since then, when the number of modules has increased again, the performance of the network has not been greatly improved.

2.2   Reconstruction Network

As shown as Fig. 1, the reconstruction network consists of three parallel branches, the first branch is three serial 3*3 convolution kernels, the second branch is a 1*1 [16] convolution kernel, and the third branch is the same as the first branch. As the name suggests, the reconstruction layer up-sampling the image to complement the image details.

In previous models, deconvolution (also known as transposed convolution) was mostly used in the reconstruction layer to up-sampled the image. The process of transposed convolutional layer is similar to the usual convolutional layer and the reconstruction ability is limited. This means that the deeper the deconvolution layer, the better the reconstruction performance, but it also means that the computational burden is increased. So we propose a parallelized CNN structure, which usually consists of 1*1 [16] and 3*3 convolutional kernels.

As stated in the model, because there are a lot of connection operations in the feature extraction layer, the input data dimension of the reconstruction layer is very large. So we use 1*1 [16] CNNs to reduce the input dimension before generating the HR pixels.

The last CNN, represented by the dark blue color in Fig. 1, compensating for the dimensional reduction caused by the parallel convolution structure.

3   Experiment

3.1   Experiment Setup

Experimental training datasets are Yang 91 [18] and BSDS200 [19]. Then we expand the training-sets by performing data augmentation operations on each image. The specific operations are that each image is vertically flipped, flipped horizontally, flipped horizontally and vertically. The total number of training images is 1,164 and the total size is 259 MB. In order to compare with existing image super-resolution algorithms, this paper converts color (RGB) images to YCbCr images and only processes Y-channels (Y-channel represents brightness). Each training image is divided into 32 steps, 32 steps, 16 steps, and 64 patches are used as the minimum batch. We use BSDS100 [19], SET 5 [20] and SET 14 [21] as test datasets.

The experiment uses an internationally accepted evaluation criterion to measure the network performance: Peak Signal-to-Noise Ratio (often abbreviated as PSNR). The unit of PSNR is dB. The larger the value is, the smaller the picture distortion is, and the better the network performance is.

We Initialize each convolution kernel using the method of He et al. All biases and PReLUs in the network are initialized to zero. The dropout rate of p = 0.8 during training. The Mean Squared Error function is used as a loss function to calculate the difference between the network output value and the true value. In addition, we used Adam [22] with an initial learning rate = 0.002 to optimize the algorithm to minimize the loss. If the loss value does not decrease after 5 training steps, the learning rate drops by 2 times. Training ends when the learning rate is less than 0.00002. An example of the results is shown in Fig. 3.

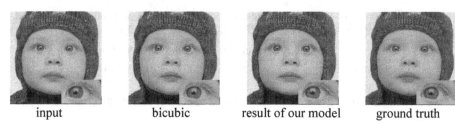

| input | bicubic | result of our model | ground truth |

Fig. 3. An example of our results of img_001 in Set5

3.2   Comparisons with State-of-the-Art Methods

Comparisons of PSNR: We use PSNR to objectively evaluate the processing result of the algorithm. Table 1 show the results of objective tests at scale = x2, scale = x3 and scale = x4 respectively. Except that the objective index of RED30 algorithm for BSD100 data set is slightly higher than our algorithm, the other test results show that the PSNR value obtained by our algorithm is higher than other algorithms, which fully demonstrates that the algorithm has better processing effect. Although the performance of our algorithm on the BSD100 dataset is slightly lower than RED30, it greatly reduces the computational complexity.

Table 1. Comparisons of PSNR with other SR Method. (scale = x2)

Dataset	SRCNN	DRCN	VDSR	RED30	CPCSC(ours)
Set5(x2)	36.66	37.63	37.53	37.66	37.75
(x3)	32.75	33.82	33.66	33.82	34.04
(x4)	30.49	31.53	31.35	31.51	31.70
Set14(x2)	32.45	33.04	33.03	32.94	33.18
(x3)	29.28	29.76	29.77	29.61	29.90
(x4)	27.49	28.02	28.01	27.86	28.24
BSD100(x2)	31.36	31.85	31.90	31.99	32.00
(x3)	28.41	28.80	28.82	28.93	28.86
(x4)	26.90	27.23	27.29	27.40	27.35

Comparison of computational complexity: Since each implementation is performed on a different hardware device or platform, it is unfair to compare test execution times. Here we calculate the computational complexity of each method. The approximate computational complexity of each method is shown in Table 2. Therefore, we can see that our DCSCN has the most advanced reconstruction performance, and the computational complexity is much smaller than VDSR [11], DRCN [12] and RED30 [14].

Table 2. Comparisons of approximate computation complexity with other SR Method. (scale = x2) For comparison, we chose f1, f2, f3, n1, n2 = (9,5,5,64,32) for SRCNN

	SRCNN (9,5,5)	DRCN	VDSR	RED30	CPCSC
CNN layers	3	20	20	30	13
CNN filters	32,64	256	64	64	32 to 96
Complexity[k]	229.5	78,083.2	2,668.5	4,152.8	80.9

4 Conclusion

This paper presents an image super-resolution method based on convolutional neural networks with parallel convolution, skip connection and ResNet. The algorithm uses parallel convolution to extract features of different scales of the image, and inputs the local and global features of each layer to the next layer by means of skip connections. In addition, the algorithm learns the residual between the low resolution image and the high resolution image. Another important point is, the model takes an image of the original size as an input, reducing image information loss. Using these methods, our model can achieve the most advanced performance with less computing resources. The experimental results also show that the model has better performance.

Acknowledgements. This work was supported by the Natural Science Foundation of China (No. 61502283). Natural Science Foundation of China (No. 61472231). Natural Science Foundation of China (No. 61640201).

References

1. Tsai, R.Y.: Multiframe image restoration and registration. Adv. Comput. Vis. Image Process. **1**, 317–339 (1984)
2. Ur, H., Gross, D.: Improved resolution from subpixel shifted pictures. CVGIP: Graph. Models Image Process. **54**(2), 181–186 (1992)
3. Komatsu, T., et al.: Signal-processing based method for acquiring very high resolution images with multiple cameras and its theoretical analysis. IEE Proc. I (Commun. Speech Vis.) **140**(1), 19–25 (1993)
4. Komatsu, T., et al.: Very high resolution imaging scheme with multiple different-aperture cameras. Sign. Proces. Image Commun. **5**(5-6), 511–526 (1993)
5. Tappen, M.F., Russell, B.C., Freeman, W.T.: Exploiting the sparse derivative prior for super-resolution and image demosaicing. In: IEEE Workshop on Statistical and Computational Theories of Vision (2003)
6. Kim, K.I., Kwon, Y.: Single-image super-resolution using sparse regression and natural image prior. IEEE Trans. Pattern Anal. Mach. Intell. **6**, 1127–1133 (2010)
7. Dai, S., et al.: Soft edge smoothness prior for alpha channel super resolution. In: IEEE Conference on Computer Vision and Pattern Recognition. CVPR 2007. IEEE (2007)
8. Yang, X., et al.: An improved iterative back projection algorithm based on ringing artifacts suppression. Neurocomputing **162**, 171–179 (2015)
9. Dong, C., et al.: Image super-resolution using deep convolutional networks. IEEE Trans. Pattern Anal. Mach. Intell. **38**(2), 295–307 (2016)

10. Dong, C., Loy, C.C., Tang, X.: Accelerating the super-resolution convolutional neural network. In: Leibe, B., Matas, J., Sebe, N., Welling, M. (eds.) ECCV 2016. LNCS, vol. 9906, pp. 391–407. Springer, Cham (2016). https://doi.org/10.1007/978-3-319-46475-6_25
11. Kim, J., Lee, J.K., Lee, K.M.: Accurate image super-resolution using very deep convolutional networks. In: Proceedings of the IEEE Conference on Computer Vision and Pattern Recognition (2016)
12. Kim, J., Lee, J.K., Lee, K.M.: Deeply-recursive convolutional network for image super-resolution. In: Proceedings of the IEEE Conference on Computer Vision and Pattern Recognition (2016)
13. Tong, T., et al.: Image super-resolution using dense skip connections. In: 2017 IEEE International Conference on Computer Vision (ICCV). IEEE (2017)
14. Mao, X.-J., Shen, C., Yang, Y.-B.: Image restoration using convolutional auto-encoders with symmetric skip connections. arXiv preprint arXiv:1606.08921 (2016)
15. Szegedy, C., et al.: Going deeper with convolutions. In: Proceedings of the IEEE Conference on Computer Vision and Pattern Recognition (2015)
16. He, K., et al.: Deep residual learning for image recognition. In: Proceedings of the IEEE Conference on Computer Vision and Pattern Recognition (2016)
17. Lin, M., Chen, Q., Yan, S.: Network in network. arXiv preprint arXiv:1312.4400 (2013)
18. Yang, J., et al.: Image super-resolution via sparse representation. IEEE Trans. Image Process. $19(11)$, 2861–2873 (2010)
19. Arbelaez, P., et al.: Contour detection and hierarchical image segmentation. IEEE Trans. Pattern Anal. Mach. Intell. $33(5)$, 898–916 (2011)
20. Bevilacqua, M., et al.: Low-complexity single-image super-resolution based on nonnegative neighbor embedding, p. 135-1 (2012)
21. Zeyde, R., Elad, M., Protter, M.: On single image scale-up using sparse-representations. In: Boissonnat, J.-D., et al. (eds.) Curves and Surfaces 2010. LNCS, vol. 6920, pp. 711–730. Springer, Heidelberg (2012). https://doi.org/10.1007/978-3-642-27413-8_47
22. Kinga, D., Ba Adam, J.: A method for stochastic optimization. In: International Conference on Learning Representations (ICLR), vol. 5 (2015)

Incorporating Description Embeddings into Medical Knowledge Graphs Representation Learning

Xi Sun[1(✉)], Yi Man[1], Yanling Zhao[2], Jin He[3], and Ningning Liu[3]

[1] School of Electronic Engineering,
Beijing University of Posts and Telecommunications, Beijing 100876, China
`sunxi9495@126.com`
[2] Instrumentation Technology and Economy, Beijing 100055, China
[3] Neusoft Corporation, Beijing 100193, China

Abstract. Representation learning of medical knowledge graphs aims to embedding entities and relations in low-dimensional vector spaces, which is beneficial to the application of medical knowledge graphs in intelligent medical systems such as intelligent guidance, disease risk prediction and question answering system of medical field. Recently, some translation-based methods including TransE, TransH and TransR built entity and relation embeddings by regarding a relation as translation from head entity to tail entity. These methods solely use the information of triplets and don't take text information into consideration. In this paper, we process a novel representation learning method by incorporating the embeddings of entity descriptions with classical translation-based methods. The embeddings of entity descriptions are built by Doc2Vec. It is easily applied for a large-scale domain-specific knowledge graphs because of its simplicity. Besides, we compare our method with classical translation-based methods to demonstrate the effectiveness of our method in medical knowledge graphs representation learning.

Keywords: Representation learning of knowledge graphs ·
Medical knowledge graphs · Translation-based methods

1 Introduction

With the development of artificial intelligence and the increasing concern about medical fields, some intelligent medical applications, such as intelligent guidance, disease risk prediction and question answering system of medical fields came out one and another. As the base of these intelligent medical applications, medical knowledge graphs are very important.

Traditionally, we use a directed graph composed of vertices and edges to represent a knowledge graph. Vertices represent entities and edges represents relations. For example, in medical knowledge graphs, entities can be drugs, disease or symptoms, relations can be curing or causing. The basic unit of knowledge graph are the triplet facts (head entity, relation, tail entity) (denoted (h, r, t)). This representation method is clear and intuitive. But these discrete symbols can't be used for calculation directly.

Y. Tang et al. (Eds.): HCC 2018, LNCS 11354, pp. 188–194, 2019.
https://doi.org/10.1007/978-3-030-15127-0_19

Recently, inspired by the idea of word embedding, researchers have proposed some knowledge graph representation methods projecting entities and relations in low-dimensional vector spaces [1–4]. Generally, knowledge graph embedding represents an entity as a k-dimensional vector h (or t) and defines a scoring function $f_r(h, t)$ to measure the plausibility of the triplet (h, r, t) [2].

1.1 Translation-Based Representation Learning

Translation-based models regard relation between head entity and tail entity in a triplet fact as translating operation in vector spaces.

TransE [1] is the first translation-based model. The main idea of TransE is that if a triplet (h, r, t) is existed, the vector embedding of head entity h plus the vector embedding of relation r is close to the vector embedding of tail entity t, otherwise they are far away. TransE use L_1-norm or L_2-norm to measure the distance between h + r and t. During the training process, TransE randomly replace head entities or tail entities with other entities to generate negative samples, and then a margin-based loss function is minimized by SGD method. TransE is effective and require fewer parameters, so it can be easily trained on large scale data.

However, TransE can't deal well with 1-to-N, N-to-1 and N-to-N relations. To address the issue, TransH [2] is proposed. Compare with TransE, it first projects the vector embedding h and t to a relation-specific hyperplane to enable an entity to have different vector embeddings when involved in different relations. And TransR [3] build entity and relation embeddings in separate entity space and relation spaces, it projects entities in the entity space into relation-specific space through a matrix.

1.2 Text-Enhanced Representation Learning

Translation-based models mentioned above solely use the information of triplets in a knowledge graph. In fact, some context information, can provide auxiliary semantic information for the representation learning of knowledge graph. The incorporation of text information can reduce the sparsity of knowledge graphs and handle 1-to-N, N-to-1 and N-to-N relations better [4].

Wang and other [5] propose a framework of jointly embedding entities and words into the same vector space. The framework consists of knowledge model, text model and alignment model. Alignment model aligns entities and words through Wikipedia anchors or names. However, Wikipedia anchors is only available on a few data source and alignment by names can be inaccurate because of the ambiguous of entity names. TEKE [4] uses an entity linking tool to label the entities and relations from text corpus to get their neighboring words, then defines entities and relations as linear transformation of the weighted average of their common neighboring words' embedding built by Word2Vec [6] model. However, entity linking tool still has data source constraints. Besides, Word2Vec has the shortage of ignoring the information of word orders. DKRL [7] adopts deep CNN to projects entity embeddings from their descriptions, then uses TransE-based method to build entity and relation embeddings. DKRL achieves high performance but require more parameters of inner layers to be tuned.

In this paper, we propose a scheme incorporating description embeddings into knowledge graphs representation learning. We use the generalized linear transformation of entities' description embeddings built by Doc2Vec as the entity representations. This single-layer model requires fewer parameters than DKRL and does not depend on entity linking tool, so it is easily applied for a large-scale domain-specific knowledge graph. Besides, we evaluate its' effectiveness in the tasks of linking prediction and triple classification on a medical knowledge graph.

2 Model

2.1 Entity Descriptions Embedding

Doc2Vec [8] is an unsupervised model widely used for representation learning of sentences and documents. Doc2Vec includes two models: DM and DBOW. DM uses document vector and context words to predict the next words. The document vector represents can act as a memory of the topic of the document. DBOW is trained by using paragraph vector to predict the words in a small window. DM takes the consideration of ordering of words and DBOW concern more about context information.

The descriptions of entities give us much information about the semantic meaning of entities. We can compute the similarity of descriptions by the distance between the description embeddings trained by Doc2Vec. So, description embeddings can help the knowledge graph representation learning. In our experiments, description embedding d_h, d_t is the concatenate of vector learned by DM and vector learned by DBOW.

2.2 Algorithm

We use the generalized linear transformation of entities' description embeddings as the entity representations $\hat{h}$ and $\hat{t}$.

$$\hat{h} = \sigma(d_h A_r + h) \tag{1}$$

$$\hat{t} = \sigma(d_t A_r + t) \tag{2}$$

Where σ is the sigmoid function, A_r is a relation-specific matrix and can be viewed as the weight of the description embeddings. h, t can be viewed as the biased vectors. Then, we use the method of translation-based models to define our score function. The score function of incorporating description embedding into TransE (DETE) is

$$f(h, r, t) = ||\hat{h} + r - \hat{t}||_{L_1/L_2} \tag{3}$$

Where L_1/L_2 is the L_1-norm or L_2-norm. The framework of TransE is illustrated by Fig. 1.

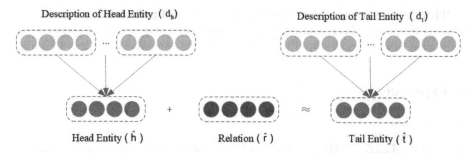

Fig. 1. The architecture of DETE. DETE incorporates the embeddings of entity description built by Doc2Vec with TransE.

This incorporating scheme can also extend to other translation-based models, such as TransH and TransR. The score function of incorporating description embedding into TransH (DETH) is

$$f(h, r, t) = ||\hat{h}_\perp + r - \hat{t}_\perp||_{L_1/L_2} \tag{4}$$

Where

$$\hat{h}_\perp = \hat{h} - w_r^T \hat{h} w_r \tag{5}$$

$$\hat{t}_\perp = \hat{t} - w_r^T \hat{t} w_r \tag{6}$$

Where w_r is a relation-specific hyperplane (the normal vector). The score function of incorporating description embedding into TransR (DETR) is

$$f(h, r, t) = ||\hat{h}_r + r - \hat{t}_r||_{L_1/L_2} \tag{7}$$

Where

$$\hat{h}_r = \hat{h} M_r \tag{8}$$

$$\hat{t}_r = \hat{t} M_r \tag{9}$$

Where M_r is a matrix projecting entities in the entity space into a relation-specific space.

To learn entity and relation embeddings, we minimize the margin-based loss function over the training set

$$L = \sum_{(h,r,t) \in S} \sum_{(h',r,t') \in S'} \max(0, \gamma + f(h, r, t) - f(h', r, t')) \tag{10}$$

Where S is the correct triplets and S' is the corrupted triplets. S' is generated by training triplets with either head entity or tail entity replaced by a random entity. $\gamma > 0$ is a margin hyperparameter. And we use SGD method to minimize the loss function.

3 Experiments

3.1 Datasets

We evaluate TETE, TETH and TETR with a medical knowledge graph. There are 302969 triplets with 77552 entities and 11 relationships. And there are 6 types of entities including disease, symptom, drug, section, exam and body part. 1-to-N, N-to-1 and N-to-N relations exist in most triplets. We randomly select 10% of triplets as validation set and 10% of triplets as testing set.

3.2 Evaluation Protocols

Researchers always evaluate the quality of the embeddings built by knowledge representation learning on the task of link prediction and triple classification.

Link Prediction. This task is to complete the missing head entity h or tail entity t in triplet (h, r, t). i.e., predict tail entity t given head entity h and relation r or predict head entity h given relation r and tail entity t. We use the similar evaluation protocols used in TransE, TransH and TransR [1–3]. For each triplet we replace the tail entity t by every entity e with the same type of t in the knowledge graph and calculate a score by function $f(h, r, t)$. Then, we rank them in descending order of the scores. And we use two evaluation metrics: (1) the averaged rank of correct entities (denoted as *Mean Rank*); (2) the proportion of ranks not larger than 10 (denoted as *Hits @ 10*). This is called the "raw" setting. Notice that if a corrupted triplet exist in the knowledge graphs, we should define it as correct. To eliminate it, we filter out those corrupted triplets which exist in training, validation or testing sets before generating the ranking list. This is called "filter" setting. In both settings, a lower *Mean Rank* and a *higher Hits @ 10* is better.

Triple Classification. This task is to judge whether a triplet (h, r, t) is correct or not. It is a binary classification task. The judge rule for classification is: if the score compute by function $f_r(h, r, t)$ is below a relation-specific threshold δ_r, it will be predicted as positive, otherwise as negative. The threshold δ_r determined by maximizing classification accuracy on validation set. In order to create a testing set for classification, we randomly replace head of tail entities with the same type of entities to generate the corrupt triplets. And the number of negative examples is equal to the number of positive examples.

3.3 Implementation

We compare TETE, TETH and TETR with baseline translation-based models: TransE, TransH and TransR. Several hyperparameters need to be tuned. We select models with learning rate λ for SGD among {0.001, 0.01, 0.1}, the margin γ among {1, 2, 10}, the embedding dimension among {20, 50, 100}, the dissimilarity measure method among {L_1, L_2}, the window of Doc2Vec among {3, 4, 5}. The best configuration is determined by the performance on validation set. We traverse all the training triplets for 1000 rounds.

3.4 Result

Table 1 shows the results of link prediction by the metric of Mean Rank and Hits @ 10. Table 2 shows the results of triple classification by the metric of accuracy. From the results we observe that: (1) On our evaluation datasets, DETE outperforms TransE, DETH outperforms TransH and DETR outperforms TransR, which proves incorporating entity descriptions built by Doc2Vec can promote knowledge representation learning. (2) TransH and TransR perform better than TransE, DETH and DETR perform better than DETE. The reason is TransH and TransR can handle 1-to-N, N-to-1 and N-to-N relations better by their relation-specific projection. (3) On WN18 and FB15 K [1–3] dataset TransR performs better than TransH, which is opposite from the result on our dataset. The reason could be there are more parameters to be tuned in TransR and DETR, it's more difficult from them to convergence.

Table 1. Experiment results of link prediction

Metric	Mean Rank		Hits @ 10	
	Raw	Filter	Raw	Filter
TransE/**DETE**	52/**46**	47/**42**	82.3/**84.6**	84.0/**85.7**
TransH/**DETH**	39/**35**	36/**33**	86.2/**87.6**	87.9/**88.6**
TransR/**DETR**	42/**38**	40/**36**	85.0/**85.3**	86.2/**86.3**

Table 2. Experiment results of triple classification

Metric	Accuracy
TransE/**DETE**	74.8/**78.3**
TransH/**DETH**	85.6/**87.2**
TransR/**DETR**	83.1/**83.2**

4 Conclusion

This paper contributes an effective and efficient representation learning method for incorporating the embeddings of entity descriptions built by Doc2Vec with classical translation-based methods. Experimental results show the entity and relation embeddings learnt by our method perform better than classical translation-based methods.

In the future, several directions are expected to explore, such as how to incorporate the text information of relations, how to incorporate other text corpus except for description.

Acknowledgement. This work is supported by the National Key R&D Program of China (2018 YFB1201500), the National Natural Science Foundation of China under (Grant No. 61471055), Beijing Major Science and Technology Special Projects under Grant No. Z181100003118012 and China Railway with project No. BX37.

References

1. Bordes, A., et al.: Translating embeddings for modeling multi-relational data. In: International Conference on Neural Information Processing Systems, pp. 2787–2795. Curran Associates Inc. (2013)
2. Wang, Z., et al.: Knowledge graph embedding by translating on hyperplanes. In: Twenty-Eighth AAAI Conference on Artificial Intelligence, pp. 1112–1119. AAAI Press (2014)
3. Lin, Y., et al.: Learning entity and relation embeddings for knowledge graph completion. In: Twenty-Ninth AAAI Conference on Artificial Intelligence, pp. 2181–2187. AAAI Press (2015)
4. Wang, Z., Li, J.: Text-enhanced representation learning for knowledge graph. In: International Joint Conference on Artificial Intelligence, pp. 1293–1299. AAAI Press (2016)
5. Wang, Z., et al.: Knowledge graph and text jointly embedding. In: Conference on Empirical Methods in Natural Language Processing, pp. 1591–1601 (2014)
6. Mikolov, T., et al.: Efficient estimation of word representations in vector space. Computer Science (2013)
7. Xie, R., Liu, Z., Jia, J., Luan, H., Sun, M.: Representation learning of knowledge graphs with entity descriptions. In: Proceedings of the 30th
8. Le, Q.V., Mikolov, T.: Distributed representations of sentences and documents, vol. 4, p. II-1188 (2014)

A Piezoelectric MEMS Harvester Suitable Adopt a New Two-Degree-of-Freedom Structure

Gaoyang Xie, Gang Tang[✉], Zhibiao Li, Xiaoxiao Yan, Bin Xu, and Xiaozhen Deng

Jiangxi Province Key Laboratory of Precision Drive and Control, Nanchang Institute of Technology, Nanchang 330099, China
tanggangnit@163.com

Abstract. In this work, we presented a piezoelectric energy harvester adopt a new two-degree-of freedom that can work at both 1st and 2nd resonant frequency. First, the inherent frequency the displacement situation were described through the Finite element simulation software. Subsequently, this device was achieved by bonding the thinned PZT onto a flexible phosphor bronze substrate, and lancing a U-shaped in order to constitute the two degree freedom structure. Two additional tungsten mass were respectively assembled at the tops of the main beam and the second beam to reduce the 1st and 2nd resonant frequency of the harvester. The experimental results show the maximum open-circuit output voltage and output power of the main beam can reach up 18.8 V when it worked at 1st resonant frequency. For the 2nd resonant frequency, the maximum open-circuit output voltage of second beam can reach up 5.44 V.

Keywords: Piezoelectric energy harvester · Two degrees of freedom · MEMS

1 Introduction

Energy harvesting plays an important role in the field of low-powered energy supply, such as wireless sensors, micro-actuators, especially in the medical, communications and military fields [1]. The use of energy harvester can reduce the use of batteries, so it is beneficial for the environment. The four main types of energy collection are triboelectric, electromagnetic, electrostatic, and piezoelectric [2]. Compared with other forms, piezoelectric devices have the advantages of small size and high energy density [3]. Meanwhile, the piezoelectric device is able to combine well with MEMS technologies [4], the output performance of piezoelectric devices will be further improved. Compared to a conventional battery, the advantages of piezoelectric MEMS harvester include miniature size, simple structure, great output performance and long lifetime [5]. Hence, many low-power devices can use it as an energy source.

The popular MEMS technologies include sputtering, epitaxial growth, sol-gel spin on, hydrothermal method, and screen printing. The screen printing of MEMS technology is different from the conventional method, because the MEMS screen printing can work at low-temperature environment. In recent years, there are more and more

Y. Tang et al. (Eds.): HCC 2018, LNCS 11354, pp. 195–200, 2019.
https://doi.org/10.1007/978-3-030-15127-0_20

researches use those MEMS technologies on the piezoelectric energy harvester, and have good output performance. Nowadays, several piezoelectric energy harvesters prepared by MEMS technology have been reported. Yi et al. reported a bimorph piezoelectric MEMS harvester via bulk PZT thick films on thin beryllium-bronze substrate, which can work under 77 Hz–88 Hz and have the output voltage of about 9.6 V-105 V, when it work at different accelerations from 0.1 g to 3.5 g [6]. Tian et al. designed a new shaped cantilever with a rectangular hole and based on bulk PZT film, which has a low natural frequency about 34.3 Hz, and the output voltage can arrive at 15.7 V [7]. But those devices only can work at the 1st resonant frequency. If one device can work at more resonant frequency, it will have better output.

In this paper, we design a two-degree-of-freedom structure cantilever for the piezoelectric MEMS energy harvesters, so that this piezoelectric MEMS energy harvesters can work under the 1st and the 2nd resonant frequency.

2   Design of the Cantilever Beam

Cantilever beam structure is the most common structure of piezoelectric energy collector, the advantages of cantilever beam structure include simple structure and easy processing, in addition, the stress produced by the cantilever beam structure under the vibration environment matches the piezoelectric material. Minimize the thickness of the cantilever beam to reduce its resonant frequency. On the other hand, in order to improve the output power, piezoelectric materials should obtain greater stress in the process of vibration. The two-degree-of-freedom structure made by cutting technology is used here. Figure 1 shows the surrogate model of two-degree-of-freedom cantilever beam. Which include piezoelectric function PZT layer, phosphor bronze substrate supporting layer, and proof masses of integrated silicon and assembled tungsten. The main beam of this two-degree-of-freedom structure is composed of the external main body of the cantilever beam, the M_1 is the proof mass of the main beam, which arranged at the free end of the main beam. Making a u-cut at the center of the main beam, and the free end is opposite to the free end of main beam. This forms the structure of the second beam. The M_2 is the proof mass of the second beam, which arranged at the free end of the second beam. The surrogate model was set up as shown in Fig. 1. The structure was simulated by ANSYS software, the result of the finite element simulation is shown in Fig. 2. Firstly, the 1st and 2nd resonant frequency of this structure are 33.463 Hz and 98.294 Hz. Secondly, for the 1st resonant frequency, the maximum displacement of the cantilever beam is at the free end of the main beam (Fig. 2a). For the 2nd resonant frequency, the maximum displacement of the cantilever beam is at the free end of the second beam (Fig. 2b). Therefore, the output performance of the main beam is significant at the 1st resonant frequency and the output performance of the second beam is significant at the 2nd resonant frequency.

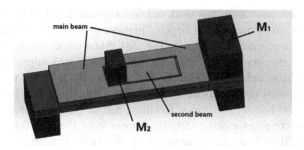

Fig. 1. The surrogate model of the device

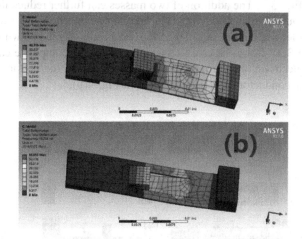

Fig. 2. (a) The first-order mode of the device, which shows the 1st resonant frequency is 33.463 Hz and the displacement situation. (b) The second-order mode of the device, which shows the 2nd resonant frequency is 98.294 Hz and the displacement situation.

3   Fabrication of the Energy Harvester

Detailed manufacturing processes are shown in the Fig. 3. The first process was on a 400-μm-thick silicon substrate, and a 2-μm-thick plasma-enhanced chemical vapor deposition (PECVD) SiO2 film was deposited on one side to provide electrical isolation of silicon (Fig. 3a). In order to improve the flexibility and fatigue resistance of the support layer, a 100 μ m-thick phosphor bronze was bonded with prepared silicon substrate (Fig. 3b). The thickness of the phosphor bronze was processed to be 90 μm. A 200-nm-thick Cr/Au was sputtered onto the polished side of the PZT, which is the bottom electrode of the PZT. Then, the conductive epoxy (provided by Shanghai Research Institute of Synthetic Resins) was coated by screen printing, which is as the intermediate layer for bonding bulk PZT with phosphor bronze substrate. Control the thickness of conductive epoxy less than 3 nm, and solidify it in a vacuum of 175 °C for 3 h. The piezoelectric material would be completely depolarized, if it is heated to its Curie temperature. The Curie temperature of the PZT used in this work is 295 °C

(Fig. 3c). The whole process of this work at low temperature. The bulk PZT is thinned down to about 40 μm by mechanical lapping, in order to reduce the resonant frequency of cantilever structure energy harvester (Fig. 3d). Meanwhile, the bulk PZT is also polished to avoid the small holes and scratches on PZT, and improve the adhesion between the PZT film and the top electrode. A 200-nm-thick Cr/Au was sputtered as the top electrode onto upper surface of PZT (Fig. 3e). Using ion beam milling to separate the top and bottom electrodes. Retained part of the silicon substrate as proof mass by deep reactive ion etching (DRIE) process from the backside (Fig. 3f). The two-degree-of-freedom structure is patterned by ultraviolet irradiation from the back-side, the specific process is to cut a U shape (Fig. 3h). Two additional tungsten masses were assembled at the free end of the main part of the cantilever beam and the free end of the U shape (Fig. 3g). The addition of two masses can further reduce the 1st and the 2nd resonant frequency of this device. Last, making a scratch on the root the second beam by ultraviolet irradiation, in order to let two beams has their own top electrode (Fig. 3h). Figure 3(a) shows the cross-sectional view of the composite cantilever. The main beam top electrode, the second beam top electrode and the bottom electrode are educed in copper conductor in PCB, as shown in Fig. 4(b).

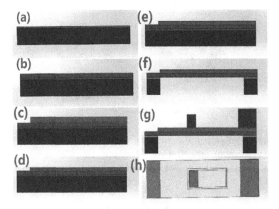

Fig. 3. Fabrication process of energy harvester.

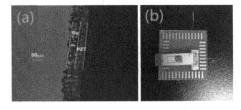

Fig. 4. (a) Cross-sectional view of the composite cantilever. (b) Micro fabricated piezoelectric device assembled with PCB

4   Testing and the Performance of Energy Harvester

The performance of the piezoelectric MEMS harvester was tested using the experimental setup given in Fig. 5, which includes a waveform generator, an amplifier, a vibrator, an accelerator monitor, and an oscilloscope. The waveform generator generated the vibrating signal and amplified by the amplifier to control the vibration of the vibrator, our device vibrated with it. While the vibration acceleration was monitored by the accelerator monitor. The output of the piezoelectric MEMS harvester under different frequency was monitored with the oscilloscope.

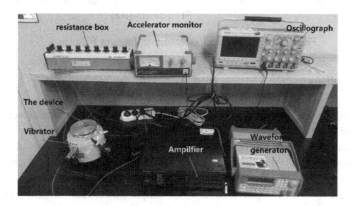

Fig. 5. Test setup of energy harvester.

According to the test, the open-circuit voltage of the main beam and the second beam are shown in Fig. 6. The device was tested at accelerations from 0.5 g to 2 g. Both beam's maximum open circuit voltages increased due to the increase of input vibration acceleration. The main beam was tested at 1st resonant frequency, the maximum open-circuit voltages were 6.08 V, 10.4 V and 18.88 V at 28.5 Hz, 28.1 Hz, 27.7 Hz. The second beam was tested at 2nd resonant frequency, the maximum open-circuit voltages were 1.64 V, 4.04 V and 5.44 V at 101.2 Hz, 99.7 Hz and 99.1 Hz. The result shows that the output voltages of both beams are increase with the increasing of acceleration. Meanwhile, the 1st resonant frequency and the 2nd resonant frequency of the energy harvester are both decreased due to the increase of the acceleration of the vibration source. Under the large stress, the damping ratio of PZT increases with the increasing of acceleration due to the nonlinearity character of PZT. Most energy harvester involved in studies can only work at 1st resonant frequency. Last, our test result suggest that the two-degree-freedom structure energy harvester can adapt to different frequency environments.

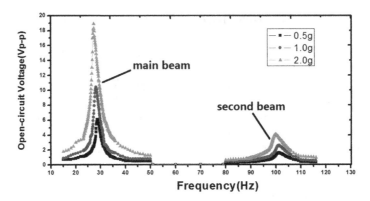

Fig. 6. The open-circuit voltage for different frequency.

5 Conclusions

In conclusion, we designed a new two-degree-of-freedom structure piezoelectric energy harvester and presented the fabrication of it. This device can operable at the 1st and 2nd resonant frequency. When the device work at 1st resonant frequency about 33 Hz, the maximum open-circuit voltage was 18.88 V. When the device work at 2nd resonant frequency about 100 Hz, the maximum open-circuit voltage was 5.44 V.

The research work of improvements in the output power and the power density for this device are under way.

Acknowledgments. This work was supported by the National Natural Science Foundation of China (51565038) and the Science Fund of Jiangxi Province Office of Education of China (GJJ151130, GJJ170986).

References

1. Erturk, A., Inman, D.J.: Piezoelectric Energy Harvesting. Wiley, Hoboken (2011)
2. Becher, P., Hymon, E., Folkmer, B., et al.: High efficiency piezoelectric energy harvester with synchronized switching interface circuit. Sens. Actuators A: Phys. **202**, 155–161 (2013)
3. Roundy, S., Wright, P.K.: A piezoelectric vibration based generator for wireless electronics. Smart Mater. Struct. **13**, 1131–1142 (2004)
4. Emad, A., Mahmoud, M.A.E., Ghoneima, M., Dessouky, M.: Testing and evaluation of stretching strain in clamped–clamped beams for energy harvesting. Smart Mater. Struct. **25**, 115006 (2016)
5. Jung, H.J., Jabbar, H., Song, Y., Sung, T.H.: Hybrid-type (d33 and d31) impact-based piezoelectric hydroelectric energy harvester for watt-level electrical devices. Sens. Actuators, A **245**, 40 (2016)
6. Yi, Z., Yang, B.: High performance bimorph piezoelectric MEMS harvester via bulk PZT thick films on thin beryllium-bronze substrate. Appl. Phys. Lett. **111**, 013902 (2017)
7. Tian, Y., Yang, B.: A low-frequency MEMS piezoelectric energy harvester with a rectangular hole based on bulk PZT film. J. Phys. Chem. Solids **117**, 21–27 (2018)

Spiking Neural P Systems
with Time Delay

Zhenxian Ma and Xiyu Liu[✉]

College of Business School, Shandong Normal University, Jinan, Shandong, China
1311065020@qq.com, sdxyliu@163.com

Abstract. Spiking neural P systems simulate the biological phenomena that the neurons cooperate to deal with spikes via synapses. In order to make the system more controllable, we introduce a new class of SNP systems, namely SNP systems with time delay (in short, TDSNP systems). In this systems, we set an initial time and a delay time for each rule. By this new way, we can use less neurons to construct each module of our system. Seen as number computing devices and number accepting devices respectively, TDSNP systems are shown to be computationally complete, both in the generating mode and the accepting mode.

Keywords: Membrane computing · Spiking neural P systems ·
Time delay · Computational completeness

1 Introduction

Membrane computing have been widely investigated since it was first proposed by pǎun in 1998 [1]. It presented a new abstract computational model inspired from the living cells. Spiking neural P systems (SNP systems, for short) is one of the youngest branches of membrane computing [2]. They are inspired from the biological phenomenon that neurons process information and communicate in the brain by sending spikes via synapses [3]. In such model, an SNP system represented by a direct graph consists of a set of neurons (placed in the nodes) and a set of synapses (represented by the directed edge of the graph) [13]. The content of each neuron is composed of a number of spikes and a series of rules. When the number of spikes present in one neuron meets certain conditions, this neuron uses its rule and sends information to other neuron. With a computation we can associate a result in various way [4,8,9].

Since SNP systems have significant potential in various application, more and more variants of SNP systems were proposed by researchers [5–7]. They were all proved to compute all Turing computable sets of natural numbers. A standard feature of those systems is the fact that when the contents of the neuron meet certain restrictive conditions, the rule fires immediately.

Real-world processes usually have the start time and the end time. In order to make the system more controllable, we introduce a new class of SNP systems,

© Springer Nature Switzerland AG 2019
Y. Tang et al. (Eds.): HCC 2018, LNCS 11354, pp. 201–208, 2019.
https://doi.org/10.1007/978-3-030-15127-0_21

namely SNP systems with time delay (in short, TDSNP systems). In TDSNP systems, we set an initial time t_0 and a delay time τ for each rule. At one step, when the number of spikes meets specified conditions, the corresponding neuron closes immediately, gets fired after t_0 step and sends spikes (or removes spikes) after $t_0 + \tau$ step. More precise definitions will be shown in Sect. 2.

Then, we prove that TDSNP systems are able to compute all Turing computable sets of numbers, both in the generating mode and the accepting mode. The proof is based on simulating register machines with TDSNP systems. In the generating mode, the number of spikes emitted to the environment are defined as the result of computation. In the accepting mode, the number of steps after receiving the two spikes from the environment are accepted.

The rest of this paper is organized as follow. Section 2 introduces the TDSNP systems in detail. In Sect. 3, we prove that TDSNP systems are computationally complete in the generating mode (by simulating register machine). In the following section, we let TDSNP systems work in the accepting mode, Sect. 5 concludes this work.

2   SNP Systems with Time Delay

Before introducing the SNP systems with time delay, we assume the reader to have some elementary knowledge in formal language theory, e.g., from [10,14] (see [11] for basic concepts and [12] for recent information about the membrane computing).

We directly introduce the type of SNP systems we investigate in this paper, such a SNP system with time delay, of degree m $\geq$ 1, is a construct of the form [15]

$$\prod = (O, \sigma_1, \ldots, \sigma_m, \text{syn}, \text{in}, \text{out}),$$

where:

1. $O = \{a\}$ represents the singleton alphabet where a is called *spikes*;
2. $\sigma_1, \ldots, \sigma_m$ are *neurons* of the form

$$\sigma_i = (n_i, R_i), 1 \leq i \leq m,$$

   where:
   (a) $n_i \geq 0$ represents the spike numbers in neuron σ_i in the initial state;
   (b) R_i represents the set of *rules* of the following two forms:
      (1) $E/a^c \longrightarrow a^p$; t_0, τ, where E is a regular expression over a and $c \geq 1, p \geq 1, c \geq p, t_0 \geq 0, \tau \geq 0$;
      (2) $a^s \longrightarrow \lambda$; t_0, τ, for s $\geq$ 1, with the restriction that $a^s \notin L(E)$ for each rule $E/a^c \longrightarrow a^p$; t_0, τ of type (1) from R_i;
3. $syn \subseteq \{1, 2, \ldots, m\} \times \{1, 2, \ldots, m\}$ are the set of synapses among neurons (for each $1 \leq i \leq m$, $\{i, i\} \notin syn$);
4. $in \in \{1, 2, \ldots, m\}$ represents the input neuron (labeled by *in*), $out \in \{1, 2, \ldots, m\}$ represents the output neuron (labeled by *out*).

We set an initial time t_0 and a delay time τ for every rule in neurons, that means when the number of spikes in a neuron meet certain conditions, rules in this neuron do not need to fire immediately.

The rules of type (1) are firing (we also say spiking) rules, we have here two important actions in the use of firing rules: getting fired and spiking. If $E/a^c \longrightarrow a^p; t_0, \tau \in R_i$ and the neuron σ_i contains k spikes such that $a^k \in L(E)$, $k \geq c$, and the neuron σ_i closes immediately. But this rule not be fired immediately. If $t_0 \neq 0$, $\tau \neq 0$, the neuron closes in the step q, then in steps $q, q+1, \ldots, q+t_0-1$, the neuron is in the preparation phase, then the rule is fired in the step $q + t_0$, and in the step $q + t_0 + \tau$, the neuron spikes and becomes open again, that is, it can receive spikes.

The rules of type (2) are forgetting rules, We also have two important actions in the use of forgetting rules: getting fired and removing (forgetting). But in using forgetting rules, we always assume the delay time $\tau = 0$, that is to say, getting fired and removing in the same step. When neuron σ_i contains exactly s spikes, the neuron σ_i closes immediately, the rule is fired after a delay of t_0 steps. After t_0 steps, s spikes are removed from the neuron σ_i.

It is possible that there are more than one rule that can be used. In this case, only one of them is chosen non-deterministically. In short, the rules are used in the non-deterministic manner, in a maximally parallel way at the level of the system: in one step, any neuron which can use a rule, of any type, spiking or forgetting, have to close the neuron, prepare for firing the rule.

3   Computational Completeness

In this section, we will prove that TDSNP systems are able to compute all Turing computable sets of numbers in the generating mode. We denote by $TDSN_m^n P$ the family of sets of number computed as above TDSNP systems with at most $m \geq 1$ neurons and at most $n \geq 1$ rules in each neuron.

Such a register machine is in the form $M = (m, H, l_0, l_h, I)$, in this construct, m is the number of registers (be labeled by $0, 1, \ldots, m-1$); I is the set of instructions, H is the set of instruction labels, note that each label from H is associated with only one instruction from I, thus there are as many as instructions as labels; l_0 is the start label, l_h is the halt label (labeling to instruction HALT), the instruction are of the following forms [16–18]:

- l_i:(ADD(r), l_j, l_k) (add 1 to register r and then go non-deterministically to one of the instructions with label l_j and l_k)
- l_i:(SUB(r), l_j, l_k) (if register r is empty, go directly to the instruction with label l_k; if register r is non-empty, then subtract 1 from register r and go to the instruction with label l_j)
- l_h:HALT (the halt instruction)

We denote by *N(M)* the set of all numbers generated by M. It is known that register machines can generate all sets of number which are Turing computable, hence register machines can characterize the family *NRE*.

Theorem 1. $TDSN_*^2P = NRE$.

Proof. We only have to prove the inclusion $NRE \subseteq TDSN_*^2P$. To achieve this aim, we construct a spiking neural p system to simulate a register machine $M = (m, H, l_0, l_h, I)$. To simulate M means to simulate each instruction of M. Thus, with each instruction we construct a module of our system. We associate a neuron σ_l with each label $l \in H$. When a neuron σ_l gets two spikes, we say, it is active, then the instruction labeled by l starts to be simulated. We also associate a neuron with each register of M. If a register r contains the number n, then the associated neuron σ_r will contains 2n spikes.

The module ADD of the system simulates an ADD instruction $l_i:(\text{ADD}(r), l_j, l_k)$, given in Fig. 1. Assume that, at step t, two spikes present in neuron σ_{l_i}, and no spike in any other neuron, with the exception of neurons associated with the registers. Having two spikes inside, neuron σ_{l_i} closes immediately and prepares the trigger for the rule. In the step $t + 1$, the rule in the neuron σ_{l_i} gets fired. The contents of neuron σ_r increased by two. And the neuron $\sigma_{l_i'}$ also receive two spikes from neuron σ_{l_i}. We use the non-determinism of the rules in neuron $\sigma_{l_i'}$. If we use the rule $a^2 \to a^2;1,1$, the instruction associated with the label l_j becomes active. However, if instead of rule $a^2 \to a^2;1,1$ we use the rule $a^2 \to a;1,0$ in the step $t + 2$. That is, the instruction associated with label l_k becomes active.

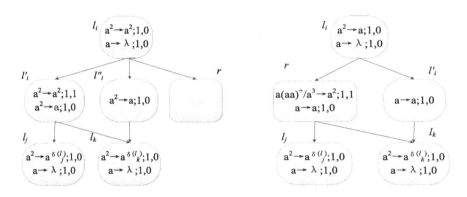

Fig. 1. Module ADD   **Fig. 2.** Module SUB

Therefore, from firing neuron σ_{l_i}, we pass non-deterministically to one of the neuron σ_{l_j}, σ_{l_k}, which also increasing by two the contents of neuron σ_r. The value of this register r of M increase by one.

Note that we use extended spiking rules in our system, thus the neurons associated with labels of ADD and SUB instructions produce different numbers of spikes. We have written the rule in the form $a^2 \to a^{\delta(l)};1,0$, with $\delta(l) = 2$ for l being the label of a ADD instruction and $\delta(l) = 1$ if l is label of an SUB instruction.

The module SUB of the system simulates an SUB instruction l_i:(SUB(r), l_j, l_k), given in Fig. 2. Assume that, at step t, two spikes present in neuron σ_{l_i}, and no spike in any other neuron except neuron σ_r. The neuron σ_{l_i} closes immediately and prepares the trigger for the rule. In the step t + 1, neuron σ_{l_i} gets fired. And neuron $\sigma_{l_i'}$ receive a spike and will be closed in the next step. The neuron σ_r also receive a spike.

Now, a similar problem appears in neuron σ_r: which rule dose it satisfy. The number of spikes in the neuron σ_r meet the first rule $(a(aa)^+/a^3 \to a^2;1,1)$ if and only if neuron σ_r contains at least three spikes, that means, there are at least two spikes existed in the neuron σ_r. In this case, after a few steps the neuron σ_{l_j} receives two spikes. Another case, if the neuron σ_r was empty, then the number of spikes meet the second rule. In this case, after a few steps the neuron σ_{l_k} receives two spikes. Therefore, from firing neuron σ_{l_i}, we ended in neuron σ_{l_j} if the register r was non-empty and decreased by one, and in neuron σ_{l_k} if the register r was empty.

The module HALT of the system simulates an HALT instruction l_h:HALT, given in Fig. 3. The process of this module is obvious, hence we omit the details. In the end, the number of spikes in neuron σ_{out} is equal to the value of register 0 of M. In conclusion, we prove that the register machine M and the TDSNP system we constructed compute the same set of numbers.

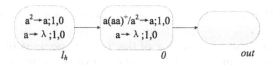

Fig. 3. Module HALT (simulating l_h:HALT)

4   SNP Systems with Time Delay Working in the Accepting Mode

In this section, we construct a TDSNP system as a number accepting device. We will also prove that it is Turing universal in the accepting mode. But in the accepting mode the non-determinism is no longer necessary. In this mode, we start the computation by introducing a number of spikes in the special neuron σ_{in}; we assume that two spikes are introduced to the system, and after receiving the two spikes, the system halts, then the number of steps is accepted. We denote by $T_{acc}DSN_mP$ the family of sets of number accepted by a P system with time delay.

Theorem 2. $T_{acc}DSN_mP = NRE$.

Proof. This proof is related to the proof of Theorem 1. That is, we start from a deterministic register machine M and we construct a TDSNP system to simulate the deterministic machine M. In the accepting mode, we must construct a further module, called INPUT module, which takes care of initializing the computation.

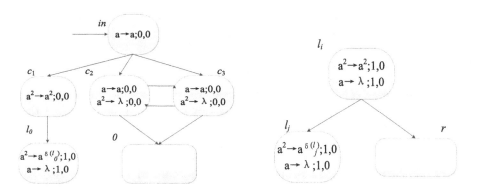

Fig. 4. Module INPUT   **Fig. 5.** Module ADD

The module INPUT is given in Fig. 4. There is a point that we have to mention that our aim is to trigger neuron σ_{l_o} as soon as possible. Thus, we set the start time t_o and the delay time τ of the first few neurons to zero to achieve this goal. When a spike enters neuron σ_{in}, the neuron σ_{in} sends a spike to neurons σ_{c_1}, σ_{c_2}, σ_{c_3}. Because the number of spikes cannot meet the conditions of the rule in the neuron σ_{c_1}, it just waits for a second spike. Neurons σ_{c_2}, σ_{c_3} get fired and exchange their spikes, the contents of neuron σ_0 increases by two. After n steps, a second spike enters the neuron σ_{in}. Neuron σ_{c_1} gets fired and sends two spikes to neuron σ_{l_0}, thus neuron σ_{l_0} triggers the simulation of a computation. While for neurons σ_{c_2}, σ_{c_3}, they will end their work. Therefore, on the one hand, the neuron σ_{l_o} starts by getting two spikes, on the other hand, the contents of the first register 0 remain 2n.

Now, we start simulating the ADD instructions and the SUB instructions. The module ADD of the system simulates an ADD instruction l_i:(ADD(r), l_j) in the deterministic case. We give it in Fig. 5. The process of this module is obvious, hence we omit the details. As for the SUB module, it remains unchanged. But the module HALT is removed, with the neuron σ_{l_h} remaining in the system, with no rule inside. Thus, the computation in register machine M stops, and the computation stops.

5   Conclude

In the real world, we always hope that the occurrence of all events can be controlled, and all processes have a start time and an end time. So we introduce a new class of SNP systems (TDSNP systems) in this paper. As an overall observation, setting an initial time and a delay time for each rule proves to be a more controlled feature. Specially, compared with previous systems, SNP systems with time delay can use smaller number of neurons on simulating a register machine. That's because we set two different times to control the system's refractory period. Inevitably, this systems take a long time to complete a particular task. We show these conclusions using simple parameters and our next goal is to combine the SNP systems with the control systems to make the potential utility of the SNP system even greater.

Acknowledgement. Project supported by National Natural Science Foundation of China (61472231, 61502283, 61876101, 61802234, 61806114), Ministry of Education of Humanities and Social Science Research Project, China (12YJA630152), Social Science Fund Project of Shandong Province, China (16BGLJ06, 11CGLJ22), China Postdoctoral Project (2017M612339).

References

1. Păun, G.: Membrane computing: an introduction. Theoret. Comput. Sci. **287**(1), 73–100 (2002)
2. Ionescu, M., Yokomori, T.: Spiking Neural P Systems and T. IOS Press, Amsterdam (2006)
3. Martín-Vide, C., Pazos, J., Păun, G., Rodríguez-Patón, A.: A new class of symbolic abstract neural nets: tissue P systems. In: Ibarra, O.H., Zhang, L. (eds.) COCOON 2002. LNCS, vol. 2387, pp. 290–299. Springer, Heidelberg (2002). https://doi.org/10.1007/3-540-45655-4_32
4. Korec, I.: Small universal register machines. Theoret. Comput. Sci. **168**(2), 267–301 (1996)
5. Cavaliere, M., Egecioglu, O., Ibarra, O.H., et al.: Asynchronous spiking neural P systems. Theoret. Comput. Sci. **410**(24), 2352–2364 (2009)
6. Păun, G.: Spiking neural P systems with astrocyte-like control. J. UCS **13**(4), 1707–1721 (2008)
7. Song, T., Pan, L., Păun, G.: Asynchronous Spiking Neural P Systems with Local Synchronization. Elsevier Science Inc., New York (2013)
8. Pan, L., Zeng, X.: A note on small universal spiking neural P systems. In: Păun, G., Pérez-Jiménez, M.J., Riscos-Núñez, A., Rozenberg, G., Salomaa, A. (eds.) WMC 2009. LNCS, vol. 5957, pp. 436–447. Springer, Heidelberg (2010). https://doi.org/10.1007/978-3-642-11467-0_29
9. Zhang, X., Zeng, X., Pan, L.: Smaller Universal Spiking Neural P Systems. IOS Press, Amsterdam (2008)
10. Rozenberg, G., Salomaa, A. (eds.): Handbook of Formal Languages. Springer, Heidelberg (1997). https://doi.org/10.1007/978-3-642-59126-6
11. Păun, G., Rozenberg, G., Salomaa, A.: The Oxford Handbook of Membane Computing. Oxford University Press, Oxford (2010)

12. The P Systems Web Page. http://psytems.disco.unimib.it
13. Păun, G., Sakakibara, Y., Yokomori, T.: P systems on graphs of restricted forms. Publicationes Math. **60**(3), 635–660 (2002)
14. Wood, D.: Theory of Computation. Harper and Row, New York (1987)
15. Zhao, Y., Liu, X., Wang, W.: Spiking neural P systems with neuron division and dissolution. PLoS ONE **11**(9), e0162882 (2016)
16. Kudlek, M.: Small deterministic Turing machines. Theoret. Comput. Sci. **168**(2), 241–255 (1996)
17. Minsky, M.L: Computation: finite and infinite machines. Am. Math. Monthly **75**(4) (1967)
18. Rogozhin, Y.: Small Universal Turing Machines. Elsevier Science Publishers Ltd., Essex (1996)

Energy Efficiency MapReduce Job Scheduling of Shuffle and Reduce Phases in Data Center

Jia Wang$^{(\boxtimes)}$, Xiaoping Li, and Xia Zhu

Southeast University, Nanjing, China
{wangjia1024,xpli,zhuxia}@seu.edu.cn

Abstract. In this paper, job scheduling of shuffle and reduce phases is considered for data center with heterogenous servers to minimize energy consumption. Constructing task list and assigning tasks to slots are designed in a job scheduling framework. The construction of task list considers jobs' deadlines and tasks' processing times. Two main steps (candidate servers construction and allocate tasks) are in the proposed assignment. The set of candidate servers is constructed in terms of data size and network topology. Allocation of tasks and slots with normalized shuffle time and data size decreases completion times of jobs, in which shuffle time is calculated by two new bandwidth allocations considering deadlines. Experimental results show that the proposed job scheduling consumes less energy than other existing adapted task scheduling strategies.

Keywords: Energy efficiency · MapReduce job scheduling · Deadline · Big data

1 Introduction

Energy consumption of compuserve and software industry made a big contribution to the increased total energy consumption of society. In China, energy consumption of data centers has reached 6.645×10^{10} KWh by 2012 [1]. And it run up to 1×10^{11} KWh in 2015, which equals to the generating capacity of three gorges dam in a year. Meanwhile, energy consumption of data center is estimated to increase by 15–20% every year [2]. For example, some Mapreduce jobs (such as the detection of anomalous traffic patterns [3] and the identification of the driver group with the same or similar behaviour features [4]) run over big data in the data center of Public Security Bureau of Jiangsu Province in China which consumes lots of energy every year. These Mapreduce jobs (CPU-intensive or IO-intensive) always massive and have given deadlines. Each Mapreduce job consists multiple map and reduce tasks. All of them are performed on heterogeneous servers in the data center. Since original and intermediate data are distributed in heterogenous servers, transmission time especially that at shuffle phase is too

© Springer Nature Switzerland AG 2019
Y. Tang et al. (Eds.): HCC 2018, LNCS 11354, pp. 209–221, 2019.
https://doi.org/10.1007/978-3-030-15127-0_22

long to neglect. Different types of tasks (CPU-intensive or IO-intensive) have different processing times. A long time containing transmission and processing time of tasks would need which results in a large energy consumption. Therefore the reduction of energy consumption by effective and efficient job scheduling is an urgent problem to solve.

Existing energy-aware task scheduling with response time always considers deadlines to minimize energy consumption [7,8] or obtains energy conservation by minimizing the (weighted) completion time [9]. While the energy-aware task scheduling considering resource utilization [10–13] minimizes energy consumption by making good use of resource fragment (CPU, memory, disk and bandwidth) of servers. Existing energy efficiency studies attach importance to task scheduling of map or reduce phases and ignore the improvement of shuffle phase, which wastes considerable performance overhead during the intermediate data transfer from map processors to reduce processors in Mapreduce [14]. Some papers considered data operators (partition and aggregation) [14,15] and transfer routers [5,16] during data transfer to reduce shuffle time. Different from the intrinsic hash function, data operators were performed in accordance with data size and the placement of map processor and reduce processor in network topology [14]. The transfer routers problem of intermediate data was converted to steiner tree problem [16]. The total transfer cost was minimized by the placement of data aggregation according to various transfer models. In a word, these papers focusing on the improvement of shuffle phase ignore the energy-aware Mapreduce job scheduling. Various strategies of shuffle phase result in different performances of reduce phase. Reduce tasks were assigned to reduce processor in accordance with data size and network distance [17] or the estimated shuffle time and network distance [18]. Tan et al. [19] improved the strategy in [17] and measured network distance by hops. These papers with job scheduling of reduce phase ignore network congestion among tasks (generally bandwidth is allocated equally), which makes longer shuffle time and larger energy consumption of data centers. Meanwhile, the incorrect assignment in [17–19] is presented because the unit of data size and network distance are different. To the best of our knowledge, there is no study on energy-aware task scheduling of shuffle and reduce phase with deadlines.

Energy consumption of the data center greatly depends on the scheduling of Mapreduce jobs. We focus on job scheduling of shuffle and reduce phases with deadline in data center. Submitted jobs are sequenced by EDF (the Earlier Deadline First) rule. BAJD (Bandwidth Allocation of Job Deadline) and BAUF (Bandwidth Allocation of Utility Function) are proposed to calculate shuffle time by proportionally allocating bandwidth among data flows with deadlines. A heuristic is developed for allocating tasks to slots with both intermediate data size and shuffle time considerations.

The rest of the paper is organized as follows. A detailed description of the under study is presented in Sect. 2. Section 3 describes a detailed job scheduling. Experimental results are given in Sect. 4. At last, conclusions is presented in Sect. 5.

2   Problem Description

2.1   System Model

In this paper, we consider energy efficient job scheduling of shuffle and reduce phases based on AIS (All-In Strategy) in terms of [20]. Let S_{ku} be the uth server in rack R_k. Assume $\mathfrak{J}^e$, $\mathfrak{J}^l$, $\mathfrak{J}^e_{ku}$ and $\mathfrak{J}^l_{ku}$ are the job with the earliest start time of shuffle phase of all jobs, with the latest finish time of reduce phase of all jobs, with the earliest start time of shuffle phase of jobs processed on S_{ku} and with the latest finish time of reduce phase of jobs processed on S_{ku} respectively. For each server, the overlap of map and shuffle phases would be presented because processing times of jobs of map phase are different. In other words, servers processing tasks of shuffle and reduce phases are already running and power-on states of servers are ignored here. With the considered AIS, all servers need to start and power off simultaneously at shuffle and reduce phases. We set time 0 and U as the start time of all servers (the start time of shuffle phase of $\mathfrak{J}^e$) and the power off time of all servers (the finish time of reduce phase of $\mathfrak{J}^l$) respectively. Since servers have the shut-down time ΔT after servers are powered off, the energy consumption is calculated from 0 to $U + \Delta T$. Let I^{Mr}_{ku} be the time to process the remained map tasks of $\mathfrak{J}^e_{ku}$ after time 0, i.e., I^{Mr}_{ku} is the start time of server S_{ku}'s shuffle phase (the start time of shuffle phase of $\mathfrak{J}^e_{ku}$). I^{Mr}_{ku} is bigger than 0 in that $\mathfrak{J}^e$ may not be processed on server S_{ku}. Let I^{Rr}_{ku} and I^d_{ku} be the running interval and idle interval of S_{ku} respectively. I^{Rr}_{ku} consists of processing times of shuffle and reduce phases in this paper. Therefore, the running interval of server S_{ku} is denoted by $I^r_{ku} = I^{Mr}_{ku} + I^{Rr}_{ku}$. Assume F^r_{ku} is the finish time of server S_{ku}, which equals the finish time of reduce phase of $\mathfrak{J}^l_{ku}$. F^r_{ku} is less than U because $\mathfrak{J}^l_{ku}$ and $\mathfrak{J}^l$ may not be the same job.

2.2   Problem Model

In order to simplify problem model, we set the arrival time of jobs as the start time of shuffle phase in this paper. Many independent jobs are start to submit from time 0. The arrival time of jobs obeys the poisson distribution P(0.5), i.e., all jobs need different time to process the remained map tasks.

A set of m jobs $J = \{J_1, J_2, \cdots, J_m\}$ share heterogeneous servers in data center. $J_i \in J$ consists of $\mathcal{N}_i$ independent reduce tasks $A_{ij} (1 \leq j \leq \mathcal{N}_i)$. With the deadline D_i of J_i, all tasks A_{ij} need to complete before D_i. In the considered data center, a set of r racks $R = \{R_1, R_2, \cdots, R_r\}$ is configured. Each rack R_k consists of N_k servers S_{ku}. N^u_k slots $L_{kuv}(1 \leq v \leq N^u_k)$ is placed on each server S_{ku}. We focus on the energy efficient job scheduling of shuffle and reduce phases for data center to minimize energy consumption with deadlines from 0 to $U + \Delta T$.

In this paper, energy consumption E consists of three types (running, idle and shut-down) energy consumption. Assume E^r, E^d and E^e be the energy consumption per unit time of running, idle and shut-down respectively,

$$E = \sum_{k=1}^{r} \sum_{u=1}^{N_k} (E^r \times I^r_{ku} + E^d \times I^d_{ku} + E^e \times \Delta T) \tag{1}$$

Since $F_{ku}^r = I_{ku}^r = I_{ku}^{Mr} + I_{ku}^{Rr}$, $U = I_{ku}^r + I_{ku}^d$ and $U = \max\limits_{1 \le k \le r} \max\limits_{1 \le u \le N_k} F_{ku}^r$,

Eq. (1) can be converted to $E = (E^r - E^d) \sum\limits_{k=1}^{r} \sum\limits_{u=1}^{N_k} F_{ku}^r + nE^d \max\limits_{1 \le k \le r} \max\limits_{1 \le u \le N_k} F_{ku}^r +$

$nE^e \Delta T$. In other words, E is determined by F_{ku}^r. Suppose $\mathcal{A}_{ku}$ is the set of tasks processing on server S_{ku}. F_{ku}^r depends on completion time C_{ij} of tasks A_{ij} in $\mathcal{A}_{ku}$. $F_{ku}^r = \max\limits_{A_{ij} \in \mathcal{A}_{ku}} C_{ij}$. C_{ij} consists of begin time t_{ij}, shuffle time $\tau_{i,j;u',u}$ and processing time p_{ij}. A decision variable $x_{i,j;k,u,v}$ is defined. $x_{i,j;k,u,v} = 1$ means task A_{ij} is allocated to slot L_{kuv}, otherwise $x_{i,j;k,u,v} = 0$. In other words, $C_{ij} = \sum\limits_{k=1}^{r} \sum\limits_{u=1}^{N_k} \sum\limits_{v=1}^{N_{ku}} (t_{ij} + \tau_{i,j;u',u} + p_{ij}) \times x_{i,j;k,u,v}$. $\tau_{i,j;u',u}$ is determined by bandwidth $B_{i,j;u',u}$. Let B_s and B_r be allocated bandwidth among servers and that among racks respectively. The allocated bandwidth of task A_{ij}'s data flow from server $S_{k'u'}$ to S_{ku} is denoted as $B_{i,j;u',u}$.

$$
B_{i,j;u',u} = \begin{cases} \propto & k' = k \text{ and } u' = u \\ B_s & k' = k \text{ and } u' \ne u \\ B_r & k' \ne k \end{cases}
$$

Generally the input data of A_{ij} need to transfer from many map processors, $\mathcal{S}_{ij}$ and $\mathcal{D}_{ij}^{ku}$ are the set of servers containing the input data of A_{ij} and the input data size of A_{ij} on S_{ku} respectively. Therefore $\tau_{i,j;u',u} = \max\limits_{S_{k'u'} \in \mathcal{S}_{ij}} \mathcal{D}_{ij}^{k'u'} / B_{i,j;u',u}$.

The considered problem is modeled as follows:

$$
\min \quad E = (E^r - E^d) \sum\limits_{k=1}^{r} \sum\limits_{u=1}^{N_k} F_{ku}^r + nE^d \max\limits_{1 \le k \le r} \max\limits_{1 \le u \le N_k} F_{ku}^r + nE^e \Delta T
$$

Subject to:

$$
F_{ku}^r = \max\limits_{A_{ij} \in \mathcal{A}_{ku}} C_{ij} \tag{2}
$$

$$
C_{ij} \le D_i \tag{3}
$$

$$
C_{ij} = \sum\limits_{k=1}^{r} \sum\limits_{u=1}^{N_k} \sum\limits_{v=1}^{N_{ku}} (t_{ij} + \tau_{i,j;u',u} + p_{ij}) \times x_{i,j;k,u,v} \tag{4}
$$

$$
\tau_{i,j;u',u} = \max\limits_{S_{k'u'} \in \mathcal{S}_{ij}} \mathcal{D}_{ij}^{k'u'} / B_{i,j;u',u} \tag{5}
$$

$$
\sum\limits_{i=1}^{m} \sum\limits_{j=1}^{\mathcal{N}_i} x_{i,j;k,u,v} \le 1 \tag{6}
$$

$$
\sum\limits_{k=1}^{r} \sum\limits_{u=1}^{N_k} \sum\limits_{v=1}^{N_{ku}} x_{i,j;k,u,v} \le 1 \tag{7}
$$

$$
x_{i,j;k,u,v} \in \{0,1\} \tag{8}
$$

where $\forall i \in \{1,2,\cdots,m\}$, $\forall j \in \{1,2,\cdots,\mathcal{N}_i\}$, $\forall k \in \{1,2,\cdots,r\}$, $\forall u \in \{1,2,\cdots,N_k\}$, $\forall k' \in \{1,2,\cdots,r\}$ and $\forall u' \in \{1,2,\cdots,N_{k'}\}$.

Equation (2) describes F_{ku}^r. Deadline constraints is implied by Formula (3). Equations (4) and (5) calculate C_{ij} and $\tau_{i,j;u',u}$ respectively. Equation (6) illustrates every task is assigned to only one slot and Equation (7) guarantees every slot is assigned to only one task. Equation (8) specifies the nature of $x_{i,j;k,u,v}$.

3 Proposed Algorithms

In this paper, energy efficient Mapreduce job scheduling considering deadlines, is proved to be a NP-hard problem [6]. Since heuristics are the most useful methods to solve NP-hard problems, a heuristic job scheduling is proposed. Initially, idle slots of all servers are managed by list W_L. Reduce tasks of each job are sequenced by LTF (the Longest Time First) rule with the best performance. In each heartbeat, arrived jobs are sequenced using EDF rule for the efficiency and simplification, which are kept by list W_J. A list W_A is constructed by job location in W_J and task location in the job. Each task in W_A is assigned to the most suitable slot in W_L by TSR (Task Scheduling of shuffle and reduce phases). W_L is updated for next heartbeat. With the simple construction of W_A and updating of W_L, task scheduling is described in detail.

3.1 Task Scheduling

In each heartbeat, tasks in W_A choose the most suitable slot in W_L to process. Different assignments result in various completion times. About the default task scheduling, the reduce task A_{ij} randomly chooses an idle slot L_{kuv} and longer completion time would be presented. Several studies [17–19] consider data size and network distance to choose a reduce processor, in which data size is considered to transfer less intermediate data and network distance is to reduce transfer time with the consideration of the placement of S_{ku} in network topology. Network distance is calculated by hops in [17–19] while hops are not accurate to measure network distance. Furthermore, the calculation of non-normalized data size and network distance in [17–19] easily results in the uncorrect assignment. In this paper, we consider the normalized intermediate data size and transfer time to assign slots in W_L to tasks in W_A.

Bandwidth Allocation. Shuffle time occupies more than half of traffics during data transfer [14]. Longer shuffle time leads to longer completion times of jobs and more energy consumption. Generally, severe bandwidth competition among data flows make shuffle time too long to violate some jobs' deadlines. In other words, shuffle time is determined by the allocated bandwidth. The larger bandwidth results in less transfer time. The existing equational bandwidth allocation between data flows in Mapreduce easily violates deadlines at reduce phase.

Assume a set of tasks T_{ku}^S are in shuffle phase on server S_{ku}, i.e., a set of tasks T_{ku}^S need to transfer intermediate data from S_{ku}. In order to ensure deadline constraints in shuffle phase, the bandwidth of S_{ku} are proportionally allocated to a set of tasks T_{ku}^S according to their deadlines (without loss of generality, deadlines of tasks are set as deadlines of corresponding jobs). Tasks in T_{ku}^S are sequenced by LDF (the Latest Deadline First), which makes the task with the latest deadline allocate the least bandwidth and transfer data of other tasks quickly. The priority of task A_{ij} represented by ω_{ij} is the location of A_{ij} in the sequenced T_{ku}^S. BAJD (Bandwidth Allocation of Job Deadline) as a bandwidth allocation is proposed firstly. The allocated bandwidth of task A_{ij} from S_{ku} to reduce processor $S_{ku'}$ by BAJD is:

$$B_{i,j;u,u'} = (\omega_{ij} / \sum_{A_{mn} \in T_{ku}^S} \omega_{mn}) \times B_s. \tag{9}$$

Considering the above assumption, the objective of allocating B_s among a set of tasks T_{ku}^S is to ensure deadline constraints of tasks by allocating reasonable bandwidth to various data flows in shuffle phase. And the sum of allocated bandwidths of T_{ku}^S is less than B_s. Various bandwidth requirements of data flows are determined by deadlines of reduce tasks. With a given allocated bandwidth, the satisfaction of each data flow can be seen as the matching degree of occupied bandwidth and its task's priority in T_{ku}^S. Therefore, the bandwidth allocation among a set of tasks T_{ku}^S is an application instance of utility function. BAUF (Bandwidth Allocation of Utility Function), another bandwidth allocation, considers utility function and deadlines to proportionally allocate bandwidth of server to various data flows. As the weighted max-min allocation is the generalization of other ones, we take the utility function of weighted max-min allocation $\mu_i \times \log(x_i)$ as the used utility function in this paper. Where μ_i and x_i are the weight of data flow and the allocated bandwidth of data flow respectively. With the consideration of deadlines in BAUF, the weight of task A_{ij}'s data flow is set as $\omega_{ij} / \sum_{A_{mn} \in T_{ku}^S} \omega_{mn}$. The calculation of ω_{ij} is same to that in BAJD. Then the bandwidth allocation from S_{ku} to $S_{ku'}$ by BAUF can be formulated as follows:

$$\max \sum_{A_{ij} \in T_{ku}^S} (\omega_{ij} / \sum_{A_{mn} \in T_{ku}^S} \omega_{mn}) \times \log(B_{i,j;u,u'}) \tag{10}$$

Subject to:

$$\sum_{A_{ij} \in T_{ku}^S} B_{i,j;u,u'} \leq B_s \tag{11}$$

Formula (11) demonstrates the allocated bandwidth of all tasks in T_{ku}^S are not larger than B_s. Since bandwidth of racks is limitation, bandwidth allocation of each rack must consider which can be obtained by the same strategies with that of each server.

Task Allocation. With the determined bandwidth allocation, an appropriate slot need select to process A_{ij}. The selection of idle slot from W_L without any heuristics is very difficult and time-consuming, we develop the set of candidate servers $\mathcal{C}_{ij}^N$ of task A_{ij} to narrow choices. With the consideration of intermediate data size, network distance and less time spent on transferring data in a rack, $\mathcal{C}_{ij}^N$ consists of the server with the biggest data size, that with the second biggest data size, that with the third biggest data size and all servers in the same rack with the server of the biggest data size.

Let C_d and C_s be the normalized intermediate data size and the normalized shuffle time respectively. C_d is calculated by the ratio of intermediate data on S_{ku} to all intermediate data of A_{ij} on a set of servers $\mathcal{S}_{ij}$ and $0 \leq C_d \leq 1$, which denotes

$$C_d = \mathcal{D}_{ij}^{ku} / \sum_{S_{k'u'} \in \mathcal{S}_{ij}} \mathcal{D}_{ij}^{k'u'}. \tag{12}$$

In order to avoid the improper assignment by different units of data size and transfer time, we need to normalize transfer time between 0 and 1. Because the codomain of $\pi/2 - \arctan(x)$ is $(0, \pi)$ and the definition domain of that is $(-\infty, +\infty)$, transfer time is normalized by $(\pi/2 - \arctan(x))/\pi$. Let $\mathcal{T}_{ij}^S$ be the shuffle time of task A_{ij}. Therefore

$$C_s = (\pi/2 - \arctan(\mathcal{T}_{ij}^S))/\pi. \tag{13}$$

where $0 < C_s < 1$. Assume $\mathbb{C}_{ku}$ is the cost of intermediate data size and transfer time on server S_{ku} $(S_{ku} \in \mathcal{C}_{ij}^N)$ for reduce task A_{ij}, which equals the weight sum of C_d and C_s. Let α be the weight of C_d.

$$\mathbb{C}_{ku} = \alpha \times C_d + (1 - \alpha) \times 1/C_s. \tag{14}$$

Since the larger C_d and the smaller C_s result in the larger $\mathbb{C}_{ku}$, we choose the server with the maximal $\mathbb{C}_{ku}$ to process reduce task A_{ij} from $\mathcal{C}_{ij}^N$. Any idle slots on server S_{ku} is randomly assigned to A_{ij}. The detailed TSR is shown in Algorithm 1. Let S^{ij} and C^{ij} be the reduce processor of A_{ij} and the cost of intermediate data size and shuffle time on server S^{ij} respectively. The assignment results are represented by AssginList. Assume T_k^S is the set of tasks at shuffle phase in rack R_k. The time complexity of Algorithm 1 is $O(|W_A||\mathcal{C}_{ij}^N||S_{ij} \max(|T_{ku}^S| \log |T_{ku}^S|, |T_k^S| \log |T_{k'}^S|))$.

Algorithm 1. TSR /*Task scheduling of shuffle and reduce phases*/

Input: W_A, W_L
Output: AssignList

1 **for** *each task A_{ij} in W_A* **do**
2   **if** *W_L is not empty* **then**
3   Obtain the set of candidate servers $\mathcal{C}_{ij}^N$;
4   **if** *$\mathcal{C}_{ij}^N$ is not empty* **then**
5           $C^{ij} \leftarrow -\infty$;
6   **for** *each server S_{ku} in $\mathcal{C}_{ij}^N$* **do**
7   Calculate C_d using Equation (12);
8              $T_{ij}^S \leftarrow -\infty$;
9   **for** *each server $S_{k'u'}$ in $\mathcal{S}_{ij}$* **do**
10   Calculate $B_{i,j;u',u}$ by BAJD or BAUF;
11   **if** $D_{ij}^{k'u'}/B_{i,j;u',u} > T_{ij}^S$ **then**
12                   $T_{ij}^S \leftarrow D_{ij}^{k'u'}/B_{i,j;u',u}$;
13   Normalize T_{ij}^S by Equation (13) and calculate $\mathbb{C}_{ku}$ using Equation (14);
14   **if** $\mathbb{C}_{ku} > C^{ij}$ **then**
15                $C^{ij} \leftarrow \mathbb{C}_{ku}$;
16                $S^{ij} \leftarrow S_{ku}$;
17   Randomly choose an idle slot of server S^{ij};
18   **else**
19   Randomly choose an idle slot of W_L;
20   Add the selected slot and task A_{ij} to AssignList;
21 **return** AssignList

---

4 Experiment

In this paper, we consider a data center to run Mapreduce jobs consisting of only reduce tasks. There are 90 servers distributing 10 racks in the data center. Each rack has different number of servers and various number of slots is deployed on each of them. The bandwidth of racks and that of servers obey the uniform distribution U(100, 300) and U(200, 400) respectively. The unit of bandwidth is MB/s. Three levels of job number are set as {50, 100, 150}. The arrival time of jobs as the start time of their shuffle and reduce phases, generates by the poisson distribution P(0.5) with the unit being second. Jobs and tasks information is generated according to the analysis performed on Yahoo! M45 production cluster [21]. The number of tasks in each job is produced with the normally distribution N(19, 145). The processing time of each task is normally distributed in N(100, 300) with the unit being second. The number of intermediate data chunks of each task obeys the uniform distribution U(5, 15). The minimum and maximum completion times of jobs are obtained by jobs process in parallel and sequentially respectively. Let $SL^s = \sum_{k=1}^{r} \sum_{u=1}^{N_k} N_k^u$ be the total number of slots. With considerations of the distribution of task number and processing time, deadline of each job is produced by the uniform distribution $U(\lceil \frac{19 \times m}{SL^s} \rceil \times 100, \lceil \frac{19}{SL^s} \rceil \times m \times 100)$ and the unit is second. In addition, E^r and E^d are set as the same values in [22] - 368 and 223 respectively. The unit of them are watt.

We run all heuristic task scheduling strategies (which are coded in Java with Eclipse Helios Release JDK1.6) on servers installed with Intel Core i5-3479 3.7 GHz processor with 4 GB of RAM and servers with Intel Core i7-3770 3.4 GHz processor with 8 GB of RAM. RPD (Residual Predictive Deviation) of E is set as the response variable. Assume E^* is the minimal value among results with the same job number. RPD of E is denoted $RPD = \frac{E-E^*}{E^*}$. A smaller energy consumption results in a less RPD of E.

4.1 Experimental Results

Parameters Tuning. There are three parameters - the construction of candidate servers, allocation of tasks and slots and bandwidth allocation in the proposed job scheduling framework to influence performance. Constructing candidate servers is to narrow choices of reduce processor for each task. Whether we construct candidate servers or not make two different methods. Let C and NC represent algorithms with and without the construction of candidate servers respectively. In this paper, tasks are allocated to slots considering both shuffle time and intermediate data size. With different units of transfer time and data size, whether we normalize them or not will result in different performances. In order to investigate the influence of weight α, we assume $N_{0.2}$, $N_{0.4}$, $N_{0.6}$, $N_{0.8}$, $N_{0.2}^N$, $N_{0.4}^N$, $N_{0.6}^N$ and $N_{0.8}^N$ designate algorithms with and without normalized shuffle time and data size when α equals 0.2, 0.4, 0.6 and 0.8 respectively. Different bandwidth allocations result in various shuffle times. Equational bandwidth allocation among tasks violates deadline constraints, two new bandwidth allocations are proposed in this paper. As aforementioned, there are $2 \times 8 \times 3 = 48$ treatments in an experimental design. Meanwhile, with the aforementioned 3 levels of job number, the number of treatments increases to 144. The 5 random instances for each level are tested so the total number of results in the calibration experiment is 720. We use the multi-factor analysis of variance (ANOVA) to analyze experimental results. The three main hypotheses (independence of residuals, homoscedasticity or homogeneity of factors levels variance and normality in residuals of the model) are checked and accepted. With the analysis of variances (the constructions of candidate servers, allocations and bandwidth allocations) for RPD of E, all p-values are less than 0.05 which indicates that these factors have a significant effect on the response variable at the 95.0% confidence level.

The means plot with 95% confidence level Tukey Honest Significant Difference (HSD) intervals for the construction of candidate servers is shown in Fig. 1. From Fig. 1, algorithm with the construction of candidate servers has the minimal RPD of E. The reason lies in that current task scheduling just happened on candidate servers, which narrows choices and simplifies computation.

The construction of candidate servers considering data size and network topology is good at the reduction of shuffle time. The means plot with 95% confidence level Tukey Honest Significant Difference (HSD) intervals for allocations is shown in Fig. 2. From Fig. 2, algorithms with normalized parameters has less RPD of E. The more exact $\mathbb{C}_{ku}$ can find the best server to process reduce task, which saves shuffle time and completion time. Especially, when α equals 0.4, algorithm with normalized parameters has the minimal RPD of E.

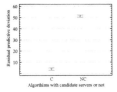

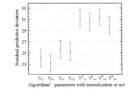

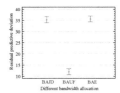

Fig. 1. Means plot of algorithms with candidate servers or not for E with 95% Tukey HSD intervals.

Fig. 2. Means plot of algorithms' different allocations for E with 95% Tukey HSD intervals.

Fig. 3. Means plot of algorithms with different bandwidth allocations for E with 95% Tukey HSD intervals.

The means plot with 95% confidence level Tukey Honest Significant Difference (HSD) intervals for bandwidth allocations is shown in Fig. 3. BAE is the equational bandwidth allocation. Algorithms with BAJD and BAE have almost the same RPD of E. BAJD considering deadlines makes jobs with small deadlines can allocate more bandwidth to reduce shuffle times. While it ignores resource constraints in the process of task scheduling (the reduce task cannot be processed because of these limited servers), which results in more waiting times of tasks and the more completion time of jobs. BAE makes longer shuffle time, which also results in longer completion time of jobs because tasks need to wait the completion of shuffle time. Algorithm with BAUF has the minimal RPD of E in Fig. 3. BAUF considers both deadlines and utility function (which represents task's current appropriate bandwidth requirement according to status of task scheduling), which allocates more reasonable bandwidth than other allocations. It balances shuffle time and resource constraints to process tasks. The measurable shuffle time and seemly task scheduling result in the less completion time and energy consumption. As aforementioned, the proposed task scheduling framework TSF with the construction of candidate servers, BAUF and normalized parameters at $\alpha = 0.4$ is to be compared with other existing algorithms in next section.

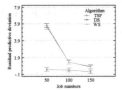
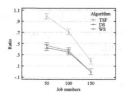

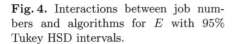

Fig. 4. Interactions between job numbers and algorithms for E with 95% Tukey HSD intervals.

Fig. 5. Interactions between job numbers and algorithms for Ratio with 95% Tukey HSD intervals.

Analysis Results. In order to investigate the effectiveness of TSF, we compare it to adapted algorithms DS (the default scheduler of Hadoop) and WS (the wait scheduling in [19]).

RPD of E as the response variable is compared firstly. Various algorithms have different completion times of jobs and total finish times of servers, which leads multiple energy consumptions. The interactions of RPD of E with 95% Tukey HSD interval is given in Fig. 4. In Fig. 4, TSF has the minimal RPD of E. Candidate servers are selected in TSF according to data size and network topology, which narrows choices and simplifies computation. BAUF considers both deadlines and utility function to balance shuffle time and processing time, which decreases completion times of jobs. Shuffle time and data size are normalized to select reduce processor, which eliminates the inaccuracy produced by different units and orders of magnitude. All of them result in less energy consumption generated by TSF. The RPD of E by WS is less than that by DS when there are 50 submitted jobs. The reason lies in that the selection of reduce processor considers data size and network topology in WS, which is better than random selection in DS. While other curves by WS and DS have no significant differences because of resource competition with more submitted jobs. Because we consider deadline constraints in this paper, we compare Ratio secondly. Ratio is the ratio of the number of jobs completed before their deadlines to the total number of submitted jobs. Smaller Ratio results in less number of jobs can finish before deadlines. Figure 5 shows the interactions of Ratio with 95% Tukey HSD interval. TSF has the maximal Ratio. Ratio generated by WS is larger than that by DS when there are 50 or 100 jobs.

5 Conclusion

In this paper, job scheduling of shuffle and reduce phases with deadlines is proposed to decrease energy consumption of data center. EDF and LTF are used to sequence jobs and tasks respectively. Task list is constructed according to job sequences and task orders. In the proposed assignment, the construction of candidate servers in terms of data size and network topology, narrows choices and simplifies computation. Two bandwidth allocations considering deadlines are designed to calculate shuffle time, which ensure jobs' deadlines. Slots are assigned to tasks according to the normalized shuffle time and data size. Experimental results show that the proposed task scheduling has less energy consumption and more jobs finished before deadlines.

References

1. Baidu builds the first solar energy cloud data center in yangquan china. http://solar.cheaa.com/2015/1130/463286.shtml
2. Koomey, J.G.: Worldwide electricity used in data centers. Environ. Res. Lett. **3**(3), 1–9 (2008)
3. Pang, L.X., Chawla, S., Liu, W., Zheng, Y.: On detection of emerging anomalous traffic patterns using GPS data. Data Knowl. Eng. **87**, 357–373 (2013)
4. Yang, J., Li, X., Wang, D., Wang, J.: A group mining method for big data on distributed vehicle trajectories in WAN. Int. J. Distrib. Sens. Netw. **11**(8), 1–9 (2015)
5. Chowdhury, M., Zaharia, M., Ma, J., Jordan, M.I., Stoica, I.: Managing data transfers in computer clusters with orchestra. ACM SIGCOMM Comput. Commun. Rev. **41**(4), 98–109 (2011)
6. Tiwari, N., Sarkar, S., Bellur, U., Indrawan, M.: Classification framework of MapReduce scheduling algorithms. ACM Comput. Surv. **47**(3), 1–49 (2015)
7. Mashayekhy, L., Nejad, M.M., Grosu, D., Zhang, Q., Shi, W.: Energy-aware scheduling of MapReduce jobs for big data applications. IEEE Trans. Parallel Distrib. Syst. **26**(10), 2720–2733 (2015)
8. Cardosa, M., Singh, A., Pucha, H., Chandra, A.: Exploiting spatio-temporal trade-offs for energy-aware MapReduce in the cloud. IEEE Trans. Comput. **61**(12), 1737–1751 (2012)
9. Bampis, E., Chau, V., Letsios, D., Lucarelli, G., Milis, I., Zois, G.: Energy efficient scheduling of MapReduce jobs. In: Silva, F., Dutra, I., Santos Costa, V. (eds.) Euro-Par 2014. LNCS, vol. 8632, pp. 198–209. Springer, Cham (2014). https://doi.org/10.1007/978-3-319-09873-9_17
10. Wang, X., Wang, Y., Cui, Y.: Energy and locality aware load balancing in cloud computing. Integr. Comput. Aided Eng. **20**(4), 361–374 (2013)
11. Patil, V.A., Chaudhary, V.: Rack aware scheduling in HPC data centers: an energy conservation strategy. Cluster Comput. **16**(3), 559–573 (2013)
12. Song, J., Liu, X., Zhu, Z., Zhao, D., Yu, G.: A novel task scheduling approach for reducing energy consumption of MapReduce cluster. IETE Techn. Rev. **31**(1), 65–74 (2014)
13. Hartog, J., Dede, E., Govindaraju, M.: MapReduce framework energy adaptation via temperature awareness. Cluster Comput. **17**(1), 111–127 (2014)
14. Ke, H., Li, P., Guo, S., Guo, M.: On traffic-aware partition and aggregation in MapReduce for big data applications. IEEE Trans. Parallel Distrib. Syst. **27**(3), 818–828 (2016)
15. Yu, W., Wang, Y., Que, X., Xu, C.: Virtual shuffling for efficient data movement in MapReduce. IEEE Trans. Comput. **64**(2), 556–568 (2015)
16. Guo, D., Xie, J., Zhou, X., Zhu, X., Wei, W., Luo, X.: Exploiting efficient and scalable shuffle transfers in future data center networks. IEEE Trans. Parallel Distrib. Syst. **26**(4), 997–1009 (2015)
17. Arslan, E., Shekhar, M., Kosar, T.: Locality and network-aware reduce task scheduling for data-intensive applications. In: International Workshop on Data-Intensive Computing in the Clouds, pp. 17–24 (2014)

18. Hammoud, M., Rehman, M.S., Sakr, M.F.: Center-of-gravity reduce task scheduling to lower MapReduce network traffic. In: International Conference on Cloud Computing, pp. 49–58 (2012)
19. Tan, J., Meng, X., Zhang, L.: Coupling task progress for MapReduce resource-aware scheduling. In: IEEE INFOCOM, pp. 1618–1626 (2013)
20. Lang, W., Patel, J.M.: Energy management for MapReduce clusters. VLDB Endowment **3**(1), 129–139 (2010)
21. Verma, A., Cherkasova, L., Campbell, R.H.: Orchestrating an ensemble of MapReduce jobs for minimizing their makespan. IEEE Trans. Dependable Secure Comput. **10**(5), 314–327 (2013)
22. Leverich, J., Kozyrakis, C.: On the energy (in)efficiency of Hadoop clusters. ACM SIGOPS Operating Syst. Rev. **44**(1), 61–65 (2010)

Research on Weibo Emotion Classification Based on Context

Weidong Huang[(⊠)], Xinkai Yao, and Qian Wang

Nanjing University of Posts and Telecommunications, Nanjing, China
huangwd@njupt.edu.cn

Abstract. In order to classify the speech and information published on the social network platform, this paper proposes an emotion classification method, based on text word vector and deep learning. According to the characteristic of weibo short text itself, the corpus is preprocessed. This paper uses word2vec to obtain the text vector of weibo short text, and classifies emotion through the classification model which is based on XGBoost. The experimental results for NLPCC corpus show that this method achieves a good emotion classification results, and can effectively improve the accuracy of sentiment classification.

Keywords: Internet public opinion · Sentiment classification · XGBoost

1 Introduction

With the popularization of mobile internet devices and the expansion of network scale, the scale of netizens has also expanded. As of December 2017, the number of Internet users in China has reached 772 million, and the penetration rate has reached 55.8% [1]. Social networking platforms have also become an indispensable part of people's lives, and people are used to getting information on these platforms and expressing their opinions. Among them, the openness, immediacy, interactivity and freedom of Weibo have made it the main platform for most people to publish public opinion. More and more government agencies and enterprises have begun to choose Weibo as the publishing platform for important information. Therefore, the information on Weibo contains a large number of social hotspots and the emotional tendencies of netizens. Through analyzing emotion of the content published by weibo users, the government can grasp the trend of public opinion, control the direction of public opinion, and rationally formulate and adjust relevant policies.

2 Background

The main research approaches of sentiment classification are emotional classification based on machine learning and logical classification based on semantic dictionary. In terms of semantic dictionaries, the commonly used sentiment dictionaries in China include Taiwan University NTUSD Simplified Chinese Emotion Dictionary, Hownet Emotion Dictionary, Dalian University of Technology Chinese Emotional Vocabulary Ontology Library, etc. Xu and others used primary school textbooks, film scripts,

© Springer Nature Switzerland AG 2019
Y. Tang et al. (Eds.): HCC 2018, LNCS 11354, pp. 222–231, 2019.
https://doi.org/10.1007/978-3-030-15127-0_23

literary periodicals and fairy tales as corpus sources to mark a large number of Chinese corpus, forming the "Emotional Vocabulary Ontology Library" [2]; On the basis of the seven emotions of the Chinese emotional vocabulary ontology library of Dalian University of Technology, Dun Xinhui enriched a kind of emotional "doubt" and constructed the emoji dictionary by using the point mutual information method, and also considered the influence of negative words and degree words on emotional expression. They used these to analyze the sentiment of weibo text [3]. Wiebe et al. added part-of-speech analysis on the basis of sentiment dictionary, and completed the subjective and objective classification of corpus based on Bootstrapping [4, 5]. In the field of machine learning, Mitchell et al. got the semantic word vector from the massive commentary corpus, and used the word vector to obtain the feature expression of the corresponding sentence or document through different semantic synthesis methods [6]. Zhang et al. used Naive Bayes (NB) and Support Vector Machine (SVM) classifiers to classify hotel reviews [7]. Jiang, He and others used SVM and EMCNN models for sentiment classification, and obtained the best known sentiment classification results on the NLPCC weibo emotional evaluation task data set [8, 9].

3   Construction of Emotional Classification Model

This paper proposes a weibo short text sentiment classification model based on word vector and XGBoost learning framework. First, we converted the weibo short text into a standardized text of the unified rule. Then we used word2vec to obtain the word vector of the microblog short text, and input the word vector into the sentiment classification model designed in this paper. We trained the XGBoost model to obtain the word vector features and classified its emotion. The specific process is shown in Fig. 1:

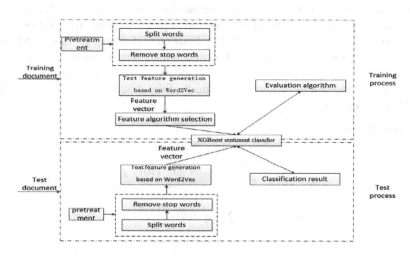

Fig. 1.

3.1   Acquisition of Word Vectors

Word2vec is a new word semantic computing technology proposed by Google in 2013. The basic idea is to map each word into a K-dimensional real number vector through training, and transform the processing problem of the text into the processing problem of the space word vector. It uses a three-layer neural network with an input layer-hidden layer-output layer. Word2vec contains two training models: CBOW and Skip-Gram. The CBOW model estimates the target word from the original sentence, while the Skip-Gram model does the opposite, and the original word is inferred from the target word. See Fig. 2 for details.

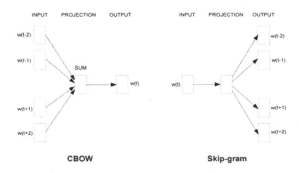

Fig. 2.

3.2   Construction of Learning Framework Based on XGBoost

XGBoost is an integrated learning framework. It's principle is to achieve accurate classification results by iteratively computing hundreds of tree models with low classification accuracy, and is capable of multi-threaded parallel computing. XGBoost uses Taylor second-order information for the loss function with higher precision and introduces regular terms besides the objective function. It obtains the optimal solution, and the gradient function is used to optimize the loss function in the iterative process to avoid over-fitting and improve the accuracy and speed of the model [11–13]. The algorithm principle is as follows:

(1) Objective function:

$$Obj = \sum_{i=1}^{n} l(y_i, \hat{y}_i) + \sum_{k=1}^{K} \Omega(f_k), f \in \mathbb{F}$$
$$\Omega(f)_t = \frac{1}{2}\lambda \sum_{j=1}^{T} \omega_j^2 + \gamma T$$

(1)

(2) Training objective function:

$$Obj^{(t)} = \sum_{i=1}^{n} l(y_i, \hat{y}_i^{(t-1)} + f_t(x_i)) + \Omega(f_t) + C \tag{2}$$

(3) Taylor second-order expansion of objective function:

$$Obj^{(t)} \approx \sum_{i=1}^{n} [l(y_i, y_i^{(t-1)}) + g_i f_t(x_i) + \frac{1}{2} h_i f_t^2(x_i)] + \Omega(f_t + C) \tag{3}$$

$$g_i = \partial_{\hat{y}^{(t-1)}} l(y_i, \hat{y}^{(t-1)}), \mathrm{h}_i = \partial^2_{\hat{y}^{(t-1)}} l(y_i, \hat{y}^{(t-1)}), \quad C \text{ is constant}$$

(4) Remove the constant term:

$$Obj^{(t)} \approx - \sum_{j=1}^{n} [g_i f_t(x_i) + \frac{1}{2} h_i f_t^2(x_i)] + \Omega(f_t) \tag{4}$$

(5) Find the optimal solution of the objective function:

$$Obj = -\frac{1}{2} \sum_{j=1}^{T} \frac{G_j^2}{H_J + \lambda} + \lambda T \tag{5}$$

It can be seen from the above steps that XGBoost performs pre-prune while optimizing the objective function, so that the optimal parameters can be obtained, and the prediction result is more accurate.

4 Experiment and Analysis

4.1 Adoption of Data Sets

This paper uses the data sets published by NLPCC in 2013 and 2014 as a word vector training corpus. First, the data set is denoised, that is, removes valueless content such as "[", "]", "#", "@", "//", tag symbols, and URL links. Then use jieba to segment the word, load the Chinese stop word dictionary, and remove the stop word processing. In order to fully test the performance of the designed sentiment classifier, this paper designs two experiments of subjective and objective emotion classification and emotional four-category experiment. Among the subjective and objective classification experiments, no emotion is considered as objective label, and happy, delight, sadness, evil, anger, fear, and surprise are considered as subjective labels. In the emotional four-category experiment, select four labels: happy, delight, sadness, and evil. The final data set is divided as shown in Tables 1 and 2.

Table 1. Partitioning with annotated data sets

	Objective	Subjective
2013	6701	12299
2014	10194	9804

Table 2. Partitioning with annotated data sets

	Happy	Delight	Sadness	Evil
2013	2153	3486	3129	2360
2014	3246	1900	1362	1781

4.2   Design of Evaluation Indicators

In order to fully evaluate the performance of the model constructed in this paper, the accuracy evaluation rate, recall rate, F1-Score, AUC and ROC curves are used as evaluation indicators to form the whole model performance evaluation system.

Accuracy rate measures the accuracy of sentiment classification; recall rate measures the success rate of sentiment classification. According to the relationship between the real emotions of the weibo short texts and the predicted emotions, the model prediction results can be represented by a confusion matrix. As shown in Table 3:

Table 3. Four cases of predictive results of sentiment classification models

Precision/Recall	Actual positive	Actual negative
Forecast positive	TP	FP
Forecast negative	FN	TN

The precision and recall rate are defined as:

$$Precision = \frac{TP}{TP + FN} \tag{6}$$

$$Recall = \frac{TP}{TP + FN} \tag{7}$$

The harmonic mean of the precision and recall rates is defined as:

$$F1-Score = \frac{2 \times Precision \times Recall}{Precision + Recall} \tag{8}$$

AUC (Area under Curve): The area under the Roc curve, and it's between 0.1 and 1. AUC can intuitively evaluate the quality of the classifier. The bigger the value, the better it is. First the AUC is a probability value. When you randomly select a positive sample and a negative sample, the probability that the current classification algorithm ranks the positive sample in front of the negative sample based on the calculated Score value is the AUC value. The larger the AUC value, the more likely the current classification algorithm is to place a positive sample in front of the negative sample, so that it can be better classified.

ROC curve is a comprehensive indicator reflecting the continuous variables of sensitivity and specific effects. It is a compositional method to reveal the relationship between sensitivity and specificity. It calculates a series of sensitivities and specificities by setting successive variables to different thresholds. The ROC curve is based on a series of different two-category methods (demarcation value or decision threshold), with the true case rate (sensitivity) (TPR) as the ordinate, and the false positive rate (1-special effect) (FPR) as the abscissa. The ROC observation model correctly identifies the trade-off between the proportion of the positive case and the proportion of the model incorrectly identifying the negative case data as a positive example. The area under the ROC curve is a measure of the accuracy of the model.

$$TPR = \frac{TP}{TP + FN} \tag{9}$$

$$FPR = \frac{FP}{FP + TN} \tag{10}$$

TPR represents: positive sample prediction result/positive sample actual number. FPR means: number of negative sample results predicted to be positive/negative sample actual number.

4.3 Comparative Experiment

This experiment uses the classic SVM model as a comparison of the designed sentiment classification model. According to the data set processed in Sect. 4.2, in the subjective and objective experiments, 20,000 short texts of processed microblogs were used as experimental data, 80% of them are used as training sets and 20% are used as test sets. SVM kernel function is rbf, parameter C is 0.8, parameter gamma is 20. And before the SVM model training, we perform PCA dimension reduction to 200-dimensional feature vectors on the extracted original feature vectors. This is equivalent to feature selection. In the emotional four-category experiment, 16,000 short texts of processed microblogs were used as experimental data. We also use 80% as a training set and 20% as a test set. Table 4 gives a comparison of the performance of the two models in subjective and objective sentiment classification. The parameter of the SVM is C, and the original feature vector is reduced to 100 dimensions. The remaining parameters are unchanged.

Table 4. Performance comparison of subjective and objective sentiment classification models.

	Precision	Recall	F1-Score	AUC	Training time
XGBoost	0.9698	0.9852	0.9706	0.9695	4025.89 (s)
SVM	0.7253	0.4994	0.6482	0.7284	282.02 (s)

It can be seen from Table 4 that the sentiment classification model designed in this paper is superior to the classical SVM sentiment classification model in the accuracy of sentiment prediction. However, the training of the model will take longer. Next, we analyze the performance of the two algorithms in more detail from the two aspects of confusion matrix and ROC curve.

Figure 3(a) shows the confusion matrix of the prediction results of the subjective and objective sentiment classification model based on XGBoost. We found that only 121 weibo short texts were misclassified in 4000 weibo short text test data. On the contrary, the classic SVM sentiment classification model has a large number of mis-classifications, as shown in Fig. 3(b). On the contrary, the classic SVM sentiment classification model has a large number of misclassifications, as shown in Fig. 3(b). Figure 4(a) shows the ROC curve of the prediction results of the subjective and objective sentiment classification model based on XGBoost. We can find that the AUC value of the subjective sentiment classification model based on XGBoost can reach 0.97, while the AUC value of the classic SVM sentiment classification model can only reach 0.73.

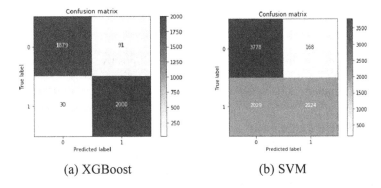

(a) XGBoost   (b) SVM

Fig. 3. Confusion matrix

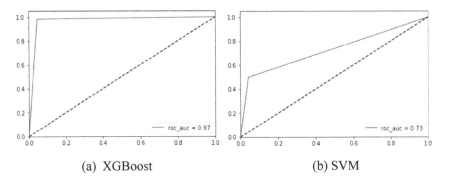

(a) XGBoost   (b) SVM

Fig. 4. ROC curve

Table 5. Performance comparison of emotional four-category experiment

	Accuracy	Training time
XGBoost	0.9956	1282.56 (s)
SVM	0.5625	52.40 (s)

In the emotional four-category experiment, the specific performance pairs are shown in Table 5:

It can be seen from Table 5 that the XGBoost-based classification model is still superior to the SVM-based classification model on the weibo short text sentiment multi-classification problem. Obviously, compared with the subjective and objective sentiment classification problem of weibo short text, SVM has a great decline in the accuracy of the emotional multi-classification problem. However, the sentiment classification model designed in this paper still maintains good performance. The experimental results show that the traditional SVM sentiment classification model can't deal with the problem of multi-classification well, and the XGBoost-based emotional multi-classification model can achieve good results. Similarly, we present the confusion matrix of two emotional multi-classification models, as follows:

Figure 5(a) shows the confusion matrix which presents the XGBoost-based sentiment four-class model prediction results. We found that only 14 weibo short texts were misclassified in 3200 weibo short text test data. On the contrary, the classic SVM emotional multi-classification model still shows a lot of misclassification, as shown in Fig. 5(b).

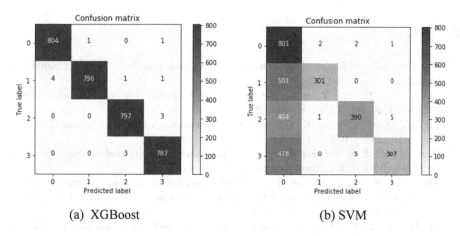

(a) XGBoost   (b) SVM

Fig. 5. Confusion matrix

The reason why the sentiment classification model designed in this paper can achieve better classification results is as follows. First of all, this paper makes full use of the published microblog corpus as the input of the word2vec word vector model. The word2vec word vector model itself has been proved to be able to obtain the contextual semantic relationship of the text well, so the feature extraction of the document vector is more reasonable. Secondly, the XGBoost learning framework is the most advanced classification learning model based on integrated learning in the field of machine learning. It has greatly improved the performance of the algorithm compared with the traditional machine learning model. However, due to the complexity of the XGBoost model, the training time required for the model is more than that of the traditional sentiment classification model.

5   Conclusion

The weibo short text contains a wealth of emotional information. By analyzing the short text of Weibo, it helps the government to control the public opinion. This paper proposes an emotion classification model based on XGBoost learning framework, which has achieved good classification results in both subjective and objective experiments and emotional four-category experiment. It effectively solves the problem of traditional classification methods neglecting word collocation and context, and improves the accuracy of sentiment classification. However, there is still no good solution to the emotional classification of more dimensions. In the future, we can deepen our research in this aspect.

Acknowledgement. Work described in this paper was funded by the National Natural Science Foundation of China under Grant No. 71671093. The authors would like to thank for the help of the college innovation team and other researchers at Nanjing University of Posts and Telecommunications.

References

1. The 41st Statistical Report on the Development in China. http://www.cnnic.net.cn/hlwfzyj/hlwxzbg/hlwtjbg/201803/t20180305_70249.htm. Accessed 5 Mar 2018
2. Xu, L., Ling, H., Pan, Y., et al.: The construction of emotional lexical ontology. J. China Soc. Sci. Tech. Inf. **27**(2), 180–185 (2008)
3. Guo, X., Zhang, Y., Yang, K.: Fine-grained sentiment analysis based on Weibo. Data Anal. Knowl. Discovery **1**(07), 61–72 (2017)
4. Wiebe, J., Riloff, E.: Finding mutual benefit between subjectivity analysis and information extraction. IEEE Trans. Affect. Comput. **2**(4), 175–191 (2012)
5. Riloff, E., Wiebe, J., Wilson, T.: Learning subjective nouns using extraction pattern bootstrapping. In: Conference on Natural Language Learning at HLT-NAACL, pp. 25–23. Association for Computational Linguistics (2003)
6. Mitchell, J., Lapata, M.: Composition in distributional models of semantics. Cogn. Sci. **34** (8), 1388–1429 (2010)
7. Ye, Q., Zhang, Z., Law, R.: Sentiment classification of online reviews to travel destinations by supervised machine learning approaches. Expert Syst. Appl. **36**(3), 6527–6535 (2009)
8. Jiang, F., Liu, Y., Luan, H., et al.: Micro-blog sentiment analysis with emotion space model. J. Comput. Sci. Technol. **30**(5), 1120–1129 (2015)
9. He, Y., Sun, S., Niu, F., Li, F.: A deep learning model of emotional semantic enhancement used in weibo emotion analysis. Chin. J. Comput. **40**(04), 773–790 (2017)

10. Mikolov, T., Chen, K., Corrado, G., et al.: Efficient estimation of word representations in vector space. Computer Science (2013)
11. Chen, T., Guestrin, C.: XGBoost: a scalable tree boosting system. In: ACM SIGKDD International Conference on Knowledge Discovery and Data Mining, pp. 785–794. ACM (2016)
12. Taylor, R.A., Moore, C.L., Cheung, K.H., et al.: Predicting urinary tract infections in the emergency department with machine learning. PLoS ONE **13**(3), e0194085 (2018)
13. Zhang, Z., Li, Y., Jin, S., et al.: Modulation signal recognition based on information entropy and ensemble learning. Entropy **20**(3), 198 (2018)
14. Liu, Y., Bi, J.W., Fan, Z.P.: A method for multi-class sentiment classification based on an improved one-vs-one (OVO) strategy and the support vector machine (SVM) algorithm. Inf. Sci. **394–395**, 38–52 (2017)
15. Liu, S., Li, F., Li, F., et al.: Adaptive co-training SVM for sentiment classification on tweets, pp. 2079–2088 (2013)

Research on Data Visualization in Different Scenarios

Xin Guo[1], Mingshu He[1(✉)], Mo Chen[2], Xiaojie Zhao[3], and Ye Tian[2]

[1] Electronic Engineering Institute,
Beijing University of Posts and Telecommunications, Beijing 100876, China
hemingshu@bupt.edu.cn
[2] Institute of Network Technology,
Beijing University of Posts and Telecommunications, Beijing 100876, China
[3] Information Science and Technology Institute,
Beijing Normal University, Beijing 100875, China

Abstract. With the development of computer science, the era of big data has arrived. Facing the new era and new challenges, the traditional analytical methods of problems in various fields have been unable to meet the needs. Data visualization is a rapidly developing discipline, it has significant advantages in analyzing problems, so data visualization shines in the era of big data. As people are very concerned about the fields of energy and environment, we choose to conduct data visualization studies in two areas, energy and the environment. According to the different characteristics of data in different fields, we propose targeted data visualization processes and design data visualization solutions. For energy data, we follow the process of data processing, visualization design, and data visualization. Based on the principle of high efficiency and intuitiveness, we add timeline and a combination of various charts to our design, and finally show a dynamic effect. We also propose a multi-dimensional visual mapping visualization scheme. The scheme can refine and enrich the visual results. For environmental data, we follow the process of goal analysis, data processing, visualization and analysis, the work shows the importance of visualization in information analysis and decision-making.

Keywords: Data visualization · Timeline · Multi-dimension ·
Decision-making · Energy · Environment

1 Introduction

1.1 Data Visualization

At the 2017 IEEE Pacific Visualization Conference, Ebert [1], a visual analysis expert from Purdue University, gave a keynote speech titled "Changing the World with Visual Analytics". In his speech, he pointed out that in order to solve the challenges in the world, we not only need to advance computer science and big data analysis, but also need a new analysis and decision-making environment. We must effectively combine human decision-making with advanced, guided analysis. And conduct human-computer cooperation discussions and decisions.

© Springer Nature Switzerland AG 2019
Y. Tang et al. (Eds.): HCC 2018, LNCS 11354, pp. 232–243, 2019.
https://doi.org/10.1007/978-3-030-15127-0_24

This method is data visualization. Data visualization uses intuitively graphical images to present data information in front of people's eyes, enabling people to effectively extract useful information from the original complex data, and perform data analysis in a more intuitive way. It can help humans find the correlation between data in order to make the right decision [2]. Data visualization analysis technology is an important method of big data analysis, which can help data analysts find the rules and patterns implied in data more quickly [3]. Data visualization technology needs the support of modern computer technologies, such as multimedia technologies, and mobile intelligent terminal technologies [4]. With the development of visualization tools, various kinds of visualization works tell us stories of various data [5]. In the field of visualization, science, technology and art are perfectly integrated. Data visualization mainly includes seven steps: acquiring data, analyzing data, filtering data, mining data, displaying data, summarizing data, and human-computer interaction [6].

1.2   Application Area

The advent of the computer age has injected new ideas for data visualization and provided performance capabilities and efficiency that cannot be achieved in the hand-painted era. Compared to various numbers, forms, texts and other information, the bright and intuitive graphical forms are more easily accepted. The science of data visualization was quickly applied to various industries.

The air pollution problem in China is becoming more and more serious. The air quality parameters can be acquired through the network. However, we still lack the visual display of air pollution data, lack the comprehensive chart of various parameter information that affects air quality over a period of time, and cannot display the distribution law of air pollution [7].

Based on the above, the data visualization work in this paper will focus on the following two areas: energy and the environment.

1.3   Our Work

This paper will basically follow the general steps of data visualization, plan the data visualization process for different application scenarios, and design a visualization scheme based on this scenario. This paper proposes visual design schemes for different data formats and presents them.

In the field of energy, as today's energy shortage, taking the path of sustainable development is an inevitable choice. Visual analysis of energy data is essential [8]. Li [9] and others proposed using the three-dimensional model CityGML to visualize urban energy consumption. Wang [10] and others proposed to visualize the urban power system by using GIS-based visualization. Both programs are based on geographic information, visualize and analyze energy data. For different emphasis, this paper focuses on the characteristics of energy data itself. To present multi-dimensional data, a visual mapping scheme for multi-dimensional data was proposed. In the case that geographic information is very important, according to the point made by Xu [11] and others, the map visualization technology has good geospatial features, which can ensure that various statistical information is better displayed. This paper will be combined with the map for visual analysis at the right time.

Brehmer [12] proposed that the timeline is often used to describe events. Events can be recorded as data. This paper combines the data and timeline in the research scenario and proposes a dynamic and efficient visualization solution.

In the field of environment, many experts and scholars at home and abroad have used statistical analysis methods on air quality, such as using Spearman rank correlation analysis. Compared to statistical analysis methods, the results of air quality visual analysis are easy to read and understand [13]. This paper will propose a different idea of visualization from the energy field. We will start from the analysis of targets, and then proceed with data processing and data visualization. This case illustrates that data visualization is not only used for the presentation of results, but also promotes research.

The chapters are arranged as follow. Section 2 introduces the visualization based on energy dataset. Section 3 introduces the visualization in environment area. Section 4 summarizes the achievements of this paper.

2   Energy

In the field of energy, energy consumption includes direct energy consumption and embodied energy consumption. Direct energy consumption refers to the energy consumed in the manufacture of products and services. Embodied energy consumption refers to the total amount of energy consumed in the process of production, transportation of products, services and their destruction [14]. At present, we have obtained terminal energy consumption, also called direct energy consumption, in 30 provinces and 6 industries over the past 8 years. We also have the input-output table, which is the embodied energy consumption among provinces in monetary expression. For the energy data, the data will be visualized according to the data processing, visualization design, visualization process.

2.1   Data Processing

From the energy balance sheet, the direct energy consumption of each of the 6 industries in 30 provinces was obtained. The energy consumption table shows the consumption of various energy sources. In order to analyze the direct energy consumption among provinces and industries, we should have a unified scale. That means, the various types of energy consumption should be uniformly converted to standard coal consumption by using the following formula.

$$d_i = \sum_k C_{ki} \times d_{ki} \tag{1}$$

where d_i represents the direct energy consumption of department i, expressed as standard coal consumption. C_{ki} represents the conversion factor of the kth energy consumption converted to standard coal. d_{ki} represents the direct energy consumption of department i's kth kind energy source.

Based on the principle of ecological input-output table, the embodied energy flow between industries is calculated. According to the principle of conservation of energy, embodied energy and direct energy flowing into an industry follow the following relationship, as shown in the Fig. 1.

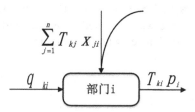

Fig. 1. Direct energy and embodied energy flowing

where q_{ki} indicates the direct consumption of the k kind energy in department i. x_{ji} denotes the intermediate input of department j to department i in the input-output table, which is the monetary expression of embodied energy. T_{kj} represents the kth energy consumption intensity of department j and is defined as the k energy amount contained in the unit output produced by department j. Therefore, $T_{kj} \times x_{ji}$ represents the amount of k energy implicit in department j's money flow to department i. p_i is the total output of department i. $T_{ki}p_i$ represents the k energy consumption contained in the total output of department i. According to the law of conservation of energy, the total input of the k energy of department i should be equal to the total output of the k energy. The formula is as follow.

$$q_{ki} + \sum\nolimits_{j=1}^{n} T_{kj} \times x_{ji} = T_{ki}p_i \qquad (2)$$

Based on the above, the implicit consumption intensity of the k energy in department j can be obtained on the basis of the economic input-output table and the energy balance table. So, we can obtain the amount of embodied energy that flow between all the departments by combing with input-output table. Similar to the problems faced by direct energy consumption, in order to have a unified measurement scale, conversion factors are used to convert various types of energy consumption into standard coal consumption and then sum up. So far, we get results similar to the input-output table, named the embodied energy table. This table shows the energy flow relationship that is expressed in standard coal.

The relationship of energy often reflects the degree of economic development between cities. We divide the provinces into several business circle by clustering algorithms. The four business circles are the Hebei Business Circle, the Guangdong Business Circle, the Anhui Business Circle and the Shaanxi Business Circle.

2.2 Visualization Design

Timeline. In the face of data visualization in a series of time, we first consider the dynamic display combined with the time axis.

In the dynamic visualization of energy data, histograms are favored for their intuitive and efficient features. In the dynamic display designing, the energy data itself is taken into consideration, and geographical information is weakened. The energy consumption of various industries in each province changes with timeline of the year through the histogram. At the same time, in order to analyze the changes of direct energy consumption from the perspective of the industry of the whole country, a multi-diagram linkage design scheme was proposed. The pie chart and the bar chart were simultaneously displayed and changes with time like the Fig. 2.

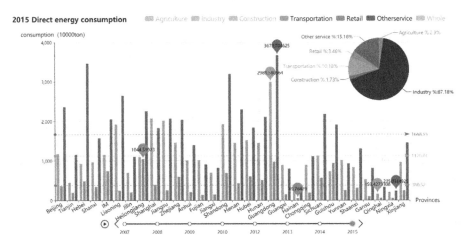

Fig. 2. Data visualization based on timeline

Combination of Multiple Charts. Multi-dimensional data information can be displayed through a variety of display methods combined. Like Fig. 3.

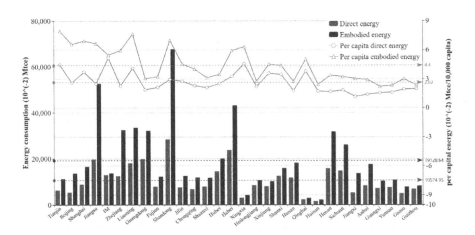

Fig. 3. Combination of bar and line charts

2.3   Multi-dimensional Visual Mapping

The visualization of bar and line charts is relatively simple. They are intuitive and efficient in some limited data. However, for data with multiple dimensions, the data visualization based on several axes only appears to be thin. The following will introduce the idea of multi-dimensional data information in visual mapping.

Scatterplot Multidimensional Visual Mapping. Although the scatter plot still only has two data axes, multi-dimensional information can be displayed in a scatter plot through visual mapping.

The Fig. 4 shows an example. The horizontal axis represents the embodied energy consumption, and the vertical axis represents the direct energy consumption. The size of the scattered dots can represent the level of GDP per capita. The labels of the scattered dots can distinguish among provinces, even the colors of scattered dots can also be used to display population density or other information (the scatter color in the Fig. 4 is only used to distinguish provinces). The auxiliary lines in the scatter plot can help users to quickly determined the level of the energy consumption of a province compared with the national average. And it's easy to see the proportion of direct energy consumption and embodied energy consumption in a province. Multi-dimensional information is mapped to a variety of visual information, then enrich the visual information.

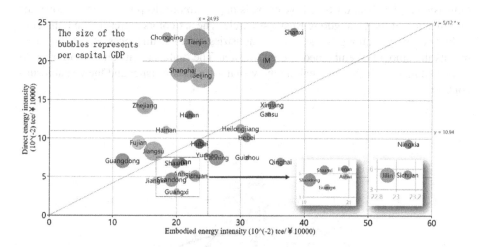

Fig. 4.   Scatter plot multi-dimensional visual mapping (Color figure online)

Map Multi-dimensional Visual Mapping. The same idea can be applied to map-based visualization.

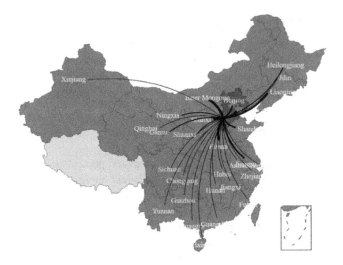

Fig. 5. Map1 multi-dimensional visual mapping

Figure 5 shows the country's embodied energy transfer. The thickness of the line represents the amount of embodied energy transfer.

In the visualization of the business circle obtained by clustering, the multi-dimensional visualization based on maps is applied more deeply. Figure 6 shows the visualization of the Guangdong Business Circle. The six colors represent the transfer of embodied energy among the six major industries, and the thickness of the lines represents the amount of embodied energy transfer. The multi-dimensional information is displayed in a single picture, making the visual information richer and the visual results concise.

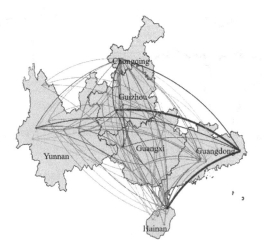

Fig. 6. Map2 multi-dimensional visual mapping (Color figure online)

3   Environmental Data on Air Pollution

3.1   Target Analysis

At present, environmental monitoring data of some environmental monitoring sites in Beijing, Hebei, Henan, Shandong, Shanxi and so on have been obtained during a certain period of time.

The environmental monitoring data includes the monitoring time, the name of the monitoring site, the concentration of various environmental pollutants and so on. It is a data set with both time and geographic information, and has a variety of data.

The visualization method can be applied to the following three stages in the analysis of environmental data: the early stage (the stage of collating, filtering and cleaning of information), the intermediate stage (information analysis activity), and the later stage (display of information analysis results). Before carrying out these three stages, the target analysis was performed on the PM2.5 data. It can help us find the ways of analyzing the internal laws. The specific process is shown in Fig. 7.

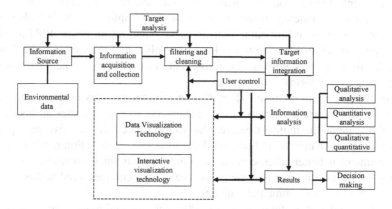

Fig. 7. Process of environmental data visualization

In this paper, the PM2.5 data from the environment monitoring is dynamically visualized to qualitatively analyze the trends of PM2.5 during this period of time. At the same time, we will quantitatively analyze the correlation of PM2.5 among regions and the relationship between the correlation and regional distance. We visual the results in order to get a prediction. We visualize a variety of environmental data in the form of multiple axes, qualitatively and quantitatively analyze the data.

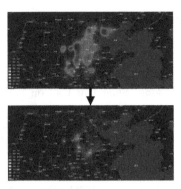

Fig. 8. Visualization of original PM2.5 data

3.2   Data Processing

In the original environment data, there is a considerable portion of redundant data, including no information data, duplicate data, contradictory data, and so on. This part of data not only drastically increases the amount of information, but also brings unnecessary trouble to information analysis. The visualization of the PM2.5 original data can be very obvious to show this kind of trouble, and through visualization, we can quickly discover this problem. Therefore, data visualization can help users quickly find a problem or abnormal situation.

From the visualization result (Fig. 8), the monitoring sites at each time in the original data have certain differences, that means, the monitoring sites at this moment are not the same as the moment before. Due to this defect, the intuitive perception of the environmental change trend becomes deceptive. The visualization result shows that the environment is better when compared to the previous time, however, this conclusion may because of the fact that many sites have not been measured at this time, not because of the real environmental changes.

We filter the data at different times of 200 environmental monitoring stations from the original data. It includes monitoring time, monitoring locations, and environmental pollutant values. For situations that with timeline and geographic information, a map-based timeline visualization scheme is proposed. In order to achieve the dynamic qualitative analysis in the target analysis and the quantitative analysis of the correlation of sites, the following processing and expansion of the data is still required.

We obtained the latitude and longitude coordinate information based on Baidu coordinate system to achieve dynamic visualization analysis with maps qualitatively. The linear distance between sites is calculated from latitude and longitude information by using the Haversine formula to analysis the relationship between the PM2.5 correlations and the sites' straight-line distances quantitatively.

3.3   Visual Display

A part of the timeline visualization based on the map is shown as Fig. 9.

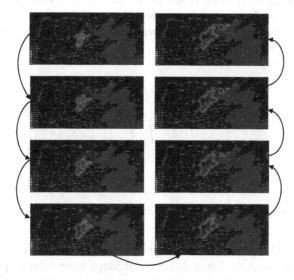

Fig. 9. Dynamic visualization based on map and timeline

Through this dynamic visualization combined with geographical location, we can easily qualitatively analyze the regularity of PM2.5 data. Dynamic visualization combined timeline with geographic information is an efficient and clear visualization solution. We discovered the law of change and further explored the underlying reasons.

We visualize the correlation between a part of the sites as Fig. 10.

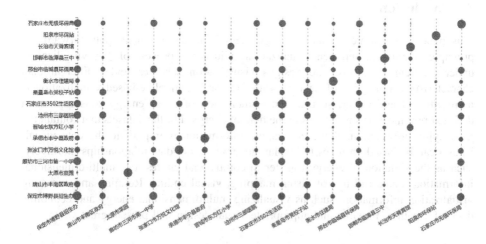

Fig. 10. Correlation between several stations (Color figure online)

In Fig. 10, the correlation between sites is mapped to the size of red circles so that the degree of correlation between a station and another station can be intuitively obtained. We can see the size are different. So, the correlation data visualization also brings us another question: what is the reason for this and what is the relationship between correlation and distance of stations.

In order to further solve this problem, we visualize the correlation coefficients and linear distances in one picture as Fig. 11.

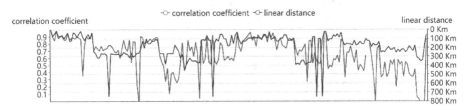

Fig. 11. Visualization based on correlation coefficients and linear distances (Color figure online)

From the Fig. 11, the red line that represents the correlation coefficients and the green line that represents the linear distances, although have different positions of the peaks, have the same trend. The visualization results indicate that most sites follow a mathematical rule that the distance between them is longer and the correlation coefficient is smaller.

Thus, in information analysis activities, visualization is not only used as the last step before the analysis of laws, or as a display of research results, but also a key tool to promote the analysis activities. Starting from the information which is presented by visualization on complicated data, new issues are further proposed to promote research.

4   Conclusion

This paper concentrate on the areas of energy and environment, puts forward the principle of efficient and rich visualization, adjusts the process of data visualization under different scenarios, and designs visualization schemes under different data characteristics. For types of datasets with time series, timeline-based visualization is more efficient and richer, such as the visualization of direct energy consumption in different provinces and industries in the field of energy. For the datasets that contain the geographic information, if the geographic information is important, the data visualization can be based on maps to display geographic and data relationships intuitively, such as the situation of energy transfer in commercial circles. For multi-dimensional information, make full use of visual mapping, visual coding. Refining and enriching visualization information, such as visualizing multidimensional energy and economic data information. Flexible and appropriate visual design will present great value.

Acknowledgments. This research is funded by The National Science Foundation of China (61601053) and National Natural Science Foundation of China (Grant No. 61602051) and the Fundamental Research Funds for the Central Universities under Grant 2017RC11.

References

1. Ebert, D.: Changing the world with visual analytics. In: 2017 IEEE Pacific Visualization Symposium, Seoul, South Korea, p. 8. IEEE Computer Society (2017)
2. Chang, Y.: The data visualization method in the era of big data. Electron. Technol. Softw. Eng. **2018**(5), 156 (2018)
3. Qiao, S., Nurbol, Su, R.: Map analysis for research status and development trend of data visual analysis. Mod. Electron. Tech. **41**(14), 161–169 (2018)
4. Chang, Y.: Visual analysis of web data under big data background. China CIO News **2018**(5), 148–150 (2018)
5. Lee, B., Riche, N.H., Isenberg, P., Carpendale, S.: More than telling a story: transforming data into visually shared stories. IEEE Comput. Graphics Appl. **35**(5), 84–90 (2015)
6. Dai, S., Dong, J., Xue, J.: Big data visualization analysis and application in scientific computing. J. Eng. Stud. **6**(3), 275–281 (2014)
7. Ma, C., Li, L., Xue, W.: Research on visualization of air pollution characteristics and distribution. J. Northwest. Polytechnical Univ. **35**(6), 1073–1078 (2017)
8. Chen, S.: Analysis of the evolution of energy consumption structure in American four states. Chem. Enterp. Manag. **2018**(11), 30–32 (2018)
9. Li, B., Mao, B.: Energy consumption analysis and visualization based on CityGML-a case study of Swedish Smart City. Geomatics World **24**(4), 48–52 (2017)
10. Wang, Y., Yu, J., Wu, F.: Data analysis and visualization based on geographic information in the background of Global Energy Internet. Electr. Power Inf. Commun. Technol. **14**(3), 49–54 (2016)
11. Xu, N., Luo, J.: Statistical analysis of data map visualization. Technol. Innov. Appl. **2018**(22), 67–68 (2018)
12. Brehmer, M., Lee, B., Bach, B., Riche, N.H., Munzner, T.: Timelines revisited: a design space and considerations for expressive storytelling. IEEE Trans. Visual Comput. Graphics **23**(9), 2151–2164 (2017)
13. Wang, R., Zhou, M., Wang, Y., Liu, Y.: Mapping spatial and temporal patterns of air condition in Northeast China. Bull. Surveying Mapp. **2017**(8), 88–91 (2017)
14. An, Q., An, H., Wang, L.: Analysis of embodied energy flow network between Chinese industries. J. Syst. Eng. **29**(6), 754–762 (2014)

Model Checking for Turn-Based Probability Epistemic Game Structure

Guocheng Zheng[1], Pin Wang[1], Xinyu Zhang[1], Chuan Zhang[1],
Mengao Li[1], Yuxiao Yang[2], and Weifeng Xu[2(✉)]

[1] Systems Engineering Research Institute, Beijing, China
zheng_guocheng@163.com, wangpin20@163.com
[2] North China Electric Power University, Baoding 071003, China
yangyuxiao_ncepu@outlook.com, weifengxu@163.com

Abstract. In this paper, a *turn-based probability epistemic game structure* (TPEGS) is proposed to model knowledge preconditions for actions of system and environment firstly, which is an extension of *turn-based synchronous game structures* with probabilistic transition. Secondly, we introduce probability operator $P_{\sim \lambda}$ into *alternating temporal epistemic logic* (ATEL) and define *turn-based probability alternating-time temporal epistemic logic* (tPATEL) for model checking the properties of TPEGS quantitatively. The probability of agents knowing some precondition before they implement an action can be expressed in tPATEL. Thirdly, we propose a method to compute probability for model checking verification problems of tPATEL based on DTMC and CTMC, and then analyze the time complexity of the method. Then, we are able to convert a part of tPATEL verification problems into the PATL ones by defining the knowledge formula $K_a\phi$, $E_{As}\phi$ and $C_{As}\phi$ as atomic propositions. Finally, we study a flight procedure in STAS using PRISM-games to demonstrate the applicability of the above model checking framework and expand the application field of model checking.

Keywords: Quantitative model checking · Game structure ·
Knowledge precondition · Flight procedure

1 Introduction

Recently, many scholars with different academic backgrounds have studied on how to make knowledge reconditions of actions from different angles. In AI planning, a major ongoing research problem is that of correctly formulating knowledge preconditions for actions and plans (Allen 1991), and alternating-time temporal epistemic logic (ATEL) (Van der Hoek 2003) was defined to express the properties such as "group *As* can cooperate to bring about ϕ iff it is common knowledge in *As* that ψ" in model checking.

In fact, it is not enough to describe people's cognition in our possible world only depending on "true" or "false", therefore probability plays an important role. In this paper, we will study the knowledge precondition for actions in quantitative model checking.

© Springer Nature Switzerland AG 2019
Y. Tang et al. (Eds.): HCC 2018, LNCS 11354, pp. 244–256, 2019.
https://doi.org/10.1007/978-3-030-15127-0_25

Ubiquitous computing is a vision of computing in which the computer disappears from view and becomes embedded in our environment, in the equipment we use, in our clothes, and even in our body (Kwiatkowska 2013). The unprecedented dependence on ubiquitous computing creates an urgent need for modelling and verification technologies to support the design process, which may improve the reliability and reduce production costs. The researchers did a lot of work to cater for ubiquitous computing, such as The VERIWARE project at the Department of Computer Science, University of Oxford. In quantitative modelling and verification approaches, probabilistic temporal logic e.g., PCTL, PTCTL and rPATL is usually used to express properties of states or paths in probabilistic system model. Recently, probability logics have successfully been applied to express properties of probabilistic system models in a variety fields successfully, from Autonomous Urban Driving to Microgrid Demand Management, (Chen et al. 2012, 2013), to DNA Computing (Dannenberg et al. 2013a, b). From the above, we will incorporate flight procedure checking in the application.

2   Preliminaries

In this section, we explain some preliminary concepts including discrete-time Markov chains (DTMC) and Continuous Markov chain (CTMC).

2.1   DTMC and CTMC

Firstly, we recall the definition of discrete-time Markov chains (DTMC) and Continuous Markov chain (CTMC), which both will be used to help us to define the new stochastic system more precisely. In the following sections, Dist(Y) is defined to represent the set of discrete probability distributions over finite set Y.

Definition 2.1. (DTMC) A Discrete-Time Markov Chain is a tuple DTMC = (S, P, L) where:

- S is a non-empty set of states;
- $P : S \times S \to [0, 1]$ is a probability transition matrix, such that $\sum_{s' \in S} P(s, s') = 1$ for all state $s' \in S$;
- $L : S \to 2^{AP}$ is a labeling function mapping each state to a set of atomic propositions, AP is a finite set of atomic propositions.

$$P_s(\omega_{\text{fin}}) \stackrel{\text{def}}{=} \begin{cases} 1 & \text{if } n = 0 \\ P(\omega_{\text{fin}}(0), \omega_{\text{fin}}(1)) * \ldots * P(\omega_{\text{fin}}(n-1), \omega_{\text{fin}}(n)) & \text{otherwise} \end{cases} \quad (1)$$

$Path_{ful}(s)$ is the set of all infinite paths starting in state s. Next, we define the cylinder of a finite path ω_{fin} as: $C(\omega_{\text{fin}}) \stackrel{\text{def}}{=} \{\omega \in Path_{ful}(s) | \omega_{\text{fin}} \text{ is a predix of } \omega\}$, and let $\sum_s$ be the smallest σ-algebra on $Path_{ful}(s)$ which contains the cylinder $C(\omega_{\text{fin}})$ for $\omega_{\text{fin}} \in Path_{\text{fin}}(s)$. Finally, $Prob_s$ on $\sum_s$ is defined as the unique measure such that $Prob_s(C(\omega_{\text{fin}})) = P_s(\omega_{\text{fin}})$ for all $\omega_{\text{fin}} \in Path_{\text{fin}}(s)$. For further details, see the paper (Kwiatkowski 2007).

Definition 2.2. (CTMC) A Continuous-Time Markov Chain (*CTMC*) is a tuple $\mathcal{C} = (S, \alpha, P, E, \mathcal{L})$ where:

- S is a finite, non-empty set of states;
- $\alpha \in \text{Dist}(S)$ is the initial distribution;
- $P : S \times S \rightarrow [0, 1]$ a probability matrix;
- $E : S \rightarrow \mathbb{R}_{\geq 0}$ is the exit rate function;
- $\mathcal{L} : S \rightarrow 2^{AP}$ is a labeling function mapping each state to a set of atomic propositions, AP is a finite set of atomic propositions.

$C(s_0, I_0, \ldots, I_{k-1}, s_k)$ denotes the cylinder set. The probability measure $Prob^{\mathcal{C}}$ on $Path(s_0)$, which is the smallest σ-algebra on $Path(s_0)$, is the unique measure defined by induction on k by $Prob^{\mathcal{C}}(C(s_0)) = \alpha(s_0)$ and for $k > 0$:

$$Prob^{\mathcal{C}}(C(s_0, I_0, \ldots, I_{k-1}, s_k)) = Prob^{\mathcal{C}}(C(s_0, I_0, \ldots, I_{k-2}, s_{k-1}))$$
$$* \int_0^{I_{k-1}} P(s_{k-1}, s_k) E(s_{k-1}) e^{-E(s_{k-1})\tau} d\tau \qquad (2)$$

In general, computing the probability of a cylinder set with k interval $I_0, \ldots, I_{k-1}$ reduces to calculate k integral over $I_0, \ldots, I_{k-1}$ (Baier et al. 1999).

3   The Role of Probability in Knowledge

There is a close link between implementing actions and their relevant knowledge. In many domains, some relevant knowledge is required in order to do actions properly, i.e. the knowledge is the preconditions of actions. A major ongoing research problem in ai planning is that of correctly formulating knowledge preconditions for actions and plans.

In some complex domains, it is very difficult or even impossible for a agent to have perfectly grasp the knowledge preconditions before implementing a action. In ATEL, this problem is expressed as the formula:

$$\langle\langle\text{professor}\rangle\rangle(K_{\text{professor}}(\text{communicate}) \cup \text{sendEmail}) \qquad (3)$$

Example 3.1. In the current WAN, if a professor in Jilin University must have known that the network is no problem before he sends an e-mail to the other professor in Oxford University, which can be expressed in ATEL as the formula:

$$\langle\langle\text{professor}\rangle\rangle(K_{\text{professor}}(\text{communicate}) \cup \text{sendEmail}) \qquad (4)$$

Example 3.2. An earthquake rescue team must have known that they need 10 types of relief supplies before they get into the disaster, which can be expressed as the formula:

$$\langle\langle volunteers\rangle\rangle(K_{volunteer}(RStypes = 10 \cup Access)) \tag{5}$$

Example 3.3. Let us analyze the Coordinated Attack problem (Gray 1978). Two generals A and B must decide whether to attack a common enemy, but we require that any attach be a coordinated attack: that is, A attacks iff B attacks and B attacks iff A attacks, which can be expressed as the formula:

$$\langle\langle A, B\rangle\rangle((K_A(Battack) \wedge K_B(Aattack)) \cup Attack) \tag{6}$$

However, the messengers may be captured by the enemy. So it is impossible for the generals to coordinate an attack under such conditions (Halpern and Moses 1990), i.e. the formula (6) is unsatisfiable in practice. Halpern and Tuttle (1993) relax this condition and require that if the generals can communicate with high probability they coordinate their attack, and new condition is that "A attacks iff B attacks and B attacks iff A attacks" holds with high probability 0.99, the knowledge of which can be expressed as the formula:

$$K_A(P_{\geq 0.99}(Battack)) \wedge K_B(P_{\geq 0.99}(Aattack)) \tag{7}$$

Similarly, we can relax the conditions for *Examples* 3.1, 3.2 respectively as follows:

$$K_{professor}(P_{\geq 0.9}communicate) \tag{8}$$

$$K_{volunteer}(P_{\geq 0.8}(Materialtypes = 10)) \tag{9}$$

The cooperation modality $\langle\langle As\rangle\rangle$ in ATL is used to express cooperation concept. Based on ATL, we can express the relation between the system and the environment in the game. And probabilistic–epistemic logic (PCTLK) was defined to specify epistemic multi-agent systems. However, there is the cooperation operator $\langle\langle\rangle\rangle$ in the formula (3), (4), (5) defined in ATEL, pre-existing logics can not be used directly in the probabilistic knowledge precondition for actions, which is the key problem that we will study in this paper.

4 Probability Epistemic Game Structures

4.1 Probability Epistemic Game Structure

Definition 4.1. A probability epistemic game structure (PEGS) is a tuple $M_{PEGS} = (\Lambda, S, A, \sim_{ag}, TStep, \mathcal{L})$, where:

- Λ is a finite, non-empty set of agents;
- S is a finite, non-empty set of states;
- A is a finite, non-empty set of actions;
- $\sim_{ag} \subseteq S \times S$ is an epistemic accessibility relation for each agent $ag \in \Lambda$.

- *TStep* : $S \times \Lambda \times (A \cup \mathbb{R}_{\geq 0}) \rightarrow \text{Dist}(S)$ is a probabilistic transition function, such that $A \cap \mathbb{R}_{\geq 0} = \varnothing$, $\theta \in \text{IDist}(S)$ is a point distribution, if $a \in A$, *TStep* is called discrete transition function, or $a \in \mathbb{R}_{\geq 0}$, *TStep* is continuous transition function *TStep*(s, Λ, a), where $s \in S, a \in A \cup \mathbb{R}_{\geq 0}$, is probability distributions over the set $S_1 \subseteq S$ which is chosed by every agent $ag \in \Lambda$ with action $a \in A \cup \mathbb{R}_{\geq 0}$.
- $\mathcal{L} : S \rightarrow 2^{AP}$ is a labeling function mapping each state to a set of atomic propositions, AP is a finite set of atomic propositions.

The agent $ag \in \Lambda$ of the PEGS M_{PEGS} is for all kinds of objects in the world, e.g. vehicle, organization, system, software process, flight procedures even people etc. For every state $s \in S$ of the PEGS M_{PEGS}, the set available actions is denoted by $A(s) \stackrel{\text{def}}{=} \{a \in A | TStep(s, \Lambda, a) \neq \perp\}$. We assume that $A(s) \neq \varnothing$ for all $s \in S$. A probabilistic transition $s \xrightarrow{a,\theta} s'$ is made by every agent $ag \in \Lambda$ and a state s based on non-deterministically selecting a distribution *TStep* $\in \text{IDist}(S)$. According to *TStep* (that $1 \geq TStep(s, \Lambda, a) > 0$), agents choose the successor state s'.

There are two same ways in which PEGS's computation may be repressed as that proposed for PS. A path $\omega = s_0 \xrightarrow{a_0,\theta_0} s_1 \xrightarrow{a_1,\theta_1} s_2 \xrightarrow{a_2,\theta_2} \cdots (\theta(s) = TStep(s, \Lambda, a))$ is a non-empty finite or infinite sequence of transition in structures, and represents nondeterminism and stochastic behavior of the transition. $|s_0 s_1 s_2 \cdots s_n|$ is used to denote the length of finite path e.g., the length of path $\omega = s_0 \xrightarrow{a_0,\theta_0} s_1 \xrightarrow{a_1,\theta_1} s_2 \xrightarrow{a_2,\theta_2} \cdots s_2$ is n which is denoted by $|s_0 s_1 s_2 \cdots s_n| = n$, and the length of an infinite path $|s_0 s_1 s_2 \cdots|$ is $+\infty$. st_ω is defined to denote $s_0 \rightarrow s_1 \rightarrow s_2 \rightarrow \cdots$, and $st_\omega(i)$ for s_i. $last(st_\omega)$ is the last state of a finite path ω. A prefix of an infinite path ω is denoted by a finite path ω_{fin} of length n if $st_{\omega fin}(i) = st_\omega(i)$ for $0 \leq i \leq n$. ω_i is denoted the suffix of the path ω starting in state s_i, e.g. $s_3 \xrightarrow{a_3,\theta_3} s_4 \xrightarrow{a_4,\theta_4} \cdots$. $Path_{fin}(M)$ is the set of all finite paths in M and $Path_{fin}(s)$ is the set of finite paths which starts in state s. Similarly, $Path_{ful}(M)$ and $Path_{ful}(s)$ are denoted to represent the set of infinite paths respectively.

Coexisting with a path, we use an adversary to resolve the non-determinism problem of the structure. Formally, an adversary Ad is defined as a function σ that maps the finite path $st_{\omega fin}$ to a distribution θ. By an adversary Ad, for every path $\omega \cdot s \in Path_{fin}(M_{PEGS})$ where $s \in S$, denotes a discrete probability distribution $\sigma(last(st_\omega), s)$ over $A(s)$. In this paper, AdvPEGS is used to denote the set of adversaries of the PEGS, which is called memoryless if $\sigma(last(st_\omega), s) = \sigma(last(st_{\omega'}), s)$ for all paths $\omega \cdot s, \omega' \cdot s \in Path_{fin}(M_{PEGS})$, and deterministic if $\sigma(last(st_\omega), s)$ is a Dirac distribution for all $\omega \cdot s \in Path_{fin}(M_{PEGS})$.

4.2   Turn-Based Probability Epistemic Game Structure

Definition 4.2. A turn-based probability epistemic game structure (TPEGS) is a tuple $M_{TPEGS} = \langle \Lambda, S, A, S_{r:(r \in \Lambda)}, \sim_{ag}, TStep, \mathcal{L} \rangle$, where:

Λ is a finite, non-empty set of agents;
S is a finite, non-empty set of states;
A is a finite, non-empty set of actions;

- $\sim_{ag} \subseteq S \times S$ is an *epistemic accessibility relation* for each agent $ag \in \Lambda$.
- $S_{r:(r \in \Lambda)}$, is a set of states of agent r and a subset of S;
- $TStep : S_r \times \Lambda \times (A \cup \mathbb{R}_{\geq 0}) \rightarrow Dist(S)$ is a probabilistic transition function, such that $A \cap \mathbb{R}_{\geq 0} = \emptyset$, $\theta \in \mathbb{I}Dist(S)$ is a point distribution, if $a \in A$, $TStep$ is called discrete transition function, or $a \in \mathbb{R}_{\geq 0}$, $TStep$ is continuous transition function $TStep(s, r, a)$, where $s \in S_r, r \in \Lambda, a \in A \cup \mathbb{R}_{\geq 0}$, is probability distributions over the set $S_r \subseteq S$ which is chosed by a agent $r \in \Lambda$ with action $a \in A \cup \mathbb{R}_{\geq 0}$.
- $\mathcal{L} : S \rightarrow 2^{AP}$ is a labeling function mapping each state to a set of atomic propositions, AP is a finite set of atomic propositions.

The probability measure $Prob_s^{Ad}$ over $Path_{ful}^{Ad}(s)$ is defined for an adversary $Ad \in Adv_{TPEGS}$. Further, for a TPEGS and state $s \in S$, under a given adversary Ad, if $a \in A$, the behaviour from state s can be described with the countable infinite DTMC (Kwiatkowska 2007): $DTMC_s^{Ad} = \left(S_s^{Ad}, P_s^{Ad}, \mathcal{L}_s^{Ad}\right)$ where $S_s^{Ad} = Path_{fin}^{Ad}(s)$, for any finite path $\omega_{fin}, \omega'_{fin} \in S_s^{Ad}$:

$$P_s^{Ad}\left(\omega_{fin}, \omega'_{fin}\right) = \begin{cases} \theta(s') & \text{if } \omega'_{fin} \text{ is of the form } \omega_{fin} \xrightarrow{a,\theta} s' \text{ and } \sigma(last(\omega_{fin}), s) = \theta(s) \\ 0 & \text{otherwise} \end{cases} \quad (10)$$

and $\mathcal{L}_s^{Ad}\left(\omega_{fin}\right) = \mathcal{L}\left(last(\omega_{fin})\right)$ for each $\omega_{fin} \in S_s^{Ad}$. Similarly, for $a \in \mathbb{R}_{\geq 0}$, we use countable infinite CTMC: $CTMC_s^{Ad} = \left(S_s^{Ad}, \alpha, P_s^{Ad}, E, \mathcal{L}_s^{Ad}\right)$. where $S_s^{Ad} = Path_{fin}^{Ad}(s)$, for any finite path $\omega_{fin}, \omega'_{fin} \in S_s^{Ad}$, and $\mathcal{L}_s^{Ad}\left(\omega_{fin}\right) = \mathcal{L}\left(last(\omega_{fin})\right)$, and $Prob^C(C(s_0)) = \alpha(s_0)$. $P_s^{Ad} : S_s^{Ad} \times S_s^{Ad} \rightarrow \mathbb{I}Dist\left(S_s^{Ad}\right)$ is a probabilistic transition function:

$$P_s^{Ad}\left(\omega'_{fin}\right) = \begin{cases} P_s^{Ad}(\omega_{fin}) * \theta(s') * \left(e^{-E(s)*infD(a)} - e^{-E(s)*supD(a)}\right) & \text{if } \omega'_{fin} \\ \quad \text{is of the form } \omega_{fin} \xrightarrow{a,\theta} s' \text{ and } \sigma_r\left(last(\omega_{fin}), s\right) = \theta(s). \\ 0 & \text{otherwise} \end{cases} \quad (11)$$

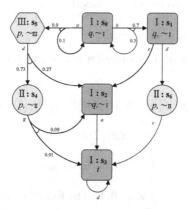

Fig. 1. Example of TPEGS

Therefore, there is a one-to-one mapping from the paths of $DTMC_s^{Ad}$ or $CTMC_s^{Ad}$ to the paths of $Path_{ful}^{Ad}(s)$. Hence it is reasonable to use the classic path probability calculation construction represented in DTMC or CTMC to define a probability measure $Prob_s^{Ad}$ over $Path_{ful}^{Ad}(s)$. Next, we give an example to describe more clearly.

Example 5.1. In the Fig. 1, there is an example of TPEGS M_{TPEGS}. In the example, the player set is $\Lambda = \{I, II, III\}$, the agent r in a state s is shown as $r : s$, i.e. $S_I = \{s_0, s_1, s_2, s_3\}$, $S_{II} = \{s_4, s_6\}$, $S_{III} = \{s_5\}$, p, q, t are atomic propositions. We have transition set$\{a, b, c, d, g\}$. Consider finite path, $\omega = s_0 \xrightarrow{a,\theta_0} s_5 \xrightarrow{g,\theta_1} s_4 \xrightarrow{c,\theta_2} s_3$, $a \in \mathbb{R}_{\geq 0}$, $\{c, d, g\} \subset A$, $\{(s_0, s_0), (s_0, s_1), (s_0, s_2), (s_1, s_1), (s_1, s_2), (s_2, s_2)\} \subseteq \sim_I$, $\{(s_4, s_4), (s_4, s_6), (s_6, s_6)\} \subseteq \sim_{II}$, $\{(s_5, s_5)\} \subseteq \sim_{III}$. We still assume that, $infD(a) = 0$, $supD(a) = 1, E(s_0) = v_0 = 10$, the initial probability $\alpha(s_0) = 1$ and player actions (transitions) are selected without the use of randomization, the probability $Prob(\omega) = \alpha(s_0) * \theta_0 * \left(e^{-E(s_0)*infD(a)} - e^{-E(s_0)*infD(b)}\right) * \theta_1 * \theta_2 = 1 * 0.9 * (e{-}10 *$ $0 - e{-}10 * 1) * 0.73 * 0.91 \approx 0.5913$ according to the formula (10), (11).

5 Turn-Based Probability Alternating-Time Temporal Epistemic Logic

Now, we represent the definition of probability alternating-time epistemic coalition structure (PAECS) based on Coalition Game (Chen 2007) as follows:

Definition 5.1. For a coalition of agents $CS \subseteq \Lambda$ of TPEGS M_{TPEGS}, we define the coalition structures of M_{TPEGS} induced by CS as two agents PTEGS $M_{CS} = \left(\{I, II\}, S, A, \left(S_I', S_{II}'\right), \sim_{ag}, TStep, \mathcal{L}\right)$, $CS = \{I, II\}$, where $S_I' = \bigcup_{r \in CS} S_r$ and $S_{II}' = \bigcup_{r \in \Lambda - C} S_r$.

For expressing the relationship among system, environment and knowledge quantitatively, we introduce probability operator $P_{\sim \lambda}$ in ATEL and substitute CS for As, with which we define the syntax of turn-based probability alternating-time temporal epistemic logic (tPATEL). The same as PCTL and PTCTL is still a CTL-style branching-time temporal logic in essence.

Definition 5.2. The syntax of alternating-time temporal epistemic logic is defined as follows:

$$\phi ::= \top \mid p \mid \neg\phi \mid \phi \wedge \phi \mid K \mid \langle\langle As \rangle\rangle P_{\sim \lambda}[\psi] \mid \langle\langle CS \rangle\rangle P_{\sim \lambda}[K]$$

$$\psi ::= X\phi \mid \phi U\phi \mid \phi U^{\leq k}\phi$$

$$K ::= K_a\phi \mid E_{As}\phi \mid C_{As}\phi$$

Where $\sim \in \{<, >, \leq, \geq\}$, $\lambda \in [0, 1]$. The formula $K_a\phi$, $E_{As}\phi$, $C_{As}\phi$, $X\phi$, $\phi U\phi$, and $\phi U^{\leq k}\phi$ are endowed with the same meaning as in ATEL. The formula $\langle\langle CS \rangle\rangle P_{\sim \lambda}[\psi]$, means that the two agents in CS can work cooperate to ensure that ψ is true with a probability $\sim \lambda$.

6   tPATEL Model Checking

In this section, we mainly present a method for model checking TPEGS, against formulae ϕ in tPATEL.

6.1   Computing Probability

We can use the classical algorithm for model checking other branching-time logics (e.g. CTL, ATEL, PCTL) formula to compute the set for atomic proposition p or logical connective e.g. $\wedge$, $\vee$ in the PATEL formula ϕ, therefore our problem is reduced to computing operators containing $\langle\langle\rangle\rangle P_{\sim\lambda}$ with knowledge operators. To simplify presentation, we use $Sat(\phi)$ to represent the state set which satisfies ϕ.

Firstly, we introduce the computation method for a probabilistic on epistemic formulas as follows:

$M, s \models \langle\langle CS \rangle\rangle P_{\sim\lambda}[K_a\phi]$ iff $Prob[M, s \models K_a\phi] \sim \lambda$, where

$$Prob[M, s \models K_a\phi] = \frac{\sum_{s \sim_{ag} s'} s' \models \phi}{|s \sim_{ag} s'|} \tag{12}$$

$M, s \models \langle\langle CS \rangle\rangle P_{\sim\lambda}[E_{CS}\phi]$ iff $Prob[M, s \models E_{CS}\phi] \sim \lambda$, where

$$Prob[M, s \models E_a\phi] = \frac{\sum_{s \sim_{CS}^E s'} s' \models \phi}{|s \sim_{CS}^E s'|} \tag{13}$$

$M, s \models \langle\langle CS \rangle\rangle P_{\sim\lambda}[C_{As}\phi]$ iff $Prob[M, s \models C_{CS}\phi] \sim \lambda$, where

$$Prob[M, s \models C_a\phi] = \frac{\sum_{s \sim_{CS}^C s'} s' \models \phi}{|s \sim_{CS}^C s'|} \tag{14}$$

6.2   Algorithm Analysis

We translate the rPATL formula for model checking the tPATEL formula. Based on PRISM-games (Chen et al. 2013), the operator knowledge operators $\sim_{ag}$, $\sim_{As}^E$, $\sim_{As}^E$, and the results of $M, s \models K_a\phi$, $M, s \models E_{cs}\phi$, $M, s \models C_{cs}\phi$ are defined as atomic propositions, which is shown in Example 5.1 partly and will be interpreted in Sect. 8 further. Therefore, model checking a tPATEL formula in PRISM-games do not have higher time complexity than model checking an rPATL formula and be stated as follows:

Theorem 7.1. Model checking a tPATEL formula where k for the temporal operator $U^{\leq k}$ is given in unary is in $NP \cap coNP$.

Proof. To prove the theorem, we will use a conversion process called polynomial time reduction algorithm. Let ξ be a PATEL formula where k for the temporal operator $U^{\leq k}$ is given in unary.

If there is no knowledge formula in ξ, i.e., $\phi ::= \top \mid p \mid \neg\phi \mid \phi \wedge \phi \mid \langle\langle CS \rangle\rangle P_{\sim\lambda}[\psi]$, let's convert $\varsigma(s, a, s')$ to $\Delta(s, a)(s')$ of stochastic multi-player game (SMG) with formula: $\Delta(s, a)(s') = \varsigma(s, a, s') / \int_0^a E(s) * e^{-E(s)\varrho} d\varrho, a \in \mathbb{R}_{\geq 0}$ or $\Delta(s, a)(s') = \varsigma(s, a, s'), a \in A$ based on formula:

$$\varsigma(s, a, s') = \begin{cases} \theta(s') & \text{if } a \in A \\ \theta(s') * \int_0^a E(s) * e^{-E(s)\varrho} d\varrho & \text{if } a \in \mathbb{R}_{\geq 0} \end{cases}$$

Then formula ξ is converted to an rPATL formula ξ'. Because the conversion process is in P, and ξ' is in NP ∩ coNP based on Theorem 1(a) in paper (Chen et al. 2012) such that the problem of deciding whether formula ξ is satisfied in s is NP ∩ coNP.

If there is knowledge formula in ξ, i.e. $\phi ::= \top \mid p \mid \neg\phi \mid \phi \wedge \phi \mid K \mid \langle\langle CS \rangle\rangle P_{\sim\lambda}[\psi] \mid \langle\langle CS \rangle\rangle P_{\sim\lambda}[K]$. Because the operator knowledge operators $\sim_{ag}, \sim_{As}^E, s \sim_{As}^E s'$ and the results of $M, s \models K_a\phi$, $M, s \models E_a\phi$, $M, s \models C_a\phi$ are defined as atomic propositions, as same as above, the conversion process by which the formula ξ can be converted to an rPATL formula ξ' is in P, such that the problem of deciding whether formula ξ is satisfied in s is NP ∩ coNP.

Therefore, Theorem 7.1 is true.

Theorem 7.2. Model checking an arbitrary tPATEL formula is in *NEXP ∩ coNEXP*.

Proof. This proof is similar to the proof for Theorem 7.1 based on the above conversion process which is a polynomial time reduction algorithm.

7   Case Study

We use PRISM-games, a model checker fool for stochastic multi-player games, to automatically verify properties of TPEGS expressed by tPATEL formulas. Based on PRISM-games, we define the operator knowledge operators $\sim_{ag}, \sim_{CS}^E, \sim_{CS}^E$, and the results of $M, s \models K_a\phi$, $M, s \models E_a\phi$, $M, s \models C_a\phi$ as atomic propositions. Thus, there are two steps to build models of TPEGS. In the first step we manually calculate if $M, s \models K_a\phi$, $M, s \models E_a\phi$ and $M, s \models C_a\phi$ is true. During building the model of TPEGS and automatically verifying properties expressed by tPATEL, we identify state $s \in S$ with above results and operators $\sim_{ag}, \sim_{CS}^E, \sim_{CS}^E$ as atomic propositions in second step.

Now, we Quantitatively analyze performances of the flight procedures in the operational concept for the small Aircraft Transportation System (SATS) with the above quantitative model checking framework. The SATS is a NASA project aimed at increasing access to small and medium sized air airports which generally lack of tower facilities and radar coverage (Muñoz and Carreño 2006). The concept is implemented by four main components: The Self Controlled Area (SCA); the Airport Management Module (AMM); data communication; and on-board navigation tools. More details about SATS and SCA, please see articles (Dowek et al. 2004; Muñoz 2006).

We model the flight procedures in the concept of operation as *Turn-Based probability epistemic game structure* (TPEGS) $M_{TPEGS} = \langle \Lambda, S, A, S_{r:(r \in \Lambda)}, \sim_{ag}, TStep, \mathcal{L} \rangle$.

In our model, we simply define state transitions which are cruise, takeoff, departure, ready, hover, descend, initialize, enter base, enter intermediate, enter final, land, taxi, enter tarmac, go around, close, re-sequence and re-assign based on twenty four rules and critical off-nominal operations.

Next, we gave a *Turn-Based probability epistemic game structure* $M_{sub} = (\Lambda, S, A, S_{r:(r\in\Lambda)}, \sim_{ag}, TStep, L)$ modeling an approach procedure that is a part of the flight procedures with three agents, one per aircraft. Agent aircraft 1, aircraft 2, aircraft 3 is represented with square, round and hexagon respectively. In the subsystem it is a real situation that before the approach procedure of the aircraft 1, aircraft 2 may have been in the holding 2 (right), and the aircraft 3 flying in the intermediate zone. From the state s_0, aircraft 1 starts a Vertical Entry (right). In the state s_0, the next state s_{10} or s_1 is selected probabilistically. Based on the data of instrument approach procedure of runway 23 at Xiamen-Gaoqi Airport (Hao 2012), we propose that Lbase (right) = 9 km (the length of the right base segment), Lbase (right) = 7 km (the length of the left base segment), Linter = 10 km (the length of the intermediate segment), Final = 11 km (the length of the final segment), the velocity, the duration and the initial location of aircrafts is reported in Table 1. In general, aircrafts do uniform variable motion in all transitions, based on which we can get the following conclusions aircrafts is reported in Table 1. In general, aircrafts do uniform variable motion in all transitions, based on which we can get the following conclusions

$$l^n = l^{n-1} + \int_0^{D(a)} v(t, l^{n-1})dt = l^{n-1} + \int_0^{D(a)} (v^{n-1} + at)dt = l^{n-1} + v^{n-1} * D(a) + \frac{1}{2} * a * D(a)^2$$

$$= l^{n-1} + v^{n-1} * D(a) + \frac{1}{2}\left(\frac{v^n - v^{n-1}}{D(a)}\right) * D(a)^2 = l^{n-1} + \frac{1}{2} * (v^n + v^{n-1}) * D(a),$$

where l^n is the location of a aircraft in n area, v^n is the velocity of a aircraft in n area, $D(a)$ is a time function for action a, $a \in A \cup \mathbb{R}_{\geq 0}$. When aircraft 1 park on the tarmac, the system M_{sub} is made into a terminal state s_{23}.

On a high level, aircrafts can land successfully and safely, which is a primary goal in the approach procedure. With tPATEL, aircraft successful landing is represented as that if a aircraft can reach a designated area, i.e. achieve $F^{<n}l_m \in$ designatedArea, where l_m is the location of aircraft m. The maximum and minimum probability of aircraft 1 and aircraft 2 collaboratively landing in the area *target Area* in n steps is represent by PATEL formula as follows:

$$<<I, II, III>> Pmax = ? [F <n (l_1 \in targetArea) \wedge (l_2 \in targetArea)] \quad (15)$$

$$<<I, II, III>> Pmin = ? [F <n (l_1 \in targetArea) \wedge (l_2 \in targetArea)] \quad (16)$$

Next, we use PRISM-games, to automatically verify those properties expressed by PATEL formulas (15, 16). In order to meet the model checker tool, we still assume that {ready, hover, descend, initialize, enter base, enter intermediate, enter final, land, taxi, enter tarmac, leave, go around} $\subset A$, i.e. all the transitions in M_{sub} is discrete.

Table 1. The data of aircrafts

Player	Velocity (m/s)	Transition duration (s)	Initial location (m)
Aircraft 1	$v_0 \leq 106$	$D(ready) = 72$	$l^0 = 61000$
	$v_1 \leq 106$	$D(descend) = 144$	
	$v_3 \leq 106$	$D(initialize) = 108$	
	$v_4 \leq 106$	$D(enter\ base) = 108$	
	$v_5 \leq 106$	$D(enter\ inter) = 150$	
	$v_6 \leq 106$	$D(enter\ final) = 226$	
	$v_7 \leq 94$	$D(land) = 30$	
	$v_8 \leq 83$	$D(taxi) = 120$	
	$v_9 \leq 17$	$D(enter\ tarmac) = 60$	
Aircraft 2	$v_{10} \leq 106$	$D(initialize) = 108$	$l^{10} = 34000$
	$v_{11} \leq 106$	$D(enter\ base) = 108$	
	$v_{12} \leq 106$	$D(enter\ inter) = 144$	
	$v_{13} \leq 94$	$D(enter\ final) = 264$	
	$v_{14} \leq 83$	$D(land) = 20$	
	$v_{15} \leq 17$	$D(taxi) = 120$	
	$v_{21} \leq 106$	$D(enter\ tarmac) = 60$	
	$v_{22} \leq 106$	$D(hover) = 144$	
		$D(go\ around) = 256$	
Player	Velocity (m/s)	Transition duration (s)	Initial location (m)
Aircraft 3	$v_{17} \leq 94$	$D(enter\ final) = 240$	$l^{17} = 15000$
	$v_{18} \leq 83$	$D(land) = 20$	
	$v_{19} \leq 17$	$D(taxi) = 120$	

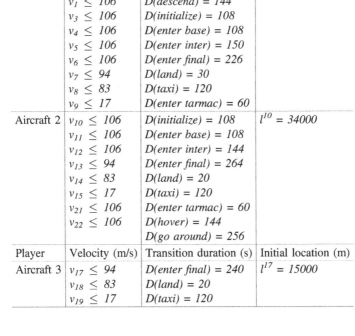

(a) Maximum probability in 100 steps   (b) Minimum probability in 100 steps

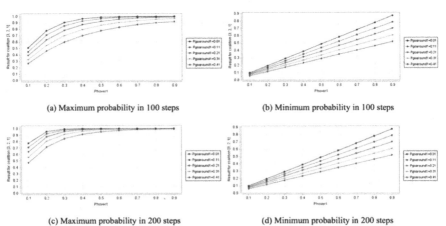

(c) Maximum probability in 200 steps   (d) Minimum probability in 200 steps

Fig. 2. Probability of aircraft 1 and aircraft 2 cooperatively landing in the target area

Figure 2(a), (c) shows, the greater the missed probability of aircraft 1, the smaller the maximum probability of aircraft 1 and aircraft 2 cooperatively landing in the designated area, and the greater the hovering probability of aircraft1 beyond the approach airspace, the greater the maximum probability of aircraft 1 and aircraft 2 cooperatively landing in the designated area. Figure 2(b), (d) shows, the greater the missed probability of aircraft 1, the smaller the minimum probability of aircraft 1 and aircraft 2 cooperatively landing in the designated area, and the greater the hovering probability of aircraft 1 beyond the approach airspace, the greater the minimum probability of aircraft 1 and aircraft 2 cooperatively landing in the designated area.

8 Conclusion

In this paper, we have successfully proposed a *turn-based probability epistemic game structure* (TPEGS), and over them syntax and semantics of *probability alternating-time temporal epistemic logic* (tPATEL) have been reasonably defined. Under this quantitative model checking framework, we verified the formulas such as formula (15, 16) using GRISM-gams, and shown the correct results. However, TPEGS can not be used to model asynchronous systems and the expression of tPATEL is not rich enough because there is not path formula like in LTL. In the future, research on the insufficient contents would be one main research direction, and another important research direction would be to analyses the environmental factors in flight procedure by quantitative model checking technology.

References

Allen, J.F.: Planning as temporal reasoning. In: KR 1991, pp. 3–14 (1991)

Alur, R., Courcoubetis, C., Dill, D.: Model-checking in dense real-time. Inf. Comput. **104**(1), 2–34 (1993)

Alur, R., Henzinger, T.A., Kupferman, O.: Alternating-time temporal logic. J. ACM (JACM) **49**(5), 672–713 (2002)

Baier, C., Katoen, J.-P., Hermanns, H.: Approximative symbolic model checking of continuous-time Markov chains. In: Baeten, J.C.M., Mauw, S. (eds.) CONCUR 1999. LNCS, vol. 1664, pp. 146–161. Springer, Heidelberg (1999). https://doi.org/10.1007/3-540-48320-9_12

Chen, T., Forejt, V., Kwiatkowska, M., Parker, D., Simaitis, A.: Automatic verification of competitive stochastic systems. In: Flanagan, C., König, B. (eds.) TACAS 2012. LNCS, vol. 7214, pp. 315–330. Springer, Heidelberg (2012). https://doi.org/10.1007/978-3-642-28756-5_22

Chen, T., Kwiatkowska, M., Simaitis, A., Wiltsche, C.: Synthesis for multi-objective stochastic games: an application to autonomous urban driving. In: 10th International Conference on Quantitative Evaluation of SysTems (QEST 2013). IEEE CS Press (2013)

Delgado, C., Benevides, M.: Verification of epistemic properties in probabilistic multi-agent systems. In: Braubach, L., van der Hoek, W., Petta, P., Pokahr, A. (eds.) MATES 2009. LNCS (LNAI), vol. 5774, pp. 16–28. Springer, Heidelberg (2009). https://doi.org/10.1007/978-3-642-04143-3_3

Dowek, G., Munoz, C., Carreno, V.A.: Abstract model of the SATS concept of operations: initial results and recommendations (2004)

Fagin, R., Halpern, J.Y., Megiddo, N.: A logic for reasoning about probabilities. Inf. Comput. **87**(1), 78–128 (1990)

Dannenberg, F., Hahn, E.M., Kwiatkowska, M.: Computing cumulative rewards using fast adaptive uniformisation. In: Gupta, A., Henzinger, T.A. (eds.) CMSB 2013. LNCS, vol. 8130, pp. 33–49. Springer, Heidelberg (2013a). https://doi.org/10.1007/978-3-642-40708-6_4

Dannenberg, F., Kwiatkowska, M., Thachuk, C., Turberfield, A.J.: DNA walker circuits: computational potential, design, and verification. In: Soloveichik, D., Yurke, B. (eds.) DNA 2013. LNCS, vol. 8141, pp. 31–45. Springer, Cham (2013b). https://doi.org/10.1007/978-3-319-01928-4_3

Gray, J.N.: Notes on data base operating systems. In: Bayer, R., Graham, R.M., Seegmüller, G. (eds.) Operating Systems. LNCS, vol. 60, pp. 393–481. Springer, Heidelberg (1978). https://doi.org/10.1007/3-540-08755-9_9

Hansson, H., Jonsson, B.: A logic for reasoning about time and reliability. Formal Aspects Comput. 6(5), 512–535 (1994)

Halpern, J.Y., Moses, Y.: Knowledge and common knowledge in a distributed environment. J. ACM (JACM) 37(3), 549–587 (1990)

Halpern, J.Y., Tuttle, M.R.: Knowledge, probability, and adversaries. J. ACM (JACM) 40(4), 917–960 (1993)

Hao, S.: The operation research of CDA based on PBN flight procedure. Civil Aviation Flight University of China (2012)

Henzinger, T.A., Nicollin, X., Sifakis, J.: Symbolic model checking for real-time systems. In: Proceedings of the Seventh Annual IEEE Symposium on Logic in Computer Science, LICS 1992, pp. 394–406. IEEE (1992)

Huth, M., Ryan, M.: Logic in computer science modelling and reasoning about system (2007)

Filar, J., Vrieze, K.: Competitive Markov Decision Processes. Springer, New York (1997). https://doi.org/10.1007/978-1-4612-4054-9

Kooi, B.P.: Probabilistic dynamic epistemic logic. J. Logic Lang. Inform. 12(4), 381–408 (2003)

Depressive Emotion Recognition Based on Behavioral Data

Yue Su[1,2], Huijia Zheng[1,3], Xiaoqian Liu[1],
and Tingshao Zhu[1(✉)]

[1] Institute of Psychology, Chinese Academy of Sciences, Beijing 100101,
People's Republic of China
tszhu@psych.ac.cn
[2] Department of Psychology, University of Chinese Academy of Sciences,
Beijing 100049, People's Republic of China
[3] St. Mark's School, Southborough, MA 01772, USA

Abstract. With the increase of pressure in people's lives, depression has become one of the most common mental illness worldwide. The wide use of social media provides a new platform for depression recognition based on people's behavioral data. This study utilizes the linguistical psychological characteristics of Weibo users to predict users' depression level. The model adopts the Gaussian process regression algorithm, sets the PUK kernel as the kernel function, applies the forward-backward search method to select feature, and uses five-fold cross-validation to evaluate performance of the model. This study finally established a prediction model with a correlation coefficient of 0.5189, which achieved a medium correlation in the psychological definition, and provided a more accurate method for the auxiliary diagnosis of depression.

Keywords: Depression recognition · Modeling · Network behavior · Machine learning

1 Introduction

Nowadays, depression has caused severe results to people's life. In February 2017, the World Health Organization [1] reported that more than 300 million people worldwide are currently living under the influence. The number of people with depression increased 18% between 2005 and 2015. These shocking numbers have forced us to pay great attention to serious harm of depression, which has become recognized as a major public health problem around the world [2].

Due to the lack of knowledge about depression, many people might miss the appropriate time for early treatment and have to suffer more pain and cost after diagnosis [3]. Diagnosis and treatment at the early stage can reduce the morbidity and mortality associated with depression and reduce costs [4]. As one of the top ten symptoms of depression, it is very necessary to investigate the depressed mood. In this paper, we start from the recognition of depression, which can make it possible to identify social media users with depression through large scale Internet data and assist in the early diagnosis of depression.

Y. Tang et al. (Eds.): HCC 2018, LNCS 11354, pp. 257–268, 2019.
https://doi.org/10.1007/978-3-030-15127-0_26

The words and expressions people use in their daily lives can reveal important aspects of their psychological states [5]. At present, the Internet provides a new platform for people expressing their opinions. Social media with various functions have emerged. On these platforms, people are engaged in making various forms of interaction and exchange such as commenting and forwarding. Users' social network behaviors are recorded in very details [6]. These behaviors express the characteristics of Internet usage of various groups of people and reflect different group traits. These records can be used as excellent data resources for analysis of different scientific uses and commercial uses.

In China, Sina Weibo occupies the prominent position in many Internet social platforms. As of September 30, 2017, the number of monthly active users of Sina Weibo has reached 376 million [7]. This enormous user provides a large amount of data, which records the behavioral characteristics of users, and provides a solid data set for this study to identify the user's depression emotional status.

This research aims to analyzing and processing users' behavioral records on Sina Weibo, and using machine learning method to build a prediction model. By building the relationship between users' online behaviors and level of depression, we try to achieve more accurate prediction of depressed mood. This research mainly combines the text information posted on Weibo with the user's depression sentiment scores obtained through the questionnaire to build prediction model and attempts different machine learning algorithms, descending dimension methods and algorithm parameters settings to make a better prediction of depression.

2 Related Work

In computer science and psychology, research on the detection of personality traits and psychological conditions of online users, especially depression, have been ongoing throughout the years.

As early as 1997, Professor Picard proposed the concept of Affective Computing [8]. One of the major research tasks of emotional computing is to perform the emotional recognition of users.

Based on people's online behavior, researchers studied many psychological traits of social media users. Quercia et al. found that the number of followings, followers, and listed counts in the user profile can accurately predict five personality traits of the user, with the predicted root mean square error capable of being below 0.88 [9].

Chen et al. used Facebook users' status and users' self-reported SWL (subjective well-being) score to predict the satisfaction with life. The prediction was moderately correlated with the self-reported score [10].

In 2015, Youyou et al. built a prediction model of personality and other psychological traits based on Facebook Likes and proved the prediction results are more accurate than those users' Facebook friends judgment [11].

When building the personality prediction model, there were improvements about feature selection in order to make the model perform better. Liu et al. utilized a deep learning algorithm to extract linguistic representation feature vector activity without supervision from text posted on Weibo, which performed well in personality prediction model [12].

Besides personality traits, researchers also dealt with some mental illness using online social media information. Guan et al. constructed suicide identification model based on profiles and linguistic features which were extracted from Weibo and proved it was possible to recognize individuals with high suicide probability [13]. McManus et al. used features from tweets to build a support vector machine prediction model of schizophrenia and achieved the performance with 92% precision [14].

Resnik et al. showed the possibility that we could estimate the severity of depression by analyzing text contents written by an individual [15].

Guntuku et al. concluded that depression and other mental illnesses could be detected in the online environment, and advances in natural language processing and machine learning have made possible large-scale filtering through social media for individuals at risk of mental illness [16].

Moreno et al. found that college students commonly displayed symptoms consistent with depression on Facebook [17].

Park et al. explored the difference of sentiment words usage on Twitter between people with and without depression, which provided evidence to recognize depression on social media platform [18]. Park et al. did similar research to explore the behavioral difference on Facebook and developed an application to evaluate depressive symptoms [19].

De Choudhury et al. obtained the released content of the social media during the year before the diagnosis of depression, and established a statistical classifier to predict the risk of depression, which could assist in early diagnosis and prevention of depression [20]. Different from the research of De Choudhury in English environment, Tsugawa et al. used various features extracted from Twitter users' activity history and web-based questionnaire of depression to construct a model in Japanese environment and managed to classify the presence of depression with an accuracy of 69% [21]. Mentioning the features used in prediction model, Tsugawa et al. found that frequency of word used in text might be useful when building depression models [22].

Mikal et al. also suggested that at least for some of the experimental participants, there were indeed questions about to what extent social media can represent the mental health of users [23].

Clearly, research on individual psychological status and emotional state recognition has been under development with combined techniques in many fields, such as computer and psychology. Studies found that online social media usage patterns are closely related to the user's personality traits. Besides, online behavioral data could be used to explore new recognition methods for suicidal tendency, schizophrenia, depression and so on. Some studies have established depression prediction models based on online social media to classify users. However, currently the prediction of depression based on online social behavior is mainly based on classification prediction, with relatively few studies on regression models.

3   Methodology

This research plans to analyze social media behavior data to identify the level of users' depression. Considering the huge users' group on Weibo, we decide to select Weibo as study platform. By API offered by Weibo, we can acquire users' behavior data required in this research, then process these data to obtain the feature dataset for modeling. Further, we select appropriate machine learning algorithm to establish a regression model to achieve a prediction of user's depression level.

The whole depression emotion recognition model primarily includes the following modules:

Data Acquisition Module: Data acquisition is mainly divided into two parts. The first part is the score of the depression level questionnaire filled out by the user. This study uses the Center for Epidemiological Survey Depression Scale (CES-D). The CES-D scale is one of the most widely used scales in the survey of depressive symptoms today [24]. The CES-D scale tends to measure the emotional experience of the subject, including some depressive symptoms such as despair and depression. The second part is the information of Sina Weibo. In the questionnaire mentioned above, we require participants to fill in their Weibo username. In the case of the user's permission, we download these users' specific Weibo contents according to their username.

Data Processing Module: Based on the detailed microblog contents obtained in the first step, we perform word segmentation, match them to the language mental feature lexicon, convert the original content into a feature set and do further feature filtering.

Model Constructing Module: Taking the feature data set processed in the data processing module as input, and the user's depression level calculated in the data acquisition module as the output, we select appropriate algorithm to establish the depression emotion recognition model, evaluate the precision of prediction through cross-validation, adjust algorithm and parameter setting with reference to the evaluation results until a predictive model of depression with the best predictive effect is obtained.

4   Behavioral Features

4.1   Data Collecting and Processing

By handing out and making users to complete questionnaires, we collected data which is needed in modeling. In order to reduce the influence of participants' subjective cognitive bias as much as possible, we conducted a questionnaire survey on two types of participants who are in different mental states. One type is ordinary person who live a normal life, another is people who have already been diagnosed with depression. By analyzing mixed data of healthy people and depressed patients, we can try to avoid the impact of cognitive bias as much as possible, and make the social media behavioral characteristics of depressed patients more prominent and the model's prediction results more accurate. After the issuance and recycling of questionnaires, we obtained a total of 2163 valid questionnaire results, which means there were 2163 records in the data set.

After obtaining the authorization of tested people, we downloaded their Weibo contents through API offered by Sina. The downloaded data included user's id, address, number of original microblogs, specific contents of microblogs and so on. In this study, we aimed at analyzing linguistic and lexical features of opinions and comments posted online. Therefore, we mainly deal with the specific content of user's microblogs. As emotions can fluctuate over time, a long period of time contains multiple emotions and would make depression features less obvious. Thus, when building model to predict the degree of depression, we only use microblogs posted within one week before users filled out the questionnaire. This period is closely related to user's mental state when filling out questionnaire and could better reflect the degree of user's depression.

The sample of Weibo contents collected is shown in Fig. 1.

Fig. 1. Example of Weibo content

According to the aim and linguistic characteristics of this research, including the exploration of mental health, Chinese context, diverse and casual linguistic traits of speeches on Weibo platform and the wide use of newly emerging popular words, we selected Simplified Chinese psychological linguistic analysis dictionary, which was improved by Gao et al. [25]. This dictionary was based on LIWC (Linguistic Inquiry and Word Count), and was added five thousand words that were most frequently used in Weibo. Also, this dictionary combined the use of emoji and punctuation on Weibo, and classified linguistic psychological vocabulary into 102 categories in total.

After matching dictionary, data features included the occurrence frequency of 102 categories of linguistic psychological vocabulary, users' id and CES-D scale scores, a total of 104 features, forming a preliminary feature matrix of 104 * 2163.

4.2   Dimension Reduction

Not all of these features are strongly related to the user's depression level. In order to eliminate irrelevant features, reduce the computational complexity and obtain a better prediction model, dimension reduction is needed.

This research mainly tried two methods. The first is to apply principal component analysis (PCA) to achieve feature extraction. The central idea of PCA is to retain the low-order principal components and ignore the high-order principal components, so as to preserve the most important features of the data to the greatest extent [26]. The second is to apply the forward-backward searching method to achieve feature selection. Its main implementation logic is to examine each possible branch of the problem separately, and examine the effect one by one [27]. The specific operation process of feature selection is as follows. At first, we use the whole feature set to build prediction model and evaluate the results. In the first round, we remove each feature in set so that every feature was removed once and re-evaluate the model. If there were new feature set working better than the larger set and the prediction result of this new feature set performed best in this selection round, we would keep this new feature set as modeling feature set and repeat the above steps. This removing process would end until there were no new feature set which performed better when modeling. At this moment, the feature set we kept worked best in the prediction of model.

4.3   Detailed Features

After feature selection of original 102 linguistic vocabulary features, the final number of linguistic psychological characteristics is 40, which would be used to build the prediction model in next part. The specific characteristics and meanings are shown in Table 1 below.

Table 1. Modeling feature table based on linguistic psychological feature set (part of the feature set)

No.	Feature	Meaning	No.	Feature	Meaning
1	Exclam	Exclamation mark	7	Leisure	Leisure word
2	SemiC	semicolon	8	Filler	Filled eulogy
3	enPast	Auxiliary verb	9	Love	Caring words
4	enPresent	Past tense	10	tNow	Current word
5	enFuture	present tense	11	tFuture	Future words
6	Quant	Negative Words	12	Colon	colon

5   Experiment

After getting the feature set needed in this research, we need to choose appropriate algorithm to build prediction model How to evaluate the performance of predictive model and optimize it based on evaluation index are an essential step in the modeling process, and in our study we select PCC to make the evaluation of prediction model.

PCC, which means the Pearson correlation coefficient, describes the ratio of covariance and standard deviation between two variables. It can measure the correlation between two variables [28]. It is generally believed in the field of psychology that when PCC reaches 0.4, there is a significant correlation between two variables, and the regression prediction model performs well. Also, the larger the value of PCC is, the better predictive model performs.

In order to make best use of the limited data, we adopt the method of cross validation. The main idea of cross validation is not to use all the original data for modeling, but to group the original data. Some of these data are used as training set to get input to train the model, and the other part acts as test set to evaluate the performance of model's prediction [29].

The modeling aim of our research is to build a depression prediction model, and in this study we will predict the specific value of depression level through users' linguistic psychological features of microblog' s detailed contents obtained on Sina Weibo. By handing out and making users to complete questionnaires, we collected data which is needed in modeling. And we got a total of 2163 valid questionnaires. After the process described in Sect. 4, we obtained a feature matrix of 103 * 2163 (including 102 linguistic vocabulary features and CES-D scale scores).

Experiment 1. In this experiment, we selected functions built in WEKA to model. Data set was the original set without dimension reduction, which contained 102 linguistic features. Then, we selected linear regression and Gaussian process regression as the prediction model algorithms. The evaluation results of five-fold cross-validation are shown in Tables 2 and 3.

Table 2. The evaluation results of five-fold cross-validation of linear regression

Index	Result
Correlation coefficient	0.3639
Mean absolute error	10.1539
Root mean squared error	12.5066
Relative absolute error	90.5309%
Root relative squared error	94.2919%
Total number of instances	2163

Table 3. The evaluation results of five-fold cross-validation of Gaussian process regression (default parameter)

Index	Result
Correlation coefficient	0.4334
Mean absolute error	9.848
Root mean squared error	11.9721
Relative absolute error	87.804%
Root relative squared error	90.2616%
Total number of instances	2163

According to the value of correlation coefficient, we found that the prediction performance of Gaussian process regression is significantly better than linear regression, and we decided to use Gaussian process regression as the modeling algorithm.

The modeling process of Gaussian process regression includes the setting of many parameters. The selection of specific parameters has a great influence on the prediction result of the model. We used the default parameters in the above modeling process. Then we need to adjust the parameters and try to accomplish better prediction performance. We mainly focus on the selection of kernel functions.

Kernel functions is an important concept in machine learning modeling. Its main purpose is to map data that cannot be divided in low dimension to high dimension to realize the division of data. Kernel function is a function to achieve the mapping process [30]. Choosing a proper kernel function could improve the predictive performance of the model effectively. After many attempts, we found that the best kernel function in the Gaussian process predictive depression model is the PUK kernel. The evaluation results of five-fold cross-validation are shown in Table 4.

Table 4. The evaluation results of five-fold cross-validation of Gaussian process regression (PUK kernel)

Index	Result
Correlation coefficient	0.4606
Mean absolute error	9.6883
Root mean squared error	11.7753
Relative absolute error	86.3799%
Root relative squared error	88.7781%
Total number of instances	2163

It can be observed that the correlation coefficient with PUK kernel is improved by about 0.027 compared to the prediction model using the default RBF kernel. PUK Kernel is short for Pearson VII Universal Kernel, which was developed on the basis of Pearson VII function [31]. PUK Kernel was developed on the use background of support vector machines, and our application to Gaussian process regression modeling also achieved good performance.

Experiment 2. In this experiment, we applied PCA to extract features from original data set. Then we used linear regression, Gaussian process regression to build model respectively, adjusted kernel function setting to explore better prediction performance and evaluated model performance by five-fold cross-validation. The evaluation results are shown in Tables 5, 6 and 7.

Table 5. The evaluation results of five-fold cross-validation of linear regression (the data set was processed by PCA)

Index	Result
Correlation coefficient	0.3406
Mean absolute error	10.3548
Root mean squared error	12.5941
Relative absolute error	92.3219%
Root relative squared error	94.9513%
Total number of instances	2163

Table 6. The evaluation results of five-fold cross-validation of Gaussian process regression with default parameter (the data set was processed by PCA)

Index	Result
Correlation coefficient	0.4153
Mean absolute error	10.0191
Root mean squared error	12.0912
Relative absolute error	89.3294%
Root relative squared error	91.1598%
Total number of instances	2163

Table 7. The evaluation results of five-fold cross-validation of Gaussian process regression with PUK kernel (the data set was processed by PCA)

Index	Result
Correlation coefficient	0.4287
Mean absolute error	9.8964
Root mean squared error	11.9805
Relative absolute error	88.2355%
Root relative squared error	90.325%
Total number of instances	2163

By comparison, we found that model performed better without feature selection than model processed by PCA in above three modeling situations. Then we chose to give up the idea of using PCA to reduce data set dimension.

Experiment 3. The forward-backward selection method is to select features of the dataset at the stage of modeling process, so we should select the modeling algorithm firstly. According to the above experiments, we found that when we applied Gaussian process regression with PUK Kernel, the prediction model performed best. Based on this, we carried out the modeling and feature selection process.

We accomplished feature selection and model building through python, and obtained the optimized model evaluation. The final model is established as follows:

Model feature: a total of 40 features, as shown in Table 1
Modeling algorithm: Gaussian process regression
Kernel function: PUK Kernel

The evaluation results of five-fold cross-validation are shown in Table 8.

Table 8. The final evaluation results of five-fold cross-validation

Index	Result
Correlation coefficient	**0.5189**
Mean absolute error	9.2469
Root mean squared error	11.3442
Relative absolute error	82.4443%
Root relative squared error	85.5279%
Total number of instances	2163

The value of correlation coefficient of prediction model built at last was 0.5189, which reached moderate correlations in psychology and achieved high prediction accuracy in regression predicting models.

The experiment results show that the specific person pronouns, non-specific person pronouns, auxiliary verbs, past tense, present tense and negative words in the text contents published on social media have a high correlation with depression, and their occurrence frequency could act as the features when recognizing the emotion of depression. And the prediction model built by Gaussian process regression worked best when using PUK Kernel, which provided new perspective for the establishment of depression level prediction model.

6   Conclusion

This study analyzed the linguistic psychological characteristics of the expressions by social media users, established the relation between these features and the corresponding depression level, and then built a depression prediction model. The model adopted the Gaussian process regression algorithm, set the PUK kernel as the kernel function, used forward-backward search method to make the further feature selection in the modeling process, and evaluated the prediction performance by the five-fold cross-validation. In this study, a prediction model with a correlation coefficient of 0.5189 was established, which achieved a medium correlation in the field of psychology and further improved the accuracy based on the existing research. The results of this study can be further examined and promoted to realize the automatic recognition of social media users' emotion of depression and to identify such emotion in advance, which will be used in the prevention of depression.

Acknowledgement. The authors gratefully acknowledge the generous support from National Basic Research Program of China (2014CB744600).

References

1. World Health Organization: Depression and other common mental disorders: global health estimates (2017)
2. Marcus, M., Yasamy, M.T., van Ommeren, M., et al.: Depression: a global public health concern (2012)
3. Wang, P.S., Simon, G., Kessler, R.C.: The economic burden of depression and the cost-effectiveness of treatment. Int. J. Methods Psychiatric Res. **12**(1), 22–33 (2003)
4. Montano, C.B.: Recognition and treatment of depression in a primary care setting. J. Clin. Psychiatry **55**(12), 18–33 (1994)
5. Pennebaker, J.W., Mehl, M.R., Niederhoffer, K.G.: Psychological aspects of natural language use: our words, our selves. Annu. Rev. Psychol. **54**(1), 547–577 (2003)
6. Pak, A., Paroubek, P.: Twitter as a corpus for sentiment analysis and opinion mining. In: Proceedings of the 7th International Conference on Language Resources and Evaluation (LREC 2010), pp. 1320–1326, May 2010
7. 钮成明, 詹国华, 李志华. 基于深度神经网络的微博文本情感倾向性分析. 计算机系统应用 **27**(11), 205–210 (2018)
8. Picard, R.W.: Affective computing (1995)
9. Quercia, D., Kosinski, M., Stillwell, D., et al.: Our Twitter profiles, our selves: predicting personality with twitter. In: 2011 IEEE Third International Conference on Privacy, Security, Risk and Trust (PASSAT) and 2011 IEEE Third International Conference on Social Computing (SocialCom), pp. 180–185. IEEE (2011)
10. Chen, L., Gong, T., Kosinski, M., et al.: Building a profile of subjective well-being for social media users. PLoS ONE **12**(11), e0187278 (2017)
11. Youyou, W., Kosinski, M., Stillwell, D.: Computer-based personality judgments are more accurate than those made by humans. Proc. Natl. Acad. Sci. **112**(4), 1036–1040 (2015)
12. Liu, X., Zhu, T.: Deep learning for constructing microblog behavior representation to identify social media user's personality. PeerJ Comput. Sci. **2**, e81 (2016)
13. Guan, L., Hao, B., Cheng, Q., et al.: Identifying Chinese microblog users with high suicide probability using internet-based profile and linguistic features: classification model. JMIR Mental Health **2**(2), e17 (2015)
14. McManus, K., Mallory, E.K., Goldfeder, R.L., et al.: Mining Twitter data to improve detection of schizophrenia. AMIA Summits Transl. Sci. Proc. **2015**, 122 (2015)
15. Resnik, P., Garron, A., Resnik, R.: Using topic modeling to improve prediction of neuroticism and depression in college students. In: Proceedings of the 2013 Conference on Empirical Methods in Natural Language Processing, pp. 1348–1353 (2013)
16. Guntuku, S.C., Yaden, D.B., Kern, M.L., et al.: Detecting depression and mental illness on social media: an integrative review. Curr. Opin. Behav. Sci. **18**, 43–49 (2017)
17. Moreno, M.A., Jelenchick, L.A., Egan, K.G., et al.: Feeling bad on Facebook: depression disclosures by college students on a social networking site. Depress. Anxiety **28**(6), 447–455 (2011)
18. Park, M., Cha, C., Cha, M.: Depressive moods of users portrayed in Twitter. In: Proceedings of the ACM SIGKDD Workshop on Healthcare Informatics (HI-KDD), pp. 1–8. ACM, New York (2012)

19. Park, S., Lee, S.W., Kwak, J., et al.: Activities on Facebook reveal the depressive state of users. J. Med. Internet Res. **15**(10), e217 (2013)
20. De Choudhury, M., Gamon, M., Counts, S., et al.: Predicting depression via social media. In: ICWSM, vol. 13, pp. 1–10 (2013)
21. Tsugawa, S., Kikuchi, Y., Kishino, F., et al.: Recognizing depression from twitter activity. In: Proceedings of the 33rd Annual ACM Conference on Human Factors in Computing Systems, pp. 3187–3196. ACM (2015)
22. Tsugawa, S., Mogi, Y., Kikuchi, Y., et al.: On estimating depressive tendencies of twitter users utilizing their tweet data. In: 2013 IEEE Virtual Reality (VR), pp. 1–4. IEEE (2013)
23. Mikal, J., Hurst, S., Conway, M.: Investigating patient attitudes towards the use of social media data to augment depression diagnosis and treatment: a qualitative study. In: Proceedings of the Fourth Workshop on Computational Linguistics and Clinical Psychology–From Linguistic Signal to Clinical Reality, pp. 41–47 (2017)
24. Radloff, L.S.: The CES-D scale: a self-report depression scale for research in the general population. Appl. Psychol. Meas. **1**, 385–401 (1977)
25. Gao, R., Hao, B., Li, H., Gao, Y., Zhu, T.: Developing simplified Chinese psychological linguistic analysis dictionary for microblog. In: Imamura, K., Usui, S., Shirao, T., Kasamatsu, T., Schwabe, L., Zhong, N. (eds.) BHI 2013. LNCS (LNAI), vol. 8211, pp. 359–368. Springer, Cham (2013). https://doi.org/10.1007/978-3-319-02753-1_36
26. Jolliffe, I.: Principal component analysis. In: Lovric, M. (ed.) International Encyclopedia of Statistical Science, pp. 1094–1096. Springer, Heidelberg (2011). https://doi.org/10.1007/978-3-642-04898-2
27. Zhou, G., Ye, G.: Forward-backward search method. J. Comput. Sci. Technol. **3**(4), 289–305 (1988)
28. Sedgwick, P.: Pearson's correlation coefficient. BMJ **345**, e4483 (2012)
29. Arlot, S., Celisse, A.: A survey of cross-validation procedures for model selection. Stat. Surv. **4**, 40–79 (2010)
30. Bergman, S.: The Kernel Function and Conformal Mapping. American Mathematical Society, New York (1950)
31. Üstün, B., Melssen, W.J., Buydens, L.M.C.: Facilitating the application of support vector regression by using a universal Pearson VII function based kernel. Chemometr. Intell. Lab. Syst. **81**(1), 29–40 (2006)

Knowledge Graph Embedding by Translation Model on Subgraph

Yifan Tan, Rongjun Li$^{(\boxtimes)}$, Jianjun Zhou, and Shuhua Zhu

Jinan University, Guangzhou 510632, China
{lrj,zsh}@jnu.edu.cn

Abstract. In this paper, we propose a translation model on subgragh representing knowledge graph. The model builds an ensemble TransE model on subgraph divided by features of relations in triplets by training the model with different parts of dataset independently. Afterwards, experimental results on link prediction show improvements on parameters compared to the state-of-the-art baselines.

Keywords: Knowledge graph · Knowledge representation ·
Representation learning · Translation models · Triplet classification

1 Introduction

Knowledge graphs such as Freebase [1] and Wordnet [2] show their application value in many fields, especially in the fields such as knowledge retrieval, knowledge storage, and intelligent question answering.

Knowledge graphs represent concepts in the physical world, and in turn, reflect the relationship between concepts in a mesh structure. Numerous triples are intertwined to form a network of knowledge structures [3].

The advantages of knowledge representation learning are mainly reflected in:

(1) Significantly increase computational efficiency. It means that representation learning effectively uses the numbers in each dimension, and many sparse data in the one-hot learning process are likely to cause effective information to be equal to 0 and disappear after the operation.
(2) Representation learning defines the dimensions of the entity and the relation vector. During the training process in which the objective function loss value is minimized, the value of the entity relationship vector is adjusted within a certain range of a certain space, even if the entity vector and relationship vector locate in totally different areas. In the same model it is expressed.
(3) Effectively solve the data sparseness problem. In the one-hot representation, the representation of each entity actually uses only one dimension, and the other dimensions are zero. However, in knowledge representation learning, each dimension of space can be fully utilized, which will help improve the utilization of data space.

© Springer Nature Switzerland AG 2019
Y. Tang et al. (Eds.): HCC 2018, LNCS 11354, pp. 269–280, 2019.
https://doi.org/10.1007/978-3-030-15127-0_27

In the process of continuous development of knowledge representation learning, predecessors proposed a variety of knowledge representation models adapted to various situations to represent the relationship between entities and relationships. When the translation model (TransE model) was proposed, due to the ease of use and accuracy of the model, the models proposed after the translation model started to be built based on the translation model.

In this paper, we first introduce the background of knowledge representation. After that, we summarize the improvements of previous work on the translation model in different directions.

Finally, we propose a translation model on subgraph. Based on the TransE model, several subspaces are formed by several subgraphs. Therefore, the training data in our model are divided according to their complexities. We ensemble these subspaces into a new TransE model called TransS. Finally, experiments are performed on the task of link prediction. The experimental data show that the TransS model proposed in this paper effectively reduces the error rate of the original model and improves the accuracy of the model. The innovation of this method is that it defactorizes massive dataset according to their complexities based on the observation that models perform better on simpler dataset.

2 Background

In recent years, with the rapid development of knowledge representation learning [4], knowledge representation learning models based on translation models have the ability to efficiently handle knowledge graph [5], and knowledge representation learning entities stored in a large amount in the knowledge graph through machine learning. The relationship is trained as a specific vector with a specific objective function. As the new model is continuously proposed and improved, the representation model based on the translation model has also been improved in performance and accuracy.

Compared with the more commonly used one-hot representation, learning representation has the advantage of possessing a low vector dimension. At the same time, it can also use the mathematical space in the entity and relation vectors to effectively solve the problem of data sparseness. Although the entity or relationship is also represented as a vector, the vector represented by this method usually has only one dimension of 1 and the values of other dimensions are 0. For example, in context [banana, apple, pear], when you need to represent a banana, the vector of the banana is [1, 0, 0]. In the same way, Apple's vector is [0, 1, 0]. This also means that in the one-hot representation, each entity as a vector in space is perpendicular to each other, and when the number of entities to be represented increases, the dimension of space also increases, resulting in dimensional curse.

The problem of representation learning mainly exists in two fields. One is that current model cannot predict entity correctly when relation in the graph is complex. Another is that the training time increases tremendously when certain improvements are applied on the benchmark model. The target of this task is to improve prediction accuracy and reduce training time.

3   Related Work

At present, compared with the previous model, the TransE model's result improves the precision and accuracy of knowledge representation learning to a level that can be applied to practical applications. Afterwards, there are two directions in its improvements.

3.1   Improvements on Loss Function

Bordes et al. [6] proposed the TransE model. The tail vector t is considered as the head vector h obtained by using the relationship vector r to offset its position in space. Compared with the previous model, the parameters and variables involved in the TransE model are very few, and the complexity is very low, so it is simple and effective. The TransH model proposed by Wang et al. [7] is a further improvement of the TransE model that can handle complex relational problems well. The TransH model preprocesses the head entity vector h and the tail entity vector t through the transformation matrix. This model projects them onto the hyperplane where the relation vector is located. In this hyperplane, the tail vector tt is considered as the head. The vector ht is obtained by using the relationship vector r to offset its position in space. The TransR model and the cTransR model map the head entity h and the tail entity t to the vector space where the relation r is located in the transformation matrix, and perform entity learning in the new space. The cTransR model proposed by Lin et al. [8] subdivides the relationship r into multiple sub-relationships r on the basis of the TransR model, and performs learning on each sub-relationship r. The TransA model proposed by Xiao et al. [9] replaces the Cartesian distance in the objective function with the Mahalanobis distance and sets different weights for training in each dimension (Fig. 1).

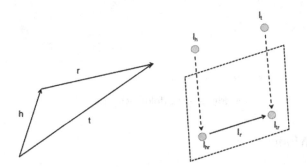

Fig. 1. TransE and TransH schematic

3.2   Improvements on Geometric Meaning

Besides improvements on numerical function, Lin et al. [10] use a path ranking algorithm to improve the TransE model. Xiao et al. [11] adopt sphere and hyperplane to solve ill-posed algebraic problem, using an overstrict geometric form problem in the TransE model. Liu et al. [12] employ analogical structure to enhance representation ability of model semantically. The TransE-DT proposed by Chang et al. [13] introduces parameter vectors to achieve flexible translation. Duan et al. [14] propose PtransW, which combines the space projection with the semantic information of relation path. The TransAH proposed by Fang et al. [15] adopts an adaptive metric, replacing Euclidean distance with weighted Euclidean distance (Figs. 2 and 3).

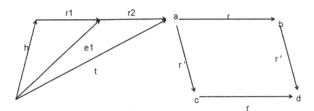

Fig. 2. PTransE and ANALOGY schematic

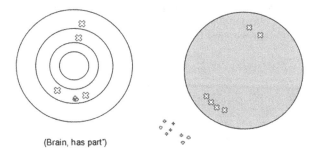

Fig. 3. ManifoldE schematic

4   TransS Model

4.1   The Motivation of Ensemble Translation Model on Subgraph

Like most of the machine learning task, there exists a problem of dataset disequilibrium. When performing a multi-classification task, the training set may have different percentage of instances on different categories. For example, to classify handwritten alphabet, the training data of "e" may be much more than that of "z". Therefore, the model may see the training data of "z" as a noise and perform better on "e" rather than "z".

The same problems are faced by knowledge graph embeddings. There exist complex relations which contain more than thousands of triplets. At the same time, simple relations which contain lesser than ten of triplets also exist. That's one of the reasons why the TransE model cannot perform well on complex relations.

4.2 Illustration of TransS Model

In TransS, when performing a link prediction, a search is performed on the relationship r. Then, the model would choose the most suitable subspace to perform its task. All the subspaces and its specific relations are stored in a list (Fig. 4).

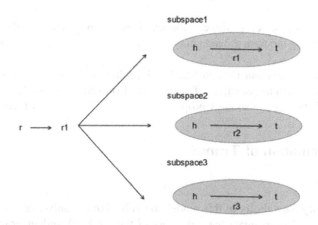

Fig. 4. TransS schematic

4.3 Dataset Division Algorithm

To solve the dataset disequilibrium problem, an intuitive way is to divide the whole triplets into several datasets where each relation has similar triplets.

Therefore, there are four steps to divide knowledge graph into subgraph.

First, count the number of triplets in specific relations, and we can get a dictionary data structure that can store all the relations and their specific numbers of triplets.

Second, define a cut-off line for relations that can divide triplets. For example, we have 100 relations which have more than 1000 triplets, and the lesser ones are the rest. So 1000 is the cut-off line that can divide the whole graph into two subgraphs.

Third, collect the related triplets of each subgraph to form their training set, valid set and test set.

Fourth, collect the entities in these subgraphs and set up a hash table of them for further use.

4.4   TransS Model Algorithm and Training

We have the same score function based on the translation model.

$$f(h,t,r) = \{y|y = |h+r-t|\} \tag{1}$$

Where h, t, and r are the head entity, tail entity, and relation vector, respectively.

Like the above model, the setting of the objective function includes positive training, artificial counterfeit partial sample counter training, and final range current training loss values. Artificial counterfeit is set for accelerating training process.

$$L = \sum_{(h,r,t)\in S} \sum_{(h',r,t')\in S'} (f(h,t,r) + b - f(h',t',r')) \tag{2}$$

Where S is the set of positive samples, S' is the negative sample obtained by randomly replacing the wrong head entity h or tail entity t within the current triple, and b is the offset.

The Stochastic Gradient Descent (SGD) algorithm is applied to adjust entities and relationship vectors to target values. The method of starting the variables is the same as the method of starting the squares proposed by the original model of TransE.

5   Implementation of TransS

5.1   Dataset

In order to verify the validity of this model, the WN18 of the universal knowledge map Wordnet was used to evaluate the indicators of the model. Wordnet provides a large amount of effective entity information and more than a dozen relationships. FB15k is another dataset which contains more than 1400 relations, and it is much more complex than WN18. The basic information of the data set is shown in Table 1:

Table 1.   Data set statistics

Dataset	#Rel	#Ent	#Train	#Valid	#Test
WN18	18	40,943	141,442	5,000	5,000
FB15k	1,345	14,951	483,142	50,000	59,071

To define the cut-off line of TransS model, we need to preprocess the dataset to find its features.

Taking WN18 as an example, we collect the basic entities and related triplets of each relation using the previous data division algorithm (Table 2).

Table 2. Statistics on relations of WN18

r	type	triplets	ent
0	12m	3341	3453
1	12m	7928	8173
2	m2m	31867	16737
3	12m	983	1042
4	121	86	82
5	m21	37221	36762
6	m21	7928	8173
7	m21	3150	3034
8	12m	675	659
9	m21	3335	3447
10	12m	37221	36762
11	12m	3150	3034
12	m21	669	654
13	12m	5142	5444
14	121	1220	1038
15	m21	5148	5455
16	m21	982	1041
17	m2m	1396	1061

In the table, we can find that r5 and r10 have 37221 triplets, which have the greatest number of triplets in the 18 relations. On the opposite, r4 has only 86 triplets. Problem of dataset disequilibrium actually exists.

5.2 Link Prediction

Link prediction first processes a complete triple, removes parts of its head entity or tail entity, and then imports the existing entities and relationships into the model to obtain the remaining removed parts. For example, the vector of the tail entity can be predicted by giving the head entity and the relation two vectors, and the tail entity vector and the relation vector can also be given to enter the trained model to calculate the head entity vector. After all the triples are processed in the same way, the loss values are calculated and ranked according to the accuracy of the prediction entity among all the entities, and finally the accuracy of the entire model is evaluated.

MeanRank is the sorting average of the correct triples. That is, through the model prediction, the value of the loss value corresponding to the entity marked as correct in the loss of all entities is sorted. After the prediction, the proportion of the correct triples is less than 10, which is denoted as hit@10. The Filt setting refers to removing a damaged triplet that is already present in the knowledge map from the training set and the verification set test set before the ranking score for each test triplet is obtained.

Specific parameters: the learning rate a is 0.01, the offset amount b is 1.0, the amount of change for each m is 0.1, and the training batch size is 1414, indicating that the dimension is 50.

In order to define how dataset disequilibrium affects the result of TransE model, we perform link prediction on relation 0–17 step by step as a method to simulate the increase of knowledge graph. The result shows in Table 3.

Table 3. Result of link prediction on 0–17 relation combination

r	epoch	mk	hit@10% (filter)	ent	triplet	itriplet	ient	r-ent
0–0	250	1501	0.03	3453	3341	3341	3453	3453
0–1	250	5398	0.02	11525	11269	7928	8072	8173
0–2	250	3155	0.71	26590	43136	31867	15065	16737
0–3	250	3222	0.7	27322	44119	983	732	1042
0–4	250	3276	0.7	27337	44205	86	15	82
0–5	250	8243	0.35	38764	81426	37221	11427	36762
0–6	250	6713	0.46	38764	89354	7928	0	8173
0–7	250	7493	0.45	40773	92504	3150	2009	3034
0–8	250	7492	0.45	40795	93179	675	22	659
0–9	250	6226	0.48	40795	96514	3335	0	3447
0–10	250	823	0.82	40795	133735	37221	0	36762
0–11	250	727	0.84	40795	136885	3150	0	3034
0–12	250	767	0.82	40795	137554	669	0	654
0–13	250	951	0.82	40854	142696	5142	59	5444
0–14	250	843	0.81	40862	143916	1220	8	1038
0–15	250	584	0.82	40862	149064	5148	0	5455
0–16	250	520	0.83	40862	150046	982	0	1041
0–17	250	568	0.83	40943	151442	1396	81	1061

To avoid overfitting in simple relation combination, we limit the training epochs as 250. Column itriplet and ient mean the increased number of triplets and entities respectively compared to the previous one relation combination. For example, itriplet 983 in relation 0–3 means relation combination [0, 1, 2, 3] has 983 triplets more than relation combination [0, 1, 2]. The number of entities related to this relation is represented as r-ent.

In Fig. 5, we can see that hit@10 increases as the number of triplets continuously increases. But there is a sudden drop on r0-4 to r0-5, in which the number of entities and triplets dramatically increases when relation 5 is added to the training model. Afterwards, hit@10 increases as the number of triplets increases without a sudden increase of entities. We can infer that the sudden drop may be caused by the fact that sparse entities are added to graph but relations are not.

Another interesting phenomenon is that hit@10f in relation 0–0 and relation 0–1 is fairly low, which reveals that TransE model may not perform well on sparse knowledge graph.

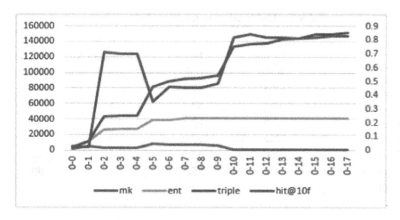

Fig. 5. Result of link prediction on 0–17 relation combination

To solve the problems of entities sparseness and data disequilibrium, we classify relations into several groups which have similar numbers of triplets and then ensemble them as a new model TransS.

6 Algorithm Evaluation

We first make link prediction on each subgraph to make contrast with TransE. The results show that TransE on each subgraph whose relations have similar training triplets performs much better when they are combined. Table 4 shows that hit@10 of each subgraph is much better than the whole graph.

Table 4. Link prediction result of relation on four subgraphs Group 1: 1, 2, 6, 5, 10. Group 2: 4, 12, 8, 16, 3. Group 3: 14, 17, 11, 7, 9, 0, 13, 15

r	epoch	MeanRank	Hit@10f
g1	1000	463	**95.7**
g2	1000	**21**	94.1
g3	1000	269	92.0
all	–	254	89.2

All graphs are divided into four groups according to their related number of their triplets. We then ensemble each subgraph model into a new model TransS. We compare the scores of different models and list the performance of each model in Mean-Rank and Hits@10.

Table 5. Link prediction results and comparisons

Method	WN18		FB15k	
	MeanRank (filter)	Hits@10 (filter)	MeanRank (filter)	Hits@10 (filter)
RESCAL	1163	52.8	683	44.1
TransE	251	89.2	125	47.1
TransH	388	82.3	84	58.5
TransR	**225**	92.0	78	65.5
TransD	230	92.3	95	75.2
TransS (our model 10 times avg)	412	**95.1**	**78**	**88.4**

As shown in Table 5, in WN18, the original method models such as TransE and TransR perform better on MeanRank, but on Hit@10, the TransS model performs better. The TransS model improves the TransE model based on the TransE model. In FB15k, TransS performs best on MeanRank and Hits@10. Therefore, TransS model has certain advantages in predicting correct entities.

From the perspective of algorithm complexity, the TransS model also has superiority in terms of parameter complexity and time complexity.

Table 6. Complexities

Model	#Parameters	#Operations (Time complexity)
TransE	$O(Nem+Nrn)$	$O(Nt)$
TransH	$O(Nem+2Nrn)$	$O(2mNt)$
TransR	$O(Nem+Nr(m+1)n)$	$O(2mnNt)$
CtransR	$O(Nem+Nr(m+d)n)$	$O(2mnNt)$
TransS	$O(Nem+Nrn+Nem')$	$O(Nt')$

Note: Ne and Nr represent the number of entities and relations, respectively; Nt represents the number of triples in the knowledge graph; m is the dimension of the entity vector; and n is the dimension of the relation vector. Parameter a is the number of subgraph. Nem' represents duplicate entities vectors caused by dividing knowledge graph into subgraphs.

Table 6 shows the following advantages of the TransS model compared to the TransE model and the improved method:

(1) The method of dividing a knowledge graph into subgraphs not only applies to the improvement of the classical TransE model, but also applies to other models such as TransR and TransH.

(2) The TransS model effectively solves problems of entities sparseness and data disequilibrium. For the first situation, we categorize relations that have similar sparseness of entities into one category and run model on it. The results show that

it performs well when relations that have different sparseness are insulated. For the second one, because all the relations have similar sparseness in one group, data disequilibrium is reduced.

(3) The TransS model builds only on the subgraphs of the whole knowledge graph. It also has the characteristics of less parameters and faster operation.

7 Conclusion

This paper proposes a TransS model for knowledge graph representation. For the problems of data equilibrium and entities sparseness in the translation model, the TransS model builds subspace of knowledge graph according to the relation-entity sparseness. Link prediction has been performed on WN18 and FB15k, and the experiment results show that the TransS model has certain advantages compared with other models.

References

1. Bollacker, K., Evans, C., Paritosh, P., et al.: Freebase: a collaboratively created graph database for structuring human knowledge. In: Proceedings of KDD 2008, pp. 1247–1250. ACM, New York (2008)
2. Miller, G.A.: WordNet: a lexical database for English. Commun. ACM **38**(11), 39–41 (1995)
3. Liu, Q., Li, Y., Duan, H., Liu, Y., Qin, Z.: Knowledge graph construction techniques. J. Comput. Res. Dev. **53**(03), 582–600 (2016)
4. Liu, Q.: Research on natural language semantic representation and reasoning based on neural networks. University of Science and Technology of China (2017)
5. Liu, Z., Sun, M., Lin, Y., Xie, R.: Knowledge representation learning: a review. J. Comput. Res. Dev. **53**(02), 247–261 (2016)
6. Bordes, A., Usunier, N., Garcia-Duran, A., et al.: Translating embeddings for modeling multi-relational data. In: Proceedings of NIPS 2013, pp. 2787–2795. MIT Press, Cambridge (2013)
7. Wang, Z., Zhang, J., Feng, J., et al.: Knowledge graph embedding by translating on hyperplanes. In: Proceedings of AAAI 2014, pp. 1112–1119. AAAI, Menlo Park (2014)
8. Lin, Y., Liu, Z., Sun, M., et al.: Learning entity and relation embeddings for knowledge graph completion. In: Proceedings of AAAI 2015. AAAI, Menlo Park (2015)
9. Xiao, H., Huang, M., Hao, Y., et al.: TransA: an adaptive approach for knowledge graph embedding. Computer Science (2015), 27 September 2016
10. Lin, Y., Liu, Z., Luan, H., Sun, M., Rao, S., Liu, S.: Modeling relation paths for representation learning of knowledge bases. arXiv preprint arXiv:150600379 (2015)
11. Xiao, H., Huang, M., Zhu, X.: From one point to a manifold: knowledge graph embedding for precise link prediction. In: Proceedings of the 25th International Joint Conference on Artificial Intelligence, pp. 1315–1321. AAAI (2016)
12. Liu, H., Wu, Y., Yang, Y.: Analogical inference for multi-relational embeddings. arXiv preprint arXiv:1705.02426 (2017)
13. Chang, L., Zhu, M., Gu, T., Bin, C., Qian, J., Zhang, J.: Knowledge graph embedding by dynamic translation. IEEE Access **5**, 20898–20907 (2017)

14. Duan, P., Wang, Y., Xiong, S., Mao, J.: Space projection and relation path based representation learning for construction of geography knowledge graph. J. Chin. Inf. Process. **32**(03), 26–33 (2018)
15. Fang, Y., Zhao, X., Tan, Z., Yang, S., Xiao, W.: A revised translation-based method for knowledge graph representation. J. Comput. Res. Dev. **55**(01), 139–150 (2018)
16. Xie, R., Liu, Z., Sun, M.: Representation learning of knowledge graphs with hierarchical types. In: International Joint Conference on Artificial Intelligence, pp. 2965–2971. AAAI Press (2016)
17. Lin, Y., Liu, Z., Sun, M.: Knowledge representation learning with entities, attributes and relations. IEEE Signal Process. Lett. **23**(4), 1 (2016)

Linguistic Signatures of Impulsive Buying Consumer Based on Microblog

Yujuan Han[1,2], Mingming Liu[1,2], and Tingshao Zhu[1,2(✉)]

[1] Institute of Psychology of the Chinese Academy of Sciences,
Beijing 100101, China
tszhu@psych.ac.cn
[2] Department of Psychology, University of Chinese Academy of Sciences,
Beijing 100101, China

Abstract. Since the systematic differences in language is stable and unique, the language in social media can be harnessed to create a valid measure of psychological traits by computer techniques. Combining computer technology and psychology theory may provide supplement for previous empirical research. This paper focused on the linguistic features of impulse buying users by using lexical analysis on microblog text. We analyzed the linguistic features of 321,610 microblog texts from 550 Weibo volunteers. The regression result indicated that there were obvious differences in the linguistic features of impulsive buyers. Systematic differences in the language use, especially in the impulse consumer, have received attention in the industry. The method of distinguishing impulsive consumers by the difference of linguistic features is convincing, especially for those who did not have any purchase records. In this paper, we proposed a new and valid strategy to study impulsive consumer.

Keywords: Impulsive buying · Linguistic signature · Microblog

1 Introduction

Microblog has become an important way for people to express attitudes and opinions. The motivation of Chinese SNS users could be divided into four types: acquiring information, entertainment, contacting with friends and conformity psychology [1]. The explosion of social media has resulted in massive data sets with hundreds of millions of users and billions of microblogs, allowing for data mining [2]. The huge dataset brings new opportunities for impulse buying research. At the same time, the methods on short text sharing in real time have been rapidly used. Some recent studies has discussed automatic personality assessment through social media language [3, 4]. The developing techniques combining multiple sources may provide better assessment of behaviors.

While impulse purchase has been extensively investigated for several decades, past research has indicated that unplanned consumption accounts for more than 60% of all purchases. The fact that impulse buying contributes to a considerable percentage of all purchases has attracted the attention of the industry company [5]. And many recent studies involved this direction. Recent research indicated that product reviews could

© Springer Nature Switzerland AG 2019
Y. Tang et al. (Eds.): HCC 2018, LNCS 11354, pp. 281–292, 2019.
https://doi.org/10.1007/978-3-030-15127-0_28

influent impulse buying intention under the network environment [6]. Another study investigated the influence of the big five personality on impulse buying in Facebook commerce [7]. However, few studies have examined the language features of the impulse buyer in social network environment. This is the innovation of this research.

2 Related Work

2.1 Framework and Influence Factors of Impulse Buying

Since DuPont Company put forward the concept of impulsive buying in 1945, impulsive purchase has attracted the attention of the vast majority of researchers. Some behavior theory has been put forward, including the stimulation theory, the emotion theory, the self-construal theory and the regulatory focus theory.

According to marketing stimulus theory, stimulation can affect consumers' purchase decision when there are significant differences at the level of time or space [8]. As 4P marketing theory, the stimulation mainly comes from four aspects: product, price, place and promotion. Impulsive buying is mainly stimulated by product stimulation and environment stimulation. The product information will induce the potential or new needs of consumer, which resulted in impulse buying.

The marketing emotion theory holds that impulse purchase is more emotional. The consumer might feel being out of control when the strong urge to buy occurs. Also impulse buying was more likely accompanied by less consideration of its consequences [9]. Thus, Impulse buying behavior was considered as the output of emotional reactions [10]. According to the study of Oliver, emotions were divided into positive and negative emotions. Consumers will not only buy impulsively because of positive emotions, but buy impulsively in order to improve their negative mood [11]. Then, scholars began to measure the emotional of consumers based on the pleasure-arousal-dominance scale (PAD) [12]. Some studies have shown that the rapidly rising pleasure feeling can significantly induce impulse buying behaviors, while in the unpleasant situation, reducing the level of arousal can increase consumer buying behavior [13].

In recent years, self-construal theory and regulation focus theory showed some new conclusions on consumer impulse buying. Studies on cultural psychology suggested that people hold different opinions on the self. When people come to aware of the relationship between the self and others, the degree that they regard themselves as an independent entity separated from others or as a part of the social relationship connected with others is important [14]. The independent self-construal is described by the opinion of oneself as a unique person, defined by the internal attributes and distinguishing characteristics. The interdependent self-construal is described by the opinion of oneself as part of a social relations, defined by others and by social relationships [15]. Hong and others have found that self-construal would influence consumers' focus toward the self or toward others in making a choice, which would conversely affect their decision making [16]. Studies form Zhang and Shrum indicated that independent self-construal showed higher impulsive buying tendency, while interdependent self-construal showed a lower impulsive buying tendency [17].

According to regulatory focus theory [18, 19], individuals tend to show specific ways of self-regulation in the process of achieving goals. One is promotion focus, which is related to motivation of achievement and development. The other is prevention focus, which is related to safety motivation such as protection and injury prevention. People with higher achievement, development, training and other motivations tend to have higher impulsive trait [20]. It is also easy to understand that people who are in the prevention focus condition tend to show less impulse buying tendency and stronger self-control ability [21]. The regulation focus can not only be shown as a long-term personality trait which is gradually formed in the process of growth, but also can be induced temporarily by scene inspiration [22]. Previous studies have shown that there are many specific methods to initiate consumer regulation focus. Self-guidance is one of the methods which allows subjects to report how their hopes and goals change with practice in order to start up promotion focus and allows subjects to report how their obligations and responsibilities change over time to start up prevention focus. Autobiographical memory is another way by encouraging participants to write short sentences supporting "success in life comes from action" or short articles about achievement and improvement to initiate focus or by encouraging participants to write short sentences supporting "prevention is the best treatment" or short articles about safety and prudence to initiate prevention focus. The third method is used by writing vocabulary related to defensive goals (such as avoidance, prevention, error, etc.) and facilitative goals (such as pursuit, success, harvest, etc.) [20, 21]. These methods provides theoretical support for the study of the linguistic signatures of impulsive consumers.

2.2 Linguistic Styles as an Individual Difference

Individuals express themselves with their own unique way, even when the content of the message is the same. The differences in expression are obvious both in spoken language and in written language from person to person. Analyses technology of word frequencies have been used to understand speaking styles of political leaders [23], to identify the anonymous author of a best-seller book and to establish the identities of the authors of literary works [24]. According to these studies, language styles can be regarded as a stable characteristic of individual differences [25].

Many of previous studies have used a similar approach—comparing word use with self-reports or behavioral assessments—to link language to personality. Previously, researchers used either historical language samples, such as scientific abstracts, literature, and other printed materials, or encouraged subjects to write new text. Now, social media provides researchers with millions of the natural language of people.

More recently, a word counting approach has been used to analyze language based on previous studies. This popular tool, Linguistic Inquiry and Word Count (LIWC) [25, 26], automatically counts word frequencies depending on predefined categories of words for over 72 psychological categories, such as negative emotional words (e.g., depression, anxiety, ugly), and causal words (e.g., because, reason, and why). Many researchers [27, 28] have tried to use this vocabulary analyses to predict a Facebook or Twitter user's personality. Social psychologists are more interested in understanding what makes people tick and why people behave the ways they do [26].

Psychologists conduct studying the words people use in their Twitter feeds, email, or Facebook status to reveal thinking styles, social relationships, motivations and personalities.

2.3   Developing the Hypothesis

Many theories elaborates the internal mechanism of impulsive buying from various aspects. Scholars have also studied various factors affecting impulsive buying behavior, such as personality traits, social impact, purchase environment and product characteristics [22]. Personality characteristics can be steadily reflected in oral and written language, which has been unanimously agreed in the current study. As a special group, the questions whether there exists difference in language expressing of these personality characteristics in the impulsive buying consumers and whether impulsive buyers can be identified automatically according to such differences have also become a topic of concern for marketing psychology researchers.

However, the current related research, especially the psychological research based on social network data on impulsive consumers' linguistic characteristics, is still relatively rare. So it has become the focus of this study. Then, we proposed the hypothesis that language features of impulse buying consumers are different from others.

3   Method

3.1   Objects and Tools

Participants

Participants were invited from users of Sina Weibo. We selected 1000 microblogs related to impulse buying by searching for keywords such as "impulse shopping" or "impulse is the devil" or "impulse buying". In order to ensure the diversity of microblog content, similar microblogs with higher frequency were removed manually. Then, we sent private letters to these 650 selected microblog users to invite them to fill in the questionnaire of the Chinese Impulse Buying Tendency which was initiated in the professional survey platform WenJuanXing. All users agreed to the anonymous use of their survey responses for research purposes. Totally 505 questionnaires were sent successfully and 340 valid questionnaires were collected among 382 reclaimed questionnaires. The efficiency is 89%. Then above 25% of score of impulse buying tendency (M = 3.02, Q3 = 3.52, SD = 0.78) were selected as the high impulse buying tendency group. There were 85 users (M = 4.03, Q3 = 4.44, SD = 0.39) in this group. 475 other general users (excluding marketing ID and enterprise ID) were randomly selected from the SINA Weibo platform as the control group. We captured every status message of the two group users who shared the microblog data to us voluntarily through the API interfaces of microblog platform. Finally, over 321,610 microblog text data could be used.

Impulse Buying Tendency Measure
This study used the Chinese Consumer Impulse Buying Tendency Scale [29], which was comprised of two subscales: cognitive subscale and emotional subscale. There were six dimensions: regardless of the future, rapid decision, unplanned buy, purchase impulse, mood adjustment, and buying experience. Subjects responded on a scale ranging from 1 (very false to me) to 5 (very true to me). The scales have shown adequate internal consistency reliability (0.90). In this study, the internal consistency of the six dimensions is between 0.75–0.83.

3.2 Data Processing

Text Cleaning
The 321,610 microblog text data was firstly cleaned by deleting the contents between @ symbol and the first blank space after that (usually as the nickname). At the same time some special symbols or meaningless words were deleted. Then the microblog text for each user was segmented by Chinese text segmentation software "Jieba" (Chinese for "to stutter"). The results of the segmentation formed the user's eigenvector.

Language Feature Extraction
We used Chinese Psychoanalytic Linguistic Inquiry and Word Count (SCLIWC) which was compiled by Gao et al. [30], on the basis of LIWC [25], to examine the cognitive, emotional and structural components behind oral and written language. Word stems are organized into 91 categories. Previous qualitative and empirical studies have related these categories to psychological phenomena of interest, such as emotional state, temporal focus, and certain cognitive processes [25]. The analysis of language features was completed by the "Text-Mind" software which was developed by the computer network psychological research group of Institute of Psychology, CAS. The "Text-Mind" program processed text files one word at a time by matching the base form of words to a dictionary of over 12176 word stems. A frequency count of the total instances of target words from each category was calculated, and this count was then divided by the total number of words in the text to represent for individual differences. Thus, scores reflected a percentage of word matched in that category.

3.3 Results and Analysis

General Situation
In the experimental group, mean user age was 25.84 (SD = 2.33), and 77 of them (88.24%) were female. Mean number of followers was 284 (SD = 181.85). Mean number of fans was 421 (SD = 539.3). Users wrote an average of 5071 (SD = 8611.0) words across all status messages (M = 1599; SD = 2564.66). In the control group, mean user age was 26.63 (SD = 4.72), and over half (68.21%) was female. Mean number of followers was 238 (SD = 181.85). Mean number of fans was 367 (SD = 539.3). Users wrote an average of 20837 (SD = 8611.0) words across all status messages (M = 1073; SD = 2564.66). The results showed that two groups used microblog widely and female were in a majority.

Linguistic Differences of Impulse Buying Tendency Consumer

To examine whether there were qualitative differences among the category variables analyzed with the "Text-Mind" program, the distribution of the language variables was carried out with the Kolmogorov-Smirnov (K-S) normality test. The K-S test result indicated that the language feature data did not conform to the normal distribution (P < 0.05). Thus, the language features were examined by Mann-Whitney U Rank Sum test. The results were partly listed in Table 1. As can be seen in Table 1, the use frequencies of family words, friends words (e.g., visitor, partner, neighbor), negative emotional words (e.g., depression, anxiety, ugly), positive emotional words (e.g., happy, exciting and beautiful), in the high impulsive buying tendency group were significantly (P < 0.05) higher than that of the control group. The use frequencies of leisure words (e.g., sports, travel, and music) and work words (e.g., resume, boss, examination), achievement words (e.g., influence, fame, reward and trophy) in the high impulsive buying tendency group were significantly (P < 0.05) higher than that of the control group. However, there was no significant difference (P > 0.05) in the use frequency of causal words (e.g., because, reason, and why), sex words, ingest words, love words, health words, money words, and religion words.

Table 1. Part results of Mann-Whitney U rank sum test of language features.

Language features	High impulsive buying tendency group (n = 85)		Control group (n = 475)		U value	W value	Z value	P value
	M	QR	M	QR				
Family	0.045	0.054	0.003	0.005	16725.0	134095.0	−3.056	0.002
Friend	0.002	0.004	0.002	0.002	15070.0	132440.0	−4.224	0.000
PosEmo	0.028	0.033	0.023	0.029	14703.5	132073.5	−4.483	0.000
NegEmo	0.013	0.018	0.009	0.016	14375.5	131745.5	−4.714	0.000
Cause	0.008	0.011	0.009	0.012	18820.0	22648.0	−1.577	0.115
Sexual	0.003	0.004	0.003	0.004	18400.0	135770.0	−1.873	0.061
Ingest	0.008	0.013	0.006	0.012	18977.0	136347.0	−1.466	0.147
Leisure	0.019	0.029	0.021	0.038	17279.5	21107.5	−2.664	0.008
Work	0.018	0.025	0.031	0.060	9702.5	13530.5	−8.012	0.000
Achieve	0.008	0.011	0.014	0.018	10887.0	14715.0	−7.176	0.000

Note: M: Median; QR: Interquartile Range; U: Mann-Whitney U; W: Wilcoxon W. PosEmo: positive emotion; NegEmo: negative emotion.

Correlations Analysis

For each of the 91 language dimensions and impulsive buying tendency (IBT) score, correlation analysis were computed. The results showed that family words (r = 0.254, P < 0.05), home words (r = 0.271, P < 0.05), work words (r = 0.293, P < 0.01), achieve words (r = 0.218, P < 0.05), space words (r = 0.223, P < 0.05), and inclusive words (r = 0.381, P < 0.01) were significantly positive correlation with impulse buying tendency (P < 0.05). Family words were significantly negative correlation with leisure words, religion words and space words, and were significantly positive correlation with work words (P < 0.01) and inclusive words (P < 0.05).

Factor Analysis

On the basis of below considerations, twenty of the original 91 variables were retained for subsequent factor analyses. Firstly, categories were included only when they did not considerably overlap with other included categories. Prepositions, cognitive process words, affect process words, social process words and so on were not included. Secondly, linguistic categories that did not stand for meanings of concrete words (e.g., number hash tag, words per sentence, or dictionary words) were excluded. Factor analysis was used to examine the most valuable dimensions of the main features variables [25]. Diagnostic tests indicated that a factor model was appropriate for the data (KMO = 0.58, Bartlett's test of sphericity = 507.63, P < 0.001). Eight common factors with eigenvalues above 1 were extracted by principal component analysis. The cumulative contribution rate of the eight common factors was 72.3%, which showed that the eight common factors could better reflect the main information of the original variables. The rotated factor loadings for component analysis were summarized in Table 2.

Table 2. Rotated factor loadings for component analysis of text mind dictionaries.

Dictionary	Fac 1:	Fac 2:	Fac 3:	Fac 4:	Fac5:	Fac 6:	Fac 7:	Fac 8:
	16.85%	12.24%	9.63%	8.21%	7.20%	6.78%	5.98%	5.45%
	Vari	Vari	Vari	Vari	Vari	Vari	Vari	Vari
Inhibit	0.704			−0.296				
Inclusive	0.666							
Space	0.623							
tfuture	0.615	−0.421						
Friend		0.735					0.409	
Description		−0.684						0.397
PosEmo	0.372	−0.600						
Family	−0.434		0.676			0.335		
Work			0.711				−0.301	
Motion			0.659			−0.554		
Achieve				0.605				
Leisure				−0.578				
NeEmo				−0.461	0.534			

Note. Only loadings of .30 or above were shown. Fac: factor; Vari: variation coefficient;

Dictionaries loading on the first factor (eigenvalue = 3.37) included the majority number of inhibitive words, part inclusive words, part space words and future words. This factor was labeled Restriction. The second factor (eigenvalue = 2.45) included such dictionaries as majority friend words, part description words and positive emotion words. This factor was termed Social Distance Emotion. The third factor (eigenvalue = 1.93) was characterized by part work words, family words and motion words. Thus, the third factor was termed Working Condition. The fourth factor (eigenvalue = 1.64) was characterized by achievement words, leisure words, and fewer negative emotion words. This factor was termed Achievement Condition. The fifth factor (eigenvalue = 1.44) was characterized by negative emotion words. This factor was termed Negative Emotion. The sixth factor (eigenvalue = 1.36) was characterized by motion words and fewer family words. This factor was termed Residence. The seventh factor (eigenvalue = 1.20) was characterized by fewer friend words and fewer work words. This factor was termed Relationship. The last factor (eigenvalue = 1.09) was characterized by fewer descriptive words. This factor was termed Description.

Principal Component Regression Analysis

An important question was whether dictionary factors can predict impulse buying. Stepwise regression was then computed, in which impulse buying tendency was used as dependent variable. The results of regression analysis were shown in Table 3. The dictionary factors significantly predicted impulse buying tendency, $F(570) = 9.03$, $P < 0.001$, $R2 = 41.0$. The Residence factor, the Work condition factor, the Relationship factor, the Restriction factor, and the Description factor finally entered the regression equation. Four factors positively affected impulse buying significantly ($P < 0.05$). The Relationship factor negatively affected impulse buying significantly ($P < 0.05$). While the Social Distance Emotion factor, the Achievement Condition factor, the Negative Emotion did not enter the regression model. And regression coefficient showed that the Residence factor (0.379), the Work Condition factor (0.265), the Restriction factor (0.212) and the Description factor (0.181) significantly positively affected impulse buying ($P < 0.05$), while the Relationship factor (-0.062) negatively affected impulse buying significantly ($P < 0.05$).

Table 3. Results of regression analysis of 8 factors and impulse buying tendency.

Model		Unstrd coefficients		Strd coefficients	t	Sig.
		B	Std. Error	Beta		
5	(Constant)	4.093	0.023			0.000
	Residence	0.104	0.024	0.397	4.410	0.000
	Work condition	0.070	0.024	0.265	2.951	0.004
	Relationship	−0.062	0.024	−0.237	−2.633	0.010
	Restriction	0.056	0.024	0.212	2.352	0.021
	Description	0.047	0.024	0.181	2.008	0.048

a. Dependent Variable: impulse buying tendency

4 Discussion

This study indicated that among the consumers with high impulsive buying tendency, females were significantly in a majority, which is consistent with other studies [8]. The words frequency in the high impulsive buying tendency group, such as family words, friends words (e.g., visitor, partner, neighbor), negative emotional words (e.g., depression, anxiety, ugly), positive emotional words (e.g., happy, exciting and beautiful), were significantly ($P < 0.05$) higher than that of the control group. The frequency of leisure words (e.g., sports, travel, and music) and work words (e.g., resume, boss, examination), achieve words (e.g., influence, fame, reward and trophy) in the high impulsive buying tendency group were significantly ($P < 0.05$) lower than that of the control group. The hypothesis was supported by the results.

Moreover, what we were most interested was whether dictionary factors can predict impulse buying. We tried to find out the potential cause of these influence of these dictionary factors. And the regression analysis results showed that the dictionary factors can significantly predict impulse buying tendency ($P < 0.001$).

Dictionaries loading on the Restriction factor included inclusive words and inhibition words (e.g., hold and restrain) which were part of the cognitive processes dimension, space words and future words which were part of the space dimension. This factor relates to the self-cognitive factors and situational factors [8], which had been proved by scholars to affect impulse buying in self-construal theory. Markus divided self-construal into independent self-construal and dependent self-construal from the perspective of self-cognition [14]. Independent self-construal positively affects impulse buying, while inter-dependent self-construal negatively affects impulse buying.

The Social Distance Emotion factor which included discrepancy words which Higgins et al. [31] have linked to health, friend words which were part of the communication dimension, and positive emotion words which were part of the emotional process dimensions had not entered the regression equation. Previous research found that impulse buying was influenced by normative evaluations from companions. The person of independent self-construal showed higher impulse buying level when they went shopping with companions who had a small difference in value than with companions who had a big difference [32].

The Working Condition factor which was characterized by high use of family and work words (current concerns dimension), and motion words, entered the regression equation. This was in agreement with the studies that consumers will buy impulsively in relatively relaxed environment [11]. The Achievement Condition factor which included achievement words (current concerns dimension), had not entered the regression equation. While this was inconsistent with regulation focus theory. Pervious research found that people who were in promotion focus tend to show higher impulsive buying tendency. The Negative Emotion factor which was characterized by negative emotion words, had failed to enter the regression equation. While previous studies showed that consumers will buy impulsively in order to improve the negative emotions [11].

Dictionaries loading on the Residence factor and Relationship factor had all entered the regression model. Studies indicated that independent self-construal people showed higher impulse buying tendency because of be less affected by the impact of others in the process of making decisions [17]. The Description factor entered the regression equation was only to be expected. Another studies showed that people in promotion focus condition who tended to be concerned with advancement, growth, and accomplishment also related to high impulsive purchase [20].

5 Conclusion

This paper focused on the linguistic features of impulse buying users, using lexical analysis on microblog text. Part of the linguistic features factors entered the regression model of impulse buying, which indicated that there exist obvious differences in the language characteristics of the impulsive buyers. Systematic differences in the language usage of impulse consumer, have received attention in industry, which becomes a relatively prospecting field because of its ramifications for a broad areas of practical application issues such as marketing, persuasive campaigns and so on. The label used by industry to depict impulsive consumers is usually by behavioral features, such as "heavy internet addicts, mobile phone dependents, fans of online shopping enthusiast". Now, we tried to distinguish impulse consumers from the perspective of language features on the social network platform. This method can be more convincing, especially for those who did not have any purchase records. In this paper, we proposed a new and valid strategy to study impulse consumer.

Since the systematic differences in the preference of language usage is stable and unique, analyzing language feature by compute techniques can be used as an effective and reliable measure method for behavior and psychological traits research. Combining these computer technologies with psychological theories may provide supplement for previous empirical research. Even though machine learning technology and natural language processing technology are now used to analyze language features, rich psychological theories are still indispensable in order to reveal the reasons behind the behavior.

Like most research in the social sciences, this studies is still affected by sampling and social expectation biases. Future works may consider collecting richer sample data to overcome these limitations. Second, we used impulse buying tendency (IBT) scores instead of measuring actual impulse buying behavior. We could not make sure that the observation subjects would actually rush to buy the corresponding product when they reported higher IBT. Therefore, it is recommended that future investigation may use actual impulse buying data instead of consumers' IBT.

References

1. Chang, Y., Zhu, D.: Motivations for the user participation in social networking site. Libr. Inf. Serv. **55**(14), 32–35 (2011)
2. Miller, G.: Social scientists wade into the tweet stream. Science **333**(6051), 1814–1815 (2011)
3. Park, G., Schwartz, H.A., Eichstaedt, J.C., Kern, M.L., Kosinski, M., Stillwell, D.J., et al.: Automatic personality assessment through social media language. J. Pers. Soc. Psychol. **108**(6), 934–952 (2015)
4. Schwartz, H.A., Eichstaedt, J.C., Kern, M.L., Dziurzynski, L., Ramones, S.M., Agrawal, M., et al.: Personality, gender, and age in the language of social media: the open-vocabulary approach. PLoS one **8**(9), e73791 (2013)
5. Amos, C., Holmes, G.R., Keneson, W.C.: A meta-analysis of consumer impulse buying. J. Retail. Consum. Serv. **21**(2), 86–97 (2014)
6. Chang, Y., Xiao, W., Qun, W., Yan, J.: The influence mechanism of third-party product reviews (TPRs) on impulse buying intent within the internet environment. Acta Psychol. Sin. **44**(9), 1244–1264 (2012)
7. Leong, L.Y., Noor, I., Sulaiman, A.: Understanding impulse purchase in Facebook commerce: does big five matter? Internet Res. **27**(4), 786–818 (2017)
8. Mattila, A.S., Wirtz, J.: The role of store environmental stimulation and social factors on impulse purchasing. J. Serv. Mark. **22**(7), 562–567 (2008)
9. Rook, D.W.: The buying impulse. J. Consum. Res. **14**(2), 189–199 (1987)
10. Baba, S., Alexander, F.: Heart and mind in conflict: the interplay of affect and cognition in consumer decision making. J. Consum. Res. **26**(3), 278–292 (1999)
11. Richard, L.: Oliver: whence consumer loyalty? J. Mark. **63**(Special Issue), 33–44 (1999)
12. Holbrook, M.B., Rajeev, B.: Assessing the role of emotions as mediators of consumer responses to advertising. J. Consum. Res. **14**(3), 404–420 (1987)
13. Donovan, R.J., Rossiter, J.R.: Store atmosphere: an environmental psychology approach. J. Retail. **58**(1), 34–57 (1982)
14. Markus, H.R., Kitayama, S.: Culture and the self: implications for cognition, emotion, and motivation. Psychol. Rev. **98**(2), 224–253 (1991)
15. Liu, Y.: Self-construal: review and prospect. Adv. Psychol. Sci. **19**(3), 427–439 (2011)
16. Hong, J., Chang, H.H.: "I" follow my heart and "We" rely on reasons: the impact of self-construal on reliance on feelings versus reasons in decision making. J. Consum. Res. **41**(6), 1392–1411 (2015)
17. Zhang, Y., Shrum, L.J.: The influence of self-construal on impulsive consumption. J. Consum. Res. **35**(5), 838–850 (2008)
18. Higgins, E.T.: Beyond pleasure and pain. Am. Psychol. **52**(12), 1280–1300 (1997)
19. Hah, J.Y., Higgins, E.T.: Regulatory concerns and appraisal efficiency: the general impact of promotion and prevention. J. Pers. Soc. Psychol. **80**(5), 693–705 (2001)
20. Sengupta, J., Zhou, R.R.: Understanding impulsive eaters choice behaviors: the motivational influences of regulatory focus. J. Mark. Res. **44**(2), 297–308 (2007)
21. Jing, F.J., Xiong, S.H.: Underlying mechanism of impulsive consumption behavior: the influence of promotion focus on impulsive consumption behavior. Adv. Psychol. Sci. **16**(5), 789–795 (2008)
22. Lockwood, P., Jordan, C.H., Kunda, Z.: Motivation by positive or negative role models: regulatory focus determines who will best inspire us. J. Pers. Soc. Psychol. **83**(4), 854–864 (2002)

23. Hart, R.P.: Verbal Style and the Presidency: A Computer-Based Analysis. Academic Press, New York (1984)
24. Foster, D.: Primary culprit: who is anonymous? N. Y. Mag. **29**, 50–57 (1996)
25. Pennebaker, J.W., King, L.A.: Linguistic styles: language use as an individual difference. J. Pers. Soc. Psychol. **77**(6), 1296–1312 (1999)
26. Pennebaker, J.W.: Expressive writing in psychological science. Perspect. Psychol. Sci. **13**(2), 226–229 (2018)
27. Golbeck, J., Robles, C., Turner, K.: Predicting personality with social media. In: Proceedings of the 2011 Annual Conference on Human Factors in Computing Systems - CHI 2011, Vancouver, BC, pp. 253–262 (2011)
28. Jing, F.J., Yue, H.L.: Studies on Chinese consumers' impulsive buying tendency scale. Public Financ. Res. **5**, 37–40 (2005)
29. Gao, R., Hao, B., Li, H., Gao, Y., Zhu, T.: Developing simplified chinese psychological linguistic analysis dictionary for microblog. In: Imamura, K., Usui, S., Shirao, T., Kasamatsu, T., Schwabe, L., Zhong, N. (eds.) BHI 2013. LNCS (LNAI), vol. 8211, pp. 359–368. Springer, Cham (2013). https://doi.org/10.1007/978-3-319-02753-1_36
30. Higgins, E.T., Vooklles, J., Tykocinski, O.: Self and health: how "patterns" of self-beliefs predict type of emotional and physical problems. Soc. Cogn. **10**, 125–150 (1992)
31. Rook, D.W., Fisher, R.J.: Normative influence on impulsive buying behavior. J. Consum. Res. **22**(3), 305–313 (1995)
32. Kramer, A.D.I.: An unobtrusive behavioral model of "gross national happiness. In: International Conference on Human Factors in Computing Systems 2010, DBLP, Atlanta, Georgia, USA, pp. 287–290. ACM (2010)

Research on the Model of Anomaly Detection of FMCG Based on Time Series

Illustrated by the Case of Cosmetics

Qiaohong Zu$^{(\boxtimes)}$ and Shaojun Nan$^{(\boxtimes)}$

School of Logistics Engineering, Wuhan University of Technology,
Wuhan 430063, People's Republic of China
zuqiaohong@foxmail.com, 458225996@qq.com

Abstract. FMCG (Fast Moving Consumer Goods) as one of the most attractive commodities in the logistics industry, its small number of consumption patterns affects all aspects of supply chain activities. FMCG is an impulse buying product, which will affect the uncertainty of demand and the bullwhip effect caused by demand fluctuation. According to the characteristics of cosmetics, this paper divides the products into two categories, then designs and improves two abnormal detection algorithms to avoid the impact of impulse consumption on demand. Aiming at a large number of common commodities, the K-iForest algorithm based on Isolation Forest (iForest) algorithm and K-means algorithm is proposed to control the sensitivity of the algorithm while ensuring the efficiency of the algorithm. For the luxury goods, the Support Vector Regression (SVR) model with sliding window is improved, which makes the algorithm more inclusive for small fluctuations.

Keywords: FMCG · Demand uncertainty · Abnormal detection

1 Introduction

With the e-commerce penetrating into all aspects of daily life rapidly and efficiently, the consumer market of FMCG has ushered in a new growth point. With the increase of urbanization rate, the change of population structure and the increase of income level, China has rapidly advanced from a simple populous country to a large consumer country. The market size of FMCG has expanded year by year and continues to climb [1]. As a typical category of fast-moving consumer goods, sales of cosmetics also rose. Online sales are more accustomed to attracting consumers at lower prices. In large online promotions, the number of cosmetics participating has shown explosive growth.

In the wake of explosive orders, the fluctuations in demand for cosmetics over time has put forward higher requirements for the control of corporate procurement and inventory. In order to make the demand stable, it is necessary to avoid the impact of impulsive consumption when making demand forecasts.

Y. Tang et al. (Eds.): HCC 2018, LNCS 11354, pp. 293–303, 2019.
https://doi.org/10.1007/978-3-030-15127-0_29

2   Research Status of Abnormal Pattern Recognition of Time Series

Yan et al. combined the sliding window model with the clustering algorithm to effectively judge the abnormal type while monitoring the real-time anomaly, then eliminate the influence of noise and sudden value on the algorithm [2]. Song and others performed sequential pattern mining with sliding window to obtain a common mode, and then determined the outliers by similarity comparison [3]. Yu et al. [4] combined the sliding window with the single-layer linear network prediction model, and made anomaly judgment by comparing the actual value with the prediction interval [4]. Liu et al. used an optimized support vector regression (SVR) model for anomaly detection. By predicting and fitting the monitoring data and establishing a confidence interval, the detection rate of outliers reached 90% [5]. Huo et al. segmented the data through the time window and built the model using the hidden Markov model. The detection accuracy was increased to 93.4% by adjusting the threshold [6].

3   Time Series Anomaly Detection for Impulsive Consumption

3.1   Question Raised

The selling trend of cosmetics is characterized by the nature of the goods, this article divides cosmetics into two categories according to their sales trends: luxury goods and general categories. It is known from experience that luxury goods have almost no sales on regular sales days. Consumer impulse spending caused abnormal data because of promotional activities. Ordinary categories have a certain amount of sales almost every day, and sales are amazing when experiencing promotions. At the same time, sales of some products have obvious seasonal trends. Therefore, for such goods, in order to pick out the impulse consumption of consumers, the algorithm needs to be relatively sensitive, and it is expected to be able to human intervention in the sensitivity of the algorithm. In contrast, there are too many days when no sales for luxury goods, the algorithm needs to have a certain degree of tolerance for small fluctuations.

3.2   Sliding Window Impulse Consumption Detection Algorithm Based on Support Vector Regression

Aiming at the anomaly detection of luxury sales data, this paper proposes a support vector machine for regression and sliding window (SVR-SW). First determine the sliding window based on the sales data. A data set consisting of window history data is predicted by the SVR algorithm. If the actual value is outside the prediction interval, it is determined to be abnormal, and the actual value is replaced, after which the window slides forward and continues to be detected according to the SVR algorithm until the end of the data. The detailed steps are as follows:

(1) Sliding window construction

The size selection of the sliding window has an impact on the efficiency and accuracy of the algorithm [7]. The larger the window size, the larger the amount of historical data used for prediction, and the higher the accuracy. On the situation, the corresponding occupied memory and the required time also increase. To start with, the e-commerce's large-scale promotion activities must be separated to avoid the situation where some of the large-scale promotions are detected in the same window, and some abnormal values are detected as normal values. At the same time, since the time interval of data collection is 1 day, the sliding window size is not very large; the size of the sliding window can be constrained between 14 and 30 days after considering the customer's repurchase characteristics.

(2) Construction of predictive model based on SVR algorithm

The core of SW-SVR is to establish a single-step prediction model based on SVR. Before the time t, the data set of the time range q is selected as the observation set of the sliding window, and the SVR model is input to estimate the predicted value at time t. The predicted value can be expressed as:

$$x_t' = M(D_t) + R \tag{1}$$

In the formula: $M(D_t)$–SVR Kernel function model; R–Empirical risk. The SVR model contains kernel functions, loss functions, and related parameters.

(1) Kernel function

The function selection steps include: first, selecting a kernel function type, and second, selecting an appropriate corresponding kernel function parameter value. There are four commonly used kernel functions: Polynomial kernel, Radial basis (RBF) core, Fourier nucleus, Sigmoid core. Among them, the RBF kernel function is the most widely used. This is because most of the actual data can be considered to approximate the Gaussian distribution, and it shows good adaptability to different situations such as dimension height and sample size. Moreover, since there is only the parameter σ in the RBF kernel function, it is relatively easy to grasp the value of the parameter.

(2) Loss function

When the SVM is applied to the regression problem and the SVR model is built, in order to preserve the important property of sparsity, a loss function needs to be introduced. Among, the ε-insensitive loss function is the only function that can make the support vector have sparsity.

$$L_\varepsilon(\xi) = \begin{cases} 0, |\xi| < \varepsilon; \\ |\xi| - \varepsilon, |\xi| \geq \varepsilon. \end{cases} \tag{2}$$

(3) Parameter selection

In this paper, the RBF kernel function is chosen to construct the model, and the polynomial kernel is used for comparative analysis.

The RBF function is a typical local kernel function. When σ tends to 0, the sample can be made linearly separable, but it can only affect the sample in a relatively small field, weakening the generalization ability. When the sample distance is much larger than σ, its value will gradually approach 0. Therefore, σ can be considered as a compromise between the local fitting ability and the generalization ability of the kernel function.

As a typical global kernel function, in the extrapolation process, if d approaches 0, the value of the corresponding kernel function is also close to 1, so that the weight control is limited. Thus, the corresponding parameter d is a reasonable choice for global fitting capability.

(3) Judgment and Processing of Abnormal Data Based on Predictive Model

After performing single-step prediction on the historical data in the sliding window, the corresponding prediction value is determined according to the prediction model, and then the residual is determined to obtain the prediction interval of p. If the probability of the newly occurring measured value falling within the prediction interval is 1-α and the corresponding residual satisfies the probability distribution requirement, the prediction interval is expressed as:

$$P_t = \overline{x_t} \pm t_{\frac{\alpha}{2}, n-1} S \sqrt{1 + 1/n} \tag{3}$$

In the formula: P_t–Prediction interval at time t, $t_{\frac{\alpha}{2}, n-1}$–is the t-distribution obeying the degree of freedom n−1, S–Standard deviation of n samples.

After determining the corresponding prediction area, the actual measurement value and the prediction interval of the sales volume at time t are compared and analyzed. If the actual measurement value at time t falls within the interval according to the comparison result, it can be judged that the value satisfies the requirement, and in the opposite case, it can be considered as abnormal data.

If the comparison determines the abnormal data, the measured value is identified, and the predicted value of the SVR model regression calculation replaces the original value. In the subsequent calculation process, the predicted value is used as the input data, and the iteration is continued.

3.3   Impulse Consumption Detection Algorithm Based on Isolated Forest Clustering

(1) Isolated forest algorithm

For the common commodities, a K-mean and Isolation Forest (K-iforest) algorithm is proposed. The isolated forest algorithm (iForest) has two stages. In the process of processing, t corresponding tree is constructed to form an isolated forest, and then the abnormal score of the sample is determined. The specific process is as follows.

The first stage: Identifying isolated forests consisting of t iTrees. The implementation steps are as follows:

(1) Randomly select Ψ sample points according to the principle of existence and put them into the root node of the corresponding tree.
(2) Randomly assign a dimension to determine the cut point p of the node data, and then specify the maximum value interval of the dimension.
(3) Generate a hyperplane at the cut point, then divide the current node data space, then put the data smaller than p in the dimension into the node on the left side, and put the larger value on the child node on the right side.
(4) Then repeat the second step and the third step continuously, and establish a new child node on this basis, and end the cutting when only one data is found in each node.

The second stage: calculate the abnormal score for the tested samples.

For a training data x, the traversal operation is carried out to determine the final level that x eventually falls on each iTree and its average depth h in each iTree. If the depth of the sample in the iTree is found to be smaller, it can be inferred that the abnormal score is higher. The sample to be tested; C's exception score is defined as shown in Eq. 4:

$$s(x) = 2^{-\frac{e[h(x)]}{c(\Psi)}}$$

(4)

Where: h(x) is the depth of the node retrieved by the detected sample in the iTree; E[ ] is the average of all t iTrees; c(Ψ) is the average path length of the binary search tree constructed by Ψ point (it is a standard term); $H(K) = \ln(k) + \xi$, ξ is the Euler's constant.

Analysis of the above results shows that the closer s(x) is to 1, the more likely it is to be abnormal data, and the opposite is more likely to be normal. In the analysis process, if the s(x) values of most of the training samples are found to be around 0.5, it can be judged that there is no abnormal value.

(2) **Algorithmic model construction of isolated forest**

The construction steps of iTree in this paper are as follows:

Firstly, randomly select one feature among the six sales curve characteristic indicators. Secondly, a value k of this feature is determined based on the random algorithm. Thirdly, according to the feature classification, all the records are processed, and then place the records in which no more than k are placed in the left child node, and the rest are placed on the right side. Fourthly, the branches on both sides of the recursive processing amount until the following requirements are met: The incoming data set has only one record or multiple identical records and the height of the tree has reached a defined height.

In most cases, the amount of abnormal data is generally small, there are also significant differences from the normal data, and thus the path length is also relatively low. In this case, as long as the portion below the average depth is considered, the requirement can be satisfied.

(3) Determination of outliers

The iForest algorithm can only evaluate one item at a time during the evaluation process. In the evaluation process, a corresponding traversal operation is needed to determine the position of the leaf node where the object falls. And then the corresponding abnormal score is determined. Based on this result, it is determined whether the sales volume on that day is abnormal or not.

This article uses the K-means algorithm here. Due to the general distinction criteria that iForest's anomaly scores have given, it is possible to exclude the special case where all abnormal scores approach 0.5. Therefore, this paper considers the K-means calculation of the abnormal score and the minimum path of the data, and divides the data into multiple normal classes and one exception class by modifying the number of cluster centers.

4 Instance Verification

4.1 Verification of Sliding Window Anomaly Detection Algorithm Based on Support Vector Regression

(1) Selection of sliding window

The sales data of one month is analyzed. Therefore, the sliding window size is set to 15, 20, 25, 30 in the experiment, and then, SVR model based on RBF kernel function is used to predict sample sales data under different window size. The trend of the evaluation results is shown in Fig. 1:

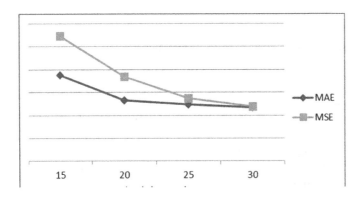

Fig. 1. Effect of window size on the model

Looking at the prediction effect of the SVR model under different window sizes in Fig. 1, as the window size increases, the average absolute error MAE and the mean square error MSE gradually decrease, while the average running time increases gradually, and when the size of the sliding window increases to 20, the running time changes begin to level off. Therefore, after considering the accuracy and time

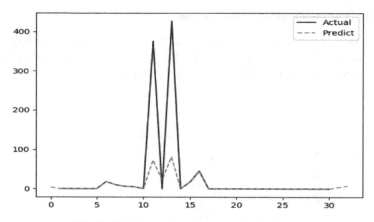

Fig. 2. Single-month forecast results for luxury

efficiency, the window size is chosen to be 30, which makes the single-step prediction model better (Fig. 2).

(2) **Analysis of prediction results based on SVR**

After determining the optimal sliding window size, the RBF kernel function and the polynomial kernel function are respectively compared and verified. The average absolute error MAE, mean square error MSE, average running time and other parameters are used to evaluate the model. When measuring, some extremely obvious abnormal dates are removed.

It can be seen from the result that the multiple parameters of the polynomial kernel function in the SVR model are relatively poor. The SVR prediction mode based on the RBF kernel function performs better in terms of prediction effect and running time. As can be seen from Fig. 3, in the case of no sales for a large amount of time, the model can well fit the curve of small fluctuations in sales while recognizing extremely high sales days.

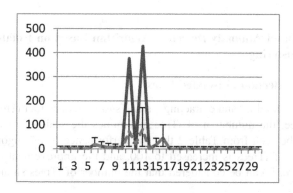

Fig. 3. Prediction interval of the 95% confidence level

(3) **Analysis of abnormal detection results**

According to the analysis and verification of the previous paper, the sliding window size is 30, and the SVR model uses the RBF kernel function to establish a single-step prediction model to forecast and test for daily sales (Fig. 4).

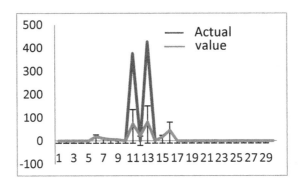

Fig. 4. Prediction interval of 99% confidence level

From the single-month test results of the luxury category in the figure, it can be seen that after the confidence level is increased from 95% to 99%, the data points on the 12th day between the two sales peaks are included in the prediction interval. From detection effect statistics in Table 1, it can also be seen that both the detection rate and the false detection rate have improved. Since the algorithm needs to weaken the sensitivity to luxury data, it is reasonable to take a 99% confidence level.

Table 1. Evaluation table of the impact of confidence level on detection results

Confidence level/PI	Detection rate	False detection rate
95%	96.63	9.35
99%	98.27	3.89

4.2   Verification of Anomaly Detection Algorithm Based on Isolated Forest Clustering

(1) **Parameter selection of isolated forest**

By randomly sampling and extracting a small subset to construct iTree to ensure the diversity of iTree, the number n of iTree determines the scale of model integration learning. It can be seen from Table 2 that the performance of the algorithm is stable after the number of iTrees is increased to 100, and the improvement of AUC is not helpful. Therefore, it can be considered that the number of iTrees should be 100.

Table 2. Calculation time and AUC of the isolated forest algorithm for different iTree numbers t

The number of iTrees n	Time/s	AUC
5	0.079	0.839
10	0.105	0.908
25	0.191	0.927
50	0.405	0.934
100	0.787	0.942
150	1.024	0.953
200	1.336	0.947
250	1.778	0.952
300	2.074	0.948

(2) **Analysis of the results of anomaly detection algorithms based on isolated forests**

When the number of iTrees is 100, the algorithm works well. When detecting the sales data of common categories, there is a small amount of misjudgment due to the accuracy problem. When the iForest algorithm is completed, its abnormal score is taken out and K-means clustering is carried out with the source data. Table 3 shows the time and effect of the algorithm when selecting different cluster numbers K for common categories.

Table 3. Influence of different cluster number K on model algorithm (general category)

Cluster number K	Time/s	Correct rate
2	2.08	0.523
3	2.06	0.531
4	2.03	0.623
5	2.11	0.983
6	2.08	0.981
7	2.11	0.983
8	2.15	0.982

It can be seen from the table that after the K-mean algorithm is combined, the accuracy rate increases with the increase of the clustering center, and the correct rate is relatively stable after reaching 5 clusters. It is shows that the clustering effect of two types of goods when K takes different values through figure.

It can be seen that when the value of K reaches 5, the abnormal scores are obviously high after K-means clustering and the sparsely distributed parts are abnormal classes. And some points with an abnormal score of around 0.6 are classified as normal due to density, which reduces the probability of misdetection, and effectively eliminates the abnormal group from some sales days with relatively small fluctuations (Fig. 5).

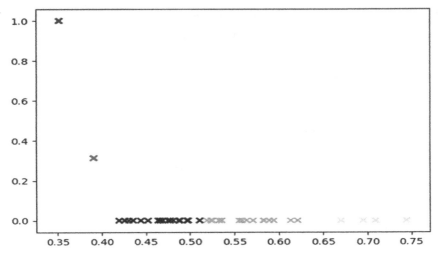

Fig. 5. k-iForest test results (k = 5)

5   Conclusion

Based on the sales characteristics of FMCG, this paper proposes an anomaly detection model, improves the anomaly detection related algorithm, and achieves the stability of demand forecast for FMCG. For luxury goods, the SW-SVR algorithm is designed and improved. Combining the advantages of window-based algorithm and model-based prediction algorithm, it has high adaptablity to regular sales days and small-scale promotion fluctuations. The results show that the algorithm can fit the sales curve well when the efficiency of the algorithm meets the requirements, and it is also good tolerance for small sales fluctuations. For common commodity products, the high efficiency of iForest (isolated forest) algorithm is combined with K-means algorithm to adjust the sensitivity of the algorithm, and the K-iForest algorithm designed and integrated has high efficiency and controllability. It is proved that the K-means algorithm can effectively reduce the sensitivity of iForest and maintain the high efficiency of the algorithm.

References

1. Liang, X.: Research on new media marketing strategy of FMCG in marketing 3.0 environment. J. Bus. Econ. (19), 61–63 2017
2. Yan, Y., Sheng, G., Liu, Y., et al.: Transformer state anomaly detection based on sliding window and clustering algorithm. High Volt. Eng. **42**(12), 4020–4025 (2016)
3. Song, H., Wei, D., Tang, G., et al.: User behavior anomaly detection algorithm based on pattern mining. J. Chin. Comput. Syst. **37**(2), 221–226 (2016)
4. Yu, Y., Zhu, Y., Wan, D., et al.: Hydrological time series anomaly detection based on sliding window prediction. J. Comput. Appl. **34**(8), 2217–2220 (2014)

5. Liu, S., Wu, Y., Che, Y.: Detection of outliers in water supply network based on interaction recognition. Water Supply Sewerage **11**, 150–154 (2015)
6. Huo, S., Zhao, J., Li, D., et al.: Illegal intrusion recognition algorithm for abnormal sequence detection. Comput. Eng. Appl. **53**(20), 68–74 (2017)
7. Chen, R., Liang, C., Xie, F.: A review of the application of nonlinear time series prediction based on SVR. J. Hefei Univ. Technol. (Nat. Sci.) **36**(3), 369–374 (2013)

Research on Font Emotion Based on Semantic Difference Method

Shuo Zhang[1]([✉]) [iD], Pengjiang Wang[1], and Wenjun Hou[2]

[1] Beijing University of Posts and Telecommunications, Beijing 100876, China
zhangshuoholo@gmail.com
[2] Beijing Key Laboratory of Network System and Network Culture, Beijing
University of Posts and Telecommunications, Beijing 100876, China

Abstract. At present, there are numerous studies on the design of Chinese character fonts, but there is few research on the emotion tendency of them. Semantic difference method and some statistical analysis methods were used to find out the common feelings of different people on the emotion conveyed by fonts. Through literature research, a large number of words were collected which describe the emotional tendencies conveyed by the fonts and two times of clustering analysis were performed using the card sorting data to complete the screening, then some participants were invited to obtain the evaluation scores of the 20 kinds of representative fonts on the remaining 16 pairs of emotion words. Subsequently, the project analysis method was used to correct the scale. Afterwards the factor analysis was used to extract the factors that could affect emotions conveyed by fonts. Finally, three factors were identified as the main influence on the font's emotional tendency: the degree of exquisiteness, the cabined feeling and the sense of order.

Keywords: Chinese character font ·
Semantic difference method (SD Method) · Emotion · Factor analysis

1 Introduction

As one of the significant carriers for the inheritance of Chinese culture, Chinese characters constantly evolve in line with the development of history. It has a history of over 3000 years, beginning with the inscriptions on bones or tortoise shells of the Shang Dynasty. After that in the early days of Qin and Han Dynasty, there have been tremendous changes in just a few decades [1]. Then, with the improvement of social productivity, people not only pay attention to the writing speed and efficiency of fonts, but also emphasize the function of self-expression. In order to satisfy these demands, the art form of calligraphy began to emerge and gradually developed into a unique system. The development of calligraphy relies heavily on the expression of emotion and consciousness, which may be the earliest application of the emotion expression of creator by the appearance of text. The development of Chinese character fonts is accompanied by the evolution of carriers. Pottery, oracle bones, bronzes, jade, stone, bamboo slips and wooden tablet are the main carriers before the invention of paper [2]. Nowadays, the universal popularity of smart devices has made screen the main carrier of information transmission, therefore more and more people choose to read on the terminals such as mobile phones instead of traditional paper media [3]. The portability of these terminals allows people to spend more time in reading, and the importance of fonts is self-evident.

There are a lot of researches on Chinese characters in China, and the categories are also various. The Chinese character has a long history, so after a long period of development and accumulation, it has formed a distinctive writing system which is not only an graphic symbol for conveying information, but also a symbol of the profound background of Chinese culture. However, there are few related studies on the emotional communication between Chinese character fonts and the audience, and major researchers still focus on the font modeling itself.

In the emotion research of fonts, foreign scholars carried out relative research earlier and proposed that the characteristics of the fonts contribute to convey specific emotions to readers, therefore they did some researches on the relationship between fonts and emotions. For example, Grohmann et al. studied how the brand impressed customers with its logo's font and what kind of brand personality was communicated through characteristics of text's typeface and color [4]. Some scholars conducted eye movement experiment to analyze the reader's emotional experience in psychology. However, there are several studies on font emotions in China, most of which explain the relationship among the structure of fonts, glyph features and emotions in the form of theoretical narrative, e.g. some researchers analyzed the different emotional performances in the aspect of the point, line and surface of Chinese character design [5]. Some proposed using emotion as the standard for font classification, and collected a large number of questionnaires for analysis, but it still belongs to the category of qualitative analysis [6]. There are also some researchers studied how to design fonts with emotional tendency [7, 8].

In summary, currently there is still a lack of research on the emotional tendency of Chinese characters. Most of the researches tend to be theoretical with a strong subjectivity. Some researchers were aware of this problem and conducted certain experiments, but the analysis of data remains in qualitative level. Meanwhile quantitative analysis was not deep enough, so hardly any inner relationship were discovered among the data. Most of the research focused on decorative fonts, which do not have sufficient versatility, instead of common fonts.

The need for Chinese characters to convey emotions makes font design not only a pictographic font modeling but also a graphic symbol design, which is an emotional bond between the reader and the text [9]. Whether in the design or application of the font, designers are required to correctly convey the emotions that they want to express to the audience. How these subjective and abstract emotions are understood and accepted is the primary problem that designers are currently facing with, and another ensuing question is what are the potential factors that influence the tendency of the emotion the readers feel.

The purpose of this study is to collect a large amount of data and perform statistical analysis of it to solve the two problems mentioned above and find out the commonality between the designers' and readers' emotional tendency towards the font. Then further research was conducted to explore potential relationships in data, and extract the main factors that affect the differences in the emotional tendency of the fonts. These factors can also be used to predict the emotional tendency of the font and as a guide for font design.

2 Method

2.1 Summary of Research Procedures

The study was carried out according to the following process: First, numerous emotion words were collected by literature review and they were screened through expert interviews and cluster analysis, then the remaining words were paired according to SD method. Second, various fonts which are frequently-used in design and reading were collected and the representative fonts were selected to establish a scale for testing combined with the emotion word pairs mentioned before. Third, project analysis of effective results was conducted to remove the words that couldn't cause obvious differences in the test so that the word pairs that were used for the subsequent analysis were obtained. Finally, the hidden factors with insignificant linear relationship were extracted by using factor analysis as the main factor that influence the emotional feeling of audience conveyed by fonts.

2.2 Data Analysis Methods

The SD method is mainly used in the process of establishing the scale. It was first applied in the field of psychology experiments and first introduced by American psychologist Osgood in 1957. The degree of emotion word in semantics is used as a yardstick for psychological experiments [10]. By scoring a pair of words with opposite meanings in Likert scale, this method makes it possible to describe the research object quantitatively. In simple terms, it is to quantify the subjective feelings of a person for a particular object, so that it is no longer a vague feeling, but a specific degree. Later, this method gradually applied in other fields, especially after the emergence of kansei engineering, its application range had been greatly expanded.

In the process of emotion word screening, clustering was mainly used, because this process required the support of statistical methods rather than relying on the judgment of individuals. Later in the data processing period, project analysis and factor analysis played an important role.

3 Experiment

3.1 Collecting Emotion Words

Emotion words should cover the feelings of different people as many as possible, in order to comprehensively evaluate different fonts. Through the literature, the network and in-depth interviews with experts, a total of 348 emotion words were collected for describing texts, fonts and calligraphy styles, which nearly contains all descriptions of the emotions conveyed by the fonts.

Preliminary Screening. The original words collected must be screened, otherwise, the test burden of participants was so heavy that a lot of invalid data would be generated. The principles of screening were as follows: Emotion words should be suitable for describing different types of fonts; The remaining words need to include as many feelings as possible that might be conveyed by the font; The meaning of emotion words themselves should be simple enough to be understood and cannot be ambiguous.

Six experts engaged in design-related fields were invited to screen words according to these principles, and selected words from the initial word collection that were not suitable for describing fonts, ambiguous or words that were difficult to understand. In the end, 191 words with a selected frequency that was equal or greater than 2 were removed, and the number of remaining words was 157.

Secondary Screening with Card Sorting. The first screening guaranteed the validity of the emotion words, but words with similar semantics still existed, so a second screening is indispensable. In this process, an experimental method of open card sorting was used, and 17 practitioners in the design field were invited to classify the remaining 157 words, and the words with similar meaning were divided into a same group. At the same time, the words that were unsuitable for describing the font emotion tendency were selected. This experiment was conducted on the Usability Test platform.

Cluster Analysis. Before cluster analysis, 76 words in the card sorting test that were considered to be unsuitable were removed, and only the 81 words and corresponding data remained for analysis.

A symmetry matrix was established by selecting the word distance (i.e., the distance between two emotion words, that was the number of participants who did not select the two words in the same group) as a data set for subsequent studies. The matrix was imported into the SPSS software for a two-stage cluster analysis: Found out the appropriate number of clusters by hierarchical clustering that would be used as the parameter K in K-Means clustering. The most representative word in each category of clustering results was selected for SD method.

Hierarchical Clustering. R-cluster analyses was used for clustering whose result showed that as the clustering progressed, the degree of similarity within the group gradually decreased. According to the data, a scree plot can be graphed, the horizontal axis is the average distance between current individual and other groups, and the vertical axis is the number of categories (Fig. 1).

As the chart shows, there is a sudden increase of the intervals when categories number is 25, 13, and 10. These points can be considered as the number of classifications. Considering that the words after clustering might become a pair, and the feeling of the fonts should also be comprehensive, 25 was decided to be the number of categories as the standard for the next clustering.

K-Means Clustering. K-Means cluster analysis was conduct on the words distance matrix and 25 was specified as the output categories amount. Finally, after 3 iterations, the results of the clustering were got. Considering the distance between the distance center of each word in each category and the representativeness of the word itself, 25 from 81 emotion words were selected to represent all the feelings, and then each word is paired with priority to matching of internal words. Finally, a total of 16 pairs of emotion words such as incompact – compact, popular – traditional were obtained (Table 1).

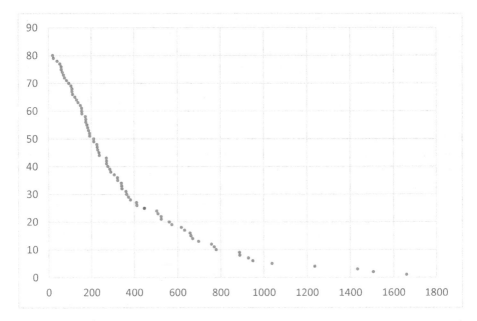

Fig. 1. Scree plot of hierarchical clustering of word distance

Table 1. 16 pairs of emotion words

Number	Emotion word pair	Emotion word pair in Chinese
w1	incompact – compact	舒展——紧凑
w2	popular – traditional	流行——传统
w3	untrammeled – steady	奔放——稳重
w4	casual – regular	随性——整齐
w5	elegant – rough	优雅——粗犷
w6	unadorned – ornate	朴实——华丽
w7	unshaped – smooth	大气——细腻
w8	mellow – hale	圆润——硬朗
w9	neoteric – classical	新颖——古典
w10	slender – plump	纤细——饱满
w11	intricate – simple	复杂——简单
w12	delicate–wild	秀气——狂野
w13	gentle – lively	文雅——活泼
w14	well-balanced – swollen	匀称——臃肿
w15	thick – acuminate	浑厚——锐利
w16	romantic – insipid	浪漫——平淡

3.2 Determine the Chinese Character Fonts for Experiment

According to the purpose of this study, 56 commonly used design fonts were collected through literature research, internet and other methods. Considering the difficulty of distinguishing fonts and testing itself, too many fonts will lead to an increase in the difficulty of distinguishing emotional tendency. A large amount of fuzzy data was not conducive to the subsequent analysis, so that some fonts with little differences or similar styles were removed, and 20 fonts of Chinese characters remained.

3.3 Semantic Difference Scale

16 pairs of emotion words extracted in the steps above were used to establish a semantic difference scale, the two ends of the scale were a pair of antonym. 7-point Likert scale was used for scoring, ranging from −3 to 3, and the scores represent the degree of feelings of the font to the emotion word at either end (Table 2).

Table 2. Example of emotion word pair scale

Emotion word		Emotion word
Incompact	−3 −2 −1 0 1 2 3	Compact
Popular	−3 −2 −1 0 1 2 3	Traditional

A online questionnaire of 20 kinds of fonts that each was along with a 16 emotion word pairs scale was used for further study. The style of each font was regular and font size was 20, moreover the same content was displayed with black text on white background. A total of 36 respondents completed the questionnaires and 32 questionnaires were valid, so there were 640 copies of scale data in total. Among the respondents, there were 17 men and 15 women, and 18 of them had experience of design related work. The above data was collected to ensure that the results of the questionnaire could best reflect the public's cognitive commonality in the emotional feeling of the font.

3.4 Correction of the Scale

This study combined the Critical Value Ratio method and the Pearson Analysis method to test the reliability of the scale consist of 16 emotion word pairs, then eliminated the less reliable word pairs to correct the scale.

Critical Value Ratio Method. The sum of the scores graded by every respondent of each font under all emotion word pairs were calculated and sorted, then the front 27% of the whole data (173 sets of data groups) were regarded as the high score group, and back 27% as the low score group, then independent sample T test was conducted on the two groups(The critical point of the high or low score group can theoretically take 25% to 33%. When the number of samples is sufficient and normal distribution, the discrimination of 27% is the most reliable) (Table 3).

Table 3. High group and low group data significance test table

Emotion words	w1	w2	w3	w4	w5	w6	w7	w8
Significance	0.000	0.000	0.000	0.000	0.000	0.296	0.496	0.000
Emotion words	w9	w10	w11	w12	w13	w14	w15	w16
Significance	0.000	0.000	0.000	0.000	0.000	0.000	0.014	0.000

Whether the two-tailed significance is less than 0.01 was the standard to decide the emotion word pair made a difference to the results or not. According to the analysis results in the above table, it showed that the three emotion word pairs (w6, w7 and w15) were not so significant that they ought to be removed.

Mean Pearson Analysis Method. Pearson correlation analysis was performed using the score of each emotion words pair of a Chinese character font and the average score of all word pairs of it, and the correlation coefficient and the P value of the corresponding probability were calculated as follows (Table 4).

Table 4. Correlation coefficient significance test table

	w1	w2	w3	w4	w5	w6	w7	w8
Pearson correlation	0.544	0.490	0.526	0.487	0.470	0.039	0.077	0.477
Significance	0.000	0.000	0.000	0.000	0.000	0.325	0.053	0.000
	w9	w10	w11	w12	w13	w14	w15	w16
Pearson correlation	0.472	0.477	0.287	0.464	0.209	0.364	0.179	0.571
Significance	0.000	0.000	0.000	0.000	0.000	0.000	0.014	0.000

Whether the two-tailed significance is less than 0.01 was selected as a significant criterion for the Pearson coefficient, if the criteria were met, it indicated that the score of the word pair was significantly correlated with the overall average, which indicated that the variable had a significant impact on the whole score. The table above shows that there were 2 pairs of emotion words (w6 and w7) whose significances were more than 0.01, therefore they should be removed.

Combining the two methods of project analysis, the three pairs of emotion words (w6, w7 and w15) were removed, and finally the semantic difference scale was corrected according to the score, also the corresponding data was valid and would be used for the analysis of the next stage.

4   Results and Discussion

The emotion words have been clustered and screened multiple times in the previous process, but there are still 13 pairs of vocabulary words remaining in the evaluation scale. Although there is a comprehensive evaluation of the emotional tendency of the font, there are two problems: the number of variables is large, if these variables are involved in mathematical modeling, the computational workload in the analysis process will be greatly increased; there may still be some correlation between these variables. Factor analysis can be used to solve these problems for dimensionality reduction.

The raw data was processed, and the average of each emotion word pairs graded by every respondent was calculated. Afterwards, the word pair was used as a variable, and the average was used as sample data to perform factor analysis in SPSS.

Whether the data met the condition of factor analysis should be checked in the first place, so KMO and Bartlett test was carried out (Table 5).

Table 5. KMO and Bartlett test results

Kaiser-Meyer-Olkin Measure of Sampling adequacy		.814
Bartlett's Test of Sphericity	Approx Chi-Square	245.824
	df	78
	Sig.	.000

The KMO value was 0.814, and the number more than 0.8 indicated that it was suitable for factor analysis according to commonly used standards. The result of significance of Bartlett spheroid test turned out to be less than 0.001, so the data proved to be suitable for factor analysis.

The correlation matrix was used as the basis for the extraction of factor, all the factors with eigenvalues more than 1 were extracted, so that 3 factors were abstracted, which explained 72.62% of the total variance of the original variables. In order to make the factors named with better interpretative, the variance maximal method was used to

Table 6. Rotated component score coefficient matrix

Emotion word pairs	1	2	3
w12: delicate–wild	.889	.226	.130
w14: well-balanced – swollen	.860	.015	−.005
w13: gentle – lively	.802	.155	.269
w5: elegant – rough	.760	.278	.221
w10: slender – plump	.705	.394	.085
w2: popular – traditional	.136	.883	.147
w3: untrammeled – steady	.200	.781	.325
w9: neoteric – classical	.360	.751	−.023
w1: incompact – compact	.183	.688	.233
w11: intricate – simple	.127	.048	.890
w16: romantic – insipid	.422	.202	.755
w4: casual – regular	−.135	.510	.657
w8: mellow – hale	.265	.477	.550

implement the orthogonal rotation of the factor loading matrix, and the variance of each factor are re-allocated to improve interpretability.

Each pair of emotion words in the table is explained by three factors. The absolute value of the load of each factor represents the contribution of the factor to the interpretation of the variable. It can be seen from the table that three factors are extracted as the main factors affecting the emotional tendency of fonts: the degree of exquisiteness, the cabined feeling and the sense of order (Table 7).

Table 7. Variables and factors naming

	Emotion words	Name of factor
Factor 1	w12: delicate–wild	The degree of exquisiteness
	w14: well-balanced – swollen	
	w13: gentle – lively	
	w5: elegant – rough	
	w10: slender – plump	
Factor 2	w2: popular – traditional	The cabined feeling
	w3: untrammeled – steady	
	w9: neoteric – classical	
	w1: incompact – compact	
Factor 3	w11: intricate – simple	The sense of order
	w16: romantic – insipid	
	w4: casual – regular	
	w8: mellow – hale	

5 Conclusions

Text as a visual graphic symbol is able to convey emotion to the readers. In addition to the content of the text itself, the font is the most intuitive trigger of the emotional tendency, which determines how the audience understands the designer's thought. From Maslow's hierarchy of needs, it can be known that people will pursue high-level needs after low-level needs are met, which is same to Chinese character fonts. After the functional and practical problems were basically solved, the perceptual needs become the main concern of people. In this study, the subjective feelings of people, no matter he or she had work experience on design, were taken into account with the purpose of making the results reflect the emotions of most people. However, there were some shortcomings in this study, such as the small number of participants, and the proportion of designers was higher.

The three factors finally extracted are the evaluation criteria that affect the emotion conveyed by fonts to the readers, so through the scoring results in these three dimensions, the emotional tendency of the font under the original 13 pairs of emotion words can be inferred to a certain extent. Take any font as an example, and invite a participant score in the three dimensions (exquisite–not exquisite, not cabined–cabined, unordered–ordered) with the range from −3 to 3. Assume that the score was −2, 1, 2 severally, it would be possible to calculate the predicted score for each emotion word pair based on the coefficients in Table 6. The score on the emotion word pair delicate–wild of the participant might be

$$-2 \times 0.889 + 1 \times 0.226 + 2 \times 0.130 = -1.292$$

so speculate that he thinks the font is delicate rather than wild. This kind of prediction can provide a certain degree of guidance for the usage of fonts in the design process, and the dimensionality reduction of multiple variables also lays a foundation for further analysis, which can be combined with the study of kansei engineering, and these factors can be associated with characteristics of fonts themselves to establish a mapping relationship in order to study the deep connection between the font features and their emotional tendencies.

References

1. Handong, D.: The Researching of the End of 20th Century of the Development of Chinese Character Design. China Academy of Art, Hangzhou (2009)
2. Li, H: The studying on the Chinese character design in digital context of the Chinese character evolution. Soochow University (2013)
3. Lin, X.: The discussion of the Chinese font design for screen display. Art Des. (2), 127–128 (2011)
4. Grohmann, B., Giese, J.L., Parkman, I.D.: Using type font characteristics to communicate brand personality of new brands. J. Brand Manage. **20**(5), 389–403 (2013)
5. Yu, R: Chinese characters font design modelling and affective image analysis. Wuhan University of Technology (2012)
6. Pei, Y: Research in Chinese and English font selection optimization based on the users' experience. Dong Hua University (2012)
7. Yang, L.: The manifestation mode of emotion of modeling in the font design about Chinese characters. Design **9**, 116–117 (2014)
8. Pei, K.: Research in emotional presentation way of decorative Chinese character fonts. Tianjin Polytechnic University (2017)
9. Jin, S.: The influence of emotional design on fonts. Design **19**, 138–139 (2017)
10. Zhu, Y., Liu, L., Zhou, W.: Evolution mode of vehicle family characteristics based on the semantic differential methods. Packag. Eng. **2**, 55–58 (2015)

What Is Semantically Important
to "Donald Trump"?

Jizheng Wan[1,2(✉)], John Barnden[1], Bo Hu[3], and Peter Hancox[1]

[1] University of Birmingham, Birmingham, UK
Jizhneg.Wan@coventry.ac.uk,
{J.A.Barnden, P.J.Hancox}@cs.bham.ac.uk
[2] Coventry University, Coventry, UK
[3] Fujitsu Laboratories of Europe Ltd., London, UK
bo.hu@fle.fujitsu.com

Abstract. In the recent years, there is a growing interest in combining explicitly defined formal semantics (in the forms of ontologies) with distributional semantics "learnt" from a vast amount of data. In this paper, we try to bridge the best of the two worlds by introducing a new metrics called the "Semantic Impact" together with a novel method to derive a numerical measurement that can summarise how strong an ontological entity/concept impinges on the domain of discourse. More specifically, by taking into consideration the semantic representation of a concept that appears in documents and its correlation with other concepts in the same document corpus, we measure the importance of a concept with respect to the knowledge domain at a semantic level. Here, the "semantic" importance of an ontology concept is two-fold. Firstly, the concept needs to be informative. Secondly, it should be well connected (strong correlation) with other concepts in the same domain. We evaluated the proposed method with 200 BBC News articles about Donald Trump (between February 2017 and September 2017). The preliminary result is promising: we demonstrated that semantic impact can be learnt: the top 3 most important concepts are Event, Date and Organisation and the least essential concepts are Substance, Duration and EventEducation. The crux of our future work is to extend the evaluation with larger datasets and more diverse domains.

Keywords: Semantic impact · Ontology Learning · XYZ model · Word2Vec

1 Introduction

As a key enabling technology of the Semantic Web, the concept of ontology has been widely used in, not only research labs but also large-scale IT projects. It is "a formal language designed to represent a particular domain of knowledge" [1], and as with other knowledge-based studies in computer science research, people have always dreamed of developing a self-learning mechanism to automate the generation of such formal representations.

Since Maedche and Staab coined the term "Ontology Learning" (OL) [2], people have experimented various learning approaches. Roughly, these approaches have been grouped into four categories: statistical approach, linguistic approach, logical approach

© Springer Nature Switzerland AG 2019
Y. Tang et al. (Eds.): HCC 2018, LNCS 11354, pp. 314–329, 2019.
https://doi.org/10.1007/978-3-030-15127-0_31

and hybrid approach [3]. However, one of the challenges among all these approaches is that, at some point of the learning process, the system needs to make a decision as to whether or not a particular concept should be included in the domain ontology. It is our contention that a method to measure the importance (or relevance) of a concept to the domain knowledge is essential in making such a decision across all OL methodologies.

Using "Harry Potter" as an example, Horrocks [4] demonstrated how to use RDF and OWL to describe the text below, which makes it possible for the software agent to discover that there is a `hasPet` relation between `HarryPotter` and `Hedwig`. Additional properties have been defined at a later stage, such that `HarryPotter` is a (`rdf:type`) `Wizard` and a `Student`, and that `Hedwig` is a `SnowyOwl`.

"Harry Potter has a pet called Hedwig."

Assuming that we need to build an ontology containing key concepts in the Harry Potter story, an immediate question is what concepts could be considered as key, in other words, what makes Harry Potter "Harry Potter"? Three concepts (or ontology classes) have been identified in the above example: `Wizard`, `Student` and `SnowyOwl`. Since the whole story is about how a young wizard studies magic at Hogwarts and fights against an evil senior wizard who graduated from the same school, it is easy to understand that `Wizard` and `Student` are more "important" than `SnowyOwl`, because without them, Harry Potter would no longer be the "Harry Potter" that we are familiar with. On the other hand, the entire story is still coherent if he has a different pet or has no pet at all. Therefore, `Wizard` and `Student` concepts have a bigger influence than `SnowyOwl` on the domain knowledge. In this paper we use the term *"Semantic Impact"* to describe such influence.

In traditional NLP or IR study, there are various ways to measure how important (or relevant) a word is with respect to a document in the corpus. However, the importance or relevance of a word to a document at the syntax level is not quite the same as the importance or relevance of a concept to the domain knowledge at the semantic level. Using TF-IDF as an example, even if people can solve the problem that in fact the `Wizard` concept contains multiple words (e.g. Harry Potter, Lord Voldemort etc.), it is still difficult to reach a high *tf-idf* weight to compute its relevance to the corpus, simply because it is almost guaranteed that this concept will exist in every chapter/document about Harry Potter and therefore will have a low, if not 0, *idf* value which suggests that it is not very informative at all. Previous research also suggested that in some cases, *idf* does not provide any improvement and therefore the *tf* (or a similar) scheme itself is sufficient [5], in which case, the more a term appears in the corpus, the more relevant it is. However, it is not necessarily true at the semantic level. As demonstrated in this paper, in the news article domain, the concept of `Date` has a low frequency compared with other concepts such as `Person` and `Place`, but it can generate a more significant semantic impact compared with the other two.

Therefore, it is unreliable to purely use the frequency or statistics-based approach to decide the "relevance" or "importance" at the semantic level. One common way to handle this issue in OL study is by relying on some pre-defined knowledge to determine what should and should not be included in an ontology. By so doing, the system will lose the ability to learn new concepts and in which case it is more likely to be an ontology populator rather than a learning approach.

As part of XYZ Model research [6], this paper will introduce a new idea called the "*Semantic Impact*" to measure how valuable a concept is to the domain knowledge at the semantic level. There is a mathematical definition at the end of Sect. 2; its textual definition is given as:

Semantic Impact (SI) represents how informative a concept is in the corpus and moreover the strength of its correlation with the other concepts in the domain.

In order to accurately measure the semantic impact, a novel approach will be discussed in this paper. For demonstration purpose, we have manually collected a set of news articles, between February 2017 and September 2017, about Donald Trump and split into two corpora: Source Corpus and Target Corpus. We then use this approach to generate some interesting results about how semantically important each concept in the "Trump" domain is.

2 Research Methodology

In traditional computational linguistics study, the idea of Distributional Semantic Models (DSM) is that the meaning of words can (at least to a certain extent) be inferred from their usage and therefore the semantics can be encapsulated in high-dimensional vectors based on the nearby co-occurrence of words [7]. There are various tools/frameworks, e.g. Word2Vec [8, 9], that have been developed to vectorise the words in the corpus so as to generate semantic representations.

By adopting and expanding DSM theory, this research is based on two assumptions: (a) a high-dimensional vector can be used to infer the semantic representation of a concept, which extensionally is a set of words that belong to the same semantic group, and (b) with sufficient data, for any concept in a domain, the distribution of its semantic representation is consistent.

Therefore, the underlying philosophy of this research is to cross-compare the semantic representation information between two corpora about the same domain. So, the system will be able to identify the patterns of distribution for the domain concepts, then train a set of neural networks to distinguish the high informative concepts from other low informative concepts.

Moreover, it is possible to use the representation of a specific concept to measure the impact or influence that a particular word (or a list of words) could bring to it. By doing so for all the domain concepts on all the vocabularies in the corpora, the system will then be able to measure the correlation between each concept-pair.

Let I_a be the *informative coefficient* for the concept a, C_a be the *correlation coefficient* it has with the other concepts and λ be a constant that adjusts the weight of the correlation, then its Semantic Impact SI_a can be calculated as follows:

$$SI_a = I_a + \lambda C_a \tag{1}$$

The value of λ is normally set empirically and depends on the individual document corpus. For example, if a domain only contains a small number of concepts, then it is highly likely that all these concepts have a strong correlation with each other and thus the informative coefficient plays a more critical role in deciding the semantic impact. A smaller value, therefore, could be assigned to λ (e.g. 0.5) to reduce the overall contribution of C_a.

2.1 System Architecture

The overall process is shown in Fig. 1. The first step is to use an existing tool/method to extract the basic concepts and relations from the source and target corpus and convert into the associated Document based Ontology (DbO) sets [6]. Then step 2 uses a normalisation and vectorisation process to generate the semantic distribution vectors for all the concepts identified in the previous step. Step 3 is designed to calculate the informative coefficient (I) and then use a Maximal Information Coefficient (MIC) [10] based approach, in Step 4, to analyse the correlations between each class/concept pair and generate the correlation coefficient (C). Finally, in Step 5, use Eq. 1 to calculate the semantic impact value. The following section will discuss these steps in detail.

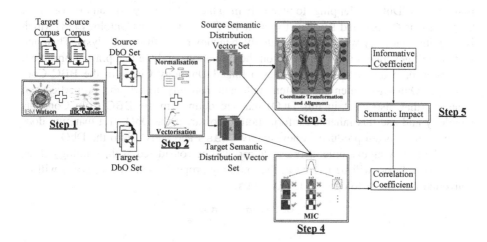

Fig. 1. Process overview

2.2 DbO Construction (Step 1)

As introduced in [6], DbO is an ontology that operates on the document level without concern for the wider context. Essentially, 200 news articles about Donald Trump were manually collected from the BBC News website and split into two corpora: Source Corpus and Target Corpus. Subsequently, the IBM Natural Language Understanding (NLU) [11] service with the default News annotation model was selected to analyse these documents and extract various semantic information (concepts and relations) from them. For example, the "Relations" information below is one of the 438 relations that have been identified from one article [12] by this process.

Relation

```
{
  "type": "agentOf",
  "firstEntityType": "Person",
  "secondEntityType": "EventCommunication",
  "secondEntity": "said",
  "firstEntity": "Sean Spicer",
  "sentence": "Before the list was published, press secretary Sean
Spicer said there were \"several instances\" of attacks that had not gained
sufficient media coverage (without specifying which fell into that
category).",
  "score": "0.99692"
},
```

After this semantic information extraction process, there is a class/property mapping process (DbO/O Mapping) to manually map the Entity Types and Relation Types to the Class and Property in the BBC Core Concepts Ontology [13] which is the initial ontology we use to generate the benchmark for further analysis. Then, the system will generate a DbO set based on these relations and the mapping information. For instance, if an Entity Type has a linked Ontology Class (in the BBC Core Concepts Ontology), then the system will automatically inherit the properties and relations (that also exist in this DbO) that are defined in the BBC Core Concepts Ontology and use this inherited information to construct the DbO. If a mapping does not exist, the system produces a new empty Class and adds it into the DbO.

In the following example, Entity Type will be considered as the ontology class; Relation Type will be treated as the ontology property, and Relations will be converted into the ontology Individuals:

Ontology Class

```
DbO:Weapon    a          owl:Class ;
        rdfs:isDefinedBy
DbO:0e0e6bf58a95f44aee0f937e33a2532b ;
        rdfs:label          "Weapon"@en .
```

Ontology Property

```
DbO:occupation   a          owl:ObjectProperty ;
        rdfs:domain        DbO:Person ;
        rdfs:isDefinedBy
DbO:0e0e6bf58a95f44aee0f937e33a2532b ;
        rdfs:label          "occupation"@en ;
        rdfs:subPropertyOf  DbO:notablyAssociatedWith .
```

```
                        Ontology Individual
DbO:15d64395d922fa5ff25ba4b01b0f9615_0.726746
        a                              DbO:NaturalEvent ,
DbO:Vehicle ;
        DbO:eventTheme              "affectedBy" ;
        Property:FirstEntity        "vehicle" ;
        Property:FirstEntityType     "Vehicle" ;
        Property:Score              "0.726746" ;
        Property:SecondEntity       "crashed" ;
        Property:SecondEntityType    "NaturalEvent" ;
        Property:Sentence           "Kuwait City, Kuwait, October
2016 What happened: An Egyptian man was detained after a bin
lorry reportedly loaded with explosives crashed into a vehicle
carrying five US soldiers." .
```

By going through all the documents in the Source and Target Corpus, the system generates 200 DbOs that are grouped into two sets: Source DbO Set and Target DbO Set. Compared with the original information extracted from the NLU, the DbO set is more convenient for us to analyse the ontological relations between each of the individuals.

2.3   Semantic Distribution Calculation (Step 2)

Vectorisation is done by the Word2Vec model in the DeepLearning4 J framework [14] with the following configuration: MinWordFrequency = 1, LayerSize = 100 and WindowSize = 5.

The semantic distribution is, in fact, a vector obtained from the vectorisation process. It is easy to get the semantic distribution for any single word in the corpus, but since a concept will contain multiple words, the challenge here is how to generate a single vector to represent the collection of individual word vectors that preserve the semantic meaning of the concept in a high-dimension space.

This is achieved by a normalisation process. The basic idea is to replace all the relevant words/entities about a specific concept from the corpus with a unique string and re-run the vectorisation process to generate a new Word2Vec model for this specific concept. Then the vector of this unique string could be considered as a projection of all the vectors of the replaced words on this newly created Word2Vec model, and considered to be tantamount to a semantic distribution vector for the original concept/class. By repeating this process, we could generate a separate Word2Vec model for all the concepts in both Source and Target Corpus respectively. We denote the new Word2Vec models created via this normalisation process W2V_<ConceptName> and the original Word2Vec model generated from the corpus Master W2V. Meanwhile, we use the Master W2V as the baseline model for aligning W2V_<ConceptName> models discussed in the next section.

There are two reasons to generate separate models instead of replacing all the relevant words from the corpus with all the unique strings in one go. Firstly, by the nature of how Word2Vec (or any word embedding method) works, replacing too many words may significantly change the grouping structure, and therefore the new model will not be able to represent the same semantic distribution as the old model does. Hence, it is essential to minimise the amount of words that need to be replaced in each model in order to maximise the consistency of the semantic representation.

Secondly, within a different context, the same word could be identified as different concepts. For example, the word "Trump" can be both Person and Place (the Trump building), and we cannot replace the same word twice with two different unique strings in one model.

2.4   Coordinate Transformation (CT) Process (Step 3 – Part 1)

By the end of the last process, the system generates two sets of the semantic distribution vector as well as the associated Word2Vec models. Essentially, the system will use these vectors to calculate the informative coefficient information. However, before giving further details, there is a more general issue that needs to be discussed here: how to compare vectors that occur in two different Word2Vec models.

Word2Vec is one of the most popular methods to vectorise words in a corpus and generate the semantic representations [8, 9], and Cosine Similarity (CS) is one of the primary methods used to compare two words/vectors inside one Word2Vec model. However, most of the vectors used by this research are in fact from different Word2Vec models and effectively projected different coordinate systems whereof CS values cannot be calculated directly. It is essential to perform coordinate transformation to align different Word2Vec models. This alignment can be anchored on common words appearing in both models.

For example, if both Word2Vec models XYZ and X'Y'Z' have words "Trump" and "President", let $\vec{V}_1^T$ and $\vec{V}_1^P$ be the vectors of the word "Trump" and "President" in the first model respectively and $\vec{V}_2^T$ and $\vec{V}_2^P$ be the corresponding vectors in the second model, the goal is to make $\vec{V}_1^T$ and $\vec{V}_2^T$ as close to each other as possible (same applies to $\vec{V}_1^P$ and $\vec{V}_2^P$). This is formally defined as follows:

$$\text{Argmax}\left(\frac{\vec{V}_1^T \cdot \vec{V}_2^T}{\|\vec{V}_1^T\|\|\vec{V}_2^T\|} + \frac{\vec{V}_1^P \cdot \vec{V}_2^P}{\|\vec{V}_1^P\|\|\vec{V}_2^P\|}\right) \tag{2}$$

We simplify the solution to the above formula to a classic supervised learning problem with neural network. Let XYZ be the master or target Word2Vec model (to be aligned against) and X'Y'Z' be the source model (aligned from). Also let $\vec{V}_2^T$ & $\vec{V}_2^P$ be the input, and $\vec{V}_1^T$ & $\vec{V}_1^P$ the labels of the associated input.

The neural network implementation consists a fully-connected feedforward neural network with 3 hidden layers as illustrated in Fig. 2. It takes a 100×1 vector (LayerSize of the W2V) as the input and outputs a 100×1 vector. We use TANH on the Output Layer forcing the output values to scale down to between $[-1, 1]$. Each of

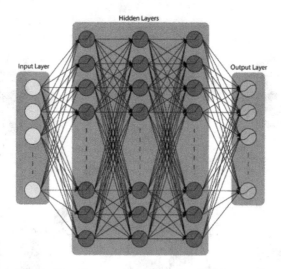

Fig. 2. Neural network structure for the CT process

the Hidden Layers contains 2000 nodes and uses ReLU as the activation function. The other configurations include:

- Using XAVIER for the weight initialisation [15].
- Using ADAM as the method for the stochastic optimisation [16].
- Set to BatchSize 100.
- Set to Number of Epochs 350.

2.5 Aligned Cosine Similarity and Informative Coefficient (Step 3 – Part 2)

As discussed before, there is a difference between the frequency-based relevance at the literal level and the informative at the semantic level. This difference is caused by the fact that the former does not take into consideration the position of a word in the sentence and its context while the second one does. Since the Semantic Distribution Vector (SDV) for a specific concept is created by the normalisation process, which is essentially a projection of all the related word vectors, their informative complexity will be inherited in the SDV which is included in the W2V_<ConceptName> model.

This section will focus only on the process we use to generate the informative coefficient (I): the reason for this will be explained in the next section.

For a specific Concept/Class a, let CS'_a be the Aligned Cosine Similarity, and $\overline{Conf}_a$ be the average confidence score. Then:

$$I_a = CS'_a \times \overline{Conf}_a \tag{3}$$

Figure 3 shows how to calculate the CS'_a. Using the Event Class as an example, Source W2V_Event is the Word2Vec model generated by the normalisation process

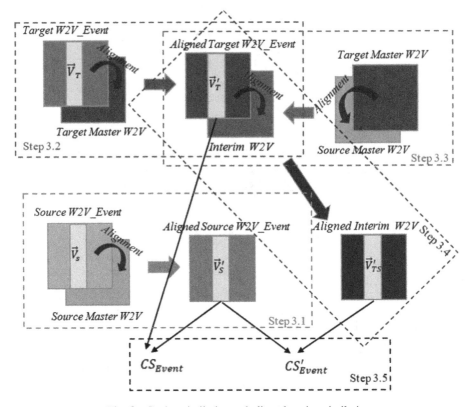

Fig. 3. Cosine similarity and aligned cosine similarity

from the source domain and $\vec{V}_S$ is the Semantic Distribution Vector for the Event concept/class in this model. So, by using the Coordinate Transformation Process discussed before, Step 3.1 aligns the Source W2V_Event model with the Source Master W2V model to create the Aligned Source W2V_Event model and the aligned distribution vector $\vec{V}'_S$. Step 3.2 applies a similar process to align the Target W2V_Event model with the Target Master W2V model, to create the Aligned Target W2V_Event model and the aligned distribution vector $\vec{V}'_T$. In Step 3.3, the system aligns the Target Master W2V model with the Source Master W2V model and creates an Interim W2V model. Then Step 3.4 will align the Aligned TargetW2V_Event model, which was created in Step 3.2, with this Interim W2V model to create an Aligned Interim W2V model which contains a new aligned distribution vector $\vec{V}'_{TS}$. Finally, the Cosine Similarity between $\vec{V}'_S$ & $\vec{V}'_T$ (CS_{Event}) and $\vec{V}'_S$ & $\vec{V}'_{TS}$ (CS'_{Event}) is calculated as:

$$CS_{Event}\left(\vec{V}'_S, \vec{V}'_T\right) = \frac{\vec{V}'_S \cdot \vec{V}'_T}{\|\vec{V}'_S\|\|\vec{V}'_T\|} \qquad CS'_{Event}\left(\vec{V}'_S, \vec{V}'_{TS}\right) = \frac{\vec{V}'_S \cdot \vec{V}'_{TS}}{\|\vec{V}'_S\|\|\vec{V}'_{TS}\|} \qquad (4)$$

By enumerating all the `Ontology Individuals` which contain at least one Event class in the DbO set, it is easy to get the sum of the score (`Property:Score` in the Individual) which is the relation confidence score obtained from the IBM NLU process and ranging from 0 (not confident) to 1 (highly confident). Let N_{Event} be the total number of such `Ontology Individual`, then $\overline{Conf}_{Event}$ will be:

$$\overline{Conf}_{Event} = \frac{\sum_{i=0}^{n} Score}{N_{Event}} \tag{5}$$

2.6 MIC-Based Correlation Analysis (Step 4) and the Final Result (Step 5)

Maximal Information Coefficient (MIC) was introduced by David Reshef to measure the strength of the linear or non-linear association between two variables [10]. It can be used to not only identify essential relationships in the dataset but also to characterise them.

Consider a Domain Ontology as a function which could be used to represent knowledge within a domain. Then the Ontology Classes will be the variables of this function. Moreover, individual words that exist in the corpus are the essential components and "material" that build the domain knowledge. Therefore, each word will have an influence on the knowledge that the Ontology represents. Hence, the individual word will have an indirect impact on the Ontology manifested through the Classes that the individuals belong to. As a result, if we could measure the impact a word could exercise on the various Classes, we could then understand the relations between these Classes. In other words, considering each word in the corpus as an independent sample, the Classes are the variables (or properties), and the value of a specific variable/property in a specific sample is the Cosine Similarity between that word and that concept/class. In this way, the system can generate a sample table with each row corresponding to a word in the Word2Vec vocabulary list, and each column corresponding to a concept/class that has been identified in the corpus.

Using the sample table as the input, the MIC algorithm generates the result that indicates the strength of the correlation between all the class pairs. The correlation coefficient for concept/class a can then be calculated as:

$$C_a = \log\left(\sum_{i=0}^{|R_a|} MIC(a,b) \times \overline{Conf}(b_i) \times \overline{Conf}(a)\right) \tag{6}$$

where $R_a = \{\langle a, b\rangle | \exists b, \langle a, b\rangle \in R\}$.

Therefore, the completed formula for the semantic impact is (Step 5):

$$SI_a = \frac{\vec{V}_S \cdot \vec{V}_{TS}}{\|\vec{V}_S\|\|\vec{V}_{TS}\|} \times \frac{\sum_{i=0}^{n} Score(a)}{N_a} + \lambda \log\left(\sum_{i=0}^{|R_a|} MIC(a,b) \times \frac{\sum_{i=0}^{n} Score(b_i)}{N_b} \times \frac{\sum_{i=0}^{n} Score(a)}{N_a}\right) \tag{7}$$

3 Results and Discussion

We carried out a preliminary evaluation wherein 35 `Entity Types` across the Source and Target Corpus have been identified using IBM NLU service. By going through manual mapping, the 35 entity types have been converted into 29 concepts (or ontology classes) in the DbO Sets as listed in Table 1.

Table 1. Concepts/Ontology classes in the DbO sets

Award	Cardinal	Crime	Date
Duration	EntertainmentAward	Event	EventBusiness
EventCustody	EventDemonstration	EventEducation	EventElection
EventPerformance	EventPersonnel	EventViolence	Facility
GeographicFeature	HealthCondition	NaturalDisaster	Organisation
Person	Place	Product	SportingEvent
Substance	Time	TitleWork	Vehicle
Weapon			

After the Normalisation and Vectorisation process (Step 2), each of them had an associated `Semantic Distribution Vector`. By going through the Step 3 in the Fig. 1, the system will be able to generate their CS, CS' and the informative coefficient.

There are four classes in the BBC Ontology: `Event`, `Organisation`, `Place` and `Person`. For now, we refer to them as `Ontology Class`. The rest of the classes in the above table are identified by the IBM NLU process and will be called `Candidate Class` in order to distinguish from the former. It is interesting to see from the above result that all the `Ontology Classes` have a high CS' value. In fact, a positive correlation between the CS' value and the informative coefficient is evidenced.

An intuitive explanation is as follows. It is easy to understand that for a class with a high informative coefficient value, such as an `Ontology Class`, it will have a more complex structure and relation (or contains more semantic information) compared to a class with a low informative coefficient value. As discussed already, this complexity will be inherited, during the normalisation process, in its semantic representation, and therefore its final *Semantic Distribution Vector* will be more "complex" (or contain more semantic information) than the distribution vector for a class with low IC value even if they have the same dimension size (100×1). Moreover, when we use the Coordinate Transformation Process to align the `W2V_<ConceptName>` Model with the Source (or Target) `Master W2V` Model (Step 3.1 and Step 3.2 in the Fig. 3), it is in fact using a neural network to predicate a vector for a word (the unique string) that never existed in the original `Master W2V` Model. As a result, the CS' value in Table 2 is essentially the degree of alignment of the predication. With this idea in mind, the above result suggests that this predication and alignment process only works well on those classes with a high informative value, otherwise, their CS' value should all be close to 1.

Table 2. Cosine Similarity, Informative Coefficient and Term Frequency (sorted by the CS')

Concept/Class	CS	CS'	I	Total TF
Event	0.0000905786	0.932403684	0.768510029	20.00957069
Organisation	−0.074791484	0.874482393	0.590578904	10.37735849
Place	0.155734465	0.838355482	0.561972973	13.93218485
Date	−0.146419838	0.816355526	0.658128843	2.782335247
Cardinal	−0.072149187	0.772277176	0.498443222	0.676784249
EventViolence	−0.067453243	0.65089637	0.406972780	0.929723817
EventPerformance	0.034194510	0.592700899	0.421278986	0.478534318
EventPersonnel	0.072302915	0.466879278	0.319738561	1.004922067
Person	−0.043701328	0.456888855	0.325094942	34.1878589
EventCustody	0.016087731	0.293029428	0.206096523	0.464861909
EventBusiness	0.012715162	0.27680552	0.205864108	0.006836205
NaturalDisaster	−0.061361331	0.190954998	0.098403009	0.116215477
Weapon	0.016315045	0.170783401	0.066954209	0.389663659
GeographicFeature	0.036188241	0.124441072	0.054562127	0.355482636
SportingEvent	−0.145656377	0.113705434	0.084004595	0.683620454
EntertainmentAward	−0.109663352	0.089443691	0.051638662	0.116215477
EventElection	0.069957979	0.087879911	0.053762208	1.196335794
Product	−0.09157607	0.080848917	0.073924671	0.006836205
EventDemonstration	−0.041675355	0.05315930	0.026291607	0.102543068
Facility	0.118815102	0.04608589	0.030943175	2.119223407
Duration	0.115452491	0.015852489	0.005636816	0.047853432
HealthCondition	−0.126119331	0.01493654	0.010940216	0.546896363
Award	−0.042326197	0.009894854	0.005742111	0.109379273
Vehicle	−0.105587758	−0.007998363	−0.004738113	0.403336068
TitleWork	0.080059260	−0.07365784	−0.047898691	0.102543068
Time	−0.196784243	−0.079296142	−0.053925132	0.129887886
Crime	0.011954751	−0.079479031	−0.055942257	0.334974022
Substance	−0.076414958	−0.092376187	−0.018941145	0.034181023
EventEducation	−0.009967238	−0.177876234	−0.086611904	0.020508614

In order to eliminate the possibility that the individual concept/class neural network had not been trained properly, we have calculated the CS and CS' for all the overlapping vocabularies in the related two Word2Vec models, which is the training data set, and then calculated its average (as shown in Table 3).

If these two models are perfectly aligned with each other, then the average value after the alignment should be equal to 1. The result clearly shows that all the neural networks are "properly" trained and work extremely well on the training dataset. This suggests that the neural network trained for a class with a low informative value may be subject to overfitting due to the simplicity of the problem it is trying to solve. However, when comes to the class with high informative value, the problem complexity helps to reduce the chance of overfitting and leads to a more accurate "good" result. A positive

Table 3. Neural Networks Evaluation Result

Neural Network	After Alignment (CS')	Before Alignment (CS)
Award	0.98031644	0.07930794
Cardinal	0.97966864	0.03505877
Crime	0.98225026	0.06569504
Date	0.98123691	0.02029948
Duration	0.97950185	0.09908933
EntertainmentAward	0.97891328	0.10126378
Event	0.98852654	0.02643711
EventBusiness	0.9786953	0.09117983
EventCustody	0.97857699	0.08236471
EventDemonstration	0.97864903	0.09305801
EventEducation	0.97960562	0.09103634
EventElection	0.97835062	0.04980883
EventPerformance	0.9785876	0.08383297
EventPersonnel	0.97807607	0.05734759
EventViolence	0.98036507	0.06654035
Facility	0.98025752	0.03982984
GeographicFeature	0.98052347	0.08840587
HealthCondition	0.97929321	0.07387248
NaturalDisaster	0.97913864	0.0762092
Organisation	0.97541052	0.0213745
Person	0.99137417	0.00545538
Place	0.98766889	0.02258751
Product	0.97633725	0.09864551
SportingEvent	0.97820687	0.08318475
Substance	0.97602775	0.08991432
Time	0.98071897	0.08842963
TitleWork	0.97798596	0.08760807
Vehicle	0.98034503	0.07701601
Weapon	0.97641792	0.07500139

side-effect is that the overfitting-ness could be used as a criterion to distinguish the low informative concepts/classes from the high informative concepts/classes.

Since there are 29 concepts/classes identified from the corpus, we have 406 class pairs in total and the correlation analysis process will generate a MIC strength value for each of the pairs. Due to the reason of size, Table 4 lists only the top 10 pairs. Based on the MIC result, it is easy to calculate the correlation coefficient value for all these 29 concepts (Formula 6) in the Source Corpus (Step 4) and then we will be able to get the final semantic impact value which is shown in Table 5 (Step 5). In this demonstration, we consider the informative and correlation are equally important and therefore $\lambda = 1$.

From the result, we can clearly see that Date (e.g. today, yesterday and next week), Event (e.g. reported, announced and promise) and Cardinal (e.g. various

Table 4. Top 10 class pairs in the Source Corpus

X var	Y var	MIC (strength)
Cardinal	Date	0.37431
Cardinal	Facility	0.35607
Date	Facility	0.32850
EventViolence	Facility	0.30661
EventViolence	Cardinal	0.30093
EventElection	Cardinal	0.28417
Crime	Cardinal	0.28044
EventViolence	Date	0.27788
EventViolence	Crime	0.27470
Organisation	Place	0.26798

Table 5. Correlation Coefficient and Semantic Impact (sorted by Semantic Impact)

Concept/Class	Correlation Coefficient	Semantic Impact
Date	0.377891836	1.036020678
Event	0.202632850	0.971142880
Cardinal	0.308715289	0.807158512
Organisation	0.147079549	0.737658452
Place	0.133240886	0.695213858
EventPerformance	0.218267737	0.639546722
EventViolence	0.231330062	0.638302842
EventPersonnel	0.259626427	0.579364988
Person	0.145728547	0.470823489
EventCustody	0.145849776	0.351946299
Facility	0.304758711	0.335701886
EventBusiness	0.100836205	0.306700312
EventElection	0.218473763	0.272235971
Crime	0.260360726	0.204418469
Product	0.125679770	0.199604442
SportingEvent	0.077800938	0.161805532
Award	0.141992566	0.147734677
HealthCondition	0.136709891	0.147650107
TitleWork	0.179914358	0.132015666
Vehicle	0.128504793	0.123766681
Time	0.079429718	0.025504586
NaturalDisaster	−0.122922189	−0.024519181
EntertainmentAward	−0.079461037	−0.027822374
Weapon	−0.123359922	−0.056405713
GeographicFeature	−0.172073896	−0.117511769
EventDemonstration	−0.155605741	−0.129314134
EventEducation	−0.176308228	−0.262920131
Duration	−0.287187334	−0.281550518
Substance	−0.384208491	−0.403149636

numbers) are the most important concepts/classes in the domain, due to the high *Informative* and *Correlation Coefficient* values. On the other hand, `EventEducation` (e.g. graduating and graduated), `Duration` (e.g. 22-min, 80-min and more than a year) and `Substance` (e.g. steel and coal) are the least important concepts.

It is interesting to see that the concept `Date` has a relatively low TF value but with a high *Semantic Impact* value as a result of both high *Informative* and high *Correlation Coefficient*. On the other hand, although the concept `Person` is still a quite an important concept (ranking 9th), it has a much higher TF value but a lower *Semantic Impact* value when compared with `Date`. This is due to its relatively small *Informative Coefficient* value even if there is a strong correlation with the other concepts. Intuitively, this is correct because all the news articles in the corpora are about Donald Trump and therefore the concept of `Person` may not be as general as the other concepts with a higher *Semantic Impact* value which leads to a small *Informative Coefficient* value as the results show.

4   Conclusion

In order to measure the importance of a particular concept to the domain knowledge at the semantic level, this paper introduced a new idea called the "Semantic Impact" which is computed from a concept's *Informative Coefficient* and *Correlation Coefficient*.

In Sect. 2, we explained the method and the process to calculate these coefficients for domain concepts. We evaluated this by using 200 BBC News articles on Donald Trump and discussed the results in Sect. 3.

We have also briefly analysed the preliminary evaluation result and explained why `Date`, `Event` and `Cardinal` have a higher *Semantic Impact* value over the others. Specifically, the concept `Person` is used as an example to explain why a high *Term Frequency* value may not necessarily result in a high *Semantic Impact* value. At this stage, we can mainly assess these results intuitively. A quantitative evaluation will be required to apply our semantic impact measure and the computation approach to other domains. This is also the crux of the future work for this research.

References

1. Zúñiga, G.L.: Ontology: its transformation from philosophy to information systems. In: Proceedings of the International Conference on Formal Ontology in Information Systems, vol. 2001, pp. 187–197. ACM, Ogunquit (2001)
2. Maedche, A., Staab, S.J.: Ontology learning for the semantic web. IEEE Intell. Syst. **16**(2), 72–79 (2001)
3. Wong, W., Liu, W., Bennamoun, M.: Ontology learning from text: a look back and into the future. ACM Comput. Surv. **44**(4), 1–36 (2012)
4. Horrocks, I.: Ontologies and the semantic web. Commun. ACM **51**(12), 58–67 (2008)
5. Beel, J., Breitinger, C., Langer, S.J.: Evaluating the CC-IDF citation-weighting scheme: how effectively can 'Inverse Document Frequency' (IDF) be applied to references. In: Proceedings of the 12th iConference (2017)

6. Wan, J., Barnden, J.: A new semantic model for domain-ontology learning. In: Zu, Q., Hu, B., Gu, N., Seng, S. (eds.) HCC 2014. LNCS, vol. 8944, pp. 140–155. Springer, Cham (2015). https://doi.org/10.1007/978-3-319-15554-8_12
7. Evert, S.: Distributional semantic models. NAACL HLT 2010 Tutorial Abstracts, pp. 15–18 (2010)
8. Mikolov, T., et al.: Efficient estimation of word representations in vector space. arXiv preprint arXiv:1301.3781 (2013)
9. Mikolov, T., et al.: Distributed representations of words and phrases and their compositionality. In: Advances in Neural Information Processing Systems. (NIPS 2013), vol. 26, pp. 3111–3119 (2013)
10. Reshef, D.N., et al.: Detecting novel associations in large data sets. Science **334**(6062), 1518–1524 (2011)
11. IBM: Natural Language Understanding. Watson Developer Cloud 2017. https://www.ibm.com/watson/developercloud/natural-language-understanding.html. Accessed 2017
12. BBC: Trump says terror attacks 'under-reported': Is that true? US & Canada (2017). http://www.bbc.co.uk/news/world-us-canada-38890090. Accessed 10 Feb 2017
13. BBC. Core Concepts Ontology (2015). https://www.bbc.co.uk/ontologies/coreconcepts. Accessed 01 Apr 2018
14. Deeplearning4J. Troubleshooting & Tuning Word2Vec. Word2Vec 2017. https://deeplearning4j.org/word2vec#trouble. Accessed 1 Feb 2017
15. Glorot, X., Bengio, Y.: Understanding the difficulty of training deep feedforward neural networks. In: Proceedings of the Thirteenth International Conference on Artificial Intelligence and Statistics (2010)
16. Kingma, D.P., Ba, J.: Adam: a method for stochastic optimization. arXiv preprint arXiv: 1412.6980 (2014)

Research on the Influence Factors of Consumer Repurchase in Dutch Auction

Weijin Jiang[1,2]([⊠]), Jiahui Chen[1], Yuhui Xu[1], Yang Wang[1], and Lina Tan[1]

[1] Institute of Big Data and Internet Innovation, Mobile E-Business Collaborative Innovation Center of Hunan Province, Hunan University of Commerce, Changsha 410205, China
jlwxjh@163.com, 18508488203@163.com, 810663304@qq.com, 363168449@qq.com
[2] School of Computer Science and Technology, Wuhan University of Technology, Wuhan 430073, China

Abstract. With the rapid development of e-commerce, the online shopping market becomes more and more competitive. How to design sales models to retain consumers and make them become continuous buying is the biggest concern of most e-commerce companies. This study was under the situation of Dutch auction model, using a large number of real auction data in Gongtianxia's "7 days auction" and "15 min auction". We use quantitative research method combining statistical analysis and panel data regression model, to study the influencing factors of consumers' repeat purchase. The results show that auction methods, consumers' perceived performance and bidding structure can affect consumers repurchase. At the same time, we established random effects regression models of the rate of consumers repurchase in 7 days and 30 days. According to the conclusion of the research in this article, we can provide some practical suggestions for Gongtianxia and a reference for other e-commerce.

Keywords: Dutch auction · Continuous purchases · Panel data · Random effects model

1 Introduction

With the rapid development of e-commerce websites, the competition among customers in China's shopping market has become increasingly fierce. In order to reflect its own characteristics and attract more consumers to pay attention to and purchase products, many companies often use a variety of innovative sales methods to attract customers. Therefore, how to obtain the attention and trust of consumers through these marketing methods and generate repeated purchases is a problem that all e-commerce companies are very concerned about.

Under the situation of Dutch auction model, this article uses a large number of real auction data in Gongtianxia, and adopts quantitative research method combining statistical analysis and panel data regression model to analyze the influencing factors of repeated purchase of consumers, thereby helping e-commerce companies to develop more reasonable sales strategies to promote the continuous purchase behavior of consumers.

2 Literature Review

2.1 Cut-Price Auction

Foreign research situation. Kotak and Alvin (2004) pointed out that goods with large transaction volume, low value and similar to daily necessities generally adopt the "Dutch auction" method [3]. Milgrom and Weber (2006) believed that the independent private value model was more suitable than the common value model for non-durable goods (such as flowers and fruits) [4]. Guerci and Kirman (2014) used mathematical analysis to analyze how to bid would maximize the buyer's interest in a continuous price reduction auction [5]. In other words, longer-lasting auctions make it easier for consumers to make purchasing decisions [6].

Some domestic scholars have also studied Dutch auction in recent years. Reduced-price auction has an auction mechanism with a bid-to-trade, which ensures the efficiency of transaction, so that it can save more time than other auction methods for the seller and the buyer [7]. Chenhui Liu (2011) used the Bayesian rule to establish the pricing of Dutch auctions on the transfer of rural land contractual management rights, and deeply analyzed how the price should be set [7].

However, most of the existing research on price reduction auctions are based on auction theory, while rarely study the impact of reduced-price auctions as a sales model on consumers of e-commerce sites. Furthermore, the research about how to influence the repeated purchase of e-commerce website consumers is very vacant.

2.2 Consumer Repeat Purchase

The theoretical research on consumer repeated purchase is more based on the theory of consumer behavior. Many research scholars believe that consumer retention is a key factor in a success of company [10].

In 1965, Cardozo introduced customer satisfaction into marketing for the first time, pointing out that customer satisfaction would increase the re-purchase behavior of customers who would buy other products of the company [12]. After that, the problem of customer satisfaction has received great attention. Michael (2000) proposed that the repeated purchase intention of consumers was affected by the combination of consumer satisfaction and conversion barriers, and the conversion obstacle was incorporated into the system of repeated purchase intention research [13].

However, for the research on the influencing factors of consumers' repeated purchases on e-commerce websites, most of the previous scholars used questionnaires to establish qualitative models from the perspective of consumer behavior or marketing. It is very vacant to use quantitative methods to study the influencing factors of the repeated purchases of consumers under real data by selecting a specific sales scenario.

3   Research Methods and Models

3.1   Data Collection and Pre-processing

The study obtained a total of 21 months of auction data from November 22, 2014 to August 23, 2016 in the "7 day" format, and 19-month auction data in the form of "15 min". Specific data collection are shown in Table 1.

Table 1. Data collection statistics

Auction	Method	Time limit	Auctions	Number of participating users	Auction record
7 day	auction	2014.11.22—2016.8.23	584	11851	40908
15 min	auction	2015.1.17—2016.8.29	1234	17826	131550

We will delete some data that is obviously different from the normal auction and the user, and only retain the auction data whose basic characteristics tend to be consistent, such as special auctions with a starting price of more than 200 yuan, only some member users can participate, and special users who have participated in the auction more than 200 times, etc.

3.2   Research Model

According to the research summary of the literature review and the analysis of the actual auction data, we established the basic model of this study as shown in Fig. 1.

The dependent variable is the consumer repurchase rate for each auction. The meanings of the respective variables and their corresponding relations in the regression equation are shown in Table 2.

3.3   Data Analysis Methods: Statistical Analysis and Panel Data Regression

A panel data regression model is further used to establish a specific quantitative model. Panel data combines the sectional data and time series data to reflect data information in more multiple directions. The reason why we choose panel data regression in this paper based on the following reasons:

(1) Individual heterogeneity can be controlled better.
(2) The changes of all variables with time factors can be controlled effectively.
(3) The panel data regression model can also avoid multi-collinearity problems better, and can provide more information, greater freedom and higher efficiency.

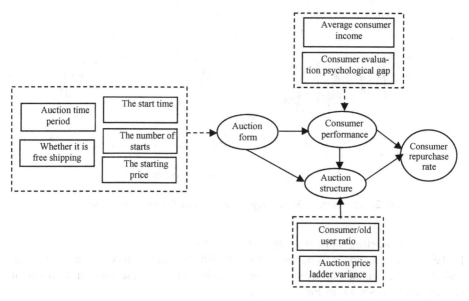

Fig. 1. The basic model of influencing factors of consumer repurchase rate

Table 2. Independent variable definition of regression model

Independent variable	Influencing factors	Variable names	Remarks
X_1	Starting price	Pl	The price of the first step of the auction
X_2	Whether it is free	postage	Yes:1 No:2
X_3	Number of starts	start_count	The starting number of the auction
X_4	Average consumer benefit	ave_profit	$Aveprofit = average((P_1 - P_{Pricetag_i})$
X_5	Average psychological gap	ave_psychoGap	$PsychoGapTag_i = Pricetag_i - Pricetag_{max}$
X_6	Average price step variance	pricetag_var	$VarPricetag = var(Pricetag_i)$
X_7	New users ratio	new_rate	$NewUser_{rate} = \frac{buytimes=1\ number\ of\ users}{Number\ of\ users\ who\ participated\ in\ this\ auction}$
X_8	Visible users ratio	sticky_rate	$StickyUser_{rate} = \frac{buytimes \geq 5 number\ of\ users}{Number\ of\ users\ who\ participated\ in\ this\ auction}$

4 Experiment Process and Conclusion

4.1 Statistical Analysis of Influencing Factors

The influence of the Auction Time Period on the Repurchase Rate

In the price reduction auction, the average repurchase rate for starting the auction every day in the time range of the entire data is shown in Fig. 2, comparing the two auction forms of "7 days auction" and "15 min auction".

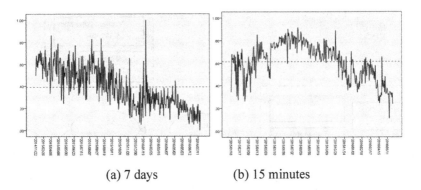

(a) 7 days   (b) 15 minutes

Fig. 2. The average 30-day repurchase rate of the two auction forms

Influence of Starting Time on Repurchase Rate
In the "15 min auction", Gongtianxia launch different auction forms at different time periods. We calculate the average user repurchase rate of the auction at different starting times as shown in Fig. 3.

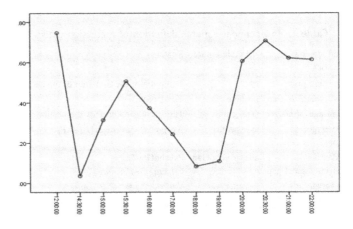

Fig. 3. Effects of different starting times on auction in "15-min auction"

Thus, for the "15 min auction", the setting of the different start time is obviously different for the overall repurchase rate of the auction. This may be because consumers have more leisure time to monitor and participate in the auction after lunch.

The influence of the Proportion of New Users Auctioned on the Repurchase Rate of Consumers
Users who participate in the auction for the first time and those who participate in the auction multiple times under the reduced price model may behave differently in the bidding process.

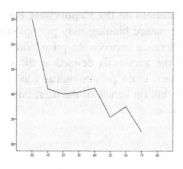

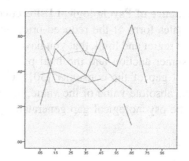

(a) the overall average repurchase
rate of new users

(b) the average repurchase rate of
new users by product

Fig. 4. The relationship between the ratio of different users and the average repurchase rate of "7 days auction"

As can be seen from Fig. 4, in the "7 days auction" and "15 min auction", the new user ratio and the average repurchase rate of the auction are basically linearly negatively correlated, In other words, the high proportion of new users is not conducive to improve the overall user repurchase rate of an auction. First-time users are more likely to be lost than repeat users.

The Influence of the Price Ladder Variance of Auction on the Repurchase Rate
In the specific purchase method of a reduced price auction, different users for the same product auction will produce the same product at different prices on different price ladders. In the auction records of "7 days auction" and "15 min auction", the proportion of each ladder auction record is shown in Fig. 5.

The variable of the price step variance of an auction reflects the concentration of the consumer bid ladder. We divide the average of the auction price step variance within 1 to average according to steps of 0.1.

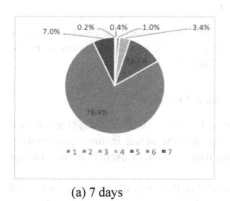

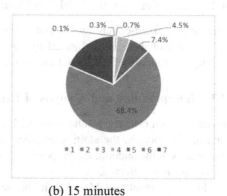

(a) 7 days

(b) 15 minutes

Fig. 5. Purchase ratio of different steps

The Influence of Psychological Difference of Consumers on the Repurchase Rate
For the sales form of the reduced-price auction, premature bidding may prompt consumers to regret and think that you have lost. The difference between the price ladder of the consumer auction and the final price ladder of the auction is defined as the psychological gap of the consumer [30]. "0" means there is no psychological gap. The larger the absolute value of the value, the earlier the bid on behalf of the user, and the greater the psychological gap generated by the auction.

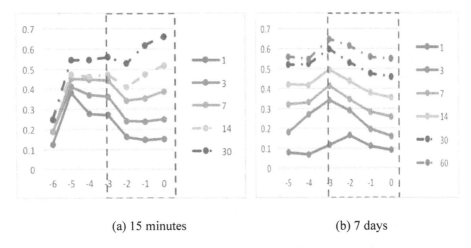

|(a) 15 minutes|(b) 7 days|

Fig. 6. Repurchase of users on the price ladder of different psychological gaps

As shown in Fig. 6, the number of users who repurchased in 1 day, 3 days, 7 days, 14 days, and 30 days on the price ladder of each psychological gap is counted as the ratio of all users. In the short term, it will inspire a greater desire to buy again in the short term. In a longer period of time, the effect of this psychological gap on repurchase is less significant.

4.2   Panel Regression Model Establishment

Similarly, a fixed-effects model and a random-effects model of the 7-day repurchase rate are established and subjected to Hausman's test. The estimated results of establishing the random effect model are shown in Table 3.

4.3   Interpretation and Analysis of Regression Model

As can be seen from the results of model 1, for the long-term (within 30 days) repurchase rate of the auction, the starting number of auctions, the actual income generated by consumers in the auction, the proportion of new users and sticky users in the auction are all significantly affected.

From the results of Model 2, it can be seen that the proportion of new users and sticky users in the auction participating users and the number of auctions started have a significant impact for the short-term (7-day) repurchase rate of the auction, and the

Table 3. Regression results of random effects model for consumer 7-day repurchase rate

7-day repurchase rate	Coefficient	Std. Err.	z	P > z
pl	−0.00127	0.001355	−0.94	0.349
postage	0.015242	0.05315	0.29	0.774
Start_count	−0.00014	8.78E−05	−1.59	0.112
ave_profit	0.002227	0.00241	0.92	0.355
ave_psychoGap.	−0.02834	0.056944	−0.5	0.619
pricetag_var	0.018456	0.110982	0.17	0.868
new_rate	−0.6588	0.298568	−2.21	0.0027
sticky_rate	−0.56373	0.156417	−3.6	0
constant term	0.900482	0.145654	6.18	0
Wald chi2(8) = 29.82				
Prob > chi2 = 0.0002				

form and coefficient of influence are basically similar to the 30-day repurchase rate of the auction. The influence of other factors on consumers' repeated purchases is more reflected in a relatively long period of time.

5   Conclusions and Prospects

This research is mainly based on the large-scale real auction data of the price reduction auction under the scene of the special sales model of the reduced price auction. The conclusion is of guiding significance to e-commerce enterprises. The following conclusions and corresponding measures and Suggestions are mainly obtained:

(1) The overall consumer repurchase rate of "15 min auction" is higher than the overall consumer repurchase rate of "7 days auction".
(2) Different start-up time has a significant impact on the consumer repurchase rate.
(3) The auction form of parcel post can improve the average repurchase rate of consumers, and this effect is more obvious in the form of "7 days auction" than "15 min auction".
(4) The proportion of new users in the auction negatively affects the repurchase rate of consumers.
(5) Excessive number of auctions are also not conducive to increasing consumers' repeated purchases, while successful bidding in a limited number of products is more conducive to consumers' satisfaction with shopping, resulting in more continuous purchases.
(6) The price ladder of consumers' bids in the auction is more dispersed, which makes some consumers have a certain psychological gap, but it is easier to enhance the entertainment of this consumption process, thus stimulating consumers' desire to try again in the short term.

In order to explain and control consumers' continuous purchasing behaviors better, it may be necessary to conduct experiments on other more effective data sets to verify our conclusions in the future.

Acknowledgement. This work was supported by the National Natural Science Foundation of China (61472136; 61772196), the Hunan Provincial Focus Social Science Fund (2016ZDB006), Hunan Provincial Social Science Achievement Review Committee results appraisal identification project (Xiang social assessment 2016JD05) The authors gratefully acknowledge the financial support provided by the Key Laboratory of Hunan Province for New Retail Virtual Reality Technology (2017TP1026). The authors gratefully acknowledge the financial support provided by the Key Laboratory of Hunan Province for Mobile Business Intelligence (2015TP1002).

References

1. Ping, H.: Wholesale market trading "new method": dutch auction. Chinese market, (Z1), 56–57 (2004)
2. Petrowski, A.: A clearing procedure as a niching method for genetic algorithms. In: Proceedings of the 3rd IEEE Conference on Evolutionary Computation, pp. 798–803. IEEE Press, Piscataway (1996)
3. Comment, R., Jarrell, G.A.: The relative signalling power of dutch-auction and fixed-price self-tender offers and open-market share repurchases. J. Finance **46**(4), 1243–1271 (1991)
4. Katok, E., Roth, A.E.: Auctions of homogeneous goods with increasing returns: experimental comparison of alternative dutch auctions. Manage. Sci. **50**(8), 1044–1063 (2004)
5. Kaida, Q., Baojian, Y., Miaohua, Z., et al.: Fresh agricultural products auction market: research hotspots and new trends. J. Kunming Univ. Sci. Technol. (Nat. Sci. Ed.) **39**(01), 98–109 (2014)
6. Guerci, E., Kirman, A., Moulet, S.: Learning to bid in sequential Dutch Auctions. J. Econ. Dyn. Control **48**, 374–393 (2014)
7. Shneyerov, A.: An optimal slow Dutch auction. Econ. Theory **57**(3), 577–602 (2014)
8. Hailiang, Z., Jiang, W.: Reflections on the way of flower auction trading. Market Res. **10**, 59–60 (2007)
9. Chenhui, L.: Study on the pricing model of "Lutch auction" for the transfer of farmland contract management rights. Central South University (2012)
10. Ampere, Q.Y.: Mathematical problems in Dutch auctions. Neijiang Technol. **29**(07), 41–42 (2008)
11. Jones, T.O.: Why satisfied customers defect. Harvard Bus. Rev. **73**(6), 11 (1995)
12. Reichheld, F.F.: Learning from customer defections. Harvard Bus. Rev. **74**(2), 56 (1996)
13. Cardozo, R.N.: An experimental study of customer effort, expectation, and satisfaction. J. Mark. Res. **2**(3), 244–249 (1965)
14. Jones, M.A., Mothersbaugh, D.L., Beatty, S.E.: Switching barriers and repurchase intentions in services. J. Retail. **76**(2), 259–274 (2000)
15. Oliver, R.L.: A cognitive model of the antecedents and consequences of satisfaction decisions. J. Mark. Res. **17**(4), 460–469 (1980)
16. Khalifa, M., Lam, R.: Web-based learning: effects on learning process and outcome. IEEE Trans. Educ. **45**(4), 350–356 (2002)

A SN P System for Travelling Salesman Problem

Hui Zhang, Laisheng Xiang, and Xiyu Liu$^{(\boxtimes)}$

Business School, Shandong Normal University, Jinan, China
1351404163@qq.com, xls3366@163.com, sdxyliu@163.com

Abstract. The Spiking Neural P system is a branch of the neuronal-like P system in the membrane system with great parallelism. However, The Travelling Salesman Problem is a long-term NP-hard problem that finds the minimum costly Hamiltonian cycles in a weighted undirected graph. In this paper, we use the rules of division and dissolution of spiking neurons, combined with the idea of point-by-point traversal, we find all Hamiltonian cycles in weighted undirected graphs. Then computing by the binary form of the spike, resulting in the minimum cost Hamiltonian cycles. A bi-directional weighted digraph is applied to prove the feasibility of the algorithm in this paper. This method takes full advantage of the great parallelism of the SN P system, using fewer neurons and simpler rules and procedures to solve Travelling Salesman Problem.

Keywords: Spiking Neural P system · Travelling Salesman Problem · Hamiltonian cycles · Neurons division

1 Introduction

Membrane system is extremely parallel, so the computational complexity can be greatly reduced. Tissue P system, cell P system and neural P system are the three basic types of membrane system [1–7], the Spiking Neural P system is a sub-branch under the neural P system. SN P system has only one object, the object is the spikes, usually expressed by a [8]. In the SN P system, the exchange of information between neurons is achieved by transmitting spikes. So far, the SN P system has been applied in many ways. For example, a SNP can perform addition, subtraction, multiplication of natural numbers using binary. In recent years, the SN P system has also been combined with clustering, once again expanding its scope of application.

The Travelling Salesman Problem asks to find the minimum costly Hamiltonian cycles in a weighted undirected graph. Scholars put forward many algorithms according to the research to solve TSP. The most classic idea in the algorithm is point-by-point traversal, many algorithms are designed based on this idea. Some scholars also applied the membrane calculation to solve the TSP, and proposed the solutions based on the P system or the cP system, and

© Springer Nature Switzerland AG 2019
Y. Tang et al. (Eds.): HCC 2018, LNCS 11354, pp. 339–346, 2019.
https://doi.org/10.1007/978-3-030-15127-0_33

so on [1,9]. Other scholars have integrated the membrane structure and convenience algorithms to form integrated technologies such as ant colony algorithm and genetic algorithm [10–12].

However, Few scholars have studied the SN P system to solve the TSP. In this paper, we use the idea of point-by-point traversal and combine the characteristics of SN P system to propose a method to solve TSP with SN P system. The first part introduces the research background, the second part introduces the foundation, the third part elaborates the method steps, The fourth part expounds an example. The fifth part summarizes the whole paper and puts forward the prospect.

2   Foundation

Our work in this paper is based on the Spiking Neural P System to solve the Travelling Salesman Problem, so we firstly give knowledge of the SN P System, and then introduce related content of the TSP.

2.1   Spiking Neural P Systems

Membrane mainly investigate three kinds of membrane system, which are cell-like P system, tissue-like P system and spiking neural P system, and SN P is one of them. Then only one object in the SN P system is the spike, which is the difference from the other membrane systems [13–16]. There are four rules in the SN P system for the evolution of the system, firing rule, forgetting rule, neurons division rule, and neurons dissolution rule, respectively.

Firing rules are used to generate pulses and pass information. Pulses consume by forgetting rules. Splitting rules are used for the division of neurons. Dissolution rules are used to ablate pulses and neurons. In addition, the SN P system can convert the number of spikes into form of binary and apply firing and forgetting rules to perform addition and subtraction of natural numbers within neurons.

2.2   Travelling Salesman Problem

Whether in graph theory or in the field of computer science, the TSP is a long-standing, well-known NP-hard problem, the full name of which is Traveling Salesman Problem [17,18]. The problem is generally reflected in the weighted graph. In essence, the TSP finds a Hamiltonian cycle that is the minimum cost in weighted digraphs. Hamiltonian cycle means starting from a vertex, passing through each vertex in the graph, and each vertex is traversed only once, finally returning to the path of the starting vertex. In a directed digraph, there will be multiple Hamiltonian cycles, so we need to find the Hamiltonian cycle with the minimum total cost (ie, the total weight). Each path has its own weight, the total weight of a path is the sum of the weights of all the subpaths that make up the path in a weighted graph. Since the discovery of the TSP, in many fields, various solutions have emerged. One of the most classic and most widely used methods is point-by-point iteration. This paper also adopted the idea of this algorithm, combined with the SN P system, re-proposed a new method of solving the TSP.

3   Description of Our SN P Systems TSP Algorithm

The algorithm uses a simple p principle, effectively a simple parallel deep exploration of digraph. So our general idea is to travel the node one by one, finally we can travel entire a weighted digraph and gain all Hamiltonian cycle. Then we could calculate total cost of every Hamiltonian cycle, total cost is expressed by total weight, and the minimum weight is selected from all Hamiltonian cycle.

The process of searching digraph is imagined as building a tree. We assume that initial neuron is regarded as the root vertex, the generating new neurons through spiking neurons division indicate children vertexes, branches are expressed by synapses between two neurons. Spiking neurons divide synchronously following expanding of the tree, finally leaves contain all information of total path, we should select the target path that has minimum cost. The weight of arc between two nodes is measured by an Euclidean distance in the digraph.

The algorithm that solves Travelling Salesman Problem with SN P systems mainly applies the firing rule and neuron division rule and neuron dissolution rule of Spiking Neural P Systems. Firstly, we provide a formal definition and four rules for the SN P system that we will use it in next context:

Rule:

r_1: $[E]_i \rightarrow [\ ]_j \| [\ ]_k$; $i \in \{1, 2, 3, \cdots, n\}$, $j \in \{1, 2, 3, \cdots, n\}$, $k \in \{1, 2, 3, \cdots, n\}$.

r_2: $a \rightarrow a$.

r_3: $E \backslash a \rightarrow a; d, \{a\} \notin E$.

r_4: $[E]_i \rightarrow \delta$, $i \in \{1, 2, 3, \cdots, n\}$.

Define:

(1) $O = \{a\}$.

(2) $\sigma_{Input} = (1, R_{Input})$, $R_{Input} = \{\ r_1, r_2, r_4, i=1\ \}$.

(3) $\sigma_2 = (2, R_2)$, $R_2 = \{\ r_1, r_3, r_4, i=2\ \}$.

(4) $\sigma_n = (n, R_n)$, $R_n = \{\ r_1, r_3, r_4, i=n\ \}$.

(5) $syn = \{(Input, 2)\} \cup \{(Input, 4)\} \cup \cdots \cup \{(i, j)\}$, $i \in \{1, 2, 3, \cdots, n\}$, $j \in \{1, 2, 3, \cdots, n\}$.

The process of this algorithm requires to be described as follows:

Input: we arbitrarily select a node as the initial vertex σ_{Input} of travelling directed graph, the number of spike indicates the sequence number of node. Every neuron has corresponding spiking that the number of spike is our pre-concerted. We make this neuron act as σ_{Input}, and it includes the weight between two vertexes. Input initial information of travelling by applying the rule r_2 in σ_{Input}.

Path construction: travel all vertexes in digraph, all legal paths in parallel could be generated as follows:

(1) Make the initial vertex to be starting vertex of path P.
(2) If there are edges from the remainder of vertexes to starting vertex, initial neuron divides to new neurons through applying division rule r_1. If there are two vertexes that they connect with starting vertex, and then the initial neuron is divided to two new neurons. The label of new neuron is the sequence number of corresponding vertex, and the spike in new neuron is generated by spikes of initial neuron, the distribution of spikes in new neurons corresponds to rule r_1. The weight of arc between father neuron and child neuron has be recursively preserved in new neuron after division.
(3) Travel remainder of vertexes by applying rule r_3 after finishing the traversal of first level vertex, if a vertex connects with several vertexes, the vertex could apply rule r_1 to divide as mentioned above in step (1). But there is a difference, after dividing, if the newly generated vertex has explored, it applies rule r_4 to dissolute itself.
(4) Repeat step (3) until all vertexes are visited.

Path detection: Many paths does not made up Hamiltonian cycle, so we should delete illegal paths. If there is an edge between last vertex and the initial vertex, add the information to the finally neuron. Otherwise we apply rule r_4, all spikes of neuron are consumed, the neuron is dissolved, then delete the path.

Path selection: the message of order and cost of path is extracted from final neurons, then recode the cost of all paths in the form of binary. We enter the binary in a pre-specified singly-linked neurons string, the number of spikes represents the cost of path. Applying rule r_3 triggers the neuron, causing the binary to enter the next neuron and subtraction, leaving a smaller binary into the subsequent neurons for comparison. Until the last neuron is fired, the minimum cost is generated. Meanwhile, the last neuron will export the minimum cost to the system. This process is end and TSP is solved.

4   Worked Example

We use the spiking neurons to represent the vertexes in the graph. The TSP is described by bidirectional graph G as shown in Fig. 1. We will use the method described in Sect. 3 to find the least cost Hamiltonian path through neurons

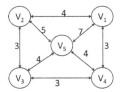

Fig. 1. The graph G

division in the SN P system. First, assume that the vertex randomly selected from the digraph is V_1, so V_1 is marked as σ_{Input}. We make the sequence number of the vertex as the subscript of the neuron, and use the initial number of spikes contained in the neuron to represent the sequence number of the vertex. Then we can get the form of the SN P system as follows:

(1)$O=\{a\}$.

(2)$\sigma_{Input}=(1,R_{Input}),R_{Input}=\{\ a \to a,[a]_1 \to [a^2]_2 \ || \ [a^4]_4 \ || \ [a^5]_5,a \to \delta\ \}$.

(3)$\sigma_2=(2,R_2),R_2=\{\ a^2 \to a,[a^2]_2 \to [a]_1 \ || \ [a^3]_3 \ || \ [a^5]_5,a^2 \to \delta\}$.

(4)$\sigma_3=(3,R_3),R_3=\{\ a^3 \to a,[a^3]_3 \to [a^2]_2 \ || \ [a^4]_4 \ || \ [a^5]_5,a^3 \to \delta\}$.

(5)$\sigma_4=(4,R_4),R_4=\{\ a^4 \to a,[a^4]_4 \to [a]_1 \ || \ [a^3]_3 \ || \ [a^5]_5,a^4 \to \delta\}$.

(6)$\sigma_5=(5,R_5),R_5=\{\ a^5 \to a,[a^5]_4 \to [a]_1 \ || \ [a^2]_2 \ ||[a^3]_3 \ || \ [a^4]_4,a^5 \to \delta\}$.

(7)$syn=\{(Input,2)\} \cup \{(Input,4)\} \cup \{(Input,5)\} \cup \{(2,Input)\} \cup \{(4,Input)\}$ $\cup \{(5,Input)\} \cup \{(2,3)\} \cup \{(2,5)\} \cup \{(3,2)\} \cup \{(5,2)\} \cup \{(3,4)\} \cup \{(3,5)\} \cup \{(4,3)\} \cup \{(5,3)\} \cup \{(4,5)\} \cup \{(5,4)\}$.

In the initial SN P system, σ_{Input} is fired first. From the digraph G, V_1 is connected with V_2, V_4 and V_5, so applying rule r_1, σ_{Input} could divide oneself and split into three neurons accordingly as shown in Fig. 2. The labels of new neurons are 2, 4 and 5, respectively. The newly generated neurons contain the serial number of the father neuron and the weight between the two neurons and their own rules.

We consider the traversal of the initial vertex as the first layer of traversal, the three neurons split by the initial vertex are the second layer of traversal, and we take the split of V_2 as an example to introduce the following traversal. Before dividing, we again emphasize the rules contained in V_2. It can be seen from the label, V_2 contains 2 initial spikes, the firing rule is $a^2 \to a$, By the syn of definition can clearly know V_2 is connected with V_1, V_3 and V_5, so the division rules is $[a^2]_2 \to [a]_1 \ || \ [a^3]_3 \ || \ [a^5]_5$, while V_2 contains the dissolution rule $a^2 \to \delta$. After firing V_2, V_2 applies the division rule and still splits into three neurons labeled 1, 3, and 5. Due to V_1 has traversed, V_2 already contains information of V_1, so the newly generated V_1 could dissolve itself by applying the dissolution rules, leaving only neurons V_3 and V_5 produced by the V_2 division as shown in Fig. 3. The remaining vertexes are traversed as the V_2 traversal method until all vertexes traversed.

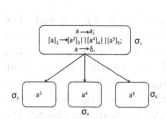

Fig. 2. The division of initial neuron

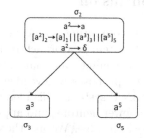

Fig. 3. The finally state of σ_2 division

After all vertexes have traversed, we need to determine whether the last vertex is connected to the initial vertex V_1, and if so, input the relevant information into the neuron. Otherwise, the corresponding neurons will dissolve using the dissolution rules to dissolve oneself. Eventually only the Hamiltonian cycle remains in the system, and the paths that do not constitute an Hamiltonian cycle have been deleted as the neurons dissolve. We employ the tree to represent the traversal of graph G, as shown in Fig. 4:

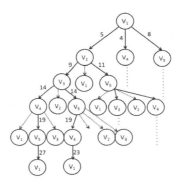

Fig. 4. The tree diagram of travelling

The traversal tree in the Fig. 4 above shows only the complete process of two Hamiltonian cycles, and the rest of the paths are similar. Finally we get seven Hamiltonian cycles. By calculating the weight of the seven paths, respectively, $21, 18, 19, 21, 19, 18, 21$. The weights are extracted from the neurons. The weights are represented by the number of vertexes. We enter the binary in a pre-specified singly-linked neurons string, and different numbers of neurons contain different path weights. Finally, we can get the minimum weight that is 18, and there are two corresponding path, which is $\{V_1 \rightarrow V_2 \rightarrow V_3 \rightarrow V_5 \rightarrow V_4 \rightarrow V_1\}$ and $\{V_1 \rightarrow V_4 \rightarrow V_5 \rightarrow V_3 \rightarrow V_2 \rightarrow V_1\}$.

5   Conclusion

Membrane computing is a new computational model inspired by biological cells, the SN P system is a branch of the membrane system. This paper proposes to solve the travelling salesman problem by applying SN P system algorithm in $O(n)$ time, proposes a new idea for solving TSP and broaden the application range of SN P system. This idea uses the maximum parallelism of the SN P system, uses fewer neurons, and applies fewer rules throughout the process, making TSP easier to solve. We use the dividing property of neurons in the SN P system to find all the Hamiltonian circuits, and apply the dissolving property of neurons to delete the paths that do not form cycles. Finally, we compare the weights with the binary form of the spikes and finally find the minimum cost

Hamilton cycle. In order to further improve the efficiency of the algorithm, the process of simplifying weight comparison is the direction of future efforts.

Acknowledgments. Project is supported by National Natural Science Foundation of China (61472231, 61502283, 61876101, 61802234, 61806114), Ministry of Eduction of Humanities and Social Science Research Project, China (12YJA630152), Social Science Fund Project of Shandong Province, China (16BGLJ06, 11CGLJ22), China Postdoctoral Project (40411583).

References

1. Guo, P., Dai, Y.: A P system for travelling salesman problem. In: 18th International Conference on Membrane Computing (CMC 2018), pp. 147–165. International Membrane Computing Society, 24–28 July 2017
2. Vitale, A., Mauri, G., Zandron, C.: Simulation of a bounded symport/antiport P system with Brane calculi. Biosystems **91**(3), 558–571 (2008)
3. Martin-Vide, C., Păun, Gh., Rodríguez-Patón, A.: On P systems with membrane creation. Comput. Sci. J. Mold. **9**(2), 134–145 (2001)
4. Mart, C., Păun, Gh., Pazos, J.: Tissue P systems. Theor. Comput. Sci. **296**(2), 295–326 (2003)
5. Freund, R., Păun, Gh., Pérez-Jiménez, M.J.: Tissue P systems with channel states. Theor. Comput. Sci. **330**(1), 101–116 (2005)
6. Zhang, X.Y., Zeng, X.X., Luo, B., Xu, J.B.: J. Comput. Theoret. Nanosci. **9**, 769 (2012)
7. Song, T., Jiang, Y., Shi, X.L., Zeng, X.X.: J. Comput. Theoret. Nanosci. **10**, 999 (2013)
8. Zhao, Y., Liu, X., Wang, W.: Spiking neural P systems with neuron division and dissolution. PLOS ONE **11**(9), e0162882 (2016)
9. Cooper, J., Nicolescu, R.: The travelling salesman problem in cP systems. In: The Sixth Asian Conference on Membrane Computing (ACMC 2017), pp. 9–21 (2017)
10. Guo, P., Liu, Z.J.: Moderate ant system: an improved algorithm for solving TSP. In: 7th International Conference on Natural Computation, pp. 1190–1196 (2011)
11. He, J., Xiao, J., Shao, Z.: An adaptive membrane algorithm for solving combinatorial optimization problems. Acta Math. Sci. **34**(5), 1377–1394 (2014)
12. Manalastas, P.: Membrane computing with genetic algorithm for the travelling salesman problem. In: Nishizaki, S., Numao, M., Caro, J., Suarez, M.T. (eds.) Theory and Practice of Computation: 2nd Workshop on Computation: Theory and Practice, pp. 116–123. Springer, Tokyo (2013). https://doi.org/10.1007/978-4-431-54436-4_9
13. Cabarle, F.G.C., Adorna, H.N., Pérez-Jiménez, M.J., Song, T.: Spiking neural P systems with structural plasticity. Neural Comput. Appl. **26**(8), 1905–1917 (2015). https://doi.org/10.1007/s00521-015-1857-4
14. Cabarle, F.G.C., Adorna, H.N., Pérez-Jiménez, M.J.: Sequential spiking neural P systems with structural plasticity based on max/min spike number. Neural Comput. Appl. **27**(5), 1337–1347 (2016). https://doi.org/10.1007/s00521-015-1937-5
15. Song, T., Zheng, P., Wong, M.L.D., Wang, X.: Design of logic gates using spiking neural P systems with homogeneous neurons and astrocytes-like control. Inf. Sci. **372**, 380–391 (2016). https://doi.org/10.1016/j.ins.2016.08.055

16. Zhang, X., Pan, L., Păun, A.: On universality of axon P systems. IEEE Trans. Neural Netw. Learn. Syst. **26**(11), 2816–2829 (2015). https://doi.org/10.1109/TNNLS.2015.2396940. PMID: 25680218
17. He, J., Zhang, K.: A hybrid distribution algorithm based on membrane computing for solving the multiobjective multiple traveling salesman problem. Fundam. Inform. **136**(3), 199–208 (2015)
18. Smith, S.L., Imeson, F.: GLNS: an effective large neighborhood search heuristic for the generalized traveling salesman problem. Comput. Oper. Res. **87**, 1–19 (2017)

Healthy and Diseased Tomatoes Detection Based on YOLOv2

Jiayue Zhao[✉] and Jianhua Qu

Shandong Normal University, Jinan, Shandong, China
zhaojiayue1994@163.com, qjh@sdnu.edu.cn

Abstract. Disease is one of the key problems that can cause serious yield lost. The effective detection of healthy and diseased tomatoes is of great significance for the development of tomato intelligent farm machinery technologies. This paper analyzed the application of YOLOv2 model on the detection of healthy and diseased tomatoes. Firstly we collected and processed the images available for study. And then the object fruits in the images were manually labeled with unified standards. Using the experimental data set, we finally obtained a YOLOv2 tomato detection network with great performance. When the threshold is 0.25, the precision rate of the network reaches up to 0.96 and the mAP reaches up to 0.91. The results indicate that YOLOv2 can be effectively applied to the detection of healthy and diseased tomatoes. This application also provides an idea for the detection of other fruits and vegetables.

Keywords: Object detection · Tomato · YOLOv2

1 Introduction

Tomato is an important cash crop, and disease is one of the key problems that restrict the output and quality of tomato [1]. The effective detection of healthy and diseased tomatoes is of great significance for the development of tomato automatic harvesting, targeted drug application and many other intelligent farm machinery technologies.

In recent years, deep learning method have gradually replaced traditional machine vision method and become the popular algorithm in the field of object detection. For the detection of tomato or tomato diseases, Zhou et al. [2] selected an 8-layer network for feature extraction and expression of tomato main organs on the basis of VGGNet, and with the network as the basic structure, they proposed a tomato main organs recognition method which relies on deep convolutional neural network. In order to enhance the proportion of directional features in classification features and improve the recognition accuracy of tomato main organs, Zhou et al. [3] designed a dual convolutional Fast R-CNN tomato organs recognition network based on RGB and grayscale image inputs. Moreover, inspired by Faster R-CNN, Zhou et al. [4] proposed a real-time recognition method of tomato main organs based on CNNs, and designed a corresponding recognition network model which can predict the object boundary and type only using the feature map. Fang et al. [5] captured the images of tomato through computer vision system, and they identified empty tomatoes applying the round value and detected abnormal tomatoes applying the variation of fruit diameter. After that,

© Springer Nature Switzerland AG 2019
Y. Tang et al. (Eds.): HCC 2018, LNCS 11354, pp. 347–353, 2019.
https://doi.org/10.1007/978-3-030-15127-0_34

artificial neural network trained with genetic algorithms was employed to conduct experimental research. Based on mathematical morphology, Computer vision technology and neural networks, Wang et al. [6] offered a fast algorithm about dilation and erosion to extract features of different tomato diseases.

However, these studies only focused on healthy tomatoes or only focused on diseased tomatoes, and did not carry out an interesting research on the both classes. Besides, the real-time and accurate recognition for tomato fruit is crucial to achieve its automated agricultural production. Well, one-stage object detection algorithm based on the regression method shows the superiority of its detection speed. Different from two-stage detection algorithm (such as Faster R-CNN) generates a series of region proposals firstly and uses a convolutional neural network to classify samples secondly, one-stage detection algorithm directly converts the object bounding box location problem into a regression problem. So that its detection speed can be greatly improved under the premise of ensuring a certain accuracy rate.

This paper selected YOLOv2 [7], one of the one-stage method with a rapid detection speed, as the experimental model, and applied it to the detection of healthy and diseased tomatoes. Tomato training set and test set will be used to train and test the YOLOv2 tomato detection network in order to demonstrate the validity of YOLOv2 for detecting the healthy condition and location information of tomato fruits.

2 Experimental Data Collection and Preprocessing

2.1 Experimental Data Collection

The data required for this experiment was collected from Baidu Image. We downloaded tomato images in Baidu Image through relevant keywords at first, and then manually selected a proper dataset to avoid the download error and the singleness of tomato images [8]. 225 images in JPEG format were finally selected. The maximum image resolution is 500 * 500 pixels and the minimum image resolution is 500 * 310 pixels. Image examples are shown in Fig. 1.

Fig. 1. Tomato images

2.2 Data Augmentation

In order to increase the sample size, ameliorate the diversity of samples and then improve the detection result [2, 9, 10], we used 5 image processing methods on the 225 tomato images, which include Highlight, Lowlight, AWB, High ISO NR and Flip Horizontal. 1125 new tomato images are obtained. Therefore a total of 1350 images compose the final dataset. 945 images were randomly selected as the training dataset, and the remaining 405 images were selected as the test dataset.

2.3 Sample Labeling and Result File Format Conversion

In this paper, the "diseased tomato" refers to the tomato fruit with physiological diseases, such as deformed fruit, hollow fruit and blotchy ripening fruit. In order to improve the detection accuracy of YOLOv2 network, the various appearances and shapes of the objects were synthetically considered during the sample labeling process [11]. 1.5.1 version of LabelImg software was used to label the healthy and diseased tomato fruits in 1350 tomato images with two object classes—"healthy tomato" and "diseased tomato". When more than 50% area of object is covered, the tomato fruit will not be labeled.

In order to facilitate the later use of data, we converted the XML format files, the result files of sample labeling, to TXT format files.

3 YOLOv2 Tomato Detection Network

3.1 YOLOv2 Network

YOLOv2 network is based on YOLO network [12]. In order to resolve the defects of YOLO in recall and position, YOLOv2 algorithm was proposed in 2017 CVPR. Using a novel, multi-scale training method the same YOLOv2 model can run at varying sizes, offering an easy tradeoff between speed and accuracy. For the time, YOLOv2 outperforms state-of-the-art methods like Faster R-CNN [13] with ResNet and SSD [14] while still running significantly faster on VOC 2007 test set.

The YOLOv2 network based on the Darknet-19 classification model is shown in Table 1 and its structure diagram is shown in Fig. 2. It uses batch normalization to stabilize training, speed up convergence and regularize the model [15]. Compared with YOLO, which predicts the coordinates of bounding boxes directly using fully connected layers, YOLOv2 introduces anchor mechanism and runs k-means clustering on the training set bounding boxes to automatically find good priors. To avoid the model instability when using anchor boxes, YOLOv2 uses direct location prediction. Since it constrains the location prediction the parametrization is easier to learn, making the network more stable. Besides, YOLOv2 network adds a pass-through layer which concatenates the higher resolution features with the low resolution features, helping localize the smaller objects.

Table 1. YOLOv2 tomato detection network

Layer	Filters	Size	Input	Output
0 conv	32	3 × 3/1	416 × 416 × 3	416 × 416 × 32
1 max		2 × 2/2	416 × 416 × 32	208 × 208 × 32
2 conv	64	3 × 3/1	208 × 208 × 32	208 × 208 × 64
3 max		2 × 2/2	208 × 208 × 64	104 × 104 × 64
4 conv	128	3 × 3/1	104 × 104 × 64	104 × 104 × 128
5 conv	64	1 × 1/1	104 × 104 × 128	104 × 104 × 64
6 conv	128	3 × 3/1	104 × 104 × 64	104 × 104 × 128
7 max		2 × 2/2	104 × 104 × 128	52 × 52 × 128
8 conv	256	3 × 3/1	52 × 52 × 128	52 × 52 × 256
9 conv	128	1 × 1/1	52 × 52 × 256	52 × 52 × 128
10 conv	256	3 × 3/1	52 × 52 × 128	52 × 52 × 256
11 max		2 × 2/2	52 × 52 × 256	26 × 26 × 256
12 conv	512	3 × 3/1	26 × 26 × 256	26 × 26 × 512
13 conv	256	1 × 1/1	26 × 26 × 512	26 × 26 × 256
14 conv	512	3 × 3/1	26 × 26 × 256	26 × 26 × 512
15 conv	256	1 × 1/1	26 × 26 × 512	26 × 26 × 256
16 conv	512	3 × 3/1	26 × 26 × 256	26 × 26 × 512
17 max		2 × 2/2	26 × 26 × 512	13 × 13 × 512
18 conv	1024	3 × 3/1	13 × 13 × 512	13 × 13 × 1024
19 conv	512	1 × 1/1	13 × 13 × 1024	13 × 13 × 512
20 conv	1024	3 × 3/1	13 × 13 × 512	13 × 13 × 1024
21 conv	512	1 × 1/1	13 × 13 × 1024	13 × 13 × 512
22 conv	1024	3 × 3/1	13 × 13 × 512	13 × 13 × 1024
23 conv	1024	3 × 3/1	13 × 13 × 1024	13 × 13 × 1024
24 conv	1024	3 × 3/1	13 × 13 × 1024	13 × 13 × 1024
25 route	16			
26 reorg		/2	26 × 26 × 512	13 × 13 × 2048
27 route	2624			
28 conv	1024	3 × 3/1	13 × 13 × 3072	13 × 13 × 1024
29 conv	35	1 × 1/1	13 × 13 × 1024	13 × 13 × 35
30 detection				

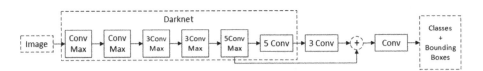

Fig. 2. YOLOv2 structure

3.2   Experimental Environment

Windows 7 Ultimate 64bit SP1, Intel Core i7-8700K @ 3.70 GHz, 64 GB RAM, Intel SSDSCKKW256H6, Microsoft Visual Studio 2015, OpenCV 3.4.1.

3.3   YOLOv2 Tomato Detection Network Training

The process of tomato detection network training is shown in Fig. 3. Firstly, collecting original data; Secondly, augmenting the amount of data; Thirdly, labeling samples; Fourthly, randomly selecting training set; Fifthly, training YOLOv2 tomato detection network; Finally, getting detection network.

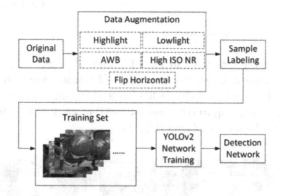

Fig. 3. Training process

We used the training set to train the YOLOv2 network and took darknet19_448.-conv.23 file as the pre-training weight file. Set batch = 64, momentum = 0.9, decay = = 0.0005, learning rate = 0.0001. During the training process, the learning rate was appropriately adjusted according to the average loss.

The changes of the average loss in iteration is shown in Fig. 4. With the increasing of iteration times, the average loss of the training set decreases gradually. As the prediction error could be reduced, the network performance was optimized during parameter updating. After 330 iterations, the average loss value was reduced to 0.165857 and no longer decreased in the following iterations, then training ended.

4   Results and Analysis

405 test images were input into the trained network for the detection of healthy and diseased tomatoes. When the IoU (Intersection over Union) of the object bounding box and the labeled bounding box is not less than 0.5, the object detection is considered to be correct.

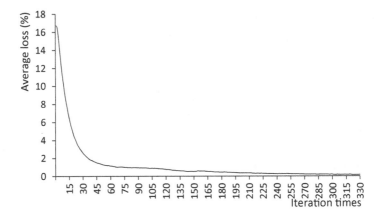

Fig. 4. Average loss curve of training

We selected precision rate, recall rate, F1-score [16] and mAP (mean average precision) as evaluation indexes. For the YOLOv2 tomato detection network, when threshold is 0.25, the precision rate reaches up to 0.96, the recall rate reaches up to 0.97, the F1-score reaches up to 0.96 and the mAP reaches up to 0.91. Examples of tomato detection results are as follows (Fig. 5).

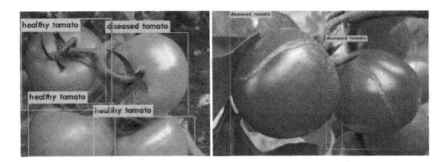

Fig. 5. Examples of tomato detection results

5 Conclusion

In this paper, we have researched and analyzed the application of YOLOv2 on the detection of healthy and diseased tomatoes. Based on the collection and augmentation of tomato images and the manual labeling of tomato fruits, YOLOv2 network was fully trained and tested by the training set and the test set. The YOLOv2 tomato detection network was finally obtained with great performances. Its precision rate is 0.96, recall rate is 0.97, F1-score is 0.96 and mAP is 0.91. These results indicate that YOLOv2 can be effectively applied to the detection of the status and location information of tomato fruits. This application also provides an idea for the detection of other fruits and vegetables.

In the future, we will improve and optimize the YOLOv2 tomato detection network according to the characteristics of tomato fruit, and further refine the disease types of object classes.

References

1. Huang, Z.H.: Research on construction method of Tomato Disease Image Database. South China Agricultural University (2016)
2. Zhou, Y.C., Xu, T.Y., Zheng, W., Deng, H.B.: Classification and recognition approaches of tomato main organs based on DCNN. Trans. Chin. Soc. Agric. Eng. 33(15), 219–226 (2017)
3. Zhou, Y.C., Xu, T.Y., Deng, H.B., Miao, T.: Recognition method of tomato key organs based on dual convolution Fast R-CNN. J. Shenyang Agric. Univ. 49(01), 65–74 (2018)
4. Zhou, Y.C., Xu, T.Y., Deng, H.B., Miao, T.: Real-time recognition of main organs in tomato based on channel wise group convolutional network. Trans. Chin. Soc. Agric. Eng. 34(10), 153–162 (2018)
5. Fang, J.L., Zhang, C.L., Pan, W., Wang, S.W.: Automated identification of tomatoes with diseases using artificial neural network trained with genetic algorithms. Trans. Chin. Soc. Agric. Eng. 03, 113–116 (2004)
6. Wang, Y.P., Dai, X.P., Huang, H., Zhang, R.: Identification of tomatoes with diseases based on mathematical morphology and neural network. J. Hunan Agric. Univ. (Nat. Sci.) 03, 344–346 (2006)
7. Redmon, J., Farhadi, A.: YOLO9000: better, faster, stronger. In: IEEE Conference on Computer Vision and Pattern Recognition, pp. 6517–6525 (2017)
8. Fu, L.S., Feng, Y.L., Elkamil, T., Liu, Z.H., Li, R., Cui, Y.J.: Image recognition method of multi-cluster kiwifruit in field based on convolutional neural networks. Trans. Chin. Soc. Agric. Eng. 34(02), 205–211 (2018)
9. Stern, U., He, R., Yang, C.H.: Analyzing animal behavior via classifying each video frame using convolutional neural networks. Sci. Rep. 5, 14351 (2015)
10. Ding, W., Taylor, G.: Automatic moth detection from trap images for pest management. Comput. Electron. Agric. 123(C), 17–28 (2016)
11. Xue, Y.J., et al.: Immature mango detection based on improved YOLOv2. Trans. Chin. Soc. Agric. Eng. 34(07), 173–179 (2018)
12. Redmon, J., Divvala, S., Girshick, R., et al.: You only look once: unified, real-time object detection. In: Computer Vision and Pattern Recognition, pp. 779–788. IEEE (2016)
13. Ren, S., He, K., Girshick, R., et al.: Faster R-CNN: towards real-time object detection with region proposal networks. In: International Conference on Neural Information Processing Systems, pp. 91–99. MIT Press (2015)
14. Liu, W., et al.: SSD: single shot multibox detector. In: Leibe, B., Matas, J., Sebe, N., Welling, M. (eds.) ECCV 2016. LNCS, vol. 9905, pp. 21–37. Springer, Cham (2016). https://doi.org/10.1007/978-3-319-46448-0_2
15. Ioffe, S., Szegedy, C.: Batch normalization: accelerating deep network training by reducing internal covariate shift, pp. 448–456 (2015)
16. Hripcsak, G., Rothschild, A.S.: Agreement, the F-Measure, and reliability in information retrieval. J. Am. Med. Inform. Assoc. 12(3), 296–298 (2005)

Research on Teaching Methods Based on Students' Autonomous Knowledge Aggregation

Xue Feng Li[✉]

Yan Shan College, Shan Dong University of Finance and Economics,
Jinan, China
lixuefengsdufe@163.com

Abstract. In information technology have driven changes in traditional educational concepts, models and methods. The main change brought about by the development of Internet and information technology in the study of teaching methods is that the method of "teaching" is more focused on the method of "learning". The development of big data provides an empirical study of teaching methods. The new data source makes the teaching method research more abundant and accurate data [1]. This paper will comprehensively consider various factors, establish a general evaluation model of student ability training effect, provide qualitative and quantitative analysis for students' ability training effect, and provide measurement standards for other similar studies.

Keywords: Knowledge aggregation · Autonomous learning ·
Teaching methods · Learning system

1 Introduction

The teaching method centered on "study" emphasizes the learning mode of student-centered, problem-oriented, student-dependent learning and collaborative inquiry. Current mainstream research includes Peer Instruction, referred to as PI [1], Problem Based Learning, referred to as PBL [2], Case Study [3], Just in Time Teaching, referred to as JiTT [4], Participatory Teaching [7, 8] and Inverted" Classroom [5, 6] and other methods. The combination of these teaching methods and these methods emphasizes the student-centered, teacher-led teaching philosophy, and strives to create a teaching context that is conducive to stimulating learning interests, and guides and encourages students to learn independently and cooperatively. A large number of studies and practices have shown that these teaching methods are active in the teaching of curriculum concepts, stimulating students' interest in learning, cultivating students' creative thinking, high-level reasoning and critical thinking skills, compared to traditional infusion-based teaching methods. The role of improving the quality of teaching and cultivating students' ability is significant. However, it should be noted that these teaching methods still have problems of applicability, limitations and effectiveness. The research also shows that different teaching methods are also presented in different teaching groups. Compared with these teaching methods, this topic will be based on "study" as the center, and the whole teaching activities will focus on students' self-completed knowledge aggregation. Students will give full play to their autonomy and initiative, and are more suitable for top-level students.

The teaching activities to cultivate students' professional ability are actually the process of helping students to establish a professional system. The professional knowledge system is a collection of professional knowledge units, including the structure of mutual relations between knowledge units. To build a professional knowledge system, it is necessary to decompose and classify professional knowledge according to professional characteristics, break down knowledge into a series of knowledge units, and then Divided into knowledge points, knowledge points can be divided into smaller knowledge points according to the details [7]. These knowledge points are distributed in various professional courses and are related to each other, which constitutes a corresponding professional knowledge system. Faced with numerous and scattered knowledge points, students often need to spend more time searching in the learning process, and it is difficult to distinguish the position of knowledge points in the whole learning process, and how to effectively organize scattered knowledge points. Contact is easy to encounter learning disabilities such as cognitive overload and information vaping [8]. Especially with the rapid development of information technology, students' access to knowledge is no longer limited to textbooks and classroom teaching, increasing the sources of Internet information, and the massive messy information on the Internet has exacerbated the seriousness of the problem. To this end, many researchers are committed to using a variety of methods and techniques such as clustering, pattern recognition, and knowledge mapping to establish knowledge associations, forming interrelated courses and even professional knowledge systems, which facilitate the navigation of students' learning processes [16, 17, 18, 19, 20, 21], further study of personalized learning path recommendations, etc. [22, 23]. These research results and practices show that the complete and systematic knowledge system provides a reasonable knowledge navigation and learning path for students to master professional knowledge. However, we believe that in these studies, the establishment of the knowledge system was pre-established under the guidance of teachers. The purpose is to systematically sort out the knowledge that is taught, centered on "teaching" rather than "learning". In addition, although there are studies on the aggregation of knowledge from the perspective of ontology, etc. [24], the source of knowledge is still limited to textbooks and other books, and does not consider other sources of knowledge, such as open literature, Internet information, etc., not adapted to current knowledge sources. Rich reality. Our research is to study the method of students' independent establishment of professional knowledge system, which is completely centered on "study", and the core of research is to complete the aggregation of multiple sources of knowledge to adapt to the fact that students' professional knowledge acquisition sources are diversified.

2 Construction of Learning System Based on the Aggregation of Students' Independent Knowledge

Change the traditional learning system to focus on classroom teaching, the passive learning mode of students, the new learning system construction will guide students to systematically complete the curriculum and even the integration of professional knowledge, supplemented by classroom teaching and give full play to students' learning initiative. Expand the professional knowledge of students, master the knowledge system, and cultivate students' scientific research ability. The main research contents of the study system construction include: knowledge system construction, knowledge aggregation method research, implementation path and performance quantification system construction.

The research goal of knowledge system construction and knowledge aggregation mode is to guide students to gradually establish the curriculum knowledge system on the basis of the existing training programs, and then establish the relationship between different curriculum knowledge, form the curriculum group knowledge system, and finally form the professional knowledge system. All knowledge system construction not only considers the basic knowledge in teaching materials and reference textbooks, but also needs to integrate relevant knowledge in open literature and public wisdom, as well as resources such as curriculum experiments, curriculum design and teaching cases completed by students. Reasonable knowledge point decomposition, association design between knowledge points, and knowledge aggregation between different knowledge sources are the key points and difficulties in the construction of knowledge systems.

The implementation path is the key to ensuring students' self-fulfilling of knowledge aggregation. The knowledge aggregation process requires full participation and self-implementation. The implementation path research includes the study of the relationship between classroom teaching and students' self-directed learning, the interaction between teachers and students; in order to screen out appropriate and correct knowledge from massive information, it is necessary to study the guiding participation method; For collective evaluation, it is necessary to study interactive participation methods.

The construction of the quantitative system of achievement is an effective means to ensure students' participation in enthusiasm and participation, and is also a measure of the effectiveness of the learning process. First, we need to study the quantitative model of achievement, and then collect and quantify the subjective and objective indicators, collective evaluation indicators, etc., and finally measure the credibility of the quantitative.

3 Construction of Student Capacity Training Effect Evaluation System

In order to test the effectiveness of the teaching method based on the aggregation of students' independent knowledge, it is necessary to establish a student capacity training evaluation system. The student's ability evaluation is a multi-faceted and multi-angle problem, involving academic achievement, employer evaluation, and Tripartite evaluation, social recognition and other factors. The main research contents include evaluation model research, determination of evaluation indicators and development of effective evaluation methods. In addition, the results of the effect evaluation will serve as effective feedback for the continuous improvement of the teaching method, and the corresponding feedback improvement mechanism should be studied. The evaluation model and evaluation method are the key points and difficulties of this part of the research.

4 Key Technology Research and Support Platform Construction

The construction of the learning system and effectiveness evaluation system will be based on relevant key technologies and supporting platforms. Key technologies include knowledge management technology, data crawling technology, text analysis technology, and behavioral data analysis technology. Since the knowledge in the knowledge system will be massive and has many kinds of related relationships, the corresponding knowledge management technology is the key content to be studied. In order to extract effective knowledge from massive Internet data, it is easy to conduct guided student participation. Data crawling technology and text analysis technology research will be necessary research content. Text analysis technology and behavioral data analysis technology will be the key technologies necessary to build a quantitative system and an effect evaluation system.

The construction of the support platform mainly follows the basic theories and methods of software engineering, and uses the agile method Scrum to organize incremental iterative development. To this end, the support platform is intended to iterate three versions, each version will be divided into several small iterative versions according to the actual situation. The completion of version 1 will support the main links of the learning system based on the aggregation of students' independent knowledge, and support knowledge management, knowledge aggregation, student self-implementation, and interactive participation. Based on the research of data crawling technology and text analysis technology, version 2 will support students' guided participation, build a quantitative system, and fully support the implementation path of knowledge aggregation. Version 3 will study the behavioral data analysis method and support the construction of the effect evaluation system based on the accumulated data in the previous version of the practice. After completing the construction of the support platform, we will continue to find problems in practice, solve problems, and continuously optimize the construction of the corresponding system to achieve the overall goal of the project.

5 Conclusion

This research will establish a new learning system, guide students to systematically complete the curriculum and even the integration of professional knowledge, supplemented by classroom teaching, give full play to students' initiative, expand students' professional knowledge, and fully grasp the knowledge system. Develop students' research capabilities. This will provide new ideas and new methods for the cultivation of college students, especially the training of top-level students in the basic disciplines. At the same time, a general evaluation model and standards will be established, which will provide qualitative and quantitative evaluation of the students' ability development effect for the research results of the subject, and can also be applied to other similar studies for effect evaluation and measurement.

References

1. Mazur, E.: Peer Instruction, "A User's Manual". Prentice Hall, Up per Saddle River (1997)
2. Barrows, H.S., Tamblyn, R.M.: Problem Based Learning: An Approach to Medical Education. Springer, New York (1980)
3. Herreid, C.F.: Case studies in science, "a novel method of science education". J. Coll. Sci. Teach. **23**, 221–229 (1994)
4. Novak, G.M.: Just in time teaching. Am. J. Phys. **118**, 63–73 (2011)
5. Bergmann, J., Sams, A.: Flip Your Classroom, Reach Every Student in Every Class Every Day, 1st edn. International Society for Technology in Education, Washington, D.C. (2012)
6. Dawson, L.A.P.: Motivation and cognitive load in the flipped classroom: definition, rational and a call for research. High. Educ. Res. Dev. **34**, 1–14 (2015)
7. El Sayed, R., El Raouf, S.: Video based lectures: an emerging paradigm for teaching human anatomy and physiology to student nurses. Alexandria J. Med. **49**, 215–222 (2013)
8. Delozie, S.J., Rrhodes, M.G.: Flipped classrooms: a review of key ideas and recommendations for practice. Educ. Psychol. Rev. 1–11 (2016)

Research and Implementation of Interactive Analysis and Mining Technology for Big Data

Li Guo[1]([✉]), Wenyuan Xu[1]([✉]), Hao Li[1], Shengxiao Zhang[1],
Dongmei Zhao[1], and Haoyang Yu[2]

[1] China Shipbuilding Industry Systems Engineering Research Institute,
Beijing, China
guolicssc@163.com, xwy0987@sina.com
[2] North China Electric Power University, Baoding 071003, China

Abstract. Leading to industry big data applications often need long-term repeated research and customized development, invisibly raising the application threshold of big data. The method to realize tasks of data preprocessing and analysis mining by user zero code will greatly reduce the complexity of data integration and big data analysis and mining [1]. Through breaking through the technologies of built-in data transformation and pre-processing, zero code analysis model development and debugging, standardized algorithm SDK interface and etc., a big data analysis application platform based on visual interaction and distributed architecture is proposed. The platform integrates data pre-processing and data analysis and mining, provides distributed computing framework, data access engine, scheduling engine, operator set, interface and visual analysis function modules, and lays a technical foundation for solving the problems of industry big data application demand variability, customized development demand variability.

Keywords: Interactive analysis and mining technology · Data integration · Data preprocessing · Distributed computing framework

1 Introduction

Big data is widely used in the fields of national security, intelligence analysis, commercial promotion, financial credit investigation, financial fraud, transportation, medical science etc. [2]. However, the wide popularity of big data is also restricted by many factors. First, isolated islands of information are widespread, and data sharing across departments and industries still needs to be improved. Policies are needed. Second, the quality of data is generally not high, and there are a lot of problems in the standardization of data, such as production data and sensor data, resulting in the low utilization rate of data and the urgent need for data governance; third, data business modeling is difficult, the application of big data needs cross-industry technology accumulation. Not only do we need to understand the data characteristics and business processes, but also we need to master professional data analysis methods. Therefore, the application of industry big data often needs long-term repeated research and customized development, which invisibly raises the application threshold of big data.

© Springer Nature Switzerland AG 2019
Y. Tang et al. (Eds.): HCC 2018, LNCS 11354, pp. 359–364, 2019.
https://doi.org/10.1007/978-3-030-15127-0_36

Based on the above background, a big data analysis application platform based on visual interaction and distributed architecture is urgently needed, which integrates data pre-processing and data analysis and mining. The task of pre-processing and analysis and mining can be arranged by user's zero code independently, which greatly reduces the difficulty and complexity of data integration and big data analysis and mining [3].

2   Research on Interactive Analysis and Mining Technology

2.1   The Characteristics of Interactive Analysis and Mining

Interactive analysis and mining applications for big data should have the following characteristics:

- Distributed architecture

For massive heterogeneous data, distributed technology architecture should be adopted to realize distributed parallel analysis and mining of massive data, and to realize iterative data analysis and mining based on DAG.

- High reliability

Large data analysis and mining, task execution cycle is long, high complexity, with the ability to deal with server, network and other failures, is essential. In terms of high reliability, the normal provision of mining services and the continuity of front-end business applications should be guaranteed from server nodes and running tasks.

- High performance

Distributed analysis and mining platform should have high performance, which is reflected in the iterative analysis and mining and fast response time. The performance of the whole platform should be guaranteed by parallel computing of multiple data nodes, parallel execution of analysis and mining tasks, and memory-based caching. Its performance is quasi-linear with the increase of cluster size.

- High scalability

Distributed analysis and mining platform should have the ability of flexible lateral expansion. With the change of business, the number of nodes can be quickly and arbitrarily increased and deleted to meet the actual business needs. The performance of the extended platform should be linearly enhanced with the increase of nodes.

2.2   Key Technologies of Distributed Analysis and Mining

- Multi-source data processing

It implements distributed analysis and mining to load and load multiple classes of data sources inside and outside clusters. During data source loading, data quality is detected and corrected.

Data quality checking adopts format matching checking, content null checking, data range checking, repeated record checking, balance checking of record number, character logic checking, rationality checking of data value range, logic expression checking, operation value checking of multiple fields, data scale checking, combination rules. Check and check the abnormal values in the data loading process [5].

Data quality correction, the use of value mapping, row-column conversion, string cutting, split fields, case conversion, missing value processing and exception value processing, etc., to achieve the detection of the problem data correction.

- Rich parallel algorithm library

Based on Spark and MapReduce framework, the algorithm includes machine learning, text analysis, numerical analysis, numerical statistics and other analysis algorithms. The algorithm implements DAG-based iterative parallel operation through distributed memory computing framework, which is suitable for efficient data mining and machine learning with large-scale data [6].

For different algorithms, algorithm evaluation operators are provided, and the results of operator-based business model analysis are analyzed. The time complexity, space complexity and correctness of operators are evaluated comprehensively and itemized, which provides guidance for selecting suitable operators for specific services and specific data types, and compares the adjustment timing of different operator parameters. Inter-complexity, spatial complexity, correctness, and other effects to facilitate are the further optimization of the selected operator.

- Construction of zero code analysis model

Provide self-help visual analysis modeling, using template componentization, modeling sketchpad, parameter setting panel and other visual interface, the modeler can easily achieve analysis business model construction, greatly reduce the technical difficulty and complexity of data analysis and mining, improve business analysis efficiency, timely response to business changes.

- Efficient analysis process

The system is based on the Spark distributed memory computing framework, which can effectively support the PB level scale and carry out efficient parallel analysis and mining. Compared with the previous MapReduce, its analysis and calculation speed will be increased by 10 to 100 times.

- Customization of custom operators

Customization of user defined operators is defined by standardized operator definitions and operator parameters. By defining the operator parameter specification, including the constraints of the operator parameters, the display mode in the foreground and other information, the definition of the operator parameter type and the data type conversion relationship is realized.

3 Implementation of Interactive Analysis and Mining Platform for Big Data

3.1 Logical Structure Design of Distributed Analysis and Mining Platform

The interactive analysis and mining platform for large data uses a distributed memory computing framework [4]. The logical framework of the distributed analysis and mining platform as shown in the following figure includes a distributed computing framework, a data access engine, a scheduling engine, an operator set, an interface, and a visual analysis interface.

In the scheduling engine layer, a configurable scheduling strategy is adopted to implement time-based periodic scheduling and manual scheduling, and the execution strategy of mining tasks is customized.

In the interface layer, JDBC, ODBC, HTTP, API and other interfaces are used to realize the unified standard access to the data in the platform, as well as the unified standard access interface for submitting, viewing, modifying, deleting and executing operators.

In the analysis of visual interface layer, visual analysis modeling interface is used to realize self-help, drag-and-drop business analysis process construction, zero code to achieve large data analysis and mining. Customized development business analysis system can realize the operation of data, algorithm, analysis task access and scheduling in the platform by calling the corresponding interface (Fig. 1).

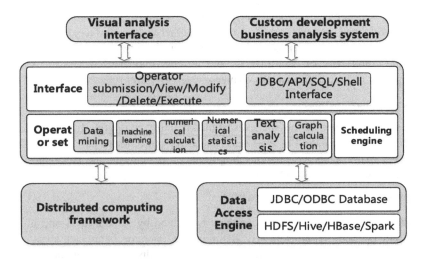

Fig. 1. Logical framework of distributed analysis and mining platform

3.2   Technology Architecture Design of Distributed Analysis and Mining Platform

In order to implement interactive distributed analysis mining, this paper presents the architecture design of distributed analysis mining technology as shown in the following figure. The architecture includes platform support layer, business layer and presentation layer.

In the platform support layer, Hadoop cluster is used as the runtime environment of the platform, which is used to store operators, SQL functions, knowledge base, user uploaded data set entities, and operator execution. HDFS and Hive components are used to realize the distributed data storage of two views of files and two-dimensional tables. HDFS API is used to realize the storage management of file data and the storage resource quota control of user directory. HCatalog API is used to realize the storage management of two-dimensional table data. The SQL operation of the underlying data is implemented by JDBC and the task submission interface is implemented. The YARN component is used to manage the computing resources in the cluster and manage and schedule the tasks of MapReduce, Spark and Samza.

At the business level, business analysis and mining logic are implemented. The model management service is used to analyze the model management, analyze the instance management and analyze the task management. Operators are used to manage

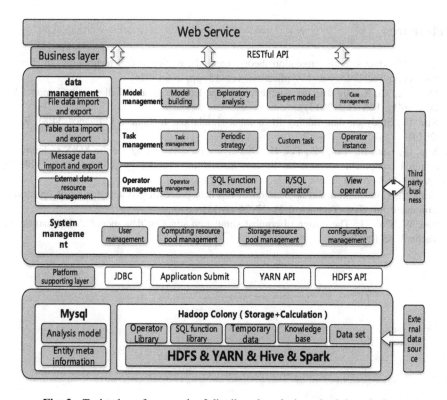

Fig. 2. Technology framework of distributed analysis and mining platform

services, and operators and operator instances are managed. Data management service is used to realize data set and knowledge base management. Use system management services to achieve user, permissions, configuration and other related system management. Using custom task management services to achieve user defined tasks.

In the display layer, Web services are mainly implemented. In the aspect of visualization, the analysis model is created and the execution status of the analysis instance is viewed. The RESTful API interface of the business layer is used to provide the calling service for the Web layer (Fig. 2).

4 Conclusions

Through breaking through the technologies of built-in data transformation and pre-processing, zero code analysis model development and debugging, standardized algorithm SDK interface, the research and development of large data analysis application platform based on visual interaction and distributed architecture is carried out. The platform provides a distributed computing framework, data access engine, scheduling engine, operator set, interface and visual analysis function modules, and provides a feasible solution to solve the problem of large data applications in the industry which need customization and development.

References

1. Li, J.: Big data science simulation theory, methods, platforms and technology. In: System Simulation Technology and Application, vol. 15, pp. 2–6 (2014)
2. Meng, X., Ci, X.: Big data management: concept, technology and challenge. J. Comput. Res. Dev. 50(1), 146–169 (2013)
3. Xue, Q.: Study on data mining for combat simulation. J. Sichuan Ordnance 34(8), 93–95 (2013)
4. Xue, Q., Cao, B., Tang, Z., Pei, H.: Discussion on data mining for the M&S technique in armored equipment support. In: System Simulation Technology & Application, vol. 13, pp. 563–567 (2011)
5. Gong, J.: Construction and application of big data processing system for naval operations. Command Inf. Syst. Technol. 6(2), 22–26 (2015)
6. Tao, W., Cui, H.: The design of combat simulation situation data model. J. Geomatics Sci. Technol. 35 (2010)

Research on Simulation Scenario Editing Method Based on Instruction Rules

Shengxiao Zhang, Hao Li, Dongmei Zhao, Wenyuan Xu[✉],
and Li Guo

Systems Engineering Research Institute, Beijing 100094, China
yangyuxiao_ncepu@outlook.com

Abstract. Simulation deduction is an important means for the army to verify the feasibility of the scheme or plan. It is necessary to provide an instruction input mode that meets the requirements of military personnel in the simulation scenario editing stage. Military scenario description language (MSDL) is analyzed, and the components of simulation scenario elements for combat plan deduction are presented. The scheme instruction system and instruction structured template are described in detail. The design method of multi-granularity combat task associated with multi-level instructions is designed based on Business Process Model and Notation. The design of scenario editing tool based on instruction description is given.

Keywords: Simulation scenario · Instruction system · Structured template · Combat mission

1 Introduction

Battle scenario deduction is the process of drilling and analyzing the situation, target or result of deployment and action at different operational stages in a combat scenario, in accordance with the operational intention, sequence and process specified in the operational plan, before or in the course of a combat operation [1]. It is an effective way to verify the feasibility of operational plan. Taking the U.S. Army as an example, JTLS, EDSIM, THUNDER and other systems is used to evaluate and deduce the combat plan before the Iraq War, and achieved good results [2].

Scenario is the basic basis for the deduction of combat plan. It is indispensable for both traditional military exercises and modern computer combat simulation. Before the simulation developer develops and runs the system, the technicians need to carry on the second description according to the military scenario and get the scenario file used in the simulation and this process is the editing of simulation scenario.

The simulation scenario editing is mainly for setting, adding and modifying the data files, including the settings of basic attributes, entity attributes and tasks, battlefield environment and operation rules, which provide important support for scenario generation and operation. In the course of deducing the combat plan, a large number of combat units need to realize the movement, situation evaluation, decision-making generation and behavior performance automatically by computer force according to their scenario content. The consistency between scenario and combat plan needs to be guaranteed, which puts forward higher requirements for the editing of combat scenario.

Y. Tang et al. (Eds.): HCC 2018, LNCS 11354, pp. 365–371, 2019.
https://doi.org/10.1007/978-3-030-15127-0_37

Based on the analysis of the content of simulation scenario, several key technologies of simulation scenario editing and formal description based on combat scenario are studied.

2 The Composition of Simulation Scenario

The composition of simulation scenario is closely related to the target and scale of simulation. For combat simulation scenario, according to the characteristics and concerns of military operations, the scenarios of different simulation systems will be very different.

To reduce tight coupling between scenario descriptions and application systems, the U.S. Army creates an XML-based mechanism for describing simulation scenario, namely the military scenario description language to initialize the simulation subsystems independently developed according to the military scenario generated by C4I system during developing OneSAF (OOS: One SAF Objective System) systems. MSDL has important theoretical and practical significance for the reuse of military scenario and the decoupling of scenario and simulation. Although it mainly refers to the US military operational rules and regulations, but refers to its development ideas and technical means, combined with the actual needs of our army combat scenario deduction, this paper refers to the MSDL standard proposed for combat scenario deduction simulation scenario elements, mainly including:

(a) Basic information: basic simulation information such as scenario name, type, background, scenario area, scenario time, simulation step, etc.
(b) Relations between belligerents: the parties involved in the scenario and their relationships, including the Red, Blue, White, Green, etc.
(c) Command sequence and military forces: the military and non-military organizations and equipment with hierarchical structure.
(d) Operational stage: the division of the operational process by time.
(e) Combat mission: information on operational action of each engagement force plan.
(f) Communication plan: communication plan for each engagement force.
(g) Tactical rules: rules of action and confrontation used in the process of engagement.
(h) Route: route information used in Scenario Planning.
(i) Region: regional information used in scenario scenario.
(j) Non war military operations: non war military operations related to operational processes.

3 Command Based Operational Plan Transformation

For the simulation of operational plans, the military orientation simulation should be accurate and no two sense mapping. Therefore, it is necessary to establish a conceptual model for the consistency description of operational plan instructions and simulation system execution instructions, and then implement the instruction transformation of the plan.

3.1 Description of Scheme Instruction System

Instruction, originally a command for computer to execute operations, is introduced into the process of scenario modeling to call a certain level of combat tasks or operations instructions. For different simulation applications, the instruction system is divided into different systems.

Instructions are numerous, diverse and complex. With the development of military theory and application, new instructions will be added or modified. XML is used to express the data model of instruction system, not only facilitating the accumulation of military instructions, but also facilitating the loosely coupled integration with scenario editing systems, which is conducive to the development and modification of instruction models. The XML description structure for the joint operational command architecture is shown in the following Fig. 1.

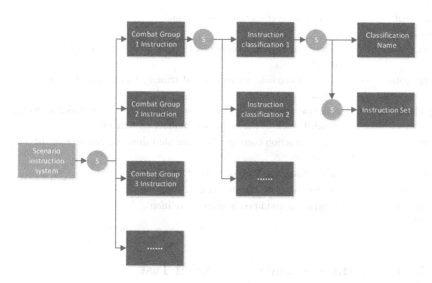

Fig. 1. Command structure of joint operations

3.2 Description of Instruction Structured Template

Instruction elements are described as conceptual models of instructions by using syntax, semantics, and diagrams that are independent of any simulation implementation. That is to extract the main elements of the instructions, such as the executing entity, the object of action, the start time, the end time, the name of the action resources, location, type of action, and so on. In this paper, EATI method is used to describe the instructions of simulation system design. Entity refers to all subjects and objects that can be identified separately in combat simulation, such as combat units, weapons and equipment, personnel, etc. Entity includes static and dynamic attributes. The state of all entities in the system constitutes the state of the system. Tasks use units with clear operational significance, which can be the smallest unit or a task group composed of multiple tasks.

Tasks consisting of one or more tasks with specific purposes and complete processes are called operational operations. An entity interacts with another entity to accomplish a task under certain conditions. This interaction is called interaction. The result is to change the state of some entity, that is, to change the state of the system.

Instructions need to drive the action of the entity, not only the executing entity, the target entity, the use of resources and other elements, but also time, space, mode of execution, execution conditions and other elements. A unified structured template for instruction is described in the following Table 1.

Table 1. Unified structured template for table instruction

Instruction name	The logo of the instruction as a sign of action
Instruction number	The ordinal number of the instruction
Functional description	Function description of instructions
Executive entity	The organization, weaponry and equipment for carrying out this directive
Target entity	The organization, weaponry and military facilities of the operational objects
Resources	The operation of a command corresponding to the operational resources needed, such as the name and number of weapons
Time	The instruction corresponds to the start time and end time of the action execution
Space	The path, area, location, etc. of an action
Execution mode	Action types of instruction execution
Conditions for execution	Beginning condition and end condition

4   Design of Multi Granularity Combat Task

The granularity and specific execution process of combat mission are closely related to domain operations. It is necessary to determine the granularity of combat mission expression according to the degree of attention of military personnel and commanders to combat process details and the available simulation resources. Tasks with large granularity need to be further decomposed according to the input of business users until the required granularity is minimized. Therefore, in the process of scenario editing and generation, tools are needed to provide the extension function of scenario task template, customizing task templates according to the task categories sorted out by the actual business.

Therefore, in view of the design requirements of multi-granularity combat task in scenario editing, it is necessary to start with the decomposition of the task, to construct the template of the basic combat action unit, and to study the composite design method of complex mission template on this basis.

4.1 Design Method of Basic Operations Unit

The basic operational unit can be understood here as a simple combination of activities to be performed by a single entity, and its basic elements of implementation are activities [4]. Activities reflect the basic business activities of an entity at a specific time. And the main information includes activity basic attribute, input parameter information, output parameter information, activation condition information and Termination condition information (Fig. 2).

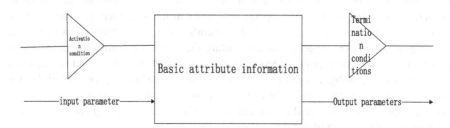

Fig. 2. Sketch map of activity elements

Activity basic attributes define activity name, type, execution entity resource and other information. Input parameters refer to all kinds of parameter information required for the execution of the activity. The output parameter mainly refers to the data generated during the operation of an activity. Activation conditions refer to the prerequisites for the execution of activities. Termination conditions refer to the prerequisites for the end of activities.

The basic operations unit can consist of one or more activities. The main information includes cell attributes, common information and node information in the cell (such as active node, logical node, event node, etc.). Unit attributes include cell name and combat unit category. Common information includes parameter interface and event interface provided by the operational unit, mainly including initialization parameters, external activation events, external output events and so on. The active node includes the element information described in the above activities. Event node (intermediate event) mainly refers to events triggered by external factors. Logical relational nodes refer to the judgment of linking active nodes, which are mainly divided into two categories. One is the yes-or node, which mainly controls the trigger relationship between the pre-order node and the post-order node, and can be subdivided into four types: connection, or connection, concurrency, or concurrency. The other is the judgment node, which mainly selects the subsequent branches according to the input data.

The above is the overall analysis of the elements of the basic operational unit, on this basis, the basic operational activities in the business field are combed and program modeling, and the above elements are formally described in XML form, to complete the formal description and modeling of the basic operational unit.

4.2   Design Method for Combat Mission Template Based on Rule

Operational mission is a higher level instantiation encapsulation of basic operational units, and the transformation of entity states between operational units is realized by conditional judgment [5]. BPMN is used for reference to establish the basic graphic elements of operational units, and then task flow modeling is carried out according to the mission requirements.

For the task flow modeling, the essence is to complete the serialization of the set of basic operational units to form a sequence of basic instructions that conform to certain rules. The sequence defines the parameters, order and conditions of execution instructions. Firstly, for the basic operational units in the sequence, it is necessary to initialize the execution parameters of the operational units according to the corresponding input command parameters, including the execution entity, the target entity, the resources used, time, space, execution mode, execution conditions and so on. These parameters can be configured in the simulation basic data, where can be quoted. On this basis, it is necessary to visually describe the execution relationship between instructions through a visual graphical configuration interface, including sequential execution, synchronous execution and conditional execution.

In the process-based combat task design, the existing combat rules can be combined and edited in the logical connector, including the addition of rules, the setting of parameters, and the editing of and or relations between rules. Through the design and serialization of the basic combat unit combination rules, the operational task templates available in the scenario can be generated.

5   Design of Editing Tools

The command-based scenario design method takes the combat mission as the traction, and realizes the configuration of entities, resources, combat targets, time and space in the combat mission by mapping the combat mission with the program command on the basis of the scenario force compilation. Using the above method, this paper designs and develops a simulation scenario editing tool. The data flow is shown as follows (Fig. 3):

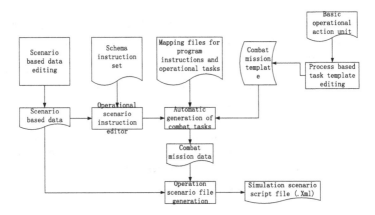

Fig. 3. Data flow diagram of scenarios edit

6 Conclusion

As the input of combat simulation scenario, the operational plan needs to ensure the accuracy and no sense mapping between the two. In this paper, the component elements of simulation scenario for scenario simulation deduction are proposed. The program instruction system, template description and multi-granularity mission design method related to mission editing are studied. The design idea and data flow of a simulation scenario editing tool are given. This paper provides a solution to the simulation scenario design based on the expression of operational instructions, and can be used in the development of scenario editing tool for combat scenario deduction.

References

1. Cheng, L.: Design and implementation of operational scheme deduction system. Ship Electron. Eng. 11 (2014)
2. Hu, X., Yang, J., Si, G., Zhang, M.: Simulation analysis and experiment of complex warfare system, p. 38. National Defense University press, Beijing (2008)
3. Military Academy of Sciences: PLA's military language, p. 179. Military Science Press, Beijing (1997)
4. Dai, J.: Research on Joint campaign Simulation Scenario edit system based on command. Master Thesis of National University of Defense Technology, May 2010
5. Li, Y.: Research on development technology of scenario scenario. Master Thesis of Harbin Institute of Technology, June 2006

Research on Social Media User Suicide Influencing Factors, Active Recognition and Intervention

Ming Guo[1,2] and Tingshao Zhu[1(✉)]

[1] Institute of Psychology, Chinese Academy of Sciences, Beijing 100101,
People's Republic of China
50990116@qq.com, tszhu@psych.ac.cn
[2] Department of Psychology, University of Chinese Academy of Sciences,
Beijing 100049, People's Republic of China

Abstract. With the development of Internet, more and more people express their own feeling on the social media, including suicide declaration, which create new opportunity for the identifying, preventing and intervening of suicide high risk group. The influencing factors of suicide behavior are numerous, mainly including internal and external factors. Internal factors include biological factors, mental disorders and psychological factors; external factors include negative life events, family factors, social environment factors and cultural factors. Systematic intervention of Internet is used to establish an automatic identification and assessment of suicide ideation in social media, and a mental psychotherapy with many other psychotherapies. Through timely intervention and referral, systematic intervention and continuous care intervention process, social media users can eliminate suicide idea and reduce suicide probability. At present, researchers pay more attention to the development process and mechanism of suicidal behavior. In the future, it will be necessary to establish Internet suicide intervention system and study the effect of different psychological intervention on Internet suicide.

Keywords: Depression recognition · Modeling · Network behavior · Systematic intervention

1 Introduction

Suicide is a serious social and public health problem, which can lead to a system of negative results. As more and more people reveal their feelings and opinions in virtual clusters, platforms such as microblog have become a way for individuals to express themselves, including expressions related to suicide. By June 30, 2018, the number of Internet users reached 802 million, and the popular rate was 57.7%. Among them, the number of mobile phone users has reached 788 million, and the proportion of Internet users accessing the Internet through mobile phones is as high as 98.3%.

Studies have shown that social media such as forums and micro-blogs have become a new field for exploring suicides. Some existing studies have also extracted suicide-related thoughts and behaviors from social networks as evidence for suicide intention

© Springer Nature Switzerland AG 2019
Y. Tang et al. (Eds.): HCC 2018, LNCS 11354, pp. 372–379, 2019.
https://doi.org/10.1007/978-3-030-15127-0_38

analysis. Therefore, the main research contents in this field at home and abroad will be reviewed, including concept definition, suicide-related model, internal motivation and influencing factors of social media users' suicide behavior, active identification and suicide intervention of high-risk groups, and future research directions for domestic colleagues.

2   Terminology and Concept Definition

Non fatal suicidal thoughts and behaviors are collectively referred to as suicidal behavior, and further refine suicidal behavior into three types: suicide ideation, suicide plan and suicide attempt. There are two main characteristics of Suicide: intentional and fatal [1].

Conceptually, suicidal ideation refers to suicidal ideas and behaviors that have lost their desire to live but have not yet caused physical injury [2]. Suicide attempt is a self-destructive act with a certain degree of suicidal intent and may result in injury or no injury [3].

Suicidal ideation is an important indicator to assess the risk of suicide. Suicidal ideation has the following characteristics: 1. concealment. 2. universality. 3. individual differences.

3   Related Research in China and Abroad

3.1   Suicide Factors

Suicide, as a result of extreme behavior, has many influencing factors. This part mainly analyzes the factors influencing suicide from two factors, namely, internal and external factors. In fact, the accumulation and interaction of these factors increase the risk of suicide. Next, use Table 1 to sort out all the factors affecting suicide.

Table 1. Suicide related factors

Suicide related factors	Illustration
Biological factors	In terms of family history, individuals with suicide were also more likely to have mental illness and to have attempted suicide or to have committed suicide than individuals without suicide [4]
Mental disorder	About 90% of suicide have mental disorder. Among them, suicide caused by depression or depressive episodes of bipolar disorder accounts for at least half of the total and is the most common mental disorder causing suicide [5–7]
Personality characteristics	A study showed that by controlling health, agreeableness, openness, responsibility and extroversion and responsibility into the regression equation, suicide risk was predicted to be 56.7% [8]

(*continued*)

Table 1. (*continued*)

Suicide related factors	Illustration
Cognitive factors	The research states clearly that the individual who has suicide attempt, has higher levels of cognitive rigidity than individuals without suicide attempts [9]
Attitude factors	The more affirmative affirmation of suicide attitude, the stronger the suicidal ideation [10]
Negative life events	Stress-the susceptible model points out that stress is one of the causes of suicidal ideation [11]
Family factors	Family factors have a huge impact on suicide. First of all, childhood abuse experiences or neglected experiences, family stability and family rearing styles can also affect individual suicide ideation [12]
Social environmental factors	There are studies showing that the Internet and the forums has great potential in knowledge popularization and intervention [13, 14]
Cultural factors	A large cultural atmosphere can influence the expression of suicidal ideation by influencing the individual's attitude toward suicide

4 Suicide Intervention Based on Social Media Suicide High Risk Groups

4.1 Suicide Intervention

Suicide intervention mainly includes drug therapy and psychotherapy. We are mainly concerned about psychotherapy. Psychotherapy can be classified according to different classification criteria. For example, according to the content, psychological intervention measures include cognitive behavioral therapy (CBT), dialectical behavioral therapy (DBT), problem solving therapy (PST), and so on. According to the form, psychological intervention measures include face-to-face intervention, Internet-based counseling, hotline counseling and so on.

From the perspective of intervention form, hotline intervention and Internet-based intervention are not to replace face-to-face intervention, but are beneficial supplements to face-to-face intervention. They not only solve the problem of shortage of medical resources, but also radiate to a wider population from time and space, providing more realistic conditions for the realization of psychological intervention.

The internal motivation and influencing factors of social media users publishing and suicide-related information on the Internet can intervene the suicide ideation of social media users to reduce and control the occurrence of social media suicide behavior, prevent the occurrence of suicide infection and improve the positive role of the Internet in suicide intervention. According to the intervention process, we divided the intervention into three stages: timely intervention and referral; systematic intervention and continuous care.

4.2   Timely Intervention and Referral of Suicide

Once a micro-blog user is identified as having a suicidal intention, immediate intervention is promptly pushed through the micro-blog private letter. This is the first step in intervention.

In addition to the private letters of intervention, we also have experienced experts and social work volunteers who can communicate with micro-blog users through private letters. The content of the exchange mainly includes six aspects, and we have standardized the process, which is a standardized language for experts or volunteers to communicate.

4.3   Systematic Suicide Intervention

Because 90% of the suicide population suffers from mental illness or mental disorders, one or a few counseling sessions can only temporarily alleviate the problem and not fundamentally solve it. Using the Internet as a tool and with the help of experts or volunteers of psychological counselors, users can receive systematic psychological intervention or treatment, which can not only solve the serious shortage of resources for our psychological counselors, but also radiate across time and space to a wider range of people to improve the national basis of psychological counseling. The role of the medical system in improving mental health and reducing suicide rate.

Recent studies have shown that in young people, CBT is the best way to reduce suicide. Internet based CBT has been widely used in the treatment of depression and anxiety, and the effect is remarkable. Although there are some network intervention systems for depression in China, there is no intervention system for suicide population, and the entry conditions for depression and other mood disorders exclude individuals with suicidal ideation. Therefore, it is urgent and necessary to establish an intervention system specifically for suicide ideation.

4.4   Continuous Suicide Intervention

Because of the recurrence and long-term nature of mental illness, or the universality of life stress, in the case of attempted suicide, many people will have repeated suicide. Therefore, continuous care and intervention are essential. By means of automatic transmission of information by computer, the relevant information can be pushed to users in time at the prescribed time point, and the problem can be solved to a great extent.

4.5   Summary

In the current research, scholars often do not confine themselves to a single intervention method or form, but combine various methods and forms in order to achieve the best results. For example, the combination of CBT, PST and PIT, the use of Internet-based interventions with the services of experts or volunteers, to maximize the role of psychological intervention, and strive to achieve maximum results in the shortest possible time.

5 Comments and Research Prospects

5.1 Active Identification of Suicide Users (Establishment of Suicide Public Opinion Monitoring System)

Existing studies have shown that most people with suicidal ideations do not seek professional help on their own initiative, especially in China. Among the reasons for not asking for help, only 20% thought it was unnecessary, and the remaining 80% were due to other reasons (such as worrying about others' opinions, worrying about others, worrying about costs, etc. Active help-seeking behavior is closely related to the use of mental health services. The lack of active help-seeking behavior may lead to the failure of individuals to help effectively in time, thus delaying the best time for suicide intervention. The popularity of the Internet can improve the current situation to a certain extent. Young people prefer not only to use the Internet for suicide declarations, but also to use Internet-based psychological intervention systems. For young people, the Internet has become an important tool for getting help. Suicide is the first cause of death among young people in China. The main user group of micro-blog is also young people.

The study of Internet as a tool has greatly expanded the scope of psychological intervention. Through the establishment of suicide public opinion monitoring system, to suicide ideation of social media (micro-blog) group users to actively identify. Two methods, keyword retrieval and machine learning modeling, are used to identify the suicide risk of microblog text (Fig. 1).

Fig. 1. Flow chart of suicide public opinion monitoring

5.2 Intervention System and Expert Collaboration (Online and Offline Intervention Mechanism)

The task of developing a self-help online counseling system based on Chinese context for suicidal ideation or behavior is in progress. Previous studies have shown that interventions with expert or volunteer assistance are significantly better than those without expert or volunteer assistance, and that assisted interventions are equivalent to face-to-face counseling. Under the guidance of this theory, we adopt the intervention system and experts to work together to intervene in suicidal ideation. The system completes the module's bad and stylized part. Experts or volunteers provide guidance to users at the beginning, provide feedback and answer to users at certain intervals (e.g. weekly), and summarize to users at the end.

5.3 Process Standardization of Manual Communication for Suicide Intervention

Private letters and micro-blog users to exchange content mainly includes six aspects, we have standardized the process, each step has a standardized language for experts or volunteers to communicate, as follows Fig. 2.

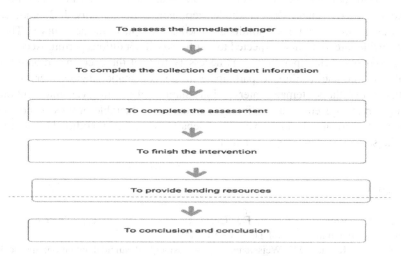

Fig. 2. Six steps for suicide intervention

The first step is to assess the immediate danger: "I'm glad you responded to our private letter and you've taken the first step in helping yourself. First of all, we want to know your state at this moment. 0 means absolutely no, 100 means the strongest. What is your current level of pain, want to die, and think life is full of hope?"

The second step is to complete the collection of relevant information: "What happened around you? When did it happen?"

The third step is to complete the assessment. First, experts or volunteers need to summarize the user's questions, then say, "To further understand the impact of this event on your mood and other aspects of life, complete the following assessment."

The fourth step is to finish the intervention. The basic principles and points of attention to solve the problem: to determine the user's rational thinking ability, problem solving initiative, help-seeking motivation(self-motivation): focus on this moment; understand the user's past coping style, effect; help users to find internal and external resources (emphasis on the importance of internal resources). Discuss with users the way to solve the problem and the obstacles that may be encountered in implementing the method.

The fifth step is to provide lending resources, such as hotline, free online consultation and hospital outpatient service.

The sixth step is to conclusion. "It seems that you already know how to solve the problem, can you summarize the contents of our private communication? Hopefully, by taking action, your situation will gradually improve."

5.4   Longitudinal Tracing Study (Cooperate with Relevant Agencies to Set up Tracking Service System)

Longitudinal follow-up studies include two parts: systematic suicide intervention and continuous suicide intervention. Because the suicide prevention system is composed of different parts of the module, in order to ensure the effect of intervention, users need to seriously learn and repeatedly experience, practice. That is, curriculum learning is a continuous process rather than a single step. For example, only 1–2 modules are learned every week, and 1–2 months are used to complete all the courses. The persistent suicide intervention is expected to last forever if conditions permit. According to the previous research experience, it is necessary to visit the user on a monthly basis after the first contact (failure to participate in the systematic suicide intervention) or after the end of the systematic intervention. Status and health level. Intervention and tracking service system is a continuous service for vulnerable groups. It cooperates with the government, relevant government departments or neighborhood committees to make the service system more systematic.

References

1. Nock, M.K., Borges, G., Bromet, E.J., Cha, C.B., Kessler, R.C., Lee, S.: Suicide and suicidal behavior. Epidemiol. Rev. **30**(1), 133–154 (2008)
2. Beck, A.T., Kovacs, M., Weissman, A.: Assessment of suicidal intention: the scale for suicide ideation. J. Consult. Clin. Psychol. **47**(2), 343 (1979)
3. O'Connor, R.C.: The relations between perfectionism and suicidality: a systematic review. Suicide Life Threat. Behav. **37**(6), 698–714 (2007)
4. Brent, D.A., Kolko, D.J., Allan, M.J., Brown, R.V.: Suicidality in affectively disordered adolescent inpatients. J. Am. Acad. Child Adolesc. Psychiatry **29**(4), 586–593 (1990)
5. Holma, K.M., et al.: Differences in incidence of suicide attempts between bipolar I and II disorders and major depressive disorder. Bipolar Disord. **16**(6), 652–661 (2014)
6. Cash, S.J., Bridge, J.A.: Epidemiology of youth suicide and suicidal behavior. Curr. Opin. Pediatr. **21**(5), 613 (2009)
7. Sareen, J., Cox, B. J., Afifi, T. O., de Graaf, R., Asmundson, G. J., ten Have, M., & Stein, M. B. (2005)
8. Voracek, M.: Big five personality factors and suicide rates in the United States: a state-level analysis. Percept. Mot. Skills **109**(1), 208–212 (2009)
9. Keilp, J., et al.: Neuropsychological function and suicidal behavior: attention control, memory and executive dysfunction in suicide attempt. Psychol. Med. **43**(03), 539–551 (2013)
10. Kwon, J.-I., et al.: Attitude toward suicide and personal experiences of suicide among doctors and health care workers in Korea. J. Korean Neuropsychiatric Assoc. **52**(4), 231–242 (2013)
11. Wasserman, D.: Suicide: An Unnecessary Death (Ming Li, Trans.). China Light Industry Press, Beijing (2003)
12. Lipschitz, D.S., Winegar, R.K., Nicolaou, A.L., Hartnick, E., Wolfson, M., Southwick, S. M.: Perceived abuse and neglect as risk factors for suicidal behavior in adolescent inpatients. J. Nerv. Ment. Dis. **187**(1), 32–39 (1999)

13. Alao, A.O., Soderberg, M., Pohl, E.L., Alao, A.L.: Cybersuicide: review of the role of the internet on suicide. CyberPsychol. Behav. **9**(4), 489–493 (2006)
14. Ikunaga, A., Nath, S.R., Skinner, K.A.: Internet suicide in Japan: a qualitative content analysis of a suicide bulletin board. Transcult. psychiatry **50**(2), 280–302 (2013)

Collaborative Filtering Recommendation Algorithm Based on AdaBoost-Naïve Bayesian Algorithm

Zuoxi Yang, Chengzhou Fu$^{(\boxtimes)}$, Ronghua Lin, Tao Peng, and Yong Tang

South China Normal University, Guangzhou 510631, China
fucz@m.scnu.edu.cn

Abstract. With the rapid development of social network and E-commence, collaborative filtering (CF) has been widely applied and studied as an effective way to alleviate the pressure of information overloading. However, the CF-based algorithms often degrade significantly in their recommendation performance due to the sparse history rating data. In view of this sparse problem, this paper presented a collaborative filtering recommendation algorithm based on AdaBoost-Naïve Bayesian algorithm. In this model, we transform the traditional algorithms of predicting ratings directly to classify unrated data. Considering that Bayesian theory is a good machine learning method that is often used for text class. Considering that machine learning method can be used for text classification, we utilize it to learn user's preferences for a certain item characteristic as a based classify. Ensemble learning, Adaboost, is also adopted to optimize and adjust adaptively the weights of the results of Bayesian classify. Finally, all of the predicted results of unrated data would be filled into the user-item matrix. The experimental results show that the sparseness of rating data has been alleviated and our proposed algorithm can improve the quality of recommender system.

Keywords: Collaborative filtering · Naïve Bayesian · AdaBoost · Movielens-100k · SCHOLAT

1 Introduction

With the rapid development of Internet, a huge variety of information is exploding at a speed over Moore's Law, causing increasingly serious pressure of information overloading [1] for users. Although traditional search engines (such as Google, Baidu) can help users better solve the requirements of retrieval, they still can not satisfy the individual requirement of different users at different time. The personalized recommendation services have come into being, where recommender system [2, 3] has been playing an increasingly significant role. Besides, collaborative filtering (CF) is one of the most widely used methods for recommender. At present, user-based CF [4] and item-based CF [5] are the two main algorithms in the proposed methods.

CF-based methods usually use the past activities or preferences from user's history feedback data and output a ranked list of items. In general, the ratings given to items by

© Springer Nature Switzerland AG 2019
Y. Tang et al. (Eds.): HCC 2018, LNCS 11354, pp. 380–392, 2019.
https://doi.org/10.1007/978-3-030-15127-0_39

different users would be the source information for recommender system to predict the unrated data. Many of them have been applied and studied widely. For example, many online services including social networks, online news and E-commerce have been using them to target their customers by recommending products or services and therefore got better economic returns. Nevertheless, these methods do have their limitations. Because of user's history data is often sparse in practice, the performance of recommender system would be dropped sharply. To address the sparse problem, many researchers try to enhance their performance. For example, setting the ratings of unrated item as fixed ratings or using similarity of different items to predict the unrated ratings simply. Both of them do have limitation. On the one hand, the former method can not fundamentally solve the problem as the ratings given to items by different users would not be exactly the same, on the other hand, the latter uses the similarities of users or item simply without considering the interaction between user and item. In general, the latter methods are more popular and effective.

In recent years, machine learning and ensemble learning methods have yield immense success on text classification, intelligent decision and manufacturing industry. Bayesian theory is one of the good supervised machine learning methods that is famous for its higher productivity and good classification performance [6, 7]. Considering the good classification ability of Bayesian method, we use it to learn to unrated data to alleviate the problem of sparsity. In other word, the task of rating prediction on user-item data is transformed to classify the unrated data based on the probabilities. It is well known that ensemble learning method can be integrated into machine learning methods to improve the performance of machine learning methods. AdaBoost (adaptive boosting) meta-algorithm is one of the best ensemble methods, which core idea is to train a number of different weak classifiers for the same training set, and then these weak classifier classification results through a certain algorithm into a strong classification algorithm [8].

In conclusion, we presented a CF recommendation algorithm based on AdaBoost-Naïve Bayesian algorithm. In our algorithm, taking into consideration of the impacts of the relationship between user attributes and item attributes on the final item rating, we select the valid user attributes and item attributes, which would be grouped together into feature attribute matrix. After having the matrix and optimizing the combination of AdaBoost algorithm and naïve Bayesian algorithm, the unrated data would be predicted and filled. Finally, the user-item rating would be denser. The experimental results show that the AdaBoost-Naïve Bayesian algorithm can achieve better prediction accuracy even with extremely sparsity of data set.

2 Related Work

Existing methods of CF for recommender system are most composed of memory-based methods [9] and model-based methods [10]. In the memory-based CF, use-based and item-based are the two CF algorithms most widely used. This two algorithms find out the sets of the target user's nearest neighbor according to the user-item rating matrix $T(m \times n)$, where $T(m \times n)$ is a matrix with size $m \times n$, m is the number of users, the n is the number of items and T_{ij} means the rating given by user i on item j. When the user-item rating matrix is dense enough, the memory-based CF always achieve good results, especially in the item-based CF, which is one of the most effective algorithms in the CF [11]. However, the user feedback data is always extremely sparse, so that the recommender system's performance declined even the results is far less than expected since it can not find out the associated neighborhood accurately.

To address the sparse problem, the methods are listed as below: (1) the ratings of unrated item would be replaced by fixed values: This is usually done by setting the user's rating for unrated items to default or average value. In the case of severe data sparsity, the reliability of this method is not high. Because the ratings of unrated items cannot be exactly the same, the problem cannot be solved fundamentally; (2) Using similarity to predict the ratings of unrated items, which would alleviate data sparsity problem by making users of the same class rating have the similar rating or setting similar items similar rating, which are both based on user similarity or item similarity simply. However, the data is usually limited to a certain aspect. For example, the user clustering method makes use of the user similarity without considering the impact of item similarity on the score, while the project clustering algorithm does the opposite. The latter method is more popular and used widely. Many researchers have been studying to enhance it. For example, the authors [12] proposed a new CF recommendation algorithm by using items categories similarity and interestingness measure to improve the performance of item-based CF. They calculated the distance between different items and used interestingness measure to analyzed the correlation degree of different items. Considering the users' information, Zhang [13] focused far more on the trust relation between users. They have proposed a user trust-based collaborative filtering recommendation algorithm by analyzing similarities between users and user trust. Besides, researchers [14] paid more attention to user preference in the local and global perspectives respectively to obtain reliable neighbors of active users. Although the above research works have further studied the item rating information, they ignored the valid attribute relationship between user and item and they are not generalizable enough, so that they would cause unstable performance of the recommender system for complex data. In this proposed algorithm, We select the user attributes and item attributes that would have impact on the users' rating to combine into feature matrix T. Each term in the matrix T element on K point scale is an abstract representation as follows:

$$(U, I)C_k \tag{1}$$

Where $U = \{U_1, U_2, ..., U_m\}$, U_k ($1 \leq k \leq m$) is a user's feature attribute, $I = \{I_1, I_2, ..., I_m\}$, I_j ($1 \leq j \leq n$) is a item's feature attribute, and the Eq. 1 indicates the probability of item rating is $C_k(C_k \in K)$ under certain users' and items' feature attributes. The results of the proposed algorithm would be filled into the original matrix for the corresponding unrated items.

3 Proposed Algorithm

In this part, we mainly introduce the algorithm process of traditional CF algorithm and AdaBoost-Naïve Bayesian CF algorithm. The general flow of user-based CF algorithm is given in Sect. 3.1. Next, in Sects. 3.2 and 3.3 describes our proposed algorithm in detail.

3.1 Similarity Method

The common similarity measures are Cosine correlation [2], Adjusted Cosine correlation [5] and Pearson correlation coefficient [15]. Their calculations are given in Eqs. 2, 3 and 4 respectively, where U is the user set who have rated for item i, j commonly, $\overline{R_u}$ is the average of all ratings of user u, R_{ui} means that the rating given by user i on item j and $\overline{R_i}$, $\overline{R_j}$ are the average of all ratings of item i, j respectively.

$$sim(i,j) = cos(i,j) = \frac{\vec{i}.\vec{j}}{\|\vec{i}\|_2 * \|\vec{j}\|_2} \tag{2}$$

$$sim(i,j) = \frac{\sum_{u \in U}(R_{ui} - \overline{R_u})(R_{uj} - \overline{R_u})}{\sqrt{\sum_{u \in U}(R_{ui} - \overline{R_u})^2}\sqrt{\sum_{u \in U}(R_{uj} - \overline{R_u})^2}} \tag{3}$$

$$sim(i,j) = \frac{\sum_{u \in U}(R_{ui} - \overline{R_i})(R_{uj} - \overline{R_j})}{\sqrt{\sum_{u \in U}(R_{ui} - \overline{R_j})^2}\sqrt{\sum_{u \in U}(R_{uj} - \overline{R_j})^2}} \tag{4}$$

3.2 Naïve Bayesian Classifier Algorithm

The prediction of uncertain knowledge is an important research area of recommender system, especially when the user requirements are not clear. Bayesian is a significant tool for dealing with uncertain recommender. In the users' feedback data, we assume that user attributes $U = (U_1, U_2, ..., U_n)$ and item attributes $I = (I_1, I_2, ..., I_n)$ are grouped into a n-dimension feature vector $X = [x_1, x_2, ..., x_n]$, and $C = [c_1, c_2, ..., c_m]$ represents m classifications/rating values. After finishing it, we can use the Bayesian theorem to calculate the probability of a sample data X being in class c_i as follows:

$$P(c_i|X) = \frac{P(X|c_i)P(c_i)}{P(X)} \tag{5}$$

Where $P(c_i)$, $P(X|c_i)$, $P(c_i|X)$ are the prior probability, likelihood, posterior probability respectively. Since the value of $P(X)$ is the same constant for each posterior probability, the selection of maximum $P(c_i|X)$ is equivalent to the selection of maximum $P(X|c_i)P(c_i)$. Then we could select the class/rating of maximum $P(X|c_i)P(c_i)$ as the class/rating of X, as follows:

$$h^*(X) = argmax\, P(X|c_i)P(c_i)(c_i \in C) \tag{6}$$

Where $P(X|c_i)$ is the coincidence probability distribution of all the attributes on sample of X. However, because it is hard to calculate the coincidence probability and the cost of the calculation is high, the Naïve Bayesian is proposed to avoid this problem, whose assumption is that attributes in the data are conditionally independent (Attribute Conditional Independence Assumption). Its core concept is that the occurrence of attributes do not depend on each other. Therefore, the Eq. 3 can be expressed as follows:

$$h^*(X) = argmax\, P(X|c_i)P(c_i) = argmax\, P(c_i) \prod_{\lambda=1}^{n} P(x_\lambda|c_i)(c_i \in C) \tag{7}$$

In the Naïve Bayesian algorithm, the prior probability can be obtained from training set. Given the training set H, H_{c_i} are the set of the training item that belongs to c_i. Since the ratings are from 1 to 5, which are discrete attributes, we assume H_{c_i,x_i} is the set of data that the value is x_i on the c_i. After that, the prior probability and $P(x_i|c_i)$ can be shown as follows respectively:

$$P(c_i) = \frac{|H_{c_i}|}{H} \tag{8}$$

$$P(x_i|c_i) = \frac{|H_{c_i,x_i}|}{|H_{c_i}|} \tag{9}$$

To avoid the probability would be 0 finally, the Eqs. 5 and 6 need to be "smoothing", so that we introduce Laplacian correction. Otherwise, the calculation of 0 would mean that information of the corresponding attribute would be erased, leading to the final results would not be accurate. The Eqs. 8 and 9 are reformulated as follows:

$$P(c_i) = \frac{|H_{c_i}| + 1}{H + n} \tag{10}$$

$$P(x_i|c_i) = \frac{|H_{c_i,x_i}| + 1}{|H_{c_i}| + n_i} \tag{11}$$

Where n is the number of the classes/ratings in the training set H, n_i is the value of the classes/ratings on the c_i.

3.3 AdaBoost-Naïve Bayesian Algorithm

Although the "Attribute Conditional Independence Assumption" has avoided the occurrence of complicated conditional probability between different attributes, which can greatly simplify the calculation of Bayesian and improve the operation efficiency, it inevitably will reduce precision. We need to further train the results of weak classifiers to improve their accuracy. Boosting, also known as reinforcement learning algorithm, is an important ensemble learning technology that is widely used and has a great impact on ensemble learning, where AdaBoost is one of the most successful representatives [16, 17]. Combining with the AdaBoost algorithm [18], the AdaBoost-Naïve Bayesian that we state above can be expressed in Algorithm 1.

4   Experimental Setup and Analysis

In out experiment, we use the public data set Movielens-100k (https://grouplens.org/datasets/movielens/) and real data set provided by scholar network Scholat (http://www.scholat.com). There are 943 users, 1682 movies and 100 000 ratings on scale of 1(dislike) to 5(like) in Movielens-100k and each user had rated at least 20 movies. We select the information about 833 users, including user's id, number of "thumbs-up", logs of exchange visits and friendships. Based on the processed data, there are 2527 ratings that are the quantification scores of logs of exchange visits from 1 to 5. After analyzing the data of Movielens-100k, we got its sparse ratio of the rating data as follows:

$$\frac{943 \times 1682 - 100000}{943 \times 1682} \times 100\% = 93.69\%$$

Obviously, the Movielens-100k data set is extremely sparse. All data sets are divided into training set (80%) and test set (20%). 5-fold cross-validation is used for evaluating the performance of our proposed algorithm by randomly selecting the different test and training set each time, and setting the average as results.

Algorithm 1. AdaBoost Algorithm

Input: Training Set $H = \{(x_1, y_1), (x_2, y_2), \dots, (x_m, y_m)\}$; Naïve Bayesian Algorithm L; Number of learning rounds T;

Process:

1 $\delta_{1,i} = \frac{1}{n}$; $Z_1 = (\delta_{1,1}, \delta_{1,2}, \dots, \delta_{1,n})$

2 **for** $t = 1, \dots, T$:

3      $h_t = L(Z, Z_t)$

4      $\varepsilon_t = (the\ number\ of\ incorrectly\ classified\ smples)/$

5          $(the\ total\ number\ of\ samples)$

6   **if** $\varepsilon_t > 0.5$ **then break;**

7      $\alpha_t = \frac{1}{2}\ln\left(\frac{1-\varepsilon_t}{\varepsilon_t}\right)$

8   **if** $h_t(x_i) = y_i$ **then**

9          $Z_i^{(t+1)} = \frac{z_i^t e^{-\alpha}}{sum(Z)}$

10   **else then**

11         $Z_i^{(t+1)} = \frac{z_i^t e^{\alpha}}{sum(Z)}$

12   **end**

Output $D(x) = sign(\sum_{t=1}^{T} \alpha_t h_t(x))$

4.1   Evaluation Measures

There are many metrics having been used for evaluating the performance of recommender system, where MAE (Mean Absolute Error) and Coverage are two common measures.

$$MAE = \frac{\sum_{i=1}^{n} |p_i - q_i|}{n} \tag{12}$$

where p_i, q_i are the predicted rating and actual ratings respectively, n is the total number of rated items. The lower MAE is, the better the accuracy will be. On the other hand, rate of Coverage is also an important measure to evaluate the ability of a recommender system to provide recommendations. The coverage can be given as follows:

$$C = \frac{K}{S} \tag{13}$$

where K, S are the number of items can be predicted and the total number of items in the test set. The higher Coverage is, the better the recommender system will be.

4.2   Similarity Measures

To find out a appropriate similarity measure for our experiment, we design experiments with Cosine correlation, Adjusted Cosine correlation and Pearson correlation coefficient in Movielens-100k data set. The results are shown in Fig. 1, 2, 3 and 4.

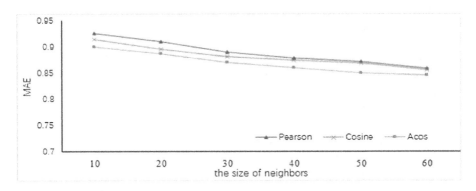

Fig. 1. MAE comparison between Cosine, Adjusted Cosine (Acos) and Pearson correlation (Pearson) measures.

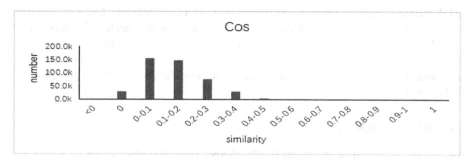

Fig. 2. Similarity distribution of Cosine correlation.

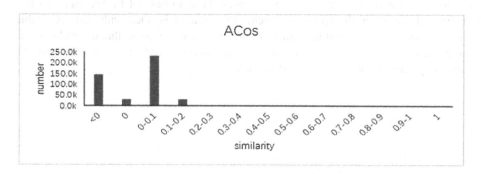

Fig. 3. Similarity distribution of Adjusted Cosine correlation.

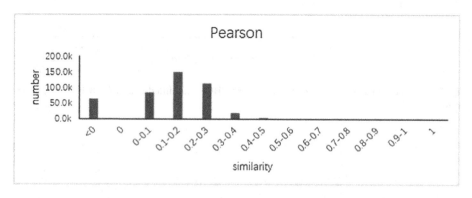

Fig. 4. Similarity distribution of Pearson correlation.

Looking into the MAE comparison between similarity measures shown in Fig. 1, although the values of Adjusted Cosine correlation is always lower than other two measures (Adjust Cosine < Cosine < Pearson), their MAEs would tend to reach a same size and the gap between them isn't so great. The similarity distributions of 3 measures is presented in Figs. 2, 3 and 4. We observe that the Pearson and Cosine correlation are more evenly distributed. However, most of the values of MAE are distributed in [0, 0.1] interval in Adjusted Cosine correlation, and there are many results of Pearson and Adjusted Cosine correlation less than 0. We get a conclusion as follows: although the values of MAE of Adjusted Cosine are lowest, there are not much different between 3 similarity measures, most of the results of Adjusted Cosine and Pearson correlation are less than 0 that are clearly contrary to realism and the values of MAE of Cosine are lower than Pearson correlation. Finally, the Cosine correlation is the best choice for the next experiments.

4.3 Filling Technology

We use the predicted rating given by AdaBoost-Naïve Bayesian classifier to fill the sparse matrix. To verify the effectiveness of the filling technology, we describe it by the comparison results shown in Figs. 5 and 6. Obviously, there are lower values of MAE on the filled matrix in user-based CF. So does the item-based CF. According to the analysis above, since the sparse user-item rating matrix is filled with more data, the recommender system would learn more information about users so that it could make a more accurate prediction. We can say that the filling is certainly useful for recommender system to improve their performance.

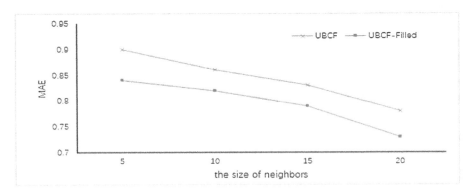

Fig. 5. MAE comparison between sparse matrix (UBCF) and filled matrix (UBCF-Filled) based on user-based CF.

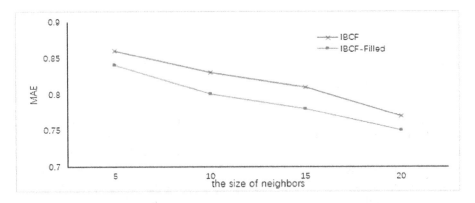

Fig. 6. MAE comparison between sparse matrix (IBCF) and filled matrix (IBCF-Filled) based on item-based CF.

4.4   Experiments with AdaBoost-Naïve Bayesian in Movielens-100k

We compare our algorithm (AdaBoost-Naïve Bayesian) with user-based CF, item-based CF and Naïve Bayesian CF algorithm. The results are shown in Figs. 7 and 8. In this section, the values of MAE in different algorithms in neighbors from 5 to 20 are shown in Fig. 7. In this picture, the values of AdaBoost-Naïve Bayesian are the lowest and the performance is that AdaBoost-Naïve Bayesian > NBCF > IBCF > UBCF. It also demonstrates Naïve Bayesian is a good classifier for recommender system. In Fig. 8, when the neighbors are in quantities, the gap between those algorithms is large where Naïve Bayesian and AdaBoost-Naïve Bayesian show better performance. The Coverage of each algorithm gradually increases and the Coverages tend to reach at 100% with increasing the amounts of neighbors. Comparing AdaBoost-Naïve Bayesian with Naïve Bayesian, we can see that AdaBoost algorithm can improve the performance of Naïve Bayesian under certain conditions.

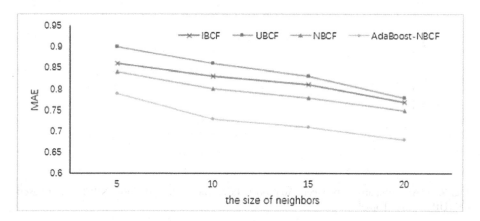

Fig. 7. MAE comparison between AdaBoost-Naïve Bayesian (AdaBoost-NBCF), user-based CF (UBCF), item-based CF (IBCF) and Naïve Bayesian CF (NBCF).

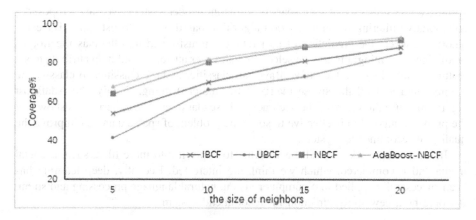

Fig. 8. Coverage comparison between AdaBoost-Naïve Bayesian (AdaBoost-NBCF), user-based CF (UBCF), item-based CF (IBCF) and Naïve Bayesian CF (NBCF).

4.5   Experiments with AdaBoost-Naïve Bayesian in Scholat

In order to verify the validity of our algorithm in real social network, we use Scholat data set provided by Scholat network. The results are shown in Fig. 9.

In this part, we have observed that the Top-N CF have reached at the lowest value of MAE when neighbors are 1, 2, 6. However, its performance is not stable enough due to the complex relationships and data in real network. Conversely, as neighbors counts increase, the values of MAE of user-based CF and AdaBoost-Naïve Bayesian gradually taper putting forward better stabilities. Our algorithm is believed to have higher accuracy. In a word, our algorithm is useful for recommender system to improve performance.

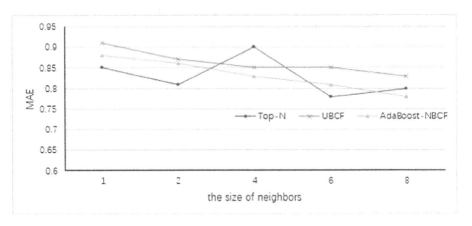

Fig. 9. MAE comparison between AdaBoost-Naïve Bayesian (AdaBoost-NBCF), user-based CF (UBCF) and Top-N.

5   Conclusion

Aiming at the difficulty of data sparsity in recommender system, we have proposed a collaborative filtering recommendation algorithm based on AdaBoost-Naïve Bayesian algorithm, where the prediction of user rating is transformed into the classification of predicting user rating. Firstly, we select the valid user attributes and item attribute to set feature vector, then use the Naïve Bayesian classifier as base classifier to classify the unrated item and fill the sparse matrix with predicted rating. Finally, the AdaBoost algorithm is used to improve the accuracy of base classifier. The experiments show that the proposed algorithm is effective to solve the problem of sparse data and improve the quality of recommender system.

In our future work, we will be working to apply it into more filed, such as social media and E-commerce, which we think are interested. Recently, deep learning has been successfully applied into computer vision, natural language processing and so on. It would be a new study direction for recommend system.

Acknowledgements. This work is supported by the National Natural Science Foundation of China (No. 61772211), the Science and Technology Project of Guangdong Province (No. 2016A030303058), the Cultivation Fund of the Science and Technology Project of the South China Normal University for Young Teachers (No. 17KJ16), the "Challenge Cup" Golden Seed Cultivation Fund of the South China Normal University (No. 18JJKA03).

References

1. Isinkaye, F.O., Folajimi, Y.O., Ojokoh, B.A.: Recommendation systems: principles, methods and evaluation. Egypt. Inform. J. **16**(3), 261–273 (2015)
2. Resnick, P., Iakovou, N., Sushak, M., et al.: GroupLens: an open architecture for collaborative filtering of netnews. In: ACM Conference on Computer Supported Cooperative Work, pp. 175–186. ACM (1994)
3. Hill, W., Stead, L., Rosenstein, M., et al.: Recommending and evaluating choices in a virtual community of use. In: SIGCHI Conference on Human Factors in Computing Systems, pp. 1994–201. ACM (1995)
4. Bellogin, A., Parapar, J.: Using graph partitioning techniques for neighbour selection in user-based collaborative filtering. In: The 6th ACM Conference on Recommender Systems, pp. 213–216. ACM (2012)
5. Sarwar, B., Karypis, G., Konstan, J., et al.: Item-based collaborative filtering recommendation algorithms. In: The 10th International Conference on World Wide Web, pp. 285–295. ACM (2001)
6. Chelly, Z., Elouedi, Z.: Hybridization schemes of the fuzzy dendritic cell immune binary classifier based on different fuzzy clustering techniques. New Gener. Comput. **33**(1), 1–31 (2015)
7. Friefman, J.S., Calvet, L.E., et al.: Bayesian inference with Muller C-elements. IEEE Trans. Circ. Syst. I-Regul. Pap. **63**(6), 895–904 (2016)
8. Purnamasari, P.D., Ratna, A.A.P., Kusumoputro, B.: EEG based patient emotion monitoring using relative wavelet energy feature and backpropagation neural network. In: The Engineering in Medicine and Biology Society, pp. 2820–2823. IEEE (2015)
9. Shi, Y., Larson, M., Hanjalic, A.: Exploiting user similarity based on rating-item pools for improved user-based collaborative filtering. In: The 3rd ACM Conference on Recommender Systems, pp. 125–132. ACM (2009)
10. Yin, H., Cui, B., Li, J., Yao, J., Chen, C.: Challenging the long tail recommendation. Proc. VLDB Endow. **5**(9), 896–907 (2012)
11. Deshpande, M., Karypis, G.: Item-based top-N recommendation algorithms. ACM Trans. Inf. Syst. **22**(1), 143–177 (2004)
12. Wei, S., Ye, N, Zhang, S.: Item-based collaborative filtering recommendation algorithm combining item category with interestingness measure. In: International Conference on Computer Science and Service System, vol. 1, no. 1, pp. 2038–2041 (2012)
13. Zhang, F., Bai, L., Gao, F.: A user trust-based collaborative filtering recommendation algorithm. In: Qing, S., Mitchell, Chris J., Wang, G. (eds.) ICICS 2009. LNCS, vol. 5927, pp. 411–424. Springer, Heidelberg (2009). https://doi.org/10.1007/978-3-642-11145-7_32
14. Zhang, J., Lin, Y., Liu, J.: An effective collaborative filtering algorithm based on user preference clustering. Appl. Intell. **45**(2), 230–240 (2016)
15. Bobadilla, J., Hernando, A., et al.: Recommender systems survey. Knowl. Based Syst. **46**(1), 109–132 (2013)

16. Freund, Y., Schapire, RE.: Experiments with a new boosting algorithm. In: Thirteenth International Conference on International Conference on Machine Learning, pp. 148–156. ACM (1996)
17. Cao, J., Kwong, S., Wang, R.: A noise-detection based AdaBoost algorithm for mislabeled data. Pattern Recogn. **45**(12), 4451–4465 (2012)
18. Zhou, Z.H.: Ensemble Methods: Foundations and Algorithms. CRC Press, Boca Raton (2012)

Anomaly Detection in the Web Logs Using Unsupervised Algorithm

Lei Jin[1], Xiao Juan Wang[1(✉)], Yong Zhang[1], and Li Yao[2]

[1] Electronic Engineering Institute, Beijing University of Posts
and Telecommunications, Beijing 100876, China
wj2718@163.com
[2] College of Information Science and Technology, Beijing Normal University,
Beijing 100875, China

Abstract. Web server in the Internet is vulnerable to be attacked. Analyzing on the web logs is one efficient method to figure out intrusion. Using unsupervised algorithm for anomaly detection is suitable for the big data situation. Therefore, the research designs a framework using unsupervised classifiers for anomaly detection in the web log. In this paper, we concentrate on the statistic features and the character features of the web logs. Using the features, we transform the web logs to vectors. We apply a suitable normalized method for our unsupervised classifiers. The principal component analysis (PCA) and the AutoEncoder (AE) are the theoretical basis for the classifiers. As we know, this paper is the first research applying PCA and AE to the web log anomaly detection combining statistic features and character features. In the simulation, we find the statistic features are efficient for the PCA. When we use the AE, character features are better. Compared with other methods, results show that our model performs better.

Keywords: Anomaly detection · Web logs · Unsupervised algorithm · Features

1 Introduction

Nowadays, the application of the web develops rapidly. However, the vulnerability in the web application is also very common. One of the most influential attacks is targeted on the web server or the web server based application. Attackers can send the special HTTP requests to the web server which help them control the web server or steal the database of the web server. Then the users of the web server will face the security threat. Analyzing the web logs is one useful way to spot the intrusion behaviors.

To observe the intrusion behaviors, there are mainly two approaches: the misuse detection and the anomaly detection [1]. In this paper, we mainly focus on the anomaly detection. In the existing researches, Zolotukhin [2] used n-gram and other statistical features to construct the normal behavior model.

© Springer Nature Switzerland AG 2019
Y. Tang et al. (Eds.): HCC 2018, LNCS 11354, pp. 393–405, 2019.
https://doi.org/10.1007/978-3-030-15127-0_40

The result shows high accuracy but consumes more time. Garcia [3] proposed a novel approach for postmortem intrusion detection which factored out repetitive behaviors. The research also combines a hidden Markov model with k-means. Dogbe [4] designed a two-tiered defense against SQL injection using the parse tree analysis and the fine grained role based access control. Joshi [5] used a supervised machine learning method based on the Naive Bayes to detect the SQL injection. Algiriyage [6] identified the web crawlers by analyzing the crawler patterns in the web logs. He developed a methodology to identify crawlers and classified them into three categories. However, the existing researches construct the normal behavior based on the normal data. In reality, the normal data is hard to get. The holder of the web server do not have the ability to provide the normal logs. Besides, using the supervised machine learning methods requests manual labels for samples. It is not suitable for the big data situation.

Thus we focus on the unsupervised methods which can be used for the anomaly detection. Several researches concentrate on using unsupervised methods to do anomaly detection. For example, we can use the iForest [7] based on the classifier trees to do anomaly detection. Williamson [8] provided a method using support vector machine to locate the abnormal samples in the data. Clustering algorithms [9] such as k-means can also be used for detecting the malicious behaviors. In this paper, we use another type of unsupervised method which uses the principal component analysis (PCA) [10] or AutoEncoder (AE) [11] as the foundation. The PCA assumes a linear system. The AE can be applied to the nonlinear mapping. For the web logs anomaly detection, we should figure out a suitable framework to use the PCA and the AutoEncoder.

The remaining of the paper is arranged as follow. In the Sect. 2, we describe our data model and the analysis framework. We focus on the feature engineering in the Sect. 3. Besides, we show the distribution of our features in the normal data and the abnormal data. In the Sect. 4, we provide the result of our model. In the Sect. 5, we summarize our work.

2   Data Model and Analysis Framework

2.1   Data Model

In this paper, we mainly focus on HTTP requests logged by most common web servers (such as Apache). Besides, our anomaly detection model analyzes GET requests which use parameters to pass values to the server. The logs set is represented as $L\{l_1, l_2 \cdots l_m\}$. A complete log includes the IP, the time stamp, the method, the referrer, the response code, the URI and so on. The URI includes most information of a log and is the main concern for anomaly detection. We use $U\{u_1, u_2 \cdots u_m\}$ to represent the URI set.

2.2   The Analysis Framework

In order to apply to the big data situation, we should use the unsupervised classifier which is no need for the manual label. We construct our framework in three steps. Figure 2 shows the processing steps of our model.

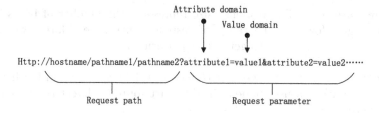

Fig. 1. The structure of an HTTP request

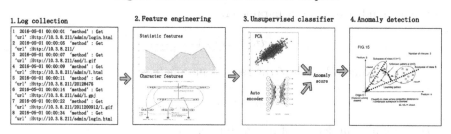

Fig. 2. The processing steps of the anomaly detection

(1) Feature extraction. In this step we transform the web logs to vectors. There are many previous researches analyzing on features extracted from the web logs. We efficiently choose useful features for anomaly detection and preprocess the features for better performance.

(2) Unsupervised classifier. We use the Principal Component Analysis (PCA) and the AutoEncoder to give the anomaly scores for each log. The reason why we choose two unsupervised classifiers is that the PCA is a linear model and the AE is a nonlinear model. So the two classifiers can be applied to different situations. For the PCA, we can first map the vectors to the low dimension, and then transform the low dimension vectors to the original dimension. We can use the L2 distance of the original data and the restored data to represent anomaly scores. The usage of the AutoEncoder is the same as the PCA. We will describe it in details in the Sect. 4.

(3) Anomaly detection. We can sort the anomaly scores and select a certain threshold percentage of data as the abnormal data.

3   The Feature Engineering

3.1   The Effective Features of the Web Logs

The Statistic Feature

a. The response code. In the web logs, many kinds of response codes exist. Besides the meanings of different response codes are various. There are about over forty kinds of response codes. In this paper, we divide the response codes into 5 classes: 200, 404, 403, 304 and others. Then we use the one-hot coding to encode the response codes.

b. The response size. The response size represents the number of bytes in the response page. However, the range of response size is very large, we should normalize the response size to get better performance.

c. The parameter features. The structure of an HTTP request can be showed in the Fig. 1. (1) We can use the length of the attribute to detect the anomaly in the attribute.

$$la_i = \mathcal{L}(q_i) \tag{1}$$

(2) The number of request parameters,

$$ln_i = \mathcal{N}(q_i) \tag{2}$$

(3) The maximum length of the request parameters,

$$lq_i = \max(q_i) \tag{3}$$

(4) The minimum length of parameter length.

$$lp_i = \min(q_i) \tag{4}$$

d. The entropy of the HTTP request. We define the entropy of the HTTP request as follow,

$$E_i = -\sum_{k=1}^{n} p_i{}^k \log_2 p_i{}^k \tag{5}$$

$p_i{}^k$ is the frequency of character k in the request. We can get the entropy of every HTTP request with formula (5).

e. The frequency of character. For an HTTP request, we can calculate the frequency of the capital letters, the lowercase letters, the digital letters and other letters.

f. The path features. (1) The number of short paths. From the Fig. 1, we can find that the request path can be divided into several short paths by the '/' character. We use the number of short paths in an URI as features of a log. (2) The maximum length of short paths. (3) The minimum length of short paths. (4) The length of request paths.

Based on the features introduced above, we can transform an original log to an 18 dimension vector. Before sending to the classifier, we should scale the vectors to the same range.

The Character Feature

n-gram. To extract character features from web logs, we use the n-gram. N-gram is widely used in speech recognition [12] and natural language processing. For example, an URL: 'abc/def.html', 2-gram is 'ab', 'bc', 'c/', '/d', 'de', 'ef', 'f.', '.h', 'ht', 'tm', 'ml'. Attackers use payloads to implement malicious behaviors. So the payloads have fixed patterns. To increase the generalization ability of the model, in the process of preprocessing the English alphabets and numbers in the logs are considered as same. In other word, 'd3' and 'z4' are same in the 2-gram. In this way, the dimension of the data decreases. The dimension of the data is equal to the number of all possible n-grams used in the dataset. We use n-grams as the character features.

3.2 The Normalized Method

For the response size, the attribute length, the path length, the number of parameters, the maximum parameter length and the extended features, the range is very large. So we must use the normalization technique to preprocess the vectors. We use the formula as follow to do normalization.

$$d_i{}^k = \log_{10} d_i{}^k / Max(\log_{10} d_i{}^k, k = 1, 2 \cdots) \tag{6}$$

Here $d_i{}^k$ is the ith dimension of log k.

Because the character features is very sparse and the maximum is not large. Using formula (7) to normalize the vectors is appropriate.

$$d_i{}^k = d_i{}^k / Max(d_i{}^k, k = 1, 2 \cdots) \tag{7}$$

4   The Result

In this paper, we use PCA and AutoEncoder to transform the vectors to lower dimensions. Then we do inverse transform to the transformed data. We use the L2 loss between the original data and the restored data to represent anomaly scores. The anomaly scores are calculated as follow,

$$A^k = \sum_i (d_i^k - d^{\iota k}_i) \tag{8}$$

Here A^k is the anomaly scores of log k. d_i^k is the ith dimension of log k and $d^{\iota k}_i$ is the ith dimension of the restored data.

4.1   The Unsupervised Classifier

The Principal Component Analysis. The PCA is the most commonly used linear dimension reduction methods. Its goal is to map the high-dimensional data to a low dimensional space with the largest variance through the linear projection. In this way, it uses fewer data dimensions, at the same time retains more of the characteristics of the original data points. The target function of the PCA is

$$\max \frac{1}{n-1} \sum_{i=1}^{n} \left(W^T (d_i - \bar{d}_i) \right)^2 \tag{9}$$

Here n is the dimension of the original data. W^T is the transformed matrix. We can get the restored data using

$$D'^k = W^T D^k W \quad st \ W^T W = I \tag{10}$$

D^k is the original data and D'^k is the restored data. The restored data keeps more information of the normal data than the abnormal data. Then we can use the anomaly scores calculated by the formula (8) to decide the abnormal logs.

The AutoEncoder. Besides the PCA, using the AutoEncoder (AE) to do anomaly detection is also effective. The AE is a special structure of the neural network. It has an input layer, an output layer and one or more hidden layers. In the training step, we set the output data the same as the input data. The input dataset experiences a dimension reduction through the hidden layer and then is reconstructed after passing output layer.

$$X'_i = X_i \tag{11}$$

Here we set the means of the jth cell in the hidden layer as

$$\hat{h}_j = \frac{1}{m} \sum_{i=1}^{m} [a_j^{(2)}(x_i)] \tag{12}$$

m is the number of samples. $a_j^{(2)}(x_i)$ is the output of the hidden layer of the sample i. Then the loss function is

$$L = \frac{1}{m} \sum_{i=1}^{m} (x' - x)^2 + \sum_{j=1}^{s_2} h \log \frac{h}{h_j} + (1 - h) \log \frac{1 - h}{1 - h_j} \tag{13}$$

Here h is a constant (usually set as 0.05). The loss function ensures that the restored data is the same as input data. Besides, it can keep the sparsity in hidden layer. We can train the AutoEncoder with the whole data. Then use the formula (8) to calculate the anomaly scores.

The Anomaly Detection. After getting the anomaly scores of each log, we should choose a threshold value to decide the anomaly logs. In this paper, we set p as the threshold. Then we sort the logs according to the anomaly scores. We choose the top p percentage of data as the abnormal data. So we can change the p to find out the best performance of the classifier. After training, we can also use the minimum score of the top p percentage of data as the threshold value. Then we use the threshold value to do online anomaly detection.

4.2   The Analysis of Our Model

We use one month data of four different web systems in this paper. The web log data is collected in the Chinese government websites. Besides, the data is provided by a web security company which cooperates with us. Table 1 shows the basic information of the web logs. We use the pattern matching to decide the labels of the web logs. (We have upload the vectorized data to https://github. com/buptjinlei/weblogs)

Table 1. The basic information of the web logs

Web logs	Number	Normal	Anomaly
Log1	174808	142329	32479
Log2	133749	112345	21404
Log3	122925	92139	30786
Log4	93221	75278	17943

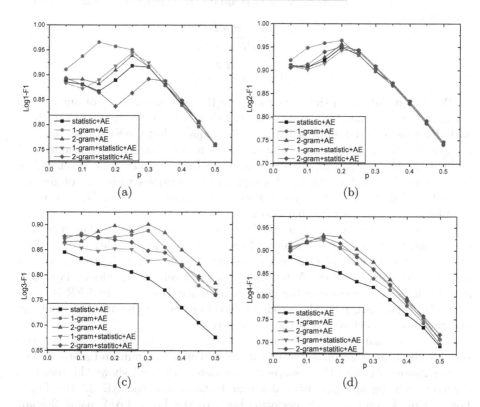

Fig. 3. The F1 score of different web log systems using the AE. (a) The F1 score of the Log1. (b) The F1 score of the Log2. (c) The F1 score of the Log3. (d) The F1 score of the Log4.

Then we compare the performance of the PCA and the AE. For the PCA, we can change the amount of the reserved information to get better performance. For the AE, we can change the number of cells in the hidden layer and the number of training iteration. Furthermore, we should use the TPR and FPR to evaluate the performance of the classifier.

$$TPR = \frac{TP}{TP + FN} \tag{14}$$

$$FPR = \frac{FP}{FP + TN} \tag{15}$$

Here TP is the number of normal samples which we predict as normal. FP is the number of abnormal samples which we predict as normal. FN is the number of normal samples which we predict as anomaly. TN is the number of abnormal samples which we predict as anomaly. TPR is the proportion of predicted normal samples in the real normal samples. FPR is the proportion of predicted normal samples in the real abnormal samples. To get high TPR and low FPR is the target. We can also use the F1 score to evaluate the overall performance in TPR and FPR.

$$F1 = \frac{2TPR * PRE}{TPR + PRE} \tag{16}$$

$$PRE = \frac{TP}{TP + FP} \tag{17}$$

We do simulation on the Log1 to decide the best parameters of our model. First, we use the statistic features. Table 2 shows the performance of the PCA with different reserved information (RI). We request that the FPR is as small as possible. It shows that when the amount of the reserved information is above 0.8 and the threshold p is 0.25, the performance is best. Besides, we can set higher p to ensure there is few abnormal samples in the samples which we predict as normal. In this way, we may get few normal samples. Then we can use these samples to construct normal behaviors.

Then we use the AutoEncoder to show the influence of the number of cells in the hidden layer to the performance of the classifier. From Table 3, we can find that when the number of cells in the hidden layer is 5 and the threshold p is 0.3, the performance is best. When the number of cells in the hidden layer is 7 and p is 0.25, the model gets the best performance. However, the FPR is 0.139 which is a little larger for the anomaly detection. So we give up this result. It also proves the effectiveness of the AutoEncoder.

Next, we analyze the influence of features on the result. We use the statistic features, 1-gram, 2-gram, statistic features + 1-gram and statistic features + 2gram. AE and PCA are both considered. Figure 3 shows the result of different web log systems using different features and the AE. In the Log1, Log2, using 1-gram and AE performs best. In the Log3, Log4, using 2-gram and AE performs best. For the AE, using 1-gram, 2-gram, or combining 1-gram and statistic features are better. Figure 4 shows the result of different web log systems using different features and the PCA. For the PCA, using

Table 2. The performance of the PCA

RI	p	0.05	0.1	0.15	0.2	0.25	0.3	0.35	0.4	0.45	0.5
0.9	TPR	0.961	0.928	0.920	0.917	**0.907**	0.847	0.789	0.730	0.675	0.614
	FPR	0.898	0.778	0.542	0.286	**0.062**	0.055	0.042	0.030	0	0
	F1	0.888	0.881	0.900	0.925	**0.944**	0.911	0.877	0.841	0.806	0.761
0.8	TPR	0.961	0.928	0.920	0.917	**0.907**	0.847	0.789	0.730	0.675	0.614
	FPR	0.898	0.778	0.542	0.286	**0.062**	0.055	0.042	0.030	0	0
	F1	0.888	0.881	0.900	0.925	**0.944**	0.911	0.877	0.841	0.806	0.761
0.7	TPR	0.939	0.878	0.831	0.792	0.731	0.670	0.615	0.604	0.603	0.600
	FPR	0.996	0.996	0.934	0.836	0.832	0.832	0.805	0.584	0.316	0.063
	F1	0.867	0.834	0.813	0.799	0.761	0.720	0.684	0.695	0.720	0.743

Table 3. The performance of the AutoEncoder

Layers	p	0.05	0.1	0.15	0.2	0.25	0.3	0.35	0.4	0.45	0.5
5	TPR	0.960	0.926	0.888	0.867	0.857	0.851	0.790	0.730	0.675	0.614
	FPR	0.905	0.785	0.685	0.507	0.283	0.038	0.038	0.031	0	0
	F1	0.886	0.880	0.869	0.875	0.892	0.915	0.872	0.841	0.806	0.761
6	TPR	0.964	0.925	0.882	0.850	0.849	0.849	0.792	0.737	0.675	0.614
	FPR	0.887	0.790	0.711	0.580	0.317	0.048	0.029	0.002	0.002	0
	F1	0.890	0.879	0.863	0.858	0.884	0.913	0.880	0.848	0.806	0.761
7	TPR	0.960	0.927	0.889	0.889	0.889	0.851	0.789	0.728	0.676	0.614
	FPR	0.905	0.783	0.677	0.409	0.139	0.040	0.039	0.037	0	0
	F1	0.886	0.880	0.870	0.897	0.924	0.914	0.878	0.839	0.806	0.761
8	TPR	0.959	0.924	0.883	0.874	0.849	0.849	0.792	0.733	0.676	0.614
	FPR	0.912	0.793	0.705	0.477	0.315	0.046	0.029	0.017	0	0
	F1	0.885	0.878	0.864	0.881	0.884	0.913	0.880	0.844	0.806	0.761

statistic features is better in the Log1 and Log2. In the Log3, it is better to use the statistic features plus 2-gram. In the Log4, using the statistic features plus 1-gram is better. Because the dimension of the vector is related to the complexity of the algorithm. In our data, the order of the dimension is $2 - gram + statistic > 2 - gram > 1 - gram + statistic > 1 - gram > statistic$. We can choose different kinds of features according to the performance and complexity. Besides, n-grams are more suitable for the AE and statistic features are more suitable for the PCA. In our model, we should set higher p to ensure lower FPR. It means the model figures out more abnormal samples. In the four web logs, we think setting p as 0.2–0.3 is better.

Here we shows the attacks which are found by our framework. Table 4 gives the attacks with its type. So we can find that our framework can detect different

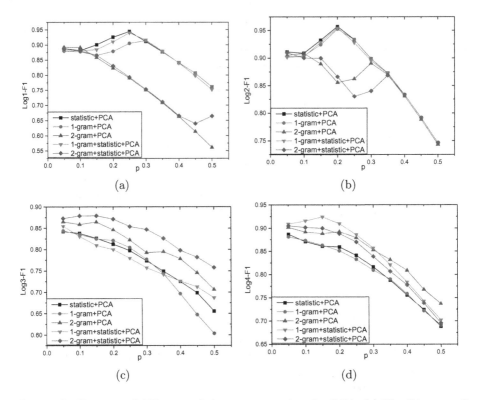

Fig. 4. The F1 score of different web log systems using the PCA. (a) The F1 score of the Log1. (b) The F1 score of the Log2. (c) The F1 score of the Log3. (d) The F1 score of the Log4.

types of web attacks such as the SQL injection, the File upload, the Trojan files and so on. These abnormal logs have different patterns compared with normal logs. Besides, it proves the effectiveness of our model.

4.3   The Comparison with Other Methods

In this section we compare our methods with the k-means [6], the robust covariance (RC) and the isolated forests (IF) in the four web logs systems. Figure 5 shows the comparison of our model with other methods. In the four web logs systems, the PCA and the AutoEncoder perform better than other methods. The AE performs better than the PCA in the FPR. We can find that our method gets higher FPR in the Log3 and the Log4 compared with the Log1 and the Log2. Because the proportion of the abnormal data is high. Our model is based on that most of the data is normal. Furthermore, we can find that the isolated forests and the robust covariance perform alike which are better than k-means. Compared with other methods, it proves the effectiveness of our model. Besides, our model is based on the unsupervised classifiers which are appropriate for

Table 4. The attacks in the logs

URL	Type
//uploadfile/userfiles/media/confg.inc.php	Sensitive documents
/oa/?s=/abc/abc/abc/${print(eval($_POST[c]))}/	Code execution
/index.php?act=../cache/adv/adv_14.cache	File download
/wp-content/themes/twentytwelve/404.php	Trojan files
/plus/search.php?typeArr[1' or '@'%3D1 and	
(SELECT 1 FROM (select count(*)	SQL injection
/admin/upload/upload.asp	File upload
/uploads/allimg/130912/8130912150509	
.jpg?bindUrl=http://sc.qq.com/qr/fx/	Redirection vulnerability

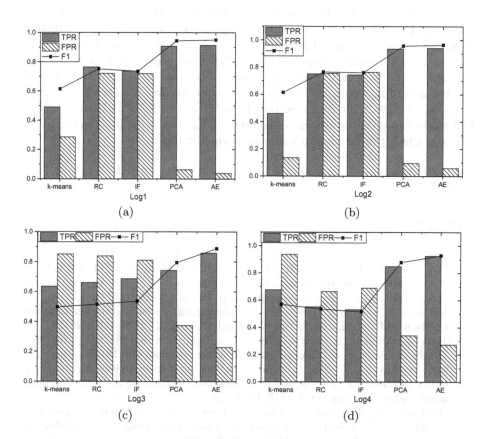

Fig. 5. The comparison with other methods in different data sets. (a) The comparison with other methods in the Log1. (b) The comparison with other methods in the Log2. (c) The comparison with other methods in the Log3. (d) The comparison with other methods in the Log4.

the big data situation. We can also use our model to do the online detection. After getting anomaly scores, we can decide a threshold value of the score. If a new log gets higher score than the threshold value, the new log is abnormal. The matrix of the PCA and the network weights of the AutoEncoder is consistent with the training.

5   The Conclusion

In this paper, we design a framework which uses the unsupervised methods to do anomaly detection in web logs. In the feature engineering, we combine the statistic features and the character features. Besides, we provide a normalized method which is suitable for web logs data. After normalization, the distribution of data is converted to the normal distribution. In our model, we use the PCA and the AutoEncoder as the unsupervised classifiers. Using statistic features, for the PCA, the amount of the restored information should be above 0.8. For the AutoEncoder, when the number of cells in the hidden layer is 5, the performance is best. Simulation reveals that when we set the PCA as the unsupervised classifier, it is better to use statistic features. When we use the AE, we should use 1-gram, 2-gram or 1-gram + statistic features. Besides, we can choose different features according to algorithm complexity requirements. Compared with other methods, our model performs better in the TPR, the FPR and the F1 score. Our model can do online anomaly detection and get the normal behaviors in the web logs which can be used to build the normal model.

Acknowledgments. This research is funded by The National Natural Science Fund (61471055).

References

1. Gollmann, D.: Computer Security, 2nd edn. Wiley, Hoboken (2006)
2. Zolotukhin, M., Hamalainen, T., Kokkonen, T.: Analysis of HTTP requests for anomaly detection of web attacks. In: IEEE, International Conference on Dependable, Autonomic and Secure Computing, pp. 406–411. IEEE (2014)
3. Garcia, K.A., Monroy, R., Trejo, L.A.: Analyzing log files for postmortem intrusion detection. IEEE Trans. Syst. Man Cybern. Part C **42**(6), 1690–1704 (2012)
4. Dogbe, E., Millham, R., Singh, P.: A combined approach to prevent SQL Injection Attacks. In: Science and Information Conference, pp. 406–410. IEEE (2013)
5. Joshi, A., Geetha, V.: SQL Injection detection using machine learning. In: International Conference on Control, Instrumentation, Communication and Computational Technologies, pp. 1111–1115. IEEE (2014)
6. Algiriyage, N., Jayasena, S., Dias, G.: Identification and characterization of crawlers through analysis of web logs. IEEE International Conference on Industrial and Information Systems, pp. 150–155. IEEE (2014)
7. Liu, F.T., Ting, K.M., Zhou, Z.-H.: On detecting clustered anomalies using SCiForest. In: Balcázar, J.L., Bonchi, F., Gionis, A., Sebag, M. (eds.) ECML PKDD 2010. LNCS, vol. 6322, pp. 274–290. Springer, Heidelberg (2010). https://doi.org/10.1007/978-3-642-15883-4_18

8. Aryal, S., Ting, K.M., Wells, J.R., Washio, T.: Improving iForest with relative mass. In: Tseng, V.S., Ho, T.B., Zhou, Z.-H., Chen, A.L.P., Kao, H.-Y. (eds.) PAKDD 2014. LNCS, vol. 8444, pp. 510–521. Springer, Cham (2014). https://doi.org/10.1007/978-3-319-06605-9_42

9. Williamson, R., Smola, A., Shawe-Taylor, J.: Support vector method for novelty detection. In: International Conference on Neural Information Processing Systems, pp. 582–588. MIT Press (1999)

10. Lin, Q., Zhang, H., Lou, J.G., Zhang, Y., Chen, X.: Log clustering based problem identification for online service systems. In: Proceedings of the 38th International Conference on Software Engineering (2016)

11. Dauxois, J., Pousse, A., Romain, Y.: Asymptotic theory for the principal component analysis of a vector random function: some applications to statistical inference. J. Multivar. Anal. 12(1), 136–154 (1982)

12. Hirsimaki, T., Pylkkonen, J., Kurimo, M.: Importance of high-order N-gram models in morph-based speech recognition. IEEE Trans. Audio Speech Lang. Process. 17(4), 724–732 (2009)

Fuzzy C-Means Clustering Problem Based on Improved DNA Genetic Algorithm and Point Density Weighting

Zhenni Jiang and Xiyu Liu[✉]

Business School, Jinan, Shandong, China
1179480219@qq.com, sdxyliu@163.com

Abstract. In the every kinds of fuzzy clustering algorithm, the fuzzy c-means clustering (FCM) is widely used in the implementation process, because of it has better local search ability and easy operation. But the fuzzy c-means clustering algorithm also has some inherent flaw and the insufficiency. In this paper, we propose some improvement for this shortcomings. Firstly, we improve the calculation method of the membership degree. Secondly, we join the density in the calculation of the membership degree. At the same time, in order to find the global optimal solution, we use the DNA genetic algorithm to assist the FCM algorithm to jump out of local optimal. In this paper, we use the Matlab2014 to realize the simulation and experiment. Firstly, we utilize the test functions and artificial datasets to prove the effectiveness of the improved DNA genetic algorithm. Secondly, we utilize the UCI data sets to validate the effectiveness of the improved fuzzy c-means algorithm. Finally, the improved fuzzy c-means algorithm was used to realize the classification for the Sogou lab corpus of text and the result proved the validity of the algorithm.

Keywords: Fuzzy c-means clustering · DNA genetic algorithm ·
Point density weighting

1 Introduction

Clustering is an unsupervised classification processing. In the process of clustering, massive objects are divided into different clusters according to certain attributes or features of the object itself. Fuzzy c-means clustering algorithm [1, 2] is typical representative for fuzzy clustering algorithm. Liu et al. [3] have used the improved fuzzy c-means algorithm in the Carbonate fluid identification. Nayak et al. [4] studied an optimization algorithm for hybrid teaching learning based on fuzzy C-means clustering algorithm is studied. Misra et al. [5] proposed an algorithm for selecting the optimal value of fuzzy parameters in fuzzy C-means clustering for the LP-Residual input function. Zhang et al. [6] added the concept of entropy to the fuzzy C-means clustering.

DNA genetic algorithm is similar to genetic algorithm, but because of it has unique coding mechanism, it can be universal applicability to many problems. Some scholars [7, 8] use the improved genetic algorithm to achieve the clustering. Broin et al. [9] combined the transcription factors with genetic k-center point clustering method to achieve the unmatched clustering. Hong He [10] proposed a clustering method based

© Springer Nature Switzerland AG 2019
Y. Tang et al. (Eds.): HCC 2018, LNCS 11354, pp. 406–415, 2019.
https://doi.org/10.1007/978-3-030-15127-0_41

on two-stage genetic algorithm. There are also have other researches about the use of DNA genetic algorithm in the clustering [11–14].

But the fuzzy c-means clustering algorithm also has some inherent flaw and the insufficiency. Firstly, the qualification of the membership degree need to be 1 will lead to the data points are more sensitive to noise and outliers. Secondly, the fuzzy c-means clustering algorithm is sensitive to the initial clustering center and easy to fall into local optimum. In this paper, we improve the fuzzy c-means algorithm. And we also optimize it by DNA genetic algorithm to improve the clustering effect. In the end, UCI dataset are used to verified the effectiveness of the improved algorithm. And then the fuzzy C-means algorithm is used to the text classification of the Sogou laboratory corpus, which has certain practical significance.

2 Improved DNA Genetic Algorithm

Traditional DNA genetic algorithm still has some defects. In order to fully utilize the optimization performance of the algorithm, we propose some improvements to the traditional DNA genetic algorithm in this section, and note as IDNA-GA.

2.1 Two-Stage Crossover Operation

Inspired by the hot spot and cold spot of DNA chromosomes, we propose a two-stage crossover strategy. We still divide chromosomes into "hot spots" and "cold spots". The crossover probability of the hot spot area was slightly larger, and the cold spot area is normal. The specific crossover operation is improved two-point intersection and described as follows:

(1) The length of the chromosome is L. The length of the hot spot is l, which l = L/2. The max generation is Gmax.
(2) When generation is less than Gmax/2, we random choose two chromosomes. And select two fragments in the hotspot area of the first chromosome. The first fragment is the length of m in the first l/2. The second fragment is the length of m in the last l/2. In the same way, select two segments in the second chromosome, and then perform crossover operation according to the probability (Fig. 1).
(3) We choose single point crossover in the cold spot area. Crossover probability is pr (pl > pr).
(4) When generation is more than Gmax/2, excessive crossover probability will disrupt existing population and lead to final convergence difficult. So we let the crossover probability in the hot spot equal to pr.

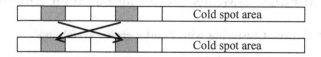

Fig. 1. Improved crossover operation

2.2 Adaptive Mutation Operation

The main idea of improving the mutation operation in this paper is that the mutation probability of the population can be adjusted according to the fitness value of the chromosome. The overall trend is: the average fitness value of the population is the dividing line. If the fitness value of the chromosome is smaller than the average fitness value, the greater the probability of mutation and vice versa. If the fitness of the chromosome is equal to the maximum fitness value, the mutation probability will be equal to 0. The specific mutation probability is updated as in formula (1):

$$P_{\mathrm{m}} = \frac{f_{\max} - f}{f_{\max} - f_{avg}} \tag{1}$$

where, pm is mutation probability, fmax is the maximum fitness in every generation. f is the fitness of the chromosome. favg is the average fitness of all chromosomes.

In order to prove the effective of the improved DNA genetic algorithm, We use a test function to verify the effective of the algorithm. The test function is $y = x + sin(x) + cos(x)$. The experimental result is compared with the common DNA genetic algorithm (Figs. 2 and 3).

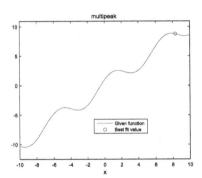

Fig. 2. Common DNA-genetic algorithm

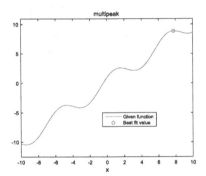

Fig. 3. Improved DNA-genetic algorithm

The experimental results show that the best fitness computed by improved DNA genetic algorithm is better than the common DNA genetic algorithm. Therefore, the improved DNA genetic algorithm can effectively avoid the algorithm falling into local optimum and achieve global optimum.

3 Improved Fuzzy C-Means Clustering Based on IDNA-GA

3.1 Fuzzy C-Means Clustering with Point Density Weighting

Fuzzy C-means algorithm (FCM) has several defects which including the number of clusters need to be specified in advance, extremely sensitive to initialization, easy to fall into local optimum, and sensitive to the independent points and noise points. In order to overcome the existing problems, we use DNA genetic algorithm to optimize the fuzzy C-means clustering. With the optimization of DNA genetic algorithm, the final clustering result can be closer to the true value.

The traditional fuzzy C-means only considers the distance between point and point, so the distance does not take into account the impact of the data point density on the cluster. So we use point density weighting when performing fuzzy C-means clustering method in this paper.

$$w_i' = \sum_{j=1, j \neq i}^{n} \frac{1}{\|x_j - x_i\|} \tag{2}$$

Where, x_i, x_j represent the ith and jth data point, respectively. It can be seen from the formula (2) that the closer the point is to the point i, the greater the density of the point i. So we normalize it:

$$w_i = \frac{w_i'}{\sum_{i=1}^{n} w_i'} \tag{3}$$

Where, w_i represent the density value of point i after normalization. Therefore, the density function of the fuzzy C-means clustering after adding the density factor is

$$J(\mu, C) = \sum_{j=1}^{C} \sum_{i=1}^{N} w_i \mu_{ij}^g d_{ij}^2 \tag{4}$$

where, d_{ij} is the Euclidean distance between the data point i and data point j. It can be calculated by the formula $d_{ij} = \|x_i - C_j\|$. C is the cluster center, which can be updated by the formula (5):

$$C = \frac{\sum_{i=1}^{N} \mu_{ij}^g x_i w_i}{\sum_{i=1}^{N} \mu_{ij}^g} \tag{5}$$

And u_{ij} is the degree of membership matrix. It can be updated by the formula (6):

$$u_{ij} = \frac{1}{\sum_{j=1}^{c} \left(\frac{d_{ij}}{d_{im}}\right)^{\frac{2}{g-1}}} \tag{6}$$

where, g is a fuzzy weighted index, and its value needs to be preset. The value of g is proportional to the degree of blurring. According to the past experiment, g is taken as 2.

3.2 Density Weighting Fuzzy C-Means Algorithm Based on IDNA-GA

FCM is actually a mapping from the cluster center to the final clustering result. But in the actual situation, there are a lot of local minimums. So in the actual clustering process, it may fall into local optimum. Therefore, we use the improved DNA genetic algorithm to optimize FCM. Here, we recorded it as IDFCM-DNAGA algorithm. The specific process is as following.

3.3 DNA Coding

Here, we use a four letter alphabet A, T, G, C to code DNA strand. And a quaternary alphabet 0, 1, 2, 3 is used to represent the characteristics of DNA nucleotide bases. The random initialization probabilities for the four bases are 0.156, 0.157, 0.344, and 0.343, respectively. The initial population is randomly generated according to probability.

3.4 Initialization

Each chromosome in the population represents the initial cluster center and also represents the number of clusters. Randomly generates M chromosomes in the initial population. The encoding length of an individual is L = n * l, and the encoding precision is equal to $(x_{maxi} - x_{mini})/4^l$.

Fitness Function

The fitness function is a mathematical representation which represent the individual adaptation to the environment. Habitually, the greater the fitness value, the stronger the individual's adaptability to the environment. Here, we defined the fitness function as following:

$$F = \frac{\delta}{J + \delta} \tag{7}$$

where, F is the value of the fitness. J is computed by the formula (4). δ is a constant which determined by yourself. Here give the statistics calculation method of δ. Firstly, we need to perform multiple clustering and record the optimal J value in each iteration process. Secondly, we can calculate the average value avg(J) of the several J values. At last, we take a same magnitude of the avg(J) and record it as δ. In this method, we can guarantee the fitness value between the interval (0, 1).

Evolutional Operation

(1) Selection operation. In the evolution process of each generation, the optimal individuals with the first 10% fitness are directly inherited to the next generation, and the rest of the individuals are selected according to the method of roulette selection. The calculation method of the selection probability is:

$$P_s(i) = \frac{F_i}{\sum\limits_{i=1}^{M} F_i} \tag{8}$$

Where, $P_s(i)$ represent the selection probability of the ith chromosome. Fi represent the fitness of the ith chromosome. M is the number of the chromosomes.

(2) Crossover and mutation operation. These two operations are described in detail in the second part.

4 Experiments Results and Analysis

4.1 Data Source and Experiments

In order to verify the performance of the improved algorithm in this paper, the IRIS dataset in the UCI database is used as a test sample. The size of sample is 150. And we compare the algorithm in this paper with the common FCM algorithm and the FCM-GA algorithm. In the experiment, the parameters are set as follows: cluster number C = 3, fuzzy index m = 2, stop threshold $\varepsilon_1 = 10^{-4}$, crossover probability $p_l = 0.9$, $p_r = 0.8$, mutation probability $p_m = 0.2$, population size Pop$size$ = 150, maximum genetic algebra T = 50, genetic algorithm stop threshold $\varepsilon_2 = 10^{-4}$.

For each algorithm, we have 10 simulations. The final clustering result is shown in Table 1. And we can see that three algorithms can converge to the final objective function value. The difference is that the average FCM algorithm requires an average of 20.4 generations to reach the objective function value, and the FCM based on the simple genetic algorithm only needs 12 generations. The proposed algorithm only needs 2 generations to converge. This shows that the optimization of the initial clustering center in the clustering process is very important. The optimization of the initial clustering center can completely reduce the number of iterations of the algorithm, and also ensure that the algorithm converges accurately to the objective function value. When the initial clustering center is selected, the IDFCM-DNAGA algorithm not only improves the operation operator of the DNA genetic algorithm, but also adds density factors to the basic FCM algorithm, which increases the problem of algorithm processing and considers more factors. So the initial cluster center is closer to the true value, and the final algorithm runs fewer algebras. At the same time, it can be seen from the above table that the performance of the proposed algorithm is more stable than the other two algorithms.

Table 1. Compared with FCM, FCM-GA and IDFCM-DNAGA

	FCM		FCM-GA		IDFCM-DNAGA	
	Converge iteration	Fitness	Converge iteration	Fitness	Converge iteration	Fitness
1	21	60.5760	13	60.5760	2	60.5760
2	15	60.5760	15	60.5760	2	60.5760
3	23	60.5760	10	60.5760	2	60.5760
4	20	60.5760	11	60.5760	2	60.5760
5	22	60.5760	12	60.5760	2	60.5760
6	19	60.5760	11	60.5760	2	60.5760
7	23	60.5760	11	60.5760	2	60.5760
8	19	60.5760	15	60.5760	2	60.5760
9	19	60.5760	12	60.5760	2	60.5760
10	23	60.5760	10	60.5760	2	60.5760
Average	20.4	60.5760	12	60.5760	2	60.5760

4.2 Experiment Analysis

In order to illustrate the validity of the IDFCM-DNAGA in this paper, we use the standard data set Iris for verification. We compare the cluster center selected by the IDFCM-DNAGA with the actual data of the Iris data set (Table 2).

Table 2. Comparison of the initial clustering center

IDFCM-DNAGA	Actual clustering center of the Iris dataset
(6.7751, 3.0524, 5.6569, 2.0536)	(6.58, 2.97, 5.55, 2.02)
(5.8892, 2.7612, 4.3643, 1.3974)	(5.93, 2.77, 4.26, 1.32)
(5.0036, 3.4030, 1.4850, 0.2515)	(5.00, 3.42, 1.46, 0.24)

From the comparison of experimental results, the initial cluster center optimized by the IDFCM-DNAGA is very close to the actual center of the data set, which naturally speeds up the running time of the whole algorithm and improves the accuracy of the algorithm. So, IDFCM-DNAGA is effective.

We also give the clustering accuracy comparison of IRIS datasets which using above three algorithms.

It can be seen from the Table 3 that the algorithm of this paper is obviously more accuracy than the FCM algorithm and the FCM-GA. At the same time, the best number of errors and the worst number of errors in the IDFCM-DNAGA are lower than the others. This indicates that the proposed algorithm is more stable.

Table 3. Compared the clustering accuracy of FCM, FCM-GA and DFCM-DNAGA

Algorithm	Best number of errors	Worst number of errors	Accuracy
FCM	16	16	89.33
FCM-GA	12	15	92.00
IDFCM-DNAGA	8	10	93.23

5 Application of Improved Fuzzy C-Means Algorithm in Text Clustering

We use the data of Sogou Lab corpus to realize the text clustering. The specific datasets are collected at http://www.sogou.com/labs/. Here, we select three classes, namely Finance, IT and Health, respectively. In each class, we select 20 samples. So, a total of 60 samples are selected to be classified.

a. Document preprocessing

   In the text clustering, the key of the preprocessing is to segment the text. Then, we need to remove some stop words against the stop word list. In this paper, the size of the stop words array is 2516 * 1.

b. Weight calculation

   After segmentation, we need to extract keywords. Here, we use the method of $IF * IDF$ algorithm to exact keywords.

   IF is the Keyword frequency, and calculated by the $TF = n/m$. Where, m is the number of the words in the dictionary and n is the number of keywords in this text. IDF is the reverse text frequency. It can be used to measure the weight index of each keyword in the text. $IDF = \log(D/D_w)$, D is the number of the articles, D_w refers to the number of articles in which the keyword appears in all articles.

c. Vector space model

   We use the vector space model (VSM) to represent the text at the process of the clustering. In short, VSM is a vector which can represent a text as the weight of a keyword. In this paper, the number of text is 60, directory is 4987 and the array dimension of the VSM is 60 * 4987.

d. The similarity of the text

   At the VSM, we usually use the cosine similarity to calculate the similarity between the texts. It should be noted that the number and order of the two text feature values to be calculated need to be the same. If a text does not have the keyword vector, the weight can be complemented by 0. The calculation formula is:

$$Sim(D_1, D_2) = \frac{\sum_{k=1}^{N} W_{1k} W_{2k}}{\sqrt{\left(\sum_{k=1}^{N} W_{1k}^2\right)\left(\sum_{k=1}^{N} W_{2k}^2\right)}} \tag{9}$$

where, D_1 and D_2 represent the first and second text, respectively. W_{1k} and W_{2k} are the value of the weight which correspond to the kth eigenvalues of the first and second text. $1 \leq k \leq N$ and N is the total number if the eigenvalues

e. Clustering

After we successfully transformed the text into the eigenvalue weight matrix, the data can be clustered according to the improved fuzzy C-means algorithm proposed in Sect. 3. The cluster of the eigenvalue vector in the final clustering result is the cluster to which the original text belongs.

f. Evaluation

The criteria for evaluation are the precision, recall and FI measures.

$$Precision = True\,positives\,/\,(True\,positives\,+\,True\,negatives) \qquad (10)$$

$$Recall = True\,positives\,/\,(True\,positives\,+\,False\,negatives) \qquad (11)$$

$$FI\,measure = 2 \times Precision \times Recall\,/\,(Precision + Recall) \qquad (12)$$

It can be seen from the above table (Table 4) that the IDFCM-DNAGA is superior to the FCM and GAFCM. This shows that the proposed algorithm can be used for text clustering and be effectively. The process of clustering involves word segmentation and keyword extraction. At present, the text processing of Chinese is not very mature, so the use of the proposed algorithm for text clustering needs further research.

Table 4. Comparison for FCM, GAFCM and IDFCM-DNAGA on precision, recall and FI

	FCM			GAFCM			IDFCM-DNAGA		
	Pre	Recall	FI	Pre	Recall	FI	Pre	Recall	FI
Finance	79.5%	80.5%	79.9%	83.5%	85.5%	84.5%	86.8%	90.6%	88.7%
IT	81.5%	80.4%	80.9%	85.8%	84.9%	85.3%	86.9%	89.2%	88.0%
Health	78.8%	78.2%	78.5%	82.4%	81.3%	81.8%	86.2%	88.6%	87.4%

6   Conclusions

In this paper, we firstly use the algorithm of Sect. 2 to optimize the initial clustering center of the fuzzy C-means algorithm. Secondly, we use the point density weighting method in the fuzzy C-means clustering. At last, we make a experiment on the UCI dataset. The experimental results show that the improved algorithm is ideal both in accuracy and optimization. Only the execution time of the algorithm will be slightly longer, because the improved algorithm needs to deal with the problem at each step. However, the number of iterations of the algorithm is also significantly reduced compared to other algorithms. In summary, the proposed algorithm is very effective. Finally, We apply IDFCM-DNAGA to text clustering and achieve better results.

Acknowledgment. The authors would like to express their thanks to the editors and the reviewers for their careful revisions and insightful suggestions. This research is supported by the National Science Foundation of China (Nos. 61876101, 61802234, 61806114, 61472231, 61502283), Social Science Fund Project of Shandong (16BGLJ06, 11CGLJ22), China Post-doctoral Science Foundation Funded Project (2017M612339) and Humanities and social sciences research projects of the Ministry of Education (12YJA630152).

References

1. Dunn, J.C.A.: Fuzzy relative of the ISODATA process and its use in detecting compact well-separated clusters. J. Cybern. **3**(3), 32–57 (1973)
2. Bezdek, J.C.: Pattern Recognition with Fuzzy Objective Function Algorithms. Plenum Press, New York (1981). 22(1171): 203–239
3. Liu, L., Sun, S.Z., Yu, H., et al.: A modified Fuzzy C-Means (FCM) clustering algorithm and its application on carbonate fluid identification. J. Appl. Geophys. **129**, 28–35 (2016)
4. Nayak, J., Naik, B., Kanungo, D.P., et al.: A hybrid elicit teaching learning based optimization with fuzzy c-means (ETLBO-FCM) algorithm for data clustering. Ain Shams Eng. J. **9**, 379–393 (2016)
5. Misra, S., Das, T.K., Choudhury, S.P., et al.: Choosing optimal value for fuzzy membership in FCM algorithm for LP-residual input features. Procedia Comput. Sci. **54**, 542–548 (2015)
6. Zhang, B., Qin, S., Wang, W., et al.: Data stream clustering based on fuzzy c-mean algorithm and entropy theory. Sig. Process. **126**, 111–116 (2016)
7. Wikaisuksakul, S.: A multi-objective genetic algorithm with fuzzy c-means for automatic data clustering. Appl. Soft Comput. **24**, 679–691 (2014)
8. Vahidi, J., Mirpour, S.: Introduce a new algorithm for data clustering by genetic algorithm. J. Math. Comput. Sci. **10**, 144–156 (2014)
9. Broin, P.Ó., Smith, T.J., Golden, A.: Alignment-free clustering of transcription factor binding motifs using a genetic-k-medoids approach. BMC Bioinform. **16**(1), 1–22 (2015)
10. Langone, R., Agudelo, O.M., De Moor, B., et al.: Incremental kernel spectral clustering for online learning of non-stationary data. Neurocomputing **139**, 246–260 (2014)
11. Haque, M.M., Nilsson, E.E., Holder, L.B., et al.: Genomic Clustering of differential DNA methylated regions (epimutations) associated with the epigenetic transgenerational inheritance of disease and phenotypic variation. BMC Genom. **17**(1), 418 (2016)
12. Dinu, L.P., Ionescu, R.T.: Clustering based on median and closest string via rank distance with applications on DNA. Neural Comput. Appl. **24**(1), 77–84 (2014)
13. Muhammad Fuad, M.M.: Hierarchical clustering of DNA microarray data using a hybrid of bacterial foraging and differential evolution. In: Dediu, A.-H., Magdalena, L., Martín-Vide, C. (eds.) TPNC 2015. LNCS, vol. 9477, pp. 46–57. Springer, Cham (2015). https://doi.org/10.1007/978-3-319-26841-5_4
14. Liu, X., Xue, J.: Spatial cluster analysis by the bin-packing problem and DNA computing technique. Discret. Dyn. Nat. Soc. **2013**(5187), 845–850 (2013)

A Hybrid Collaborative Filtering Recommendation Algorithm Using Double Neighbor Selection

Wenan Tan[1,2(✉)], Xiaofan Qin[1], and Qing Wang[1]

[1] School of Computer Science and Technology,
Nanjing University of Aeronautics and Astronautics, Nanjing 211100, China
{wtan, qinxfan}@foxmail.com
[2] School of Computer and Information, Shanghai Polytechnic University,
Shanghai 201209, China

Abstract. The traditional collaborative filtering algorithms are more successful used for personalized recommendation. However, the traditional collaborative filtering algorithm usually has issues such as low recommendation accuracy and cold start. Aiming at addressing the above problems, a hybrid collaborative filtering algorithm using double neighbor selection is proposed. Firstly, according to the results of user's dynamic similarity calculation, the similar interest sets of the target users may be dynamically selected. Analyzing the dynamic similar interest set of the target user, we can divide the users into two categories, one is an active user, and the other is a non-active user. For the active user, by calculating the trust degree of the users with similar interests, we can select the user with the trust degree of TOP-N, and recommend the target user. For the non-active user, the neighbor user may be found according to the similarity of the user on some attributes, and them with high similarity will be recommend to the target user. The experimental results show that the algorithm not only improves the recommending accuracy of the recommendation system, but also effectively solves the problem of data sparseness and user cold start.

Keywords: Collaborative filtering · Similarity · Dynamic neighbor selection · Trust · Hybrid model · Cold start

1 Introduction

As data growing rapidly, how to recommend products or information services to customers has attracted widespread attention from researchers at home and abroad. In order to help users quickly and accurately find what they want in larger amounts of data, the recommendation algorithms have appeared. The collaborative filtering recommendation algorithm is one of the most widely used [1]. The main principle is to find similar user sets or similar item sets for the recommended target users or target items based on the user-item rating in datasets. However, there are some drawbacks in the traditional collaborative filtering method. Firstly, the recommendation efficiency is declined when the data is extremely sparse or new users and new items appear due to excessive dependence on the user's rating of the item. Secondly, the recommendation

© Springer Nature Switzerland AG 2019
Y. Tang et al. (Eds.): HCC 2018, LNCS 11354, pp. 416–427, 2019.
https://doi.org/10.1007/978-3-030-15127-0_42

is assumed that the users are independent and equally distributed, and it ignores the credibility of users based on item ratings.

Thus, we propose a hybrid collaborative filtering algorithm by using double neighbor selection for addressing these problems. There are contributions: One is to improve the quality of neighbors by using user similarity and trust between users as the basis of neighbor selection. The second is to use the similarity between user attributes to solve the problems that users rating data is extremely sparse or cold start in some recommendation system.

2   Related Works

In order to improve the accuracy of the recommendation and address how to deal with sparse data and cold start in massive data, Researchers have developed various methods to overcome these problems and given some high quality recommendations. For example, in order to solve the cold start problem of the project, the literature [2] proposed a recommendation algorithm which mainly combines the user time information with the attributes of the item itself and the label information to obtain a personalized prediction score. When the item is cold-started, the item rating can be obtained directly from the item information. However, this method improved the novelty of the algorithm, the accuracy is still slightly lacking. The literature [3] proposed a new collaborative filtering algorithm to mitigate the impact of user cold start in prediction. The literature [4] proposed an attribute-to-feature mapping method to solve the cold start problem in collaborative filtering. The literature [5] proposed a combined recommendation method to solve the sparse data and cold start problems, but the algorithm only produces good recommendations for some and sparse data and it does not achieve an improved effect for a complete cold start. Moreover, the recommendation accuracy for the target user is not high. The literature [6] proposed two recommendation models to solve the complete cold start and incomplete cold start problems for new items, which are based on a framework of tightly coupled CF approach and deep learning neural network. The literature [7] proposed a collective Bayesian Poisson factorization (CBPF) model for handling the new problem of cold-start local event recommendation in Event-based social networks (EBSNs).

The literature [8] proposed a collaborative filtering recommendation algorithm based on multi-social trust. The article quantifies the credibility, reliability and self-awareness in social science to form trust, which improves the accuracy of recommendation and recall rate, although the performance of the algorithm has been reduced. The literature [9] proposed a combination recommendation algorithm based on clustering and collaborative filtering. This algorithm reduces the sparsity of the original matrix by clustering and improves the recommendation accuracy. The literature [10] proposed a general collaborative social ranking model to rank the latent features of users extracted from rating data based on the social context of users. The literature [11] proposed an improved clustering-based collaborative filtering recommendation algorithm. The literature [12] proposed a collaborative filtering algorithm based on the double neighbor selection strategy. Though this algorithm can improve the recommendation accuracy when the data reaches a certain sparsity, almost half of the users' neighbors cannot reach the expected number in the experiment.

In summary, we propose a combined recommendation algorithm based on collaborative filtering of double neighbor selection strategy. Firstly, the target users are divided into two categories according to the maximum number of neighbors of the target users. Secondly, different recommendation methods are adopted for these two types of users to achieve good recommendation results.

3   Hybrid Collaborative Filtering Recommendation Algorithm

3.1   Dynamic Selection of User Interest Similarity Set

In the traditional collaborative filtering system, the user ratings is a set of users consisting of m users $U = \{u_1, u_2, \ldots, u_m\}$ and the item ratings is set of items consisting of n items $T = \{t_1, t_2, \ldots, t_n\}$ which is composed of the rating matrix R of m × n order, where R_{ij} represents the score value of the user u_i for t_j, and if the user u_i has no evaluation for t_j, the $R_{i,j} = 0$.

In the recommendation system, if you want to, how to choose high-quality neighbors is the key to ensuring high recommendation quality. In general, we select neighbors through KNN (K-Nearest Neighbors) [13] in the traditional collaborative filtering algorithm and some of them contain a few users with minor similarity to the target users that leads to the recommendation system accuracy is inefficient. Therefore, in order to reduce the impact of target users' predictions on some their neighbors due to interest bias. The existing work is to define similarity threshold to filter out higher quality neighbors. However, defining a fixed threshold in different environments makes the algorithm less scalable. Thus, Dongyan et al. [12] proposed a method towards dynamic selection of the user's interest similarity set for the target user. The method may set the similarity threshold according to the similarity between the target user u_i and its neighbor candidate users.

Before getting a similar set of interests for the target user and the target item, we first need to obtain its candidate neighbor set. Let C denotes candidate neighbor set, Given a historical rating matrix R, the current target user $u_i \in U$ and the current target item $t_j \in T$, the $R_{ij} = 0$ at this time. If exits $u_k \in U$ that makes $R_{kj} \neq 0$, we can see that u_k is the candidate neighbor of the current target user u_i and the current target item t_j. Then the candidate neighbor set $C_j(u_i)$ of u_i on t_j can be expressed as follows:

$$C_j(u_i) = \left\{u_k | R_{ij} = 0 \wedge R_{kj} \neq 0, u_i \in U, t_j \in T\right\} \tag{1}$$

We regard $u_k \in C_j(u_i)$ as the recommended neighbor of the target user u_i corresponding to the target item t_j, using the Pearson correlation coefficient formula to calculate the similarity between each candidate neighbor and the target user. Let CI be the common item rating set, CI_{ik} denotes the set of items which had been rated by user u_k and u_i, the similarity calculation between u_k and u_i is given by

$$sim(i, k) = \frac{\sum_{t_x \in CI_{ik}} \left(R_{i,x} - \overline{R_i}\right) \left(R_{k,x} - \overline{R_k}\right)}{\sqrt{\sum_{t_x \in CI_{ik}} \left(R_{i,x} - \overline{R_i}\right)^2} \sqrt{\sum_{t_x \in CI_{ik}} \left(R_{k,x} - \overline{R_k}\right)^2}} \qquad (2)$$

where $sim(i, k)$ represents the similarity between user u_i and candidate neighbor user u_k; $R_{i,x}$, $R_{k,x}$ respectively represent the ratings of user u_i and user u_k for item t_x; $\overline{R_i}$ and $\overline{R_k}$ represent the average scores of users u_i and u_k on the common item rating set respectively.

For the traditional recommendation algorithm, it is generally necessary to select some neighbors with higher similarity by setting the threshold. In order to improve the scalability of the recommendation algorithm, this paper measures it by the similarity between the target user u_i and the recommended user u_k. Its calculation formula is as follows

$$V_{sim} = \frac{\sum_{k=1}^{x} sim(i, k)}{x}, x = \left|C_j(u_i)\right| \qquad (3)$$

If $sim(i, k) \geq V_{sim}$, u_k is a similar user of the target user u_i. We set the interest similarity set to S, and then the similar interest set of the target user u_i corresponding to the target item t_j can be expressed as $S_j(u_i)$

$$S_j(u_i) = \left\{u_k | sim(i, k) > V_{sim}, i \neq k, u_k \in C_j(u_i)\right\} \qquad (4)$$

3.2 The Calculation of Trust Degree Between Users

We also use the trust degree between users as one of the criteria for measuring neighbors to improve the accuracy and reliability of the recommendation algorithm.

The $u_k \in S_j(u_i)$ is taken as the recommended target user of user u_i and target item t_j. Referring to the literature [14], the target user u_i can be predicted based on the following formula

$$P_{i,j} = \overline{R_i} + \frac{\left(R_{k,j} - \overline{R_k}\right) \times sim(i, k)}{|sim(i, k)|} \qquad (5)$$

Where $P_{i,j}$ indicates that the user u_i is based on the predicted score of the input user u_k on the item t_j; $R_{k,j}$ represents the score of the user u_k on the item t_j; $\overline{R_i}$ and $\overline{R_k}$ respectively represent the average score of u_i and u_k; $sim(i, k)$ represents the similarity between user u_i and user u_k.

Based on the deviation of the predicted score $P_{i,j}$ from the actual score $R_{i,j}$, we can calculate the predictive power of the user u_k for the target user u_i, which is measured as follows:

$$sat_j(i, k) = \begin{cases} 1, & |P_{i,j} - R_{i,j}| < \varepsilon \\ 0, & \text{else} \end{cases} \tag{6}$$

Where $sat_j(i, k)$ is the estimated capability value of the user u_k predicting the user u_i score, and $P_{i,j}$ and $R_{i,j}$ respectively represent the predicted score and the actual score based on the (5) formula. ε is a constant, which can be randomly selected. In this paper, $\varepsilon = 1.2$ is selected by reference [12].

Through $sat_j(i, k)$ of the formula (6), the calculation formula of the trust degree between the user u_i and the user u_k is as follows

$$T(i, k) = \frac{\sum_{x=1}^{|CI_{ik}|} sat_j(i, k)}{|CI_{ik}|} \tag{7}$$

3.3 Calculating Similarity Between Users Using User Attributes

In traditional collaborative filtering, the recommendation accuracy and the recall rate decreased owing to the cold start or sparse data. For these reasons, we utilize the user attribute similarity [15] to find the nearest neighbor of the target user and consider the similar user with higher similarity as the recommended user. In this way, we could predict the target user's rating of the target item based on the recommended user's rating of the target item.

In general, user attributes include the user's gender, age, occupation, etc. We can classify users by these attributes and assign specific values. For example, gender can be divided into male and female, we could set "male = 1" and "female = 0"; we segment the age with numerical attributes, for example, the age can be divided by a period of 10 years, so that 0 to 10 years old is 1, 10–20 years old is 2 and so on, other attributes are also assigned in this method. Thus, We can obtain the item attribute matrix *Attr* by processing these raw data information, Table 1 depicts an example user–attribute rating matrix.

Table 1. A sample user-attribute matrix

	Gender	Age	Profession
User1	0	1	4
User2	1	5	3
User3	1	3	6
User4	0	1	4
User5	1	2	8

We can obviously represent the feature attribute vector of the user u_i as $Attr_{u_i} = \{a_{u_{i1}}, a_{u_{i2}}, \ldots, a_{u_{in}}\}$, if the target user u_i and the user u_k has the same r attribute values, we let $|a_{u_{ir}} \cap a_{u_{kr}}| = 1$, otherwise $|a_{u_{ir}} \cap a_{u_{kr}}| = 0$. Then users u_i and u_k can be measured by the following formula.

$$sim_{attr}(i, k) = \alpha \times |a_{u_{i1}} \cap a_{u_{k1}}| + \beta \times |a_{u_{i2}} \cap a_{u_{k2}}| + \ldots + \delta \times |a_{u_{in}} \cap a_{u_{kn}}| \quad (8)$$

Where α, β and δ are weighting factors, and $\alpha + \beta + \ldots + \delta = 1$.

Let N as the nearest neighbor set selected based on the user attribute, the $N(u_i)$ can be expressed as the nearest neighbor set of the current target user u_i.

3.4   Prediction Computation

In the recommendation system, while many scholars recommend items to the target users, they might add some elements improving the quality of the target users' neighbors as selecting neighbors to improve the recommendation accuracy and reduce the dimension of matrix to improve the efficiency of the algorithm and solve the problem of sparseness of rating matrix. However, in terms of adding elements or reducing matrix dimensions, the following two major problems arise when recommending.

- If we only add some elements that can improve the quality of the neighbors to improve the accuracy of the algorithms. This not only makes the performance of the algorithm decrease, but also makes some users with sparse scoring unable to find a suitable neighbor. Because the neighbors of some data-sparse users cannot meet the requirements of high-quality neighbors. At the same time, the more elements that are added, the longer the algorithm takes, which greatly reduces the efficiency of the algorithm and the recall rate.
- If the matrix dimension reduction method or some user attributes are applied to recommend, the algorithm efficiency and the recall rate can be improved, but the recommended accuracy is greatly reduced.

Therefore, we combine the neighbors selected by the dynamic neighbor with the trust factor which is more important to the recommendation algorithm to improve the neighbor quality. Moreover, for some users with sparse data or cold start, they are recommended based on the similarity of user attributes and then a hybrid collaborative filtering algorithm based on double neighbor selection is proposed, and named HCF-DNS. The core ideas of HCF-DNS are as follows

1. Selecting a neighbor candidate set $C_j(u_i)$ of the target user u_i for the target item t_j;
2. Obtaining a dynamic similar interest set $S_j(u_i)$ of the target user u_i based on the target item t_j according to the method of dynamically selecting a neighbor user in Sect. 3.2;
3. If the similar interest set $S_j(u_i)$ of the target user u_i is an empty set, the predicted score of the current target user u_i on the item t_j is calculated based on the similarity of the user attributes. The predicted score formula is as follows

$$P_{i,j} = \frac{\sum_{k \in N(u_i)} sim_{attr}(i, k) \times R_{kj}}{\sum_{k \in N(u_i)} |sim_{attr}(i, k)|} \quad (9)$$

If a cold start item occurs, the target user's rating for the cold start item is set to 3;

4. If the similar interest set $S_j(u_i)$ of the target user u_i is not an empty set, the trust degree of each of the target user u_i and the user similar interest set $S_j(u_i)$ is obtained in using the inter-user trust degree calculation method in Sect. 3.3. And select the Top-N users with the most trust as the trusted neighbor user set of the target user u_i;
5. Calculating the score of the target user u_i on the target item t_j based on the rating information of the neighbor of the trusted user and using the following equation.

$$P_{i,j} = \overline{R}_i + \frac{\sum_{u_k \in T(u_i)} (R_{k,j} - \overline{R}_k) \times T(i,k)}{\sum_{u_k \in T(u_i)} |T(i,k)|} \tag{10}$$

According to the above algorithm idea, the hybrid collaborative filtering algorithm based on double neighbor selection is given as follows

---

Algorithm: HCF-DNS
Input: Score matrix R, user-attribute matrix $Attr$, target user u_i, target item t_j
Output: The target user u_i predicts the score of the target item t_j, $P_{i,j}$

Begin:
$T(u_i) \leftarrow \emptyset$, $N(u_i) \leftarrow \emptyset$
$S_j(u_i) = \{u_k | sim(i,k) > V_{sim}, \ i \neq k, \ u_k \in C_j(u_i)\}$
if $S_j(u_i) = \emptyset$
   for each $u_k \neq u_i$
      utilize equation (8) to obtain similarity with u_i
   end for
   sort the similarity of u_i and each u_k to get $N(u_i)$
   select Top-N users with the highest similarity in $N(u_i)$
   utilize the equation (9) to get the predicted rating $P_{i,j}$ of the target user u_i
else
   for each $u_k \neq u_i$
      utilize equation (8) to obtain trust degree with u_i
   end for
   $T(u_i) \leftarrow Top -N$ *users with the most trust degree*
   utilize the equation (10) to get the predicted rating $P_{i,j}$
return $P_{i,j}$
End

---

4 Experiment and Results Analysis

4.1 Test Data Set

We adopt the dataset (https://movielens.umn.edu/) provided by the MovieLens [16]. In this data set, users rate on the movies they have seen with a rating range of 1 to 5. The higher the score, the higher the user's satisfaction with the movie. The data set also provides some basic information about the user, such as the user's gender, age and occupation. In this experiment, this information is collected as user attribute information. In the course of the experiment, we will collect 3 attributes and the weighting factors were set to $\alpha = 0.45$ (gender weighting factor), $\beta = 0.35$ (age weighting factor), $\delta = 0.2$ (occupation weighting factor) after referencing the literature [17]. There are 943 users in the dataset with 100,000 ratings for 1,682 movies and the its sparsity of approximately 93.7%.

In the experiment, we divided the data set into training sets and test sets according to the ratio of 80% and 20%.

4.2 Performance Evaluation Index

We make use of the mean absolute error (MAE) [1, 18], root mean square error (RMSE) [1, 18] and recall rate (RL) [1] to evaluate the prediction accuracy and prediction ability of the recommendation algorithm. The MAE and the RMSE are important parameters for evaluating the accuracy of the recommendation algorithm and the RL is the precision of the evaluation recommendation system. The MAE is obtained by the deviation between the predicted score and the actual score. The lower the MAE value, the higher the accuracy of the recommendation algorithm; the RMSE with high sensitivity to the error is the square root of the ratio of the squared deviation of the predicted rating to the actual rating and the predicted number of times. The lower the RMSE, the more accurate the prediction score. The RL is also called the recovery rate, which is used to measure the recall and precision of the algorithm. The measurement equation is as follows.

$$\text{MAE} = \frac{\sum_{k=1}^{n} |P_k - R_k|}{n} \tag{11}$$

Where P_k represents the rating prediction value, R_k represents the real evaluation value and n represents the number of evaluations.

$$\text{RMSE} = \sqrt{\frac{\sum_{k=1}^{n} (P_k - R_k)^2}{n}} \tag{12}$$

Among them, the definition of the variable appearing in the formula is the same as the Eq. (11).

$$RL = \frac{m}{n} \tag{13}$$

Where m represents the number of ratings that can be predicted by the algorithm, and n represents the number of ratings to be tested in the test set.

4.3 Experimental Method

Since the algorithm proposed in this paper is for various users and the number of corresponding neighbors is relatively scattered when selecting the similar interest sets of the target users, in order to make the comparison between the proposed algorithm and other algorithms more obvious during the test. We select 1/10, 2/10, ... , k/10, 1 of the number of neighbors to test. The following is the distribution of similar neighbors of the target users in the test set. The following is the distribution of similar neighbors of the target users in the test set one.

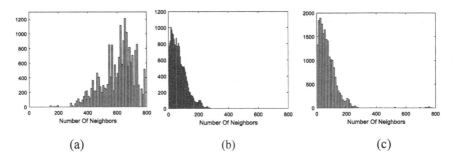

| (a) | (b) | (c) |

Fig. 1. The distribution of neighbor numbers for three algorithms

In the above figures, Fig. 1(a) is the neighbor distribution of the target user corresponding to the target item in using the CF algorithm. From this figure, we can see that the number of selected neighbors is concentrated in the interval of 400 to 800. Thus, the neighbors are not filtered when making predictions. Although the number of neighbors is large, the accuracy is not high; Fig. 1(b) is the number of neighbors selected after using the CF-DNC algorithm. As can be seen from the figure, the number of neighbors is mostly distributed in the interval of 0 to 200 and there are target users with zero neighbors. Although the quality of neighbors has improved, the cold start problem is still inevitable when the we use the measures in [20]. Figure 1(c) is the improved algorithm HCF-DNS algorithm, the distribution of neighbors is also 0 to 200 but it can be seen that the algorithm has screened the number of neighbors for high quality after comparing with the Fig. 1(a). At the same time, combined with the recall rate of Table 2 and Fig. 1(b) the comparison, it can be seen that the cold start problem has also been alleviated.

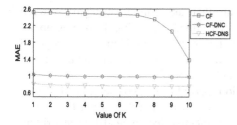
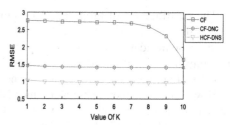

Fig. 2. The results of using MAE   **Fig. 3.** The results of using RMSE

Table 2. Comparison of recall rates

	CF	CF-DNC	HCF-DNS
Recall	1	0.91	1

4.4 Experimental Results and Analysis

In order to evaluate the accuracy of the algorithm proposed in this paper, we compare it with the traditional collaborative filtering algorithm (CF) [19] and the previously mentioned double neighbor choosing strategy based on collaborative filtering (CF-DNC) in the same experimental environment, the experimental results are as follows.

It can be seen from Figs. 2 and 3 that the algorithm HCF-DNS is better than CF-DNC and traditional CF. As can be seen from Table 2, the recall rate of the algorithm is relatively high. It can be seen that in the case of extremely sparse data sets, we not only improves the recommendation accuracy, but also improves the recall rate.

The reason why our recall rate is higher than other experiments is that we refer to the literature [20] to do some processing on some sparse data of CF and CF-DNC during the experiment. And after processing, we find that our proposed algorithm still has an advantage in recall rate.

The main reason is that in the recommendation process, we add the trust degree factor, which is more important among users to improve the recommendation accuracy. When the user starts coldly, the user attribute-based recommendation is adopted to avoid the decrease of the recall rate. Thus avoiding the problem of the drop in the recall rate. It can be seen that HCF-DNS proposed in this paper achieves better results than CF and CF-DNC.

5   Conclusion and Further Work

With the wide application of recommendation technology in e-commerce, both the accuracy of the recommendation system and the recall rate are the main research directions to improve QoS of recommendation. In this paper, a hybrid collaborative filtering algorithm based on double neighbor selection, which named HCF-DNS, is proposed for addressing these two problems.

The contributions of proposed algorithm are that taking the trust degree between users as the basis of the target user's neighbors selection, and mixing the demographic-based recommendation algorithm to prevent the cold start problem. The proposed algorithm has effectively addressed the problem of user cold start in traditional collaborative filtering and improved Recommended accuracy.

However, the demographic recommendation algorithm is a coarse-grained recommendation algorithm. The accuracy of the recommended algorithm is not high when solving the cold start. So how to design a more effective recommendation algorithm, making the recommendation efficiency of cold start users higher is the next research work of this paper.

Acknowledgments. The paper is supported in part by the National Natural Science Foundation of China under Grant No. 61672022, and Key Disciplines of Computer Science and Technology of Shanghai Polytechnic University under Grant No. XXKZD1604.

References

1. Ricci, F., Rokach, L., Shapira, B.: Recommender Systems: Introduction and Challenges. Recommender Systems Handbook, pp. 1–34. Springer, Boston (2015). https://doi.org/10.1007/978-0-387-85820-3
2. Yu, H., Li, J.: Algorithm to solve the cold-start problem in new item recommendations. J. Softw. **26**(6), 1395–1408 (2015)
3. Bobadilla, J., et al.: A collaborative filtering approach to mitigate the new user cold start problem. Knowl. Based Syst. **26**, 225–238 (2012)
4. Gantner, Z., et al.: Learning attribute-to-feature mappings for cold-start recommendations. In: 2010 IEEE 10th International Conference on Data Mining (ICDM). IEEE (2010)
5. Sobhanam, H., Mariappan, A.K.: Addressing cold start problem in recommender systems using association rules and clustering technique. In: International Conference on Computer Communication and Informatics (ICCCI) (2013)
6. Wei, J., et al.: Collaborative filtering and deep learning based recommendation system for cold start items. Expert. Syst. Appl. **69**, 29–39 (2017)
7. Zhang, W., Wang, J.: A collective bayesian poisson factorization model for cold-start local event recommendation. In: Proceedings of the 21th ACM SIGKDD International Conference on Knowledge Discovery and Data Mining. ACM (2015)
8. Ruiqin, W., et al.: A collaborative filtering recommendation algorithm based on multiple social trusts. J. Comput. Res. Dev. **53**(6), 1389–1399 (2016)
9. Napoleon, D., Ganga Lakshmi, P.: An efficient K-Means clustering algorithm for reducing time complexity using uniform distribution data points. In: Trendz in Information Sciences & Computing (TISC). IEEE (2010)
10. Forsati, R., et al.: Pushtrust: an efficient recommendation algorithm by leveraging trust and distrust relations. In: Proceedings of the 9th ACM Conference on Recommender Systems. ACM (2015)
11. Xiaojun, L.: An improved clustering-based collaborative filtering recommendation algorithm. Clust. Comput. **20**(2), 1281–1288 (2017)
12. Dongyan, J., et al.: A collaborative filtering recommendation algorithm based on double neighbor choosing strategy. J. Comput. Res. Dev. **50**(5), 1076–1084 (2013)

13. Herlocker, J.L., et al.: An algorithmic framework for performing collaborative filtering. In: ACM SIGIR Forum, vol. 51, no. 2. ACM (2017)
14. O'Donovan, J., Smyth, B.: Trust in recommender systems. In: Proceedings of the 10th International Conference on Intelligent User Interfaces. ACM (2005)
15. Raad, E., Chbeir, R., Dipanda, A.: User profile matching in social networks. In: 2010 13th International Conference on Network-Based Information Systems (NBiS). IEEE (2010)
16. Miller, B.N., et al.: MovieLens unplugged: experiences with an occasionally connected recommender system. In: Proceedings of the 8th International Conference on Intelligent User Interfaces. ACM (2003)
17. Rydén, F., et al.: Using kinect and a haptic interface for implementation of real-time virtual fixtures. In: Proceedings of the 2nd Workshop on RGB-D: Advanced Reasoning with Depth Cameras (in conjunction with RSS 2011) (2011)
18. Karafyllis, I.: Finite-time global stabilization by means of time-varying distributed delay feedback. SIAM J. Control Optim. **45**(1), 320–342 (2006)
19. Herlocker, J.L., et al.: Evaluating collaborative filtering recommender systems. ACM Trans. Inf. Syst. (TOIS) **22**(1), 5–53 (2004)
20. Schein, A.I., et al.: Methods and metrics for cold-start recommendations. In: Proceedings of the 25th Annual International ACM SIGIR Conference on Research and Development in Information Retrieval. ACM (2002)

Energy-Efficient Independent Task Scheduling in Cloud Computing

Xia Zhu$^{(\boxtimes)}$, Mehboob Hussain, and Xiaoping Li$^{(\boxtimes)}$

Southeast University, Nanjing, China
{zhuxia,Mehboob,xpli}@seu.edu.cn

Abstract. The high scientific applications which contain thousands of tasks are usually executed in virtulized cloud for many benefits. With the increment of the processing capability of the cloud system, the computation energy is significantly consumed along. Thus efficient energy consumption methods are quite necessary to save the energy cost. In this paper, the independent task scheduling problem in a cloud data center is considered. It is a big challenge to achieve the tradeoff between the minimization of computation energy and user-defined deadlines. A heuristic is proposed which consist of an energy efficient task sequencing method and a virtual machine searching strategy. Experimental results show that the proposed heuristic clearly outperforms the other algorithms.

Keywords: Virtualized cloud · Energy consumption · Task scheduling

1 Introduction

Cloud computing has started to dominate the computing environment in current days. It provides high and scientific applications being executed in the cloud for various benefits. With increase of processing capabilities, the energy consumption has also increased significantly [1]. Energy efficient execution manner for these tasks in cloud system become very essential. During the past decades, the power consumption of computing resources accounted 45% of the total power consumption in a data center [2]. Many operations have been conducted to reduce the energy consumption of the computing resources [3]. It is usually adopted to assign tasks to the slower virtual Machines (VMs) while meeting the deadline and ensuring quality of services. However, the requirements of cloud users and cloud service provider (CSP) are conflicted. The goal of CSP is to schedule tasks submitted by the cloud users in an optimal way such that it should meet the deadline and quality of service (QoS) with minimum computation energy.

Independent task scheduling problem focuses on the scheduling of a set of independent tasks to be run on heterogeneous VMs. By using the modern virtualization techniques, a large scale of user tasks can be simultaneously executed in cloud. The main goal is to properly schedule these tasks in a way that minimizes the computing resource energy and determines suitable resources for tasks

© Springer Nature Switzerland AG 2019
Y. Tang et al. (Eds.): HCC 2018, LNCS 11354, pp. 428–439, 2019.
https://doi.org/10.1007/978-3-030-15127-0_43

under deadlines. This problem extensively exists in the practical environment such as relational database queries, parametric studies and image processing, where parallel tasks are submitted to cloud provider for execution.

In order to resolve the energy problem in cloud system, many researches tried to improve energy efficiency of computing resources. A common data center usually composes of different types of VMs, which have different process capacity and power consumption characteristics [4]. An effective way is to use energy-aware scheduling approach which leverages computing resources heterogeneity. It was proved in [5] that a hybrid data center with low power VMs and high performance VMs can achieve the efficient energy purpose.

In this paper, the independent tasks scheduling problem with the purpose of minimizing resource energy consumption under user deadlines is considered. The contributions of this paper include:

- Firstly, all tasks are arranged in energy efficient order by three rules.
- Secondly, a novel scheduling algorithm is proposed for searching heterogeneous VMs to effectively reduce energy consumption and finish all tasks before deadlines.
- Lastly, experiments are performed in cloud system simulation environment to validate the proposed algorithm.

The rest of this paper is organized as follows: In Sect. 2, the related works in literature are summarized. Problem definition and mathematical model are given in Sect. 3. In Sect. 4, energy-efficient scheduling approaches for tasks in cloud are proposed. Section 5 contains the simulation experiments and performance analysis. Finally, the paper is concluded in Sect. 6.

2   Related Work

With the expanded demand of cloud computing infrastructures and the explosion in data center sizes, energy efficiency becomes a crucial issue and most of them are proved to be NP-hard. Many researchers have begun to study about energy-efficient policies for reducing energy from computing resources. There are several ways to reduce computer resource energy such as maximize resource utilization, minimize VM migration, dynamic voltage and frequency scaling (DVFS) technique and efficient task scheduling.

From the perspective of energy-efficient task scheduling, most existing works focus on task scheduling and suitable resource allocation in cloud to minimize energy. Garg et al. [6] proposed a technique for task scheduling problem in a heterogeneous data center to get the minimum energy consumption for all type of tasks. But there was an assumption that an upper bound of total jobs arrive rate is known ahead, which may not be available in a real cluster. Yigitbasi et al. [7] proposed some heuristics to get a worst energy saving of workload in Hadoop clusters. But they only focused on the MapReduce workload and neglected other types of workload in current clusters. Liu et al. [8] analyzed a heterogeneous cluster with parallel tasks and proposed an energy consumption model. They also presented an energy-aware task scheduling strategy based on clustering to shorten scheduling lengths while keeping energy consumption minimal. Li et al. [9] proposed an energy-aware task scheduling algorithm for heterogeneous clusters based on Min-Min heuristic. Its goal was to get a best time energy tradeoff. Mukherjee et al. [10] depicted three thermal-aware energy-saving job scheduling techniques to reduce the energy consumption of the data center under some performance constraints.

Some researches considered reducing power consumption in virtual computing environments. Liu et al. [11] proposed the GreenCloud architecture to reduce data center power consumption, while guaranteing the performance and leveraging live VM migration technology. Beloglazov et al. [12] proposed and evaluated heuristics for dynamic reallocation of VMs to minimize energy consumption while providing reliable QoS. Verma et al. [13] presented several approaches to capture the cost-aware application placement problem. Li et al. [14] implemented and validated a dynamic resource provision framework for virtualizing server environments. Kusic et al. [15] examined and evaluated three local resource allocation policies based on shortest queue in a heterogeneous cluster.

In contrast to previous work, the energy efficient task scheduling problem with the constraint of user deadlines is considered in this paper. The proposed scheduling algorithm focuses on reducing the energy consumption by assinging the independent tasks to the lower energy consumption VMs under the consideration of deadlines.

3   Problem Description

In this paper, the problem that a set of independent tasks being scheduled in a cloud data center which comprised of heterogeneous VMs is considered. Each task should be effectively assigned to an appropriate VM with certain quantity of computing capacity to execute. For simplicity, several assumptions for practical environments are used:

- Each task is only executed on one VM, neither task migration nor interruption is allowed.
- VMs reside in single data center, break-down is not concerned.
- Data transmission time of each task is neglected. Only the power consumption of VMs is considered.

Notations to be used in this paper are listed in Table 1. The system model, application and resource models of the considered problem are given in the following sections.

Table 1. Notations used in the paper

Notation	Explanation
N	Number of task v
M	Number of virtual machine $\mathcal{V}$
$\mathcal{V}_j$	The jth virtual machine in a data center
v_i	The ith independent task
W_i	The workload of the task v_i
d_i	The deadline of the task v_i
ζ_j	The speed of virtual machine $\mathcal{V}_j$
p_j	The power of virtual machine $\mathcal{V}_j$
T_i^e	The execution time of the task v_i
$x_{i,j}$	The determined placement of task v_i and virtual machine $\mathcal{V}_j$
B_i	The begin time of task v_i
F_i	The finish time of task v_i
T_i^{slack}	The slack time of task v_i

3.1 System Model

Figure 1 presents the system architecture of the considered problem, which includes two components: the Master Node and the Data Center Component (DCC). Tasks are first submitted to the Master Node by the cloud user. The Master Node is responsible for scheduling tasks to the appropriate VMs to minimize the energy consumption. The Master Node also plays the role as a connector between the cloud user and the DCC. Three types of VMs are configured in advance in the DCC: Small, Medium and Large.

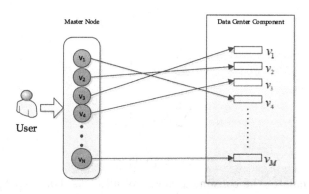

Fig. 1. The system model architecture

3.2   Application and Resource Model

Tasks are represented by $\{v_1, v_2, v_3, \ldots, v_N\}$, and VMs in a data center are defined by $\{\mathcal{V}_1, \mathcal{V}_2, \ldots, \mathcal{V}_M\}$. Each task has a workload W_i ($i = 1, \ldots, N$) and a deadline d_i, which is defined by the cloud user when tasks are submitted. The computation speed and power of VM $\mathcal{V}_j$ ($j = 1, \ldots, M$) are denoted as ζ_j and p_j, respectively. To minimize the energy consumption, each task is assigned to the slower VM under deadline d_i, which needs less energy. $x_{i,j} \in \{0, 1\}$ is a decision variable, where $x_{i,j} = 1$ only if task v_i is assigned to $\mathcal{V}_j$. The energy consumption of task v_i is determined by the power p_j and the execution time T_i^e, i.e., $T_i^e = \sum_{j=1}^{M} x_{i,j} \times \frac{W_i}{\zeta_j}$.

The considered problem can be mathematically modeled as below:

$$\min Z = \sum_{i=1}^{N} \sum_{j=1}^{M} x_{i,j} \times p_j \times T_i^e \tag{1}$$

s.t.

$$T_{j,0} = 0 \tag{2}$$

$$T_{j,k} = T_{j,k-1} + \sum_{k=1}^{N} x_{k,j} T_k^e \tag{3}$$

$$T_i^e = \sum_{j=1}^{M} x_{i,j} \times \frac{W_i}{\zeta_j} \tag{4}$$

$$F_i = \sum_{j=1}^{M} x_{i,j} \times T_{j,k} \tag{5}$$

$$F_i \leq d_i \tag{6}$$

$$\sum_{i=1}^{N} x_{i,j} = 1 \tag{7}$$

$$\sum_{j=1}^{M} x_{i,j} = 1 \tag{8}$$

$$x_{i,j} \in \{0, 1\} \tag{9}$$

Equation 1 calculates the consumed energy of all tasks. For each $\mathcal{V}_j$, the finish time $T_{j,0}$ is initialized to 0 in Eq. 2. The finish time $T_{j,k}$ of $\mathcal{V}_j$ when executing task v_k is determined by the finish time of previous task v_{k-1} and the execution time $\sum_{k=1}^{N} x_{k,j} T_k^e$ of the current task v_k, $T_{j,k}$ is formulated in Eq. 3, and T_i^e is defined in Eq. 4. The finish time F_i of v_i is presented in Eq. 5. Equation 6 reveals that the finish time F_i of v_i should be less than its deadline d_i. Equations 7 and 8 illustrates that each VM is assigned to one task and each task is assigned to one VM. Equation 9 implies that v_i is allocated to $\mathcal{V}_j$ or not.

4   Proposed Algorithm

In this section, an energy efficient heuristic scheduling algorithm is proposed to find an optimal or near optimal solution to complete all N tasks on M VMs with minimum or near minimum execution energy Z while meet the deadline d_i. As illustrated in Algorithm 1, the proposed heuristic contains two phases: (i) a task sequence is generated according to three rules inspired by HEFT algorithm [16], (ii) tasks are iteratively selected and assigned to the most energy efficient VMs.

Algorithm 1. The Algorithm Framework

   Input: Q_v; Q_d; $\{\mathcal{V}_{j,1}, \ldots, \mathcal{V}_{j,m_j}\}$

1 **begin**
2 $\quad$ $Z \leftarrow 0$;
3 $\quad$ **Task sequencing**;
4 $\quad$ **foreach** $v_i \in Q_v$ **do**
5 $\quad\quad$ $Z_i \leftarrow$ **VM Searching**;
6 $\quad\quad$ $Z \leftarrow Z + Z_i$
7 $\quad$ **return** Z;

4.1   Task Sequencing

After being submitted by users, a set of independent tasks are sequenced according to their deadlines, workloads and slack time. The slack time T_i^{slack} is determined by the actual finish time F_i and d_i of v_i, i.e., $T_i^{slack} = d_i - F_i$. F_i is decided by the execution time T_i^e and the available time of the assigned VM. However, since the heterogeneity of the VMs, the T_i^e is undetermined before scheduling and the average execution time $\overline{T_i^e} = \frac{\sum_{j=1}^{M} W_i}{\sum_{j=1}^{M} \zeta_j}$ is employed to estimate the task execution time. Assumed that the available time of each VM is 0, the T_i^{slack} can be calculated as:

$$T_i^{slack} = d_i - \overline{F_i} \tag{10}$$

$$\overline{F_i} = \sum \overline{T_i^e} \frac{\sum_{j=1}^{M} W_i}{\sum_{j=1}^{M} \zeta_j} \tag{11}$$

Since the deadline, size and slack time are critical to task sequencing, three different rules are developed as follows.

(1) Earliest Deadline First (EDF): Tasks are sequenced based on the ascending order of their deadlines. If the deadlines of two tasks are the same, the one with the smaller size will be ranked with a higher priority.
(2) Smallest Slack Time First (SSF): Tasks are sorted based on the ascending order of their slack time. If the slack time is same of any tasks, the one with the smallest total workload will be arranged first.
(3) Smallest Workload First (SWF): Tasks are sequenced based on the ascending order of their sizes.

4.2   VM Searching

The allocation of each task v_i to a VM is to make a decision on the $x_{i,j}$, i.e., $x_{i,j} = 1$ if the v_i is assigned to $\mathcal{V}_j$. As depicted in Eq. 1, the energy consumption is decided by the power p_j of the assigned VM and the task execution time T_i^e. Performance per Watt PpW_j is used in this paper to characterize the energy efficiency of $\mathcal{V}_j$, which is defined as

$$PpW_j = \frac{\zeta_j}{p_j}. \tag{12}$$

VMs are sorted according to the ascending order of PpW_j, and the available time $T_{j,0}$ of each $\mathcal{V}_j$ is initialized to 0. To energy efficiently assign each task to the optimal VM, the sorted VMs are traversed from the head to tail. If $T_{j,i-1} + T_i^e < d_i$, then $\mathcal{V}_j$ for v_i is identified, and the available time $T_{j,i}$ is dynamically updated. The details of the VM searching algorithm is described in Algorithm 2.

Algorithm 2. VM Searching (VMS)

Input: v_i: task to schedule

1 **begin**
2 | $Q_{vm} \leftarrow$ Sort the VMs by PpW_j with the ascending order;
3 | $\mathcal{V} \leftarrow NULL$;
4 | **foreach** $\mathcal{V}_j \in Q_{vm}$ **do**
5 | | $T_{j,0} \leftarrow 0$;
6 | **foreach** $\mathcal{V}_j \in Q_{vm}$ **do**
7 | | Calculate the T_i^e of $\mathcal{V}_j$ by Eq.4;
8 | | **if** $T_{j,i-1} + T_i^e < d_i$ **then**
9 | | | Calculate the $T_{j,i}$ of $\mathcal{V}_j$ by Eq.3;
10 | | | $\mathcal{V} \leftarrow \mathcal{V}_j$;
11 | | | break;
12 | Calculate the consumed energy Z_i by Eq.1;
13 | **return** $Z_i, \mathcal{V}$;

In line 2, VMs are sorted by PpW_j with the ascending order, and put into sequence Q_{vm} in which VMs are iteratively traversed. In line 3, the result sequence of VMs $\mathcal{V}$ are initialized as null. The available time $T_{j,0}$ of each VM in Q_{vm} is initialized to 0. From line 7 to 11, if the sum of the available time of $\mathcal{V}_j$ and the execution time of v_i is less than the deadline d_i, then v_i is assigned to $\mathcal{V}_j$, and the new available time $T_{j,i}$ of $\mathcal{V}_j$ is dynamically updated.

In Algorithm 2, VMs are swapped at least $M \times log(M)$ times in the sorting process. Besides, the traverse of the sorted VMs consumes M times. Totally, the time complexity of Algorithm 2 is $O(M \times log(M))$.

5   Performance Evaluation

In this section, the parameters and performance of the proposed algorithm are investigated. Different components of the proposed method are analyzed to find the best combination. All algorithms are implemented in Java and configured in the same (Intel (R) Core (TM) i5-3475 CPU @ 3.30GHz, 10G Memory). The operation software of the machine is Windows 10 for carried out our experiment.

5.1   Simulation Setup

In the experiment, five different numbers of task nodes $Q_t \in \{50, 100, 200, 400, 500\}$ are generated. The deadline of each task is defined on basis of the following equation:

$$d_i = F_i + \gamma \times F_i \tag{13}$$

The deadline d_i of a task is the sum of the earliest completion time and a certain percentage of the earliest completion time. γ is used as a parameter to control the tightness of the task deadline with value range $\gamma \in \{.2, .4, .6, .8, 1\}$. So each task can get five different size deadlines noted as $D1, D2, D3, D4, D5$.

In the algorithm comparison phase, three existing algorithms Energy Aware Rolling-Horizon (EARH) [17], Earliest Deadline First (EDF) [18] and First Come First Serve (FCFS) [19] are selected to verify the effectiveness of the proposed heuristic. For fairness, all the compared algorithms are executed with the same tasks collection and the same setting of the number of tasks and parameter verification.

In order to measure the performance of the algorithms, the Relative Percentage Deviation (RPD) in Eq. 14 is adopted:

$$RPD(\%) = \frac{Z - Z^*}{Z^*} \times 100\% \tag{14}$$

Z represents the value of the objective function obtained when executing the tasks according to the proposed algorithm. Z^* represents the minimum consumption of energy that all algorithms consume when executing the tasks. All the experimental results will be performed by the Analysis of Variance (ANOVA) technique.

The VM configuration of the performance evaluation is presented in the Table 2. There are five types of VMs, different VMs have different processing speed and the different power. The VMs of each VM type are randomly generated, during the parameter calibration and the algorithm comparison, the VM configuration remains the same.

Table 2. VM Specifications.

VM	VM1	VM2	VM3	VM4	VM5
Core	1	1	1	1	1
MIPS/Core	200	400	600	800	1000
Power/Core	50 W	100 W	150 W	200 W	250 W

5.2 Parameter Calibration

Figure 2 shows the multi factor 95.0% Turkey HSD confidence interval of RPD impact on parameter γ. As we can see, γ has a significant influence on the results of the algorithm. When γ from .2 to .6 a value RPD decreased obviously, and when $\gamma > .6$ RPD values has stabilized. Therefore, .6 is selected in this algorithm.

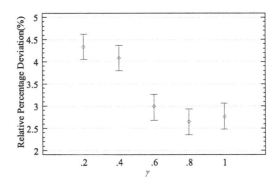

Fig. 2. The parameter γ have 95.0% confidence interval Tukey HSD Mean interval chart

5.3 Task Sequencing Methods

When the tasks are submitted into the scheduling system, three task sequencing rules (EDF, SSF and SWF) are proposed to generate the task scheduling sequence. Three sequencing rules are calibrated to select the most appropriate one. Figure 3 presents the mean plot of three task sequence rules with 95.0% Tukey HSD intervals, the RPD value of SSF is obviously lower than EDF and SWF. It is concluded that the task is scheduled by the task sequence which is generated by the SSF rule leads to smaller energy consumption. Therefore, SSF is selected for the task sequencing component.

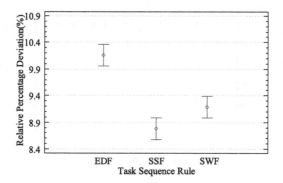

Fig. 3. The mean plot of task sequence rules with 95.0% Tukey HSD intervals

5.4   Algorithm Comparison

To evaluate the performance of the proposed algorithm, three existing task scheduling algorithm: EARH [17], EDF [18] and FCFS [19] are selected as the benchmark algorithms.

Figures 4 and 5 depict the RPD values between the proposed algorithm and the compared algorithms under different deadlines and task numbers. Figure 4 illustrates that the proposed algorithm is not as good as EDF algorithm under the tight deadline, however, with the deadline becoming loose, the RPD values of the proposed EEITS is gradually lower than the compared algorithms. The performance of each algorithm evaluated under various task number is presented in the Fig. 5. The proposed algorithm is obviously better than the compared algorithms.

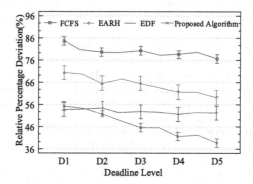

Fig. 4. Comparison of algorithms under different deadlines

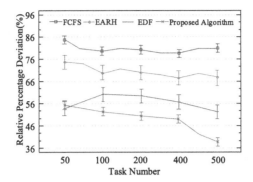

Fig. 5. Comparison of algorithms under different task numbers

6   Conclusion

In this paper, energy-efficient scheduling problem for independent task with user defined deadline in virtulized cloud is investigated. The goal of the paper is to schedule the tasks in an energy efficient way. An energy efficient independent task scheduling heuristic is proposed, which consists of a task sequencing process and an energy efficient VM selection strategy. Experimental results shows that the proposed algorithm outperforms the others in most cases.

References

1. Kizza, J.M.: Guide to Computer Network Security. Springer, London (2017). https://doi.org/10.1007/978-1-4471-4543-1
2. Filiposka, S., Mishev, A., Juiz, C.: Balancing performances in online VM placement. In: Loshkovska, S., Koceski, S. (eds.) ICT Innovations 2015. AISC, vol. 399, pp. 153–162. Springer, Cham (2016). https://doi.org/10.1007/978-3-319-25733-4_16
3. Ebrahimi, K., Jones, G.F., Fleischer, A.S.: A review of data center cooling technology, operating conditions and the corresponding low-grade waste heat recovery opportunities. Renew. Sustain. Energy Rev. **31**, 622–638 (2014)
4. Nathuji, R., Isci, C., Gorbatov, E.: Exploiting platform heterogeneity for power efficient data centers. In: Fourth International Conference on Autonomic Computing, 2007, ICAC 2007, p. 5. IEEE (2007)
5. Chun, B.-G., Iannaccone, G., Iannaccone, G., Katz, R., Lee, G., Niccolini, L.: An energy case for hybrid datacenters. ACM SIGOPS Oper. Syst. Rev. **44**(1), 76–80 (2010)
6. Garg, S., Sundaram, S., Patel, H.D.: Robust heterogeneous data center design: a principled approach. ACM SIGMETRICS Perform. Eval. Rev. **39**(3), 28–30 (2011)
7. Yigitbasi, N., Datta, K., Jain, N., Willke, T.: Energy efficient scheduling of mapreduce workloads on heterogeneous clusters. In: Green Computing Middleware on Proceedings of the 2nd International Workshop, p. 1. ACM (2011)
8. Liu, W., Li, H., Du, W., Shi, F.: Energy-aware task clustering scheduling algorithm for heterogeneous clusters. In: 2011 IEEE/ACM International Conference on Green Computing and Communications (GreenCom), pp. 34–37. IEEE (2011)

9. Li, Y., Liu, Y., Qian, D.: An energy-aware heuristic scheduling algorithm for heterogeneous clusters. In: Proceedings of the 15th International Conference on Parallel and Distributed Systems (ICPADS) (2009)
10. Mukherjee, T., Banerjee, A., Varsamopoulos, G., Gupta, S.K.S., Rungta, S.: Spatio-temporal thermal-aware job scheduling to minimize energy consumption in virtualized heterogeneous data centers. Comput. Netw. **53**(17), 2888–2904 (2009)
11. Liu, L., et al.: Greencloud: a new architecture for green data center. In: Proceedings of the 6th International Conference Industry Session on Autonomic Computing and Communications Industry Session, pp. 29–38. ACM (2009)
12. Beloglazov, A., Buyya, R.: Energy efficient allocation of virtual machines in cloud data centers. In: 2010 10th IEEE/ACM International Conference on Cluster, Cloud and Grid Computing (CCGrid), pp. 577–578. IEEE (2010)
13. Verma, A., Ahuja, P., Neogi, A.: pMapper: power and migration cost aware application placement in virtualized systems. In: Issarny, V., Schantz, R. (eds.) Middleware 2008. LNCS, vol. 5346, pp. 243–264. Springer, Heidelberg (2008). https://doi.org/10.1007/978-3-540-89856-6_13
14. Li, Z., Ge, J., Hu, H., Song, W., Hu, H., Luo, B.: Cost and energy aware scheduling algorithm for scientific workflows with deadline constraint in clouds. IEEE Trans. Serv. Comput. **11**, 713–726 (2015)
15. Kusic, D., Kephart, J.O., Hanson, J.E., Kandasamy, N., Jiang, G.: Power and performance management of virtualized computing environments via lookahead control. Cluster Comput. **12**(1), 1–15 (2009)
16. Topcuoglu, H., Hariri, S., Min-you, W.: Performance-effective and low-complexity task scheduling for heterogeneous computing. IEEE Trans. Parallel Distrib. Syst. **13**(3), 260–274 (2002)
17. Zhu, X., Yang, L.T., Chen, H., Wang, J., Yin, S., Liu, X.: Real-time tasks oriented energy-aware scheduling in virtualized clouds. IEEE Trans. Cloud Comput. **2**(2), 168–180 (2014)
18. Chetto, H., Chetto, M.: Some results of the earliest deadline scheduling algorithm. IEEE Trans. Softw. Eng. **10**, 1261–1269 (1989)
19. Schwiegelshohn, U., Yahyapour, R.: Analysis of first-come-first-serve parallel job scheduling. In: SODA, vol. 98, pp. 629–638. Citeseer (1998)

An Improved IFP-growth Algorithm
Based on Tissue-Like P Systems
with Promoters and Inhibitors

Ning Wang and Xiyu Liu[✉]

Business School, Shandong Normal University, Shandong, Jinan, China
1127278856@qq.com, sdxyliu@163.com

Abstract. The FP-growth is an effective method of mining frequent itemsets to find association rules. But this algorithm scans the database twice to create a FP-tree. This process reduces the efficiency of the algorithm. An improved method, the TPPIIFP-growth algorithm, is presented and uses two-dimensional vector table and tissue-like P systems with promoters and inhibitors to improve the original algorithm. While reducing the scanning, using the flat maximally parallel reduces the time complexity. And this method can be applied to other similar algorithms.

Keywords: Data mining · FP-growth algorithm ·
IFP-growth algorithm · Frequent itemsets · Tissue-like P systems

1 Introduction

The frequent itemsets mining is an important part of data mining, where its purpose is to find items that appear frequently in the database. Many unexpected interesting connections can be found in these frequent itemsets and these connections help managers make more correct decisions. There is a well-known example: beer and diapers [1,2]. Supermarket managers mine these two frequent itemsets in numerous data. When sellers place goods, putting beer and diapers together will increase and boost sales [3].

The FP-growth Algorithm is a classical algorithm to improve Apriori Algorithm and the method can be used to mine the frequent itemsets [4]. However, the FP-growth algorithm still needs to scan the database twice during mining. Then we can improve the algorithm to save the scanning time of the large database, and parallel processing can greatly improve the efficiency of the calculation [7]. In the study, we use the tissue-like P system to deal with the improved algorithm in parallel to improve the efficiency of the algorithm [6,8].

The P system is a new computability model of membrane structure [10]. The tissue-like P system is a parallel structure and the cells contain only objects and evolutionary rules that do not include other basic membranes. The strong distributed and parallel computing power of P systems can greatly improve the computational efficiency of the algorithm and many fields can apply to P systems [12].

© Springer Nature Switzerland AG 2019
Y. Tang et al. (Eds.): HCC 2018, LNCS 11354, pp. 440–451, 2019.
https://doi.org/10.1007/978-3-030-15127-0_44

In this study, an improved FP-growth algorithm based on a tissue-like P system with promoters and inhibitors (TPPIIFP-growth) use the parallel mechanism in P systems. Cells can communicate with each other, all itemsets are searched in parallel, regulated by a set of promoters and inhibitors. Compared with other FP-growth algorithms, the TPPIIFP-growth saves time in the time complexity [11,13].

The paper structure is as follows. Section 2 introduces some basic concepts and notions about the improved FP-growth (IFP-growth) algorithm and tissue-like P system with promoters and inhibitors. Section 3 introduces the rules that TPPIIFP-growth is performed. Section 4 introduces an illustrative example to apply the above rules. Section 5 introduces the conclusion.

2 Preliminaries

2.1 The IFP-growth Algorithm

Definitions [14]

1. Item: an item is a field in a transactional database.
2. Itemset: an itemset is a set of items.
3. m-itemset: if an itemset contain m items, it is called m-itemset.
4. Transaction: a record in a transactional database is called a transaction. The transaction is nonempty itemset.
5. Support count: the number of transaction that contains a certain itemset appears in the transactional database. It is called the support count of the itemset.
6. Frequent itemset: we usually set a minimum support count threshold s. If an itemset's support count is greater than or equal to the support count threshold s, this set of item is called a frequent itemset.

Here is the usual procedure for the IFP-growth [4,7].

Input The transactional database with D transactions and the minimum support count threshold k.

Step 1. Create a two-dimensional vector table that is used to count the two-two combination items support counts. When scanning the transactional database for the first time obtains frequent 1-itemsets and sorts the frequent 1-itemsets in the descending order, we record the support counts for he two-two combination items in the transactional database. For example, if items a, b and c are in the first record of transactional database, the 2-itemsets {a, b}, {a, c} and {b, c} are in the two-dimensional vector table.

Step 2. According to the sorted frequency 1-itemsets, we obtain the conditional pattern bases by the above two-dimensional vector scale. And delete some items that their support counts are less than the minimum support count threshold k.

Step 3. According to the conditional pattern bases, the conditional FP-tree is obtained.

Output All frequent itemsets is generated.

2.2   The Tissue-Like P Systems with Promoters and Inhibitors

A membrane structure with objects in its membranes, with evolution rules for objects, and with specified input-output prescriptions is called P system. The Tissue-Like P Systems with Promoters and Inhibitors (TPPI) also contain the above characteristics [15]. In the TPPI, cells can communicate with each other, and there are evolutionary rules, objects, promoters and inhibitors within the cell [9].

The TPPI with m elementary membranes is described as follows:

$$\Pi = (O, \sigma_1, \sigma_2, \cdots, \sigma_m, syn, \rho, i_{out}) \tag{1}$$

In the first formula, O represents is an alphabet, where include all objects in the system. σ_1, $\cdots$, σ_m represents m cells in the system. syn represents all synapses between cells. ρ represents the priority order between evolutionary rules. i_{out} represents the subscript of the output cell that store the result. Each cell is of the form

$$\sigma_k = (w_{k,0}, R_k), for 1 \leqslant k \leqslant m \tag{2}$$

In the second formula, $w_{k,0}$ represents the original objects in the cell k. The $w_{k,0} = \lambda$ represents the objects that is deleted in the k cell. The R_k represents the rules in the k cell. The rule is executed under circumstance of promoters and is stopped under circumstance of inhibitors $\neg\beta$. New objects generated in the rules are transmitted through the synapses. The $r_k p$ represents the subrule in the cell k and the symbol $\cup$ connect the same subset.

3   The TPPIIFP-growth Algorithm

3.1   Rules

Suppose the transactional database includes D transactions and t fields. The object a_{ij} refers to the ith transaction and the jth item I_j in the ith transaction. The cell structure that is used to run the TPPIIFP-growth algorithm including cell 1, $\cdots$, 6 is shown Fig. 1.

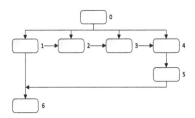

Fig. 1. Cell structure of the TPPIIFP-growth algorithm

The tissue-like P system with promoters and inhibitors for TPPIIFP-growth is as follow:

$$\Pi_{IFP-growth} = (o, \sigma_0, \cdots, \sigma_6, syn, \rho, i_{out}) \tag{3}$$

In the third formula, $o = \{a_{ij}, \beta_j, \beta_{j_1 j_2}, \delta_j, \delta_{j_1 j_2}, \eta_j, \gamma_j, \mu, \nu_j, D^{U[j]}, \alpha_j, \alpha_{j_1 j_2}, \cdots, \alpha_{j_1 \cdots j_t}\}$; $syn = \{\{0, 1\}, \{0, 2\}, \{0, 3\}, \{0, 4\}; \{1, 6\}, \{5, 6\}; \{1, 2\}, \{2,3\}, \{3,4\}, \{4, 5\}\}$; $\sigma_0 = (w_{0,0}, R_0)$, $\sigma_1 = (w_{1,0}, R_1)$, $\sigma_2 = (w_{2,0}, R_2)$, $\sigma_3 = (w_{3,0}, R_3)$, $\sigma_4 = (w_{4,0}, R_4)$, $\sigma_5 = (w_{5,0}, R_5)$, $\sigma_6 = (w_{6,0}, R_6)$; $\rho = \{r_i > r_j - i < j\}$; $i_{out} = 6$.

In $\sigma_0 = (w_{0,0}, R_0)$, $w_{0,0} = \lambda$ and
R_0:
$r_{01} = \{a_{ij} \to a_{ij,go}\} \cup \{\theta^k \to \theta^k_{,go}\} \cup \{\eta^{k'} \to \eta^{k'}_{,go}\} \cup \{\mu \to \mu_{,go}\}$
for $1 \leqslant i \leqslant D$ and $1 \leqslant j \leqslant t$

In $\sigma_1 = (w_{1,0}, R_1)$, $w_{1,0} = \lambda$ and
R_1:
$r_{11} = \{\theta^k \to \beta^k_1, \ldots, \beta^k_t\}$
$r_{12} = \{\delta^p_j \beta^{k-p}_j \to \lambda\} \cup \{(\delta^k_j)_{\neg \beta_j} \to \alpha_{j,go}\}$
for $1 \leqslant j \leqslant t$ and $1 \leqslant p \leqslant k$
$r_{13} = \{a_{ij}\beta_j \to \delta_j\}$
for $1 \leqslant i \leqslant D$ and $1 \leqslant j \leqslant t$

In $\sigma_2 = (w_{2,0}, R_2)$, $w_{2,0} = \lambda$ and
R_2:
$r_{21} = \{()_{\alpha_j \eta^{k'}} \to \gamma^{k'}_j\}$
for $1 \leqslant j \leqslant t$ and $k' = k+1, k+2, \ldots, k+n$
$r_{22} = \{\alpha_j \to \lambda\}$
for $1 \leqslant j \leqslant t$
$r_{23} = \{()_{\alpha_j a_{ij}} \to \lambda\}$
for $1 \leqslant i \leqslant D$ and $1 \leqslant j \leqslant t$
$r_{24} = \{\alpha_j \to \alpha_{j,go}\}$
for $1 \leqslant j \leqslant t$
$r_{25} = \{b^{k'}_j \to \lambda\} \cup \{b^{p'}_j \gamma^{k'-p'}_j \to \alpha_j\}$
for $1 \leqslant i \leqslant D, 1 \leqslant j \leqslant t, k' = k+1, k+2, \ldots, k+n$ and $1 \leqslant p' \leqslant k'$
$r_{26} = \{()_{a_{ij}\gamma_j} \to b_j\}$
for $1 \leqslant i \leqslant D$ and $1 \leqslant j \leqslant t$

In $\sigma_3 = (w_{3,0}, R_3)$, $w_{3,0} = \lambda$ and
R_3:
$r_{31} = \{()_{\alpha_{j_1}\alpha_{j_2}\theta^k \neg \beta^k_{j_1 j_2}} \to \beta^k_{j_1 j_2}\}$
for $1 \leqslant j_1 < j_2 \leqslant t$
$r_{32} = \{\delta^p_{j_1 j_2}\beta^{k-p}_{j_1 j_2} \to \lambda\} \cup \{(\delta^k_{j_1 j_2})_{\neg \beta_{j_1 j_2}} \to \alpha_{j_1 j_2,go}\}$
for $1 \leqslant j_1 < j_2 \leqslant t$ and $1 \leqslant p \leqslant k$
$r_{33} = \{(\beta_{j_1 j_2})_{a_{ij_1} a_{ij_2}} \to \delta_{j_1 j_2}\}$
for $1 \leqslant i \leqslant D$ and $1 \leqslant j_1 < j_2 \leqslant t$
$r_{34} = \{\alpha_j \to \alpha_{j,go}\}$
for $1 \leqslant j \leqslant t$

In $\sigma_4 = (w_{4,0}, R_4)$, $w_{4,0} = \lambda$ and
R_4:
$r_{41} = \{(\alpha_{j_1 j_2})a_{ij_1}a_{ij_2} \rightarrow a_{j_1 j_2}\}$
for $1 \leqslant i \leqslant D$ and $1 \leqslant j_1 < j_2 \leqslant t$
$r_{42} = \{a_{j_1 j_2}U_{j_1}[j_2] \rightarrow U_{j_1}[j_2]\}$
for $1 \leqslant j_1 < j_2 \leqslant t$
$r_{43} = \{\alpha_j U[j] \rightarrow U[j]\}$
for $1 \leqslant j \leqslant t$
$r_{44} = \{U_{j_1}[j_2]U[j] \rightarrow D_{,go}^{U[j]}\}$
for $1 \leqslant j \leqslant t$ and $1 \leqslant j_1 < j_2 \leqslant t$
$r_{45} = \{\mu \rightarrow \mu_{,go}\} \cup \{\theta^k \rightarrow \theta_{,go}^k\}$
In $\sigma_5 = (w_{5,0}, R_4)$, $w_{5,0} = \lambda$ and
R_5:
$r_{51} = \{\mu \rightarrow \nu_1 \cdots \nu_t\}$
$r_{52} = \{D^{U[j]}\nu_j \rightarrow a_{jj_1 j_2 \cdots j_t}\}$
for $1 \leqslant j \leqslant t$ and $1 \leqslant j_1 < j_2 < \cdots < j_t \leqslant t$
$r_{53} = \{()_{a_{jj_1 j_2 \cdots j_t}\theta^k}\neg\beta_{jj_1 j_2 \cdots j_t}^k \rightarrow \beta_{jj_1 j_2 \cdots j_t}^k\}$
for $1 \leqslant j \leqslant t$ and $1 \leqslant j_1 < j_2 < \cdots < j_t \leqslant t$
$r_{54} = \{\delta_{jj_1 j_2 \cdots j_t}^p \beta_{jj_1 j_2 \cdots j_t}^{k-p} \rightarrow \lambda\} \cup \{(\delta_{jj_1 j_2 \cdots j_t}^k)\neg\beta_{jj_1 j_2 \cdots j_t} \rightarrow \alpha_{jj_1 j_2 \cdots j_t,go}\}$
for $1 \leqslant j \leqslant t$ and $1 \leqslant j_1 < j_2 < \cdots < j_t \leqslant t$
$r_{55} = \{(\beta_{jj_1 j_2 \cdots j_t})a_{jj_1 j_2 \cdots j_t} \rightarrow \delta_{jj_1 j_2 \cdots j_t}\}$
for $1 \leqslant j \leqslant t$ and $1 \leqslant j_1 < j_2 < \cdots < j_t \leqslant t$
In $\sigma_6 = (w_{5,0}, R_4)$, $w_{5,0} = \lambda$ and
$R_6 = \varnothing$

3.2 Computing Process

Input. The input cell is cell 0. In the transactional database, the data is encoded into objects a_{ij}. And objects θ^k, $\eta^{k'}$ and μ are used to activate the computation, where k represents the support count threshold and k' is a integer that is greater than k. We send objects a_{ij}, θ^k, $\eta^{k'}$ and μ into cell 1, 2, 3, 4 by executing rule r_{01}.

In the cell 1, frequent 1-itemsets are generated. Rule r_{11} is executed to turn objects θ^k into β_j^k for $1 \leqslant j \leqslant t$, where β_j^k are the auxiliary objects to activate the next rule execution. Rule r_{13} is executed to find all frequent 1-itemsets by using the internal flat maximally parallel mechanism in the P system. We take the detection process of the frequent 1-itemset $\{I_1\}$ as an example. If object a_{i1} is the ith record in the item, the subrule $\{a_{i1}\beta_1 \rightarrow \delta_1\}$ can be executed to generate δ_1 in the cell 1. There are k copies of β in the cell 1. So, the first item needs to appear in at least k records for the itemset $\{I_1\}$ to be a frequent 1-itemset. And at most k subrules of the form $\{a_{i1}\beta_1 \rightarrow \delta_1\}$ can be executed. The detection process is executed until all objects a_{i1} or k copies of β_1 have been consumed. If there are surplus β_1 that haven't been consumed, p copies of δ_1 are generated for $1 \leqslant p \leqslant k$. Therefore, objects a_{i1} is not a frequent 1-itemset. The subrule $\{\delta_1^p\beta_1^{k-p} \rightarrow \lambda\}$ is executed to delete surplus $\delta_1^p\beta_1^{k-p}$ in the r_{12}. If all k copies of β_1 are consumed, k copies of δ_1 are generated. Therefore, objects a_{i1} have at least k copies and are frequent 1-itemset. The subrule $\{(\delta_1^k)_{\neg\beta_1} \to \alpha_1, go\}$ is executed to generate frequent 1-itemset and object α_1 is sent into cell 2 and 6 to activate the computation in cell 2. If no 1-itemset is a frequent 1-itemset, the computation halts. The detection process of other frequent itemsets is the same as frequent 1-itemset.

In the cell 2, we generate sorted frequent 1-itemsets. Rule r_{21} is executed to obtain all auxiliary objects $\gamma_j^{k'}$ that are used to obtain sorted frequent 1-itemsets, where k' is an integer that is greater than the support threshold k. The rule r_{22} is executed to delete all redundant frequent 1-itemsets α_j. Rule r_{23} and r_{25} are not executed because of missing object α_j and b_j in the cell 2. We take the sorted process of the smaller frequent 1-itemset as an example. Because the auxiliary objects $\gamma_j^{k'}$ are generated by frequent 1-itemsets. Therefore, if the objects a_{ij} are not frequent items, the rule r_{26} will not be executed. When object γ_j and frequent item a_{ij} appear both in the cell 2, the rule r_{26} will be executed to generate objects b_j. If the frequent items a_{ij} is k copies, objects γ_j will be surplus and k copies of b_j is generated. The subrule $\{b_j^{p'}\gamma^{k+1-p'} \to \alpha_j\}$ is executed to generate the smallest number of frequent 1-itemset in the rule r_{25}. Rule r_{23} is executed to delete objects a_{ij} that is the same as objects α_j. The pair of empty parentheses in the rule indicates that objects α_j will not be consumed when the rule is executed. Rule r_{24} is executed to send the smaller frequent 1-itemset into cell 3 and activate the computation in cell 3. If all objects γ_j are consumed, the number of frequent items a_{ij} is greater or equal to $k+1$. So, the subrule $\{b_j^{k+1} \to \lambda\}$ is executed to delete redundant objects b_j^{k+1}. The sorted process continues to be executed until no frequent items. The sorted process of other frequent 1-itemsets is the same as the smaller frequent 1-itemset. We obtain next frequent 1-itemset by $k' = k+2, \cdots$.

In the cell 3, we delete itemsets combined with each other that are less than the support threshold in order to construct the two-dimensional vector table. Rule r_{31} is executed to obtain all candidate two-dimensional vector table using the internal flat maximally parallel mechanism in the P system. The candidate two-dimensional vector table is the same as any two terms that intersect each other. Therefore, the detection process of the candidate two-dimensional vector table takes the itemset $\{I_1, I_2\}$ as an example. Multiple subrules working on objects with different subscripts form the rule r_{31}. If the cell 2 stores frequent 1-itemsets α_1 and α_2, the subrule $\{()_{\alpha_1\alpha_2\theta^k_{\neg\beta^k_{12}}} \to \beta_{12}^k\}$ is executed to generate β_{12}^k. So the itemset $\{I_1,I_2\}$ is candidate two-dimensional vector.

Rule r_{33} is executed to detect all two-dimensional vector that satisfy the support threshold. The detection process that items less than the support threshold are deleted takes the itemset $\{I_1, I_2\}$ as an example. Multiple subrules working on objects with different subscripts form the rule r_{33}. If the cell 3 stores objects a_{i1} and a_{i2}, the subrule $\{(\beta_{12})_{a_{i1}a_{i2}} \to \delta_{12}\}$ is executed to generate object δ_{12}. Because the k copies of β_{12} is in the cell 3, the subrule is executed k times. If all objects β_{12} are consumed to generate k copies of δ_{12}, the subrule $\{(\delta_{12}^k)_{\neg\beta_{12}} \to \alpha_{12}, go\}$ is executed to generate α_{12} that meets the support threshold in the rule r_{32}. Meanwhile, the itemsets that satisfy the support threshold are sent into cell 4. If all objects $a_{i1}a_{i2}$ are consumed and p copies of β_{12} are consumed to generate p copies of δ_{12} for $1 \leqslant p \leqslant k$, objects $a_{i1}a_{i2}$ don't meet the support threshold. Therefore, the subrule $\{\delta_{12}^{p}\beta_{12}^{k-p} \rightarrow \lambda\}$ is executed to delete those objects that don't meet the support threshold in the rule r_{12}. The detection process of two-dimensional vectors is the same as the itemset $\{I_1, I_2\}$.

Rule r_{34} is executed to send frequent 1-itemsets α_j into cell 4.

In the cell 4, we generate two-dimensional vector table and obtain the conditional pattern base. Rule r_{41} is executed to obtain objects $a_{j_1j_2}$ that satisfy the support threshold by using the internal flat maximally parallel mechanism in the P system. We take the detection process of object a_{12} as an example. If object α_{12} meets the support threshold and is in the cell 4, the subrules $\{(\alpha)_{a_{ij_1}a_{ij_2}} \rightarrow a_{12}\}$ can be executed to generate object a_{12}. All objects s_{12} will meet the support threshold. If object α_{12} doesn't satisfy the support threshold and isn't in the cell 4, the subrules can't be executed. The detection processes of objects a_{ij} are performed in the same way.

Rule r_{42} and rule r_{43} is executed to transform objects $a_{j_1j_2}$ and α_j into $U_{j_1}[j_2]$ and $U[j]$. Objects $U_{j_1}[j_2]$ and $U[j]$ can make container to transform the form of $a_{j_1j_2}$ and α_j. We take the conversion process of objects a_{12} and α_1 as an example. If objects a_{12} and α_1 is in the cell 4, Rule r_{42} $\{a_{12}U_{j_1}[j_2] \rightarrow U_1[2]\}$ and rule r_{43} $\{\alpha_1 U[j] \rightarrow U[1]\}$ is executed to generate objects $U_1[2]$ and $U[1]$.

Rule r_{44} is executed to classify $U_{j_1}[j_2]$ by $U[j]$. The rule r_{46} is actually composed of multiple subrules working on objects with different subscripts. We take the classification process of objects $U[1]$ and $U_{j_1}[1]$ as an example. If objects $U[1]$ and $U_{j_1}[1]$ are in the cell 4, the subrule $\{U_{j_1}[1]U[1] \rightarrow D_{,go}^{U[1]}\}$ is executed to classify $U_{j_1}[1]$ into $U[1]$ and to send the result into the cell 5. The classification processes of the other objects $U_{j_1}[j_2]$ are performed in the same way. Therefore, we obtain the conditional pattern base.

Rule r_{45} is executed to send objects μ and θ^k into cell 5.

In the cell 5, we obtain all frequent itemsets. Rule r_{51} is executed to generate ν_j for $1 \leqslant j \leqslant t$. The auxiliary objects ν_j represent the adhesive between objects. Rule r_{52} is executed by using the internal flat maximally parallel mechanism in the P system. We take the bonding process of the conditional pattern base $D^{[1]}$ as an example. If the conditional pattern base $D^{[1]}$ is in the cell 5, $D^{[1]}: D^{U_{j_1}[1]}$ is stored. Therefore, the subrule $\{D^{[1]}\nu_1 \rightarrow a_{1j_1...j_t}\}$ is executed to generate frequent itemsets $a_{1j_1...j_t}$ that represent any combination of object a_1 with other objects in the conditional pattern base. The bonding process of other conditional pattern bases is the same as the conditional pattern base $D^{[1]}$.

Rules r_{53}, r_{54}, r_{55}, r_{56} are executed to check all results that are generated by rule r_{52}. The 4 rules are similar to those in cells 1 and 3. First, we use rule r_{53} to generate all auxiliary objects $\beta_{jj_1j_2...j_t}^k$, that is, all candidate itemsets. Second, the redundant objects $a_{jj_1j_2...j_t}$ is deleted by rule r_{54}. Then, objects $a_{jj_1j_2...j_t}$ and $\beta_{jj_1j_2...j_t}$ consume each other to see who is completely consumed by rule r_{56}. Finally, rule r_{55} is executed to delete redundant objects or generate frequent itemsets.

All the results, that is to say, objects meeting the support threshold, are stored in the cell 6.

3.3   Time Complexity

The time complexity of TPPIIFP-growth algorithm in the worst case is analyzed. In cell, rule can use the flat maximally parallel. Firstly, 1 computational step in needed to send copies of a_{ij}, θ^k, $\eta^{k'}$ and μ into cells 1, 2, 3, 4.

The frequent 1-itemsets are generated to need 3 computational steps. First step generates candidate frequent 1-itemset. Second step finds the support threshold of frequent 1-itemsets. Third time sends frequent 1-itemsets into cells 2 and 6.

The frequent 1-itemsets are sorted to need $2 + 4n$ computational steps. First step generates auxiliary objects γ_j by frequent 1-itemsets. Second step deletes frequent 1-itemsets. The sorted processes need 4 computational steps, but they need to repeat n times. First step finds the number of objects a_{ij}. Second step generates corresponding frequent 1-itemsets. Third step deletes objects a_{ij} by the smaller frequent 1-itemsets. Forth step sends sorted frequent 1-itemsets into cell 3.

The two-dimensional vector table are generated to need 4 computational steps. First step generates candidate 2-itemsets. Second step finds the support threshold of 2-itemsets. Third step sends frequent 2-itemsets into cell 3. Forth step filters objects a_{ij} by frequent 2-itemsets to generate two-dimensional vector table.

The conditional pattern base is generated to need 3 computational steps. First step transforms the form of objects a_{ij}. Second step transforms the form of frequent 1-itemset. Third step generates the conditional pattern.

The frequent itemsets are generated to need 2 computational steps. First step generates the auxiliary objects ν_j. Second step generates frequent itemsts. The frequent l-itemsets are checks to need 4 computational steps. These steps are similar with generating frequent 1-itemsets' steps.

Therefore, the time complexity of TPPIIFP-growth algorithm is $1 + 3 + 2 + 4n + 4 + 3 + 2 + 4 = 19 + 4n$.

We compare the time complexity of TPPIIFP-growth algorithm with improved FP-growth [7] that the time complexity is $a*a*n(n-1)/2$. The time complexity of TPPIIFP-growth algorithm is batter.

4   An Illustrative Example

Table 1 shows the transaction database in the example. And the database includes 10 transactions and 7 fields. The support count threshold is k = 2. The computational processes are as follows.

Input The database is encoded as the objects a_{11}, a_{12}, a_{15}, a_{22}, a_{24}, a_{26}, a_{32}, a_{33}, a_{41}, a_{42}, a_{44}, a_{51}, a_{52}, a_{62}, a_{63}, a_{67}, a_{71}, a_{73}, a_{81}, a_{82}, a_{85}, a_{91}, a_{92}, a_{93}, a_{101}, a_{103} in a new form that the P system can recognize. The above objects

Table 1. The transactional database

TID	Items
T1	I_1, I_2, I_5
T2	I_2, I_4, I_6
T3	I_2, I_3
T4	I_1, I_2, I_4
T5	I_1, I_2
T6	I_2, I_3, I_7
T7	I_1, I_3
T8	I_1, I_2, I_5
T9	I_1, I_2, I_3
T10	I_1, I_3

a_{ij} and objects θ^2, $\eta^{k'}$ and μ are sent to the cell 0 and activate the calculation, where the objects θ^2 represent the support threshold k = 2 and the objects $\eta^{k'}$ represent an integer that is greater than k = 2. The rule r_{01} is executed to sent the all objects a_{ij} and objects θ^2, $\eta^{k'}$ and μ to the cells 1 $\cdots$ 4.

In the cell 1, we obtain the frequent 1-itemsets. Rule r_{11} is executed to generate the auxiliary objects $\beta_1^2 \cdots \beta_t^2$. If a item is a frequent item, it needs to appear in at least 2 records. Rule r_{13} is executed to find all frequent 1-itemsets in the flat maximally parallel. We take the detection process of the frequent 1-itemset $\{I_1\}$ as an example. According to the transaction database, the cell 1 stores the objects a_{11}, a_{41}, a_{51}, a_{71}, a_{81}, a_{91}, a_{101} that means the first, the forth, the fifth, the seventh, the eighth, the ninth, the tenth records contain I_1. Therefore, the subrules $\{a_{11}\beta_1 \rightarrow \delta_1\}$, $\{a_{41}\beta_1 \rightarrow \delta_1\}$, $\cdots$, $\{a_{101}\beta_1 \rightarrow \delta_1\}$ are executed. Meanwhile, 2 copies of β_1 are in the cell 1. So, the itemset $\{I_1\}$ needs to appear at least twice and is a frequent itemset. Every subrule is executed to consume a object β_1. Therefore, the itemset $\{I_1\}$ consumes 2 copies of β_1 to generate 2 copies of δ_1. Rule r_{12} can be executed to process the results obtained by rule r_{13}. Then, the subrule $\{(\delta_1^2)_{\neg\beta_1} \rightarrow \alpha_{1,go}\}$ is executed to send the object α_1 into cells 2 and 6 to demonstrate that the itemset $\{I_1\}$ is a frequent 1-itemset and to activate the computation in the cell 2. The detection process of other frequent itemsets $\{I_2\}$, $\{I_3\}$, $\{I_4\}$, $\{I_5\}$ is the same as the frequent itemset $\{I_1\}$. In the transaction database, items $\{I_6\}$ and $\{I_7\}$ only appear one time. So, they don't consume all objects β_6 and β_7 and aren't frequent 1-itemsets. The subrules $\{\delta_6^1\beta_6^1 \rightarrow \lambda\}$ and $\{\delta_7^1\beta_7^1 \rightarrow \lambda\}$ is executed to delete redundant objects.

In the cell 2, we obtain the sorted frequent 1-itemsets. Objects α_1, α_2, α_3, α_4, α_5 are in the cell 2. Rule r_{21} can be executed to generate the auxiliary objects $\gamma_1^{k'}$, $\gamma_2^{k'}, \gamma_3^{k'}, \gamma_4^{k'}, \gamma_5^{k'}$, where k' is a integer that is greater than k=2. Rule r_{22} is executed to delete the redundant objects α_j. We take the detection process of the smaller frequent 1-itemset as an example. When the k' is 3, the subrules $\{()a_{11}\gamma_1 \rightarrow b_1\}$, $\{()a_{41}\gamma_1 \rightarrow b_1\}$ and $\{()a_{51}\gamma_1 \rightarrow b_1\}$ are executed to consume all objects γ_1 and to generate 3 copies of b_1. So, the subrule $\{b_1^3 \to \lambda\}$ is executed in the rule r_{25}. Then, the subrules $\{()_{a_{12}\gamma_2} \to b_2\}$, $\{()_{a_{22}\gamma_2} \to b_2\}$ and $\{()_{a_{32}\gamma_2} \to b_2\}$ are executed to consume all objects γ_2 and to generate 3 copies of b_2. The subrule $\{b_2^3 \to \lambda\}$ is executed to delete redundant objects b_2. Meanwhile, all objects γ_3 are consumed to generate 3 copies of b_3 and to delete the objects b_3 in the rule r_{25}. The items $\{I_4\}$ and $\{I_5\}$ appear twice in the cell 2. So, the subrules are executed to consume 2 copies of objects γ_4 and γ_5 and to generate 2 copies of objects b_4 and b_5. Therefore, the subrules $\{b_4^2\gamma_4^1 \to \alpha_{4,go}\}$ and $\{b_5^2\gamma_5^1 \to \alpha_{5,go}\}$ are executed to generate the smaller frequent 1-itemsets $\{I_4\}$ and $\{I_5\}$. The sorted process of other frequent 1-itemsets is performed at the same way.

Next, frequent 1-itemsets α_4 and α_5 is in cell 2. Therefore, rule r_{23} can be executed to delete redundant a_{ij} that their frequent 1-itemsets have been selected. We take the α_4 as an example. The frequent 1-itemset α_4 is the smaller frequent 1-itemset. The subrules $\{()_{\alpha_4}a_{24} \to \lambda\}$ and $\{()_{\alpha_4}a_{44} \to \lambda\}$ are executed to delete objects a_{24}, a_{44}. The objects a_{ij} is updated in cell 2. The other screening processes are the same as frequent 1-itemset α_4. Rule r_{24} is executed to send frequent 1-itemsets α_j into cell 3 and activate the computation. Then, we execute rule r_{26} from $k' = k+2$, that is, k+2 copies of γ_j and the process is similar to the above. Finally, we obtain sorted frequent 1-itemset $\{I_4\}$, $\{I_5\}$, $\{I_3\}$, $\{I_1\}$, $\{I_2\}$ and they are sent into cell 3 and activate the computation in cell 3.

In the cell 3, we delete itemsets combined with each other that are less than the support threshold in order to construct the two-dimensional vector table. Rule r_{31} is executed to generate all candidate itemsets combined with each other. We take the detection process of the candidate itemset $\{I_1, I_2\}$ as an example. So, the subrule $\{()_{\alpha_1\alpha_2\theta^2\neg\beta_{12}^2} \to \beta_{12}^2\}$ is executed to generate 2 copies of objects β_{12}. Objects β_{12}^2 are in cell 3, which illustrates itemset $\{I_1, I_2\}$ is a candidate itemset and itemset $\{I_1, I_2\}$ must appear at least 2 times to make $\{I_1, I_2\}$ to satisfy the support threshold. The detection processes of other candidate itemset are the same as itemset $\{I_1, I_2\}$. Therefore, objects β_{12}^2, β_{13}^2, β_{14}^2, β_{15}^2, β_{23}^2, β_{24}^2, β_{25}^2, β_{34}^2, β_{35}^2, β_{45}^2 are generated in cell 3.

Rule r_{33} is executed to delete candidate itemsets that don't satisfy the support threshold. We take the detection process of itemsets $\{I_1, I_2\}$ as an example. The items $\{I_1\}$ and $\{I_2\}$ are stored in the first, the forth, the fifth, the eighth, the ninth records. The subrules $\{(\beta_{12})_{a_{11}a_{12}} \to \delta_{12}\}$ and $\{(\beta_{12})_{a_{41}a_{42}} \to \delta_{12}\}$ can be executed to consume all objects β_{12} and to generate 2 copies of object δ_{12}. Therefore, itemsets $\{I_1, I_2\}$ satisfy the support threshold. The subrule $\{(\delta_{12}^2)\neg\beta_{12} \to \alpha_{12,go}\}$ is executed and the itemset α_{12} is sent into cell 4. The detection processes of other itemsets are performed in the same way. Therefore, the objects δ_{12}^2, δ_{15}^2, δ_{25}^2, δ_{24}^2, δ_{23}^2, δ_{13}^2 are generated in cell 3. The objects α_{12}, α_{15}, α_{25}, α_{24}, α_{23}, α_{13} are generated and sent into cell 4 to activate the computation in cell 4. However, itemset $\{I_1I_4\}$ only appears once in the forth record. The subrule $\{(\beta_{14})_{a_{41}a_{44}} \to \delta_{14}\}$ is executed to generate a object δ_{14}. So, itemset $\{I_1I_4\}$ don't satisfy the support threshold. The subrule $\{\delta_{14}^1\beta_{14}^1 \to \lambda\}$ is executed to delete redundant objects in rule r_{32}.

Rule r_{34} is executed to send the objects α_j into cell 4.

In the cell 4, we obtain two-dimensional vector table and conditional pattern base. Rule r_{41} is executed to generate every frequent item a_{ij}. We take the detection process of object a_{12} as an example. Object α_{12} is a frequent 2-itemset and is in the cell 4. The subrules $\{(\alpha_{12})_{a_{11}a_{12}} \rightarrow a_{12}\}, \{(\alpha_{12})_{a_{41}a_{42}} \rightarrow a_{12}\}, \cdots, \{(\alpha_{12})_{a_{91}a_{92}} \rightarrow a_{12}\}$ is executed to generate 5 copies of a_{12}. The detection processes of other objects a_{ij} are performed in the same way. Therefore, 5 copies of a_{12}, 2 copies of a_{15}, a_{25} and a_{24} and 3 copies of a_{13} and a_{23} are in cell 4 and are frequent items to obtain two-dimensional vector table that is deleted item that don't satisfy the support threshold.

Rules r_{42} and r_{43} are executed to transform objects $a_{j_1j_2}$ and α_j into $U_{j_1}[j_2]$ and U[j]. We take the objects a_{12} and α_1 as an example. The subrules $\{a_{12}U_{j_1}[j_2] \rightarrow U_1[2]\}$ and $\{\alpha_1 U[j] \rightarrow U[1]\}$ are executed to generate objects $U_1[2]$ and U[1]. Objects 5 copies of $U_1[2]$, 2 copies of $U_1[5]$, $U_2[5]$, $U_2[4]$, 3 copies of $U_2[3]$, $U_1[3]$, U[5], U[4], U[3], U[1] and U[2] are in cell 4.

Rule r_{44} is executed to obtain the conditional pattern bases. We execute rule r_{44} by the order of frequent 1-itemsets α_j. The object U[5] is the smaller. So, We take the detection process of the conditional pattern base of U[5] as an example. The subrules $\{U_1[5]U[5] \rightarrow D^{U[5]}_{,go}\}$ and $\{U_2[5]U[5] \rightarrow D^{U[5]}_{,go}\}$ are executed to generate the conditional pattern base, U[5] = $\{U_1[5]: 2, U_2[5]: 2\}$. The detection processes of other conditional pattern bases are performed in the same way. Therefore, we obtain all conditional pattern bases, $D^{U[5]} = \{U_1[5]: 2, U_2[5]: 2\}$, $D^{U[4]} = \{U_2[4]: 2\}$, $D^{U[3]} = \{U_1[3]: 3, U_2[3]: 3\}$ and $D^{U[2]} = \{U_1[2]: 5\}$.

Rule r_{45} is executed to send auxiliary objects μ and θ^k into the cell 5.

In the cell 5, we obtain all frequent itemsets. Rule r_{51} is executed to generate all auxiliary objects ν_j for $1 \leqslant j \leqslant t$. Rule r_{52} is executed to obtain all itemsets by conditional pattern bases. We take the bonding process of conditional pattern base U[5] = $\{U_1[5]: 2, U_2[5]: 2\}$ as an example. The conditional pattern base The subrules $\{D^{U[5]}\nu_5 \rightarrow a_{15}\}$, $\{D^{U[5]}\nu_5 \rightarrow a_{25}\}$ and $\{D^{U[5]}\nu_5 \rightarrow a_{125}\}$ is executed to generate 2 copies of objects a_{15}, a_{25} and a_{125}. The bonding processes of other conditional pattern bases are performed in the same way. So, 2 copies of objects a_{15}, a_{25} and a_{125}, 4 copies of objects a_{24}, 3 copies of objects a_{13}, a_{23} and object a_{123} and 5 copies of objects a_{12} are generated in cell 5.

Rule r_{53}, r_{54}, r_{55} and r_{56} are executed to check all objects and to see who doesn't satisfy the support threshold. These rules are similar to those in cells 1 and 3. Therefore, we obtain frequent itemsets $\alpha 15$, $\alpha 25$, $\alpha 125$, $\alpha 24$, $\alpha 13$, $\alpha 23$ and they are sent into cell 6 to be stored.

Output All frequent itemsets are stored in the cell 6.

5   Conclusions

This paper presents the TPPIIFP-growth algorithm, to mine frequent itemsets. The TPPIIFP-growth algorithm uses the two-dimensional vector table to reduce a scan of the conditional pattern base and applies to the tissue-like P system with promoters and inhibitors. This method reduces the time complexity to mine frequent itemsets. Although, time complexity of the TPPIIFP-growth algorithm isn't compared with a lot of algorithms. For larger databases, using the flat maximally parallel and saving a scan can improve a lot of efficiency. This result also gives other similar algorithms a suggestion that parallel mechanisms can be used to improve algorithmic efficiency.

Acknowledgments. Project is supported by National Natural Science Foundation of China (61472231, 61502283, 61876101, 61802234, 61806114), Ministry of Eduction of Humanities and Social Science Research Project, China (12YJA630152), Social Science Fund Project of Shandong Province, China (16BGLJ06, 11CGLJ22), China Postdoctoral Project (2017M612339).

References

1. Agarwal, S.: Data mining: data mining concepts and techniques. In: International Conference on Machine Intelligence and Research Advancement. IEEE, pp. 203–207 (2014)
2. Han, J.: Data Mining: Concepts and Techniques. Morgan Kaufmann Publishers Inc., Waltham (2005)
3. Berry, M.J., Linoff, G.: Data Mining Techniques: For Marketing, Sales, and Customer Support. Wiley, Indianapolis (1997)
4. Grahne, G., Zhu, J.: Fast algorithms for frequent itemset mining using FP-trees. IEEE Trans. Knowl. Data Eng. **1710**, 1347–1362 (2005)
5. Borgelt, C.: An implementation of the FP-growth algorithm. In: Proceedings of the 1st International Workshop on Open Source Data Mining: Frequent Pattern Mining Implementations, pp. 1–5 (2005)
6. Zhang, D., et al.: Pfp: parallel fp-growth for query recommendation. In: ACM Conference on Recommender Systems, pp. 107–114. ACM (2008)
7. Yang, Y., Luo, Y.: Improved alogrithm based on FP-Growth. Comput. Eng. Des. **31**(7), 1506–1509 (2010)
8. Agrawal, R., Srikant, R.: Fast Algorithms for Mining Association Rules in Large Databases. In: International Conference on Very Large Data Bases, pp. 487–499. Morgan Kaufmann Publishers Inc., Burlington (1994)
9. Păun, G.: Computing with Membranes. J. Comput. Syst. Sci. **61**(1), 108–143 (2000)
10. Păun, G.: On the power of membrane division in P systems. Theoret. Comput. Sci. **324**(1), 61–85 (2004)
11. Bottoni, P., et al.: Membrane systems with promoters/inhibitors. Acta Informatica **38**(10), 695–720 (2002)
12. Pan, L., Song, B.: Flat maximal parallelism in P systems with promoters. Elsevier Science Publishers Ltd., Amsterdam (2016)
13. Song, B., Pan, L., Prez-Jimnez, M.J.: Tissue P systems with protein on cells. Fundamenta Informaticae **144**(1), 77–107 (2016)
14. Liu, X., Zhao, Y., Sunb, M.: An Improved apriori algorithm based on an evolution-communication tissue-like P system with promoters and inhibitors. Discrete Dyn. Nat. Soc. **2017**(1), 1–11 (2017)
15. Martn-Vide, C., et al.: Tissue P systems. Theoret. Comput. Sci. **296**(2), 295–326 (2003)

Veca: A High-Performance Consensus Algorithm for State Machine Replication

Xiangguang Yan[1,2] and Yun Wang[1,2(✉)]

[1] Southeast University, Nanjing, China
ywang_cse@seu.edu.cn
[2] Key Lab of Computer Network and Information Integration, MOE, Nanjing, China

Abstract. Consensus algorithm is fundamental for distributed systems. It requires high throughput, low latency and high availability. However, to the best of our knowledge, these requirements cannot be satisfied well at the same time in any existing consensus algorithm. In this paper, we propose Veca, a consensus algorithm for state machine replication that tries to satisfy the requirements at the same time as well as possible. Veca is a leaderless consensus algorithm for which all replicas can commit commands concurrently at any time, and each command can be committed after just one round of communication with a majority of replicas in the normal case. Veca separates agreement from ordering and execution, which allows all replicas to commit commands concurrently without determining their order, but to track their dependencies using vector clocks. Then a subsequent replay phase assigns an order to the commands and executes them in that order. Commands are committed out of order and then be executed in the same order by all replicas. A replica can take the initiative to learn the decision for an instance using a failure recovery protocol. The leaderless design makes the systems built with Veca provide continuous service as long as more than half of the replicas are available. We show that Veca has higher throughput, lower latency and higher availability than several typical consensus algorithms. The correctness of Veca has been proved theoretically and its advantages are demonstrated through an experimental evaluation in LAN and WAN.

Keywords: Consensus algorithm · State machine replication · Leaderless

1 Introduction

Distributed consensus is essential to building high available systems, which allows a collection of machines to work as a coherent group that can tolerate the failures of some of its members. Paxos [1,2] and Raft [3] consensus algorithm are widely adopted in modern large distributed systems such as MegaStone, Spanner, CockroachDB, OceanBase, and TiDB for data replication and fault tolerance. Many large-scale distributed systems like GFS, HDFS, and RAMCloud typically use a replicated state machine [4] such as Chubby, Boxwood, and ZooKeeper

© Springer Nature Switzerland AG 2019
Y. Tang et al. (Eds.): HCC 2018, LNCS 11354, pp. 452–463, 2019.
https://doi.org/10.1007/978-3-030-15127-0_45

for activities including operation sequencing, coordination, leader election, and resource discovery.

Distributed systems place three main demands on consensus algorithms: (1) High throughput for replication inside a distributed system; (2) Low latency for replication across data centers; (3) High availability for services.

The typical consensus algorithms or replication protocols like Multi-Paxos, Raft, Zab, and Viewstamped Replication have a common limitation that all clients communicate with a single leader server at all times when it is available. When the leader fails, additional consensus mechanisms are required to do leader election. Leader leases [5] are usually used as a failure detector for other replicas to discover a failed leader in time.

However, leader-based consensus algorithms still exist several problems. For example, leader bears higher load and may easily become a bottleneck of the distributed system. To solve these problems, MegaStone, Spanner, CockroachDB, OceanBase and TiDB partition data into multiple overlapping consensus groups. Mencius [6] shares the leader load by distributing the leader responsibilities round-robin among the replicas.

Egalitarian Paxos (EPaxos) [7] abandons the leader and it exploits commutativity in state machine commands. However, if concurrently proposed commands interfere with each other, EPaxos requires an additional round of communication. EPaxos can reach high availability but cannot achieve low latency and high throughput due to command interference.

To achieve low latency, high throughput and high availability at the same time as well as possible, we design a consensus algorithm Veca for state machine replication. Veca requires no particular leader replica. Instead, all replicas can commit commands concurrently at any time, and each command can be committed after just one round of communication with a majority of replicas in the normal case. Veca separates agreement from ordering and execution, which allows all replicas to commit commands concurrently without determining their order, but to track their dependencies using vector clocks. Then a subsequent replay phase assigns an order to the commands and executes them in that order. Commands are committed out of order and then be executed in the same order by all replicas. A replica can take the initiative to learn the decision for an instance using a failure recovery protocol. The leaderless design makes the systems built with Veca provide continuous service as long as more than half of the replicas are available. Veca has several advantages of load balancing, high availability and high performance.

The remainder of this paper is organized as follows. Section 2 provides Paxos background. Section 3 demonstrates the intuition of Veca. In Sect. 4, we present the detail design of Veca consensus algorithm. Section 5 implements and evaluates Veca.

2 Paxos Background

Paxos [1,2] is the most famous consensus algorithm, it decides a value with two phases. When receiving a command from a client, a replica will try to become the leader of a new instance by creating a *proposal* identified with an incremental *proposalid* and sending *Prepare* messages to a *quorum* of *acceptors* (possibly including itself). If the proposal id is higher than any previously received proposal, the acceptor replies a *Promise* message to ignore all future proposals with a less proposal id. If the acceptor accepted a proposal at some point in the past, it must include the previous proposal in its response to the proposer. If the proposer receives enough *Promise* messages from a quorum of acceptors, it successfully becomes a temporary leader of this instance and can run the second phase. If any acceptor had previously accepted any proposal, the proposer must set the value of its proposal to the value associated with the highest proposal id reported by the acceptors. If none of the acceptors had accepted a proposal up to this point, the proposer could choose any value for its proposal. Then the proposer sends *Propose* messages to a quorum of acceptors with the chosen value for its proposal. If an acceptor receives a *Propose* message, it must accept it if and only if it has not already promised to any prepared proposal with a greater proposal id. In this case, it should accept the corresponding value and send an *Accept* message to the proposer. When the proposer receives enough Accept messages from a quorum of acceptors, it commits the command locally, and asynchronously notifies all other replicas and the client.

3   Veca Intuition

Veca is a consensus algorithm for state machine replication in which all replicas in the replicated state machine run their own instances of Paxos concurrently and independently. The commands produced by the instances led by all replicas are committed without determining their order, but to track their dependencies using vector clocks [8]. Then a subsequent replay phase assigns an order that guarantees *serializability* and *linearizability* to the commands and executes them in that order.

An instance in Veca runs roughly as follows. Every replica maintains a vector clock. When a replica starts a new instance, it skips the prepare phase in Paxos and runs the accept phase directly. The replica increases its vector clock and sends *Propose* messages to at least a majority of replicas with its vector clock. When a replica receives a *Propose* message, it updates its vector clock with the received vector clock and replies an *Accept* message to the proposer with its vector clock. When the proposer receives an *Accept* message, it updates its vector clock with the received vector clock. When the proposer receives *Accept* messages from a majority of replicas, it commits the instance with its vector clock and sends *Commit* messages to all other replicas. When a replica receives a *Commit* message, it updates its vector clock with the received vector clock and commits the instance. When a replica commits an instance, it starts a replay phase to assign an order to the command produced by that instance and relevant commands according to their dependencies tracked by vector clocks and finally executes them in that order.

Figure 1 presents a simple example of how Veca works. A replicated state machine consists of replica A, B and C. Replica A and C are running two

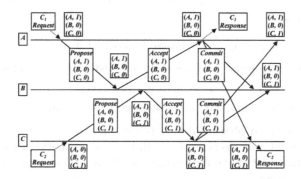

Fig. 1. Veca message flow in a simple example.

instances concurrently. Command C_1 did not discover command C_2, but C_2 discovered C_1, which makes C_2 depend on C_1. Thus, C_1 can be executed as soon as committed, but $C - 2$ must wait for the commit of C_1 and be executed after C_1.

4   Design

4.1   Assumptions

Non-Byzantine Failure. A replica may crash, or it may fail to respond to messages from other replicas indefinitely, but it cannot respond in a way that does not conform to the protocol.

Asynchronous Distributed System. Replicas are connected by a network. The network may fail to deliver messages, delay them, duplicate them, or deliver them out of order.

Totally Ordered Replicas. Each replica has a unique identifier, and all identifiers can be totally ordered.

Majority Requirement. At least a majority of replicas are available.

4.2   Design Goals

Load Balance. No leader, all replicas can commit commands concurrently at any time.

Low Latency. Each command can be committed after one round of communication with a majority of replicas.

High Throughput. A lot of requests can be handled concurrently and efficiently.

High Availability. Systems can provide continuous service as long as more than half of replicas are available.

4.3   Guarantees

Non-triviality. Only proposed commands can be learned.

Durability. Once a command has been committed, it will remain so at any later time.

Safety. At most one command can be learned in an instance.

Liveness (with High Probability). If a command has been proposed in an instance, then eventually every replica will learn a command in the instance, as long as more than half of the replicas are available and messages eventually go through before timeout.

Consistency. All committed commands will be executed in the same order by all replicas.

Linearizability. If committed commands are serialized by clients, they will be executed in the serialized order. Furthermore, The correctness of Veca is provable.

4.4   The Veca Consensus Algorithm

Veca is an algorithm for managing the replication, ordering and execution of commands inside a replicated state machine. The replicated state machine comprises $N = 2F + 1$ replicas, where F is the maximum number of tolerated failures. For every replica R there is an unbounded sequence of numbered instances $(R.id, 1)$, $(R.id, 2)$, $(R.id, 3)$, ..., $(R.id, i)$, ... that replica R owns, where $R.id$ is the identifier of replica R, and i is the incremental clock of replica R. The state of each replica includes all instances owned by every replica in the system. At most one command will be chosen in an instance. The order of instances is not pre-determined, instead, it is determined dynamically by the algorithm, as commands are chosen.

Veca comprises (1) the commit protocol for replicas to commit commands concurrently without determining their order but to track their dependencies; (2) the replay algorithm for replicas to assign an order to the commands according to their dependencies and to execute them in that order; and (3) the failure recovery protocol for replicas to take the initiative to learn the decision for an instance.

To access the service of a replicated state machine, a client sends a *Request* message to a replica of its choice. A *Response* message from that replica will notify the client the execution result of the command. If a client time out waiting the *Response* message after sending a *Request* message, it resends the *Request* message to another replica. When a replica receives a Request message from a client, it runs the commit protocol to commit a command. Then, the replica calls the replay algorithm to assign an order to the command and relevant commands and executes them in that order. Finally, the replica gets the execution result and replies to the client. If a replica times out waiting for the commit of a relevant command, it will run the failure recovery protocol to learn it.

The Commit Protocol. The commit protocol is for replicas to commit commands concurrently without determining their order, but to track their dependencies using vector clocks. Due to the page limitation, the pseudocode of these protocols are not listed here.

When a replica receives a *Request* message from a client, it becomes the leader of a new instance. The replica increases its vector clock. Then, it initializes a new command in a new instance and saves it in the instance. Finally, the replica sends *Propose* messages to all other replicas.

When a replica receives a *Propose* message, if the ballot number of the received command is smaller than the previously received largest one in the instance, the *Propose* message will be ignored. Then, the replica updates its own vector clock with the vector clock of the received command. Finally, the replica saves the received command and replies an *Accept* message to the proposer.

When a replica receives an *Accept* message, the replica updates its own vector clock with the received vector clock. Then, if the replica receives at least $\lfloor N/2 \rfloor$ *Accept* messages of the command, it updates the vector clock of the command, marks the state of the command as committed, and sends *Commit* messages to all other replicas. Finally, the replica calls the replay algorithm to execute commands, waits the command executed and gets the execution result, and replies a *Response* message to the client. It is a deterministic mechanism for replicas to start replaying commands from the same command proposed by the replica with the smallest identifier.

When a replica receives a *Commit* message, the replica updates its own vector clock with the vector clock of the received command. Then, the replica saves the received command. Finally, the replica calls the replay algorithm to execute commands.

As in Paxos, every message contains a *ballotnumber* to indicate message freshness. The difference is that the ballot number in Paxos is global, but in Veca, every instance has its own independent ballot number. Replicas disregard messages with a ballot number that is smaller than the largest they have seen for a certain instance. For correctness, ballot numbers used by different replicas must be distinct, so they include a replica identifier. A replica increases only the incremental number of the ballot number when trying to initiate a new ballot. Each replica is the default leader of its own instances, so the ballot number is initialized to the replica identifier at the beginning of every instance.

The Replay Algorithm. The replay algorithm is for replicas to assign an order to the unordered commands according to their dependencies and to execute them in that order. An instance starts at the *proposedvectorclock* and ends at the *committedvectorclock*. All concurrent instances during this period are tracked in the proposed vector clock and the committed vector clock.

Figure 2 puts the concurrent instances during the running of instance (R, S) in timeline. The proposed vector clock of instance (R, S) is $[(R_0, S_0), (R_1, S_1), \ldots, (R, S), \ldots, (R_N - 1, S_N - 1)]$, and the committed vector clock of instance (R, S) is $[(R_0, T_0), (R_1, T_1), \ldots, (R, T), \ldots, (R_N - 1, T_N - 1)]$. Note that

S and T do not necessarily have the same value, because a replica can start the next instance before the last one completed and the next instance may run faster than the first one. The two vector clocks do not only record the logical start and end time of instance (R, S), but also record the concurrent instances during the running of instance (R, S). The commands produced by those concurrent instances must be reordered according to their dependencies.

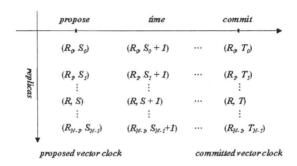

Fig. 2. The concurrent instances during the running of instance (R, S) in timeline.

Figure 3 shows the directed acyclic graph (DAG) of the dependencies between command $C(R, S)$ and its concurrent commands. Every command has the same dependencies with its concurrent commands. If we put all commands together, we can get a complete dependency graph. The self dependencies are monotonic, and they cannot form a cycle. However, the concurrent dependencies are not monotonic, and they can form a cycle. As a result, the complete dependency graph is no longer a DAG but a directed cyclic graph (DCG). Fortunately, the concurrent dependencies can only happen between the commands produced by the instances owned by different replicas. Thus, the cycles in the complete dependency graph can be simply broken using the order of the replicas.

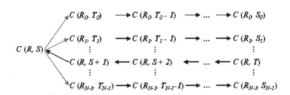

Fig. 3. The directed acyclic graph (DAG) of the dependencies between command C(R, S) and its concurrent commands. The dark arrows signify self dependencies and the gray arrows signify concurrent dependencies.

The replay algorithm is a topological sorting of a DCG. We extend the depth-first search algorithm for topological sorting. When a cycle in the DCG is detected, it is simply broken by the ascending order of the replicas. To execute a command, the replay algorithm first saves the state of the command and marks the state of the command as replaying. The first for loop recursively replays the commands that are in self dependency with, then the next for loop recursively replays the commands that are in concurrent dependency with in ascending order of replicas. Any failure will roll back the state of the replaying command and return false immediately during this period. Finally, if there is no error and the command has been committed, the command will be executed and its state will be marked as replayed.

The replay algorithm can fail to execute a command in the following cases: (1) the command has not been committed; (2) the command depends an uncommitted or missing command and the dependency cannot be broken. If the dependency can be broken, the command can be executed without waiting the dependent command committed. If a replica times out waiting for the commit of a command, the replica will take the initiative to recover the instance of the command with the failure recovery protocol.

The Failure Recovery Protocol. The failure recovery protocol is for replicas to take the initiative to learn the decision for an instance. If a replica times out waiting for the commit of an instance, the replica will try to take ownership of that instance by running the failure recovery protocol, at the end of which the replica will either learn what command was proposed in this instance then finalize committing it, or, if no replica has seen a command, will commit a no-op command to finalize the instance.

When a replica is going to recover an instance of a potentially failed replica, the replica increases the incremental number of the ballot number of this instance. Then, the replica concatenates the increased number and the identifier of the replica to generate a new ballot number. Finally, the replica sends a *Prepare* message to all replicas (including itself) with the new ballot number.

When a replica receives a *Prepare* message, and if the ballot number in the *Prepare* message is not larger than the previously received largest ballot number of the instance, the replica ignores the message. Then, the replica updates the ballot number of the instance with the received ballot number. Finally, the replica replies a *Promise* message to promise to ignore all future *Prepare* messages with a less or equal ballot number and *Propose* messages with a less ballot number in the instance.

When a replica receives a Promise message, and if the replica receives at least $\lfloor N/2 \rfloor + 1$ *Promise* messages of the instance, there are four situations according to the set of replies with the highest ballot number: (1) if the set of replies contains a command whose state is committed, which indicates the instance is committed, then the replica just saves the command and sends *Commit* messages to all other replicas; (2) if the set of replies contains at least $\lfloor N/2 \rfloor + 1$ commands whose states are accepted, which indicates the instance could be committed, then the replica updates the vector clock of the command with all the vector clocks of accepted commands, marks the state of the command as committed, and sends *Commit* messages to all other replicas; (3) if the set of replies contains a command whose state is accepted, which indicates the command has not yet been replicated to a majority of replicas, then the replica saves the command locally and sends *Propose* messages to all other replicas to start the commit protocol; (4) if the set of replies is not in any situation above, the replica initializes a no-op command and sends *Propose* messages to all other replicas to start the commit protocol.

Veca allows multiple replicas to run the failure recovery protocol concurrently for the same instance. As in Paxos, if multiple replicas propose commands at almost the same time for the same instance, there may be no command accepted by a majority of replicas, then the instance will fail again. When this happens, each replica will time out again and restart the failure recovery protocol with a larger ballot number. However, without extra measures there can be a livelock in which failed failure recovery repeats indefinitely. Like the leader election in Raft, Veca uses randomized timeouts to ensure that failed failure recovery is rare and that they are resolved quickly.

5   Evaluation

We evaluated Veca on LAN and WAN using three replicas (tolerating one failure) and five replicas (tolerating two failures). Replicas in LAN are located in our laboratory, using Gigabit Ethernet, running Windows 10. Replicas in WAN are located in Amazon EC2 datacenters in California (CA), Virginia (VA) and Ireland (EU), plus Oregon (OR) and Japan (JP) for the five-replica experiment, running Amazon Linux AMI 2017.03. Veca, EPaxos and Multi-Paxos are implemented with the optimization of pipelining while Raft and Mencius are not allowed this optimization.

5.1   Latency in WAN

We validate that Veca has low latency in WAN using three replicas. In the latency experiment, at each server there are also ten clients co-located with each replica. They generate requests simultaneously, and measure the latency for each request. In Veca, Mencius and EPaxos, clients send requests to their local replicas, while in Multi-Paxos and Raft, clients send requests to the leader. Figure 4 shows the average and 99% ile latency for Veca, EPaxos, Mencius, Multi-Paxos and Raft in WAN using three replicas.

Veca has lower latency than EPaxos. Mencius performs relatively well in the balanced experiment that all replicas receive request messages at the same aggregate rate. However, Mencius experiences latency corresponding to the round trip time to the replica that is farthest away from the client, which brings additional latency than Veca. Multi-Paxos and Raft have high latency in non-leader replicas. Multi-Paxos can be implemented with the optimization of pipelining while Raft cannot. This makes Multi-Paxos have lower latency than Raft.

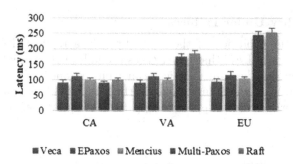

Fig. 4. Average latency (99% ile indicated by lines atop the bars) at each 3 replicas in WAN. The Multi-Paxos and Raft leader is in CA.

5.2   Throughput in LAN

Veca has been evaluated to have higher throughput than EPaxos, Mencius, Multi-Paxos and Raft. A client on a separate server sends requests in an open loop, and measures the rate at which it receives replies. For Veca, EPaxos and Mencius, the client sends each request to a replica chosen uniformly at random. Figure 5 shows the throughput for Veca, EPaxos, Mencius, Multi-Paxos and Raft in LAN using three replicas and five replicas.

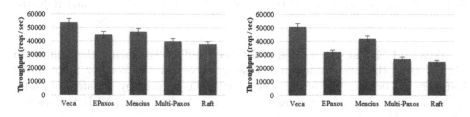

Fig. 5. Throughput in LAN for 3 replicas (left) and 5 replicas (right) (with 95% CI).

Veca has higher throughput than EPaxos, the improvements are even bigger using five replicas. Mencius cannot compare with Veca in this experiment. Because Mencius introduces a lot of overhead on synchronizing and coordinating the pre-partitioned instances, and the influences are even bigger under high concurrency. Besides, the leader rotation for every instance makes the replicated state machine run at the speed of the slowest replica, which also reduces throughput.

Multi-Paxos and Raft have low throughput because the leader becomes bottlenecked by its CPU and network. Multi-Paxos and Raft cannot achieve load balance among replicas since the leader must process more messages than other replicas.

5.3   Availability Under Failures

Figure 6 shows the evolution of the throughput in a replicated state machine using three replicas in LAN that experiences the failure of one replica. For Veca, EPaxos and Mencius, the failure replica is an arbitrary one in the system, and for Multi-Paxos and Raft, the failure replica is the leader. A client sends requests at the same appropriate rate of approximately 10,000 requests per second for every system.

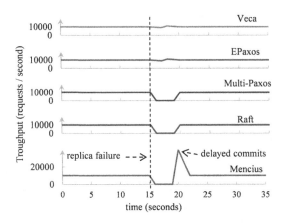

Fig. 6. Throughput when one of three replicas fails. For Multi-Paxos and Raft, the leader fails.

Veca and EPaxos have almost not been influenced during the failure. In Multi-Paxos, the new leader may take a lot of time to recover amounts of instances left behind. In Raft, the recovery time is reduced by its safety of leader election. The failure of a non-leader replica in Multi-Paxos and Raft does not affect availability.

In contrast, any replica failure disrupts Mencius. Every replica in Mencius must hear from all other replicas before committing a command. If any replica fails to respond, no other replica can make progress until the failure is detected and another replica commits no-ops on behalf of the possibly failed replica. At this point, the delayed commands are committed, which causes the throughput spike depicted. Live replicas commit no-ops periodically until the failed replica recovers, or until a membership reconfiguration.

6   Conclusion

We have presented the design of Veca, a high-performance consensus algorithm for state machine replication. With the high-efficiency and leaderless design, Veca achieves higher load balance, lower latency, higher throughput and higher availability at the same time, which are approximately optimal in existing consensus algorithms. Veca has important theoretical and practical benefits for state machine replication in LAN and WAN.

References

1. Lamport, L.: Paxos made simple. ACM SIGACT News **32**(4), 18–25 (2001)
2. Chandra, T.D., Griesemer, R., Redstone, J.: Paxos made live: an engineering perspective. In: Proceedings of the ACM Symposium on Principles of Distributed Computing, (PODC 2007), Portland, Oregon, USA, August 2007, pp. 398–407. ACM (2007)
3. Ongaro, D., Ousterhout, J.: In search of an understandable consensus algorithm. In: Proceedings of the USENIX Annual Technical Conference, (ATC 2014), pp. 305–320. USENIX (2014)
4. Schneider, F.B.: Implementing fault-tolerant services using the state machine approach: a tutorial. ACM Comput. Surv. **22**(4), 299–319 (1990)
5. Gray, C., Cheriton, D.: Leases: an efficient fault-tolerant mechanism for distributed file cache consistency. ACM SIGOPS Oper. Syst. Rev. **23**(23), 202–210 (1989)
6. Mao, Y., Junqueira, F.P., Marzullo, K.: Mencius: building efficient replicated state machines for WANs. In: Proceedings of the USENIX Symposium on Operating Systems Design and Implementation, (OSDI 2008), San Diego, CA, December 2008, pp. 369–384. USENIX (2008)
7. Moraru, I., Andersen, D.G., Kaminsky, M.: There is more consensus in egalitarian parliaments. In: Proceedings of the ACM Symposium on Operating System Principles, (SOSP 2013), pp. 358–372. ACM (2013)
8. Lamport, L.: Time, clocks, and the ordering of events in a distributed system. Commun. ACM **21**(7), 558–565 (1978)

A K-Means Clustering Algorithm: Using the Chi-Square as a Distance

Luis Ariosto Serna[1(✉)], Kevin Alejandro Hernández[1],
and Piedad Navarro González[2]

[1] Universidad Tecnologica de Pereira, Pereira, Colombia
{luarserna,kevin_loco}@utp.edu.co
[2] Corporación Instituto de Administración y finanzas (CIAF), Pereira, Colombia
investigacion@ciaf.edu.co

Abstract. The recurrent use of databases with variables of the categorical type in different fields of science. Demands new approaches when using cluster analysis techniques on this type of database. For this reason, in this article we compare the function *kmeans()* of Matlab with a function K-Means implemented by us, with the addition that it has integrated a measure of similarity that the function of Matlab does not have, the distance chi-square, both algorithms were tested in databases with quantitative and categorical variables. The experimental results showed a higher level of classification success in favor of the function implemented by us, explaining the correct functioning of the implemented algorithm and demonstrating that the chi-square distance is the measure of appropriate similarity for categorical type databases.

Keywords: Database · Cluster · K-means · Metric ·
Qualitative variable

1 Introduction

The rapid increase and integration of databases require researchers and engineers in the area of data science, data mining, etc. They are trained to provide society which requires data processing, a great contribution to scientific discoveries, optimize industrial processes and find relationships or patterns between data sets. Researchers have established algorithms and adopted new methods for the processing of large amounts of data (data mining) allowing to summarize the information in a much smaller set preserving the structure of the data and highlighting the most relevant characteristics of the same [1].

One of the most common and efficient grouping methods is the algorithm K-Means [2–4] however the choice of a measure of similarity (metric) is usually done at convenience and depends on the application, the type. The data of the variables contained in the database can influence the choice, since it is not appropriate to calculate the arithmetic mean of a data set with nominal, categorical or qualitative variables as if they were quantitative variables [1].

© Springer Nature Switzerland AG 2019
Y. Tang et al. (Eds.): HCC 2018, LNCS 11354, pp. 464–470, 2019.
https://doi.org/10.1007/978-3-030-15127-0_46

Besides, the increasing use of databases with qualitative variables demands new approaches when making cluster analyzes, such as Ralambondrainy [5]. Also, the increasing use of databases with qualitative variables demands new approaches when making cluster analyzes, such as Ralambondrainy. Which presents an approach using the K-Means algorithm to group categorical data. It converts multiple categorical attributes into binary attributes (1 for presence and 0 for the absence of that category), then treats these binary attributes as numeric in the K-Means algorithm. However, this algorithm needs a large number of binary attributes when the data sets have attributes with many more categories. By increasing the computational cost and memory storage of the algorithm. Other algorithms such as the Gower similarity coefficient [6], dissimilarity measures [7], the PAM algorithm [8], statistic fuzzy algorithms [9] and conceptual clustering methods [10] have been reported. All of them have limited performance when are they are applied to extensive data of type categorical.

With the aim of addressing this problem that arises when doing cluster analysis on categorical data. We decided to implement the K-Means algorithm in Matlab and in addition to include the standard similarity measures (sqeuclidean, city block, cosine, and correlation) the algorithm was equipped with chi-square distance since it is suitable for use over categorical data. In this article, we will compare the K-Means clustering algorithm implemented by us, and Matlab *kmeans()* algorithm, based on the level of accuracy of the labeling of quantitative and qualitative databases.

2   Materials and Methods

Let $\mathbf{X} \in \mathbb{Z}^{N \times P}$ a categorical data set with N inputs and P features, our aim is to find k groups (clusters) using the standard K-means method and the Chi-Square distance, which is similar to the Euclidean, but in this case, it is weighted. This distance is a suitable metric for analysis of qualitative, categorical, nominal and redundant data. Also, it compares the counting of corresponding categorical variables to two or more independent features [12]. We construct the distance matrix with the following expression:

$$d_{ij} = \sqrt{\sum_{n=1}^{P} \frac{1}{\widetilde{w}_n} \left(\widetilde{x}_{in} - \widetilde{x}_{jn} \right)^2}, \tag{1}$$

where: $\widetilde{x}_{in} = \frac{x_{in}}{\sum_{n=1}^{P} x_{in}}$, $a_n = \sum_{i=1}^{N} x_{in}$ and $\widetilde{w}_n = \frac{a_n}{\sum_{i=1}^{P} a_n}$. In this case $x_{in} \in \mathbb{Z}$ and $\mathbf{x}_i = \{x_{i1}, ..., x_{iP}\} \in \mathbb{Z}^P$ represents the initial form of the categorical sample, and $\mathbf{d}_i = \{d_{i1}, ..., d_{iN}\} \in \mathbb{R}^N$ is the new sample in the Euclidean space, $\widetilde{w}_i \in \mathbb{R}$ can be interpretate as a i-th feature weight, in this way the original dataset $\mathbf{X}$ is transform in a new data set $\mathbf{D} \in \mathbb{R}^{N \times N}$. Then we use the K-means algorithm applying on $\mathbf{D}$, this is method commonly used for partitioning a dataset in k groups (clusters). This is done by minimizing the distance among samples of the same cluster and it is maximized the distance among objects belonging to other groups [11]. The cluster assignment is based on the distance matrix, which is calculated with a similarity measure $\nu\left(\mathbf{d}_n, \boldsymbol{\mu}_k\right)$, and its form depends on the employed metric, being $\mathbf{d}_n \in \mathbb{R}^N$ the n-th sample and $\boldsymbol{\mu}_k \in \mathbb{R}^N$ the k-th centroid. The basic algorithm for K-means is given as follows:

1. Initialize cluster centroids $\boldsymbol{\mu}_1, \boldsymbol{\mu}_2, ..., \boldsymbol{\mu}_k \in \mathbb{R}^N$, randomly.
2. Repeat until convergence:
For every i, set:
$$c^{(i)} = \operatorname{argmin}_j \left\|\mathbf{d}^{(i)} - \boldsymbol{\mu}_j\right\|^2$$
For each j, set:
$$\mu_j = \frac{\sum_{i=1}^{m_k} 1\{c^{(i)}=j\}\mathbf{d}^{(i)}}{\sum_{i=1}^{m_k} 1\{c^{(i)}=j\}}$$
Being m_k the number of data points belonging to the k-th group c_k

2.1   Algorithm Implementation

The function $kmeans()$ Matlab in its latest version (R2018a at the time of writing this article) this has five different similarity measurement options, which can be specified using the 'Distance' parameter. However, none of these metrics is appropriate for the use of databases with categorical variables [13]. For this reason, we decided to implement the K-Means algorithm in Matlab with the standard metrics, with the addition that a measure of similarity was incorporated that the $kmeans()$ function did not have, the chi-square distance. The main advantage offered by the algorithm implemented by us on the $kmenas()$ function of Matlab is the addition of the chi-square distance which opens up new possibilities when it comes to cluster databases with qualitative variables, As it will be seen in the results, the function has a correct functioning exceeding the percentage of success in some cases. On the other hand, the main disadvantage may be that, unlike Matlab's $kmeans()$ function, our function uses the number of iterations, as well as the centroids, change as the convergence criterion, while the function $kmeans()$ of Matlab uses the algorithm k-means++, making it converge faster (Fig. 1 and Table 1).

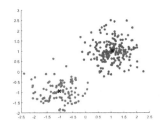

Fig. 1. Operation of the K-Means algorithm implemented, with the parameters of 10 iterations and 2 clusters, applied on a synthetic database.

Table 1. Databases used in the comparison of the grouping algorithm.

Database	Samples	Features	Classes	Variable type
Glass Identification	214	10	6	Quantitative
Iris	150	4	3	Quantitative
Blood Transfusion Service Center	748	5	2	Quantitative
Breast Cancer Wisconsin (Diagnostic)	569	30	2	Quantitative
Tic-Tac-Toe Endgame	958	9	2	Categorical
Car Evaluation	1728	6	4	Categorical
Congressional Voting Records	435	16	2	Categorical
Balance Scale	625	4	3	Categorical

2.2 Databases

To make the comparison between the K-Means algorithm implemented by us and the *kmeans()* function. The use of the databases of the UCI Machine Learning Repository website was made available [14], from there four databases were downloaded with quantitative variables (where two of them are biological databases) and four databases with variables of categorical type, which are specified in the following table.

3 Experimental Results and Discussion

The experiment had two stages, in the first stage it was tested on the quantitative databases, the algorithm implemented by us was on par with the Matlab function, showing the even better percentage of classification success although the same measures of similarity were used. In the second stage, the results of the tests carried out a show, as a general result a better classification accuracy when grouping with the chi-square distance on the databases with variables of categorical type in comparison with the other distances.

3.1 Stage 1

Both the function implemented by us and the *kmeans()* function of Matlab were approved based on their classification success rate when making the grouping, for this we used only the databases with variables of quantitative type and a number of 10 iterations as convergence criteria, the similarity measures used were the sqeuclidean, city block, cosine, and correlation.

3.2 Stage 2

Finally, both functions are compared using the criterion of percentage of success in the classification making use of the databases with variables of categorical type

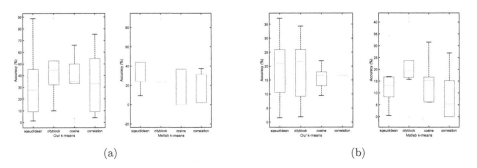

(a)   (b)

Fig. 2. Success percentage in the classification (a) based on Iris data. (b) Based on Glass Identification data.

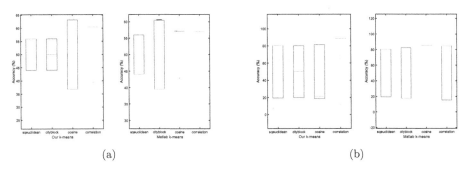

(a)   (b)

Fig. 3. Success percentage in the classification (a) based on Blood Transfusion Service Center data. (b) Based on Breast Cancer Wisconsin (Diagnostic) data.

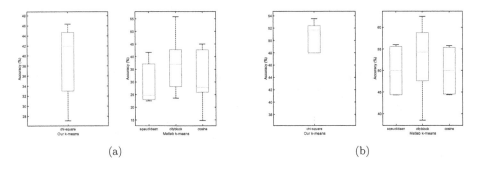

(a)   (b)

Fig. 4. Success percentage in the classification (a) based on Balance Scale data. (b) Tic-Tac-Toe Endgame data.

and a number of 10 iterations as criterion of convergence, in this comparison the function implemented by us made the grouping based on in the measure of chi-square similarity, whereas the *kmeans()* function of Matlab used the standard distances except for the 'correlation' which presented problems because some points of the databases presented relatively small standard deviations.

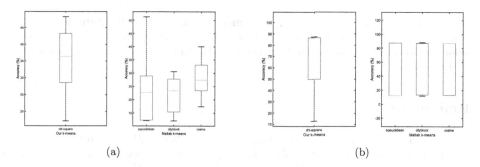

(a)   (b)

Fig. 5. Success percentage in the classification (a) based on Car Evaluation data. (b) Congressional Voting Records data.

3.3   Discussion

The differences in the percentage of success between both functions can be attributed to the way in which the algorithms assign the centroids. The *kmeans()* function of Matlab selects the centroids using the k-means++ algorithm. That is, select the centroid for the next iteration with a probability proportional to the distance from itself to the nearest centroid of the previous iteration [13]. While the algorithm implemented by us finds the centroids in the first iteration randomly and in the others, it makes use of the equation of *kmeans()*.

The chi-square distance obtains better results in the percentages of correct classification compared to the distances sqeuclidean, cityblock, and cosine and this can be attributed to the fact that the chi-square distance has an advantage over the others and that it calculates some weights assigning a relevance to each attribute, which makes this measure of similarity appropriate for use on databases with variables of categorical type.

4   Conclusions and Future Work

The K-Means algorithm implemented by us shows a percentage of success of classification that is quite similar or better in some cases than the function *kmeans()* of Matlab (Figs. 2 and 3), Thus demonstrating that the implemented one is a reliable function to do cluster analysis with databases with variables of quantitative type. While on the other hand, the incorporation of the chi-square distance gives an added value to this function.

The function implemented by us obtains a higher percentage of success in classification in each of the databases with variables of categorical type in comparison with the function *kmeans()* of Matlab (Figs. 4 and 5) and this is due to the convenient use of the metric chi-square and that it calculates the "distance" using weights which can be interpreted as the relevance of each characteristic, highlighting the essential attributes while the cluster is performed.

Acknowledgments. We would like to thank the Corporacion Instituto de Administracion y Finanzas (CIAF) and the research group of organizations and innovation belonging to the same institution. Who supported us in the development and financing of the article.

References

1. Hand, D.J.: Principles of data mining. Drug Saf. **30**(7), 621–622 (2007)
2. Anderberg, M.R.: Cluster Analysis for Applications: Probability and Mathematical Statistics: A Series of Monographs and Textbooks, vol. 19. Academic Press, Cambridge (2014)
3. Ball, G.: A clustering technique for summarizing multivariate data. Behav. Sci. **12**(2), 153–155 (1967)
4. MacQueen, J.: Some methods for classification and analysis of multivariate observations. In: Proceedings of the Fifth Berkeley Symposium on Mathematical Statistics and Probability, vol. 14, no. 1, pp. 281–297 (1967)
5. Ralambondrainy, H.: A conceptual version of the K-means algorithm. Pattern Recogn. Lett. **16**(11), 1147–1157 (1995)
6. Gower, J.C.: A general coefficient of similarity and some of its properties. Biometrics **27**, 857–871 (1971)
7. Gowda, K.: Symbolic clustering using a new dissimilarity measure. Pattern Recogn. **24**(6), 567–578 (1991)
8. Kaufman, L.: Finding Groups in Data: An Introduction to Cluster Analysis, vol. 344. Wiley, New York (2009)
9. Woodbury, M.A.: Clinical pure types as a fuzzy partition. J. Cybern. **4**(3), 111–121 (1974)
10. Michalski, R.S.: Automated construction of classifications: conceptual clustering versus numerical taxonomy. IEEE Trans. Pattern Anal. Mach. Intell. **4**, 396–410 (1983)
11. Ghosh, S., Dubey, S.K.: Comparative analysis of K-means and fuzzy C-means algorithms. Int. J. Adv. Comput. Sci. Appl. **4**(4), 35–38 (2013)
12. Mohanavalli, S.: Precise distance metric for mixed data clustering using chi-square statistics. Res. J. Appl. Sci. Eng. Technol. **10**(12), 1441–1444 (2015)
13. Mathworks.com: K-means clustering - MATLAB kmeans (2018). https://www.mathworks.com/help/stats/kmeans.html. Accessed 26 June 2018
14. Asuncion, A., Newman, D.J.: UCI Machine Learning Repository Irvine. University of California, School of Information and Computer Science (2013)
15. Martinez, T.: Improved heterogeneous distance functions. J. Artif. Intell. Res. **6**, 1–34 (1997)

Simulation Research on Aircraft Anti-collision Algorithm

Aijie Guan[1], Wenhao Xiang[2], Suhuan Jiang[1], Jun Peng[1(✉)], Jun Xu[1], and He Huang[1]

[1] College of Computer Science and Technology, Jilin University, No. 2699 Qianjin Street, Changchun City, China pengjun@jlu.edu.cn
[2] Research Institute of China Shipbuilding Industrial Systems Engineering, Beijing, China

Abstract. How to prevent collision problems on the aircraft need frequent detection for flight conflicts, therefore, it is necessary to design a good aircraft anti-collision algorithm. This paper analyzed the existing flight conflict detection algorithm paielli algorithm and carried out error analysis. On the basis of the original algorithm, this paper eliminated non-conflicting and coordinate processing, the error rate calculated by the improved algorithm was smaller. At the same time, the algorithm was verified by simulation and comparison with the conflict warning tools used at present, which proved that the new algorithm is more effective in detecting flight conflicts when the aircraft is turning.

Keywords: Aircraft · Anti-collision · Algorithm · Paielli algorithm

1 Introduction

Safety and efficiency are two key issues in aviation activities. Vehicle collision detection and avoidance are important means to solve these two problems [1]. Vehicle collision detection can be divided into three categories: long term, medium term and short term. The medium-term uses the information of the vehicle's flight plan, current position and status to predict the flight route and status of the vehicle in the next 20 min, so as to judge whether there is a possibility of conflict between the aircraft [2]. There are two kinds of mid-term conflict detection methods: one is geometric method, which can judge whether the two aircraft will conflict by linear extrapolation of their positions according to flight plans. This method is simple, but does not consider the possibility of conflict, resulting in a large number of false alarms in practical application. The other is probabilistic methods. At present, the main research on conflict probability calculation is paielli algorithm [3, 4] and prandini's Brownian motion-based conflict detection algorithm [5, 6].

This paper analyzed the existing flight conflict detection algorithm paielli algorithm and carried out error analysis. On the basis of the original algorithm, this paper eliminated non-conflicting and coordinate processing, the error rate calculated by the improved algorithm was smaller. At the same time, the algorithm was verified by simulation and comparison with the conflict warning tools used at present, which proved that the new algorithm is more effective in detecting flight conflicts when the aircraft is turning.

© Springer Nature Switzerland AG 2019
Y. Tang et al. (Eds.): HCC 2018, LNCS 11354, pp. 471–480, 2019.
https://doi.org/10.1007/978-3-030-15127-0_47

2 Parelli Analytic Algorithm

2.1 Prediction of Error Joint Covariance

If there is an aircraft performing a mission, assuming that the actual position at which it is located at any given moment is expressed as k, and the position of the track predicted by the air traffic control system is expressed as k'. Because the prediction error is zero mean distribution and the track prediction error and the covariance matrix are in accordance with the normal distribution of the diagonal matrix. Therefore, the error of the track prediction can be calculated by the difference between k and k', and d is expressed to represent the track prediction error. The calculation formula is as follows:

$$d = k - k' \tag{1}$$

The covariance matrix of d is as follows:

$$B = \mathrm{cov}(d) \tag{2}$$

The real-time position data of the aircraft is relative to the heading coordinate system. However, when designing the algorithm model, it is necessary to convert the heading coordinates of the aircraft into Cartesian coordinates. Suppose there is a matrix Z that can be used for conversion. The form of matrix Z is as follows:

$$Z = \begin{bmatrix} \cos \beta & -\sin \beta \\ \sin \beta & \cos \beta \end{bmatrix} \tag{3}$$

In the formula, it represents the heading angle of the aircraft. The heading angle refers to the angle of flight of the aircraft in a Cartesian coordinate system.

After the above formula conversion, the above formula is subjected to secondary processing, and the calculation data is converted into a form of the Cartesian coordinate system.

(1) The predicted position is:

$$k' = Zk' \tag{4}$$

(2) The track prediction error is:

$$D = Zd \tag{5}$$

(3) Error covariance matrix:

$$M = \mathrm{cov}(d) = ZBZ^T \tag{6}$$

Aircraft anti-collision technology is to reduce the aircraft collision event during the mission. Each aircraft is relatively independent, which means that each aircraft has its own prediction error of the track point. In order to perform flight collision detection, it is necessary to perform a separate calculation on the prediction error of the track point of each aircraft, which increases the number of algorithm runs.

Here, the independent error of the two aircrafts can be placed on any of the two aircraft, and the aircraft selected as the error transfer target is called a random aircraft, denoted by m. Similarly, the independent flight speeds of the two aircrafts are placed on an aircraft that has not been selected as a random aircraft. It is called a reference aircraft and is denoted by n [7, 8].

Some relationships between the aircraft m and the aircraft n are calculated as follows:

(1) The actual distance p between m and n:

$$P = P_m - P_n \tag{7}$$

(2) The distance p' between the predicted positions of m and n:

$$P' = P'_m - P'_n \tag{8}$$

(3) Track prediction error α:

$$\alpha = P - P' \tag{9}$$

(4) Joint error covariance matrix G:

$$G = \mathrm{cov}(\alpha) = M_m + M_n - M_{mm} \tag{10}$$

In the formula, the covariance matrix of the aircraft m is represented as M_m, similarly, we can know the meaning of M_n. Because there is no correlation between the prediction errors of m and n, so $M_{mn} = 0$.

2.2 Collision Probability of Aircraft Calculation

Assumed that the relative distance vector of the aircraft m and the aircraft n is $\lambda(q)$, the initial value is expressed as λ_0, where m represents a random aircraft and n represents a reference aircraft. The initial positions of m and n are represented by coordinates f_m and f_n. The relative velocity vector is denoted by v. $\lambda(q)$ can be expressed as:

$$\lambda(q) = f_m + v_m q - f_n - v_n q = \lambda_0 + vq \tag{11}$$

When $\lambda(q)$ and v are vertical, the aircraft m and n are closest. According to this property, it can be calculated when the two aircrafts are closest to each other. This time is q_m, and its calculation formula:

$$q_m = \frac{\lambda_0 * v}{v * v} \tag{12}$$

The formula of collision probability of the aircraft m and n [9, 10]:

$$P_{paielli} = \int_{-\infty}^{\infty} \int_{y_1}^{y_2} \frac{1}{2\pi} 1 e^{-\frac{1}{2}(x^2+y^2)} dxdy = \int_{-\infty}^{\infty} \frac{1}{\sqrt{2\pi}} e^{-\frac{1}{2}x^2} dx * \int_{y_1}^{y_2} \frac{1}{\sqrt{2\pi}} e^{-\frac{1}{2}y^2} dy \quad (13)$$

2.3   Analyzing Algorithm Error

Assuming that there are aircraft m and n, the errors of the aircraft m and the aircraft n are transferred to the aircraft m, and the speeds of the aircraft m and the aircraft n are transferred to the aircraft n. According to the contents of the previous section, m is a random aircraft, n is the reference aircraft.

In the ellipse, the two-dimensional normal function is integrated and evaluated. The calculation results are approximate values. In order to obtain the flight collision probability of two aircrafts, the paielli algorithm expands the protection area when calculating. The calculated value is an approximation. If the speed difference between the aircraft m and the aircraft n is relatively large, the relative distance of the flight in a time period is relatively long, so that the gap between the original protection zone and the extended protection zone can be ignored. The probability of flight collision is closer to the true value, but if the relative speed between the aircraft m and the aircraft n is small, then a small gap in the area of the extended area will have a greater impact on the final conflict value [11, 12].

3   Algorithm Improvement

3.1   Eliminating Non-conflicting Pairs

When conducting collision detection for a flight area, if we calculate the collision probability between any two aircraft who entering the area, this will be a very large amount of calculation. Therefore, if screening in advance, the aircraft pair that does not have a flight conflict will be removed. For example, the flight collision probability calculation is not required for the aircraft flying in the opposite direction in a short time. This can do non-conflicting removal operations.

The vector λ represents the relative distance of the two aircraft at the beginning, and the vector v represents the relative speed of the two aircraft. Assume that there is an aircraft m and an aircraft n, if $\lambda * v < 0$, that the aircraft m and the aircraft n are flying in opposite directions. Then the track pair composed of the aircraft m and the aircraft n that requires collision detection can be removed. If $\lambda * v > 0$, it is said that the aircraft m and the aircraft n are flying in the same direction, such conflicts need to be retained.

3.2 Position Change

Rotate the three-dimensional Cartesian coordinate system along the y-axis so that the direction indicated by the x-axis in the rotated coordinate system is on the same line as the major axis of the ellipse shown in Fig. 1. The rotation transformation matrix used in this process is denoted by F, and the rotation process is shown in Fig. 1:

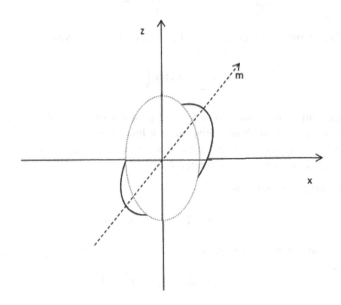

Fig. 1. Schematic diagram of coordinate rotation

As shown in Fig. 1, the black ellipse is the original position of the elliptical protection zone, and the blue ellipse indicates the new position after the coordinate rotation. In the figure, the angle between the vector m and the z-axis is expressed θ and the rotation transformation matrix F can be expressed as:

$$F = \begin{pmatrix} 1 & 0 & 0 \\ 0 & \cos\theta & -\sin\theta \\ 0 & \sin\theta & \cos\theta \end{pmatrix} \tag{14}$$

In Fig. 1 the long axis of the black ellipse is in line with the vector m, the coordinates of the intersection point in the first Cartesian coordinate system are represented by x_m and y_m, and the long axis length of the black ellipse is represented by a, and the short axis length is represented by b, c represents the elliptical focus. Let there is a variable k and assign value to it. The formula is as follows:

$$ax_m^2 + 2bx_my_m + cy_m^2 = k^2 \tag{15}$$

Utilizing the above formula, x_m and y_m can be calculated as:

$$y_m = \left(k\sqrt{\frac{a}{(ac - b^2)}} \right) \tag{16}$$

$$x_m = \frac{\left(\sqrt{b^2 y_m^2 - acy_m^2 + ak^2} - by_m \right)}{a} \tag{17}$$

The angle θ between the vector m and the z-axis in the figure can be expressed as:

$$\theta = \frac{\pi}{2} - \arctan\left(\frac{y_m}{X_m} \right) \tag{18}$$

The black ellipse is rotated to reach the position of the blue ellipse. After the rotation, the long axis of the blue ellipse and the line where the vector m is located, the coordinates of the intersection point in the first quadrant are represented by x_n and y_n, and the whole coordinates are represented by ω:

(1) The conversion equation is:

$$\omega = Fy_m, y_m = F^{-1}\omega \tag{19}$$

(2) The equation after conversion is:

$$rx_m^2 + 2ex_my_m + uy_m^2 = k^2 \tag{20}$$

(3) Deriving the z and x axes:

$$x_{max} = \frac{k}{r}, z_{max} = \frac{k}{u} \tag{21}$$

(4) Calculate the collision probability of 3 axes:

$$\begin{aligned} P(x) &= \phi(-x_b + x_{max}) - \phi(-x_b - x_{max}) \\ P(y) &= \phi(-y_b + y_{max}) - \phi(-y_b - y_{max}) \\ P(z) &= \phi(-z_b + d/2) - \phi(-z_b - d/2) \end{aligned} \tag{22}$$

(5) The whole collision probability:

$$\begin{aligned} P &= \int_{-z_b-\frac{d}{2}}^{-z_b+\frac{d}{2}} \int_{-y_b-y_{max}}^{-y_b+y_{max}} \int_{-x_b+x_{max}}^{-x_b+x_{max}} P(x,y,z)dx\,dy\,dz \\ &= \int_{-z_b-\frac{d}{2}}^{-z_b+\frac{d}{2}} P(z)dz \int_{-y_b-y_{max}}^{-y_b+y_{max}} P(y)dy \int_{-x_b-x_{max}}^{-x_b+x_{max}} P(x)dx \\ &= P(z)P_d(x,y) \end{aligned} \tag{23}$$

(6) Bring $P(x), P(y), P(z)$ into the whole collision probability equation:

$$P_d(x,y) = \int_{-y_b-y_{max}}^{-y_b+y_{max}} \int_{-x_b-x_{max}}^{-x_b+x_{max}} P(x,y)dxdy - \int_{-y_b+\frac{y_{max}}{\sqrt{2}}}^{-y_b+y_{max}} \int_{-x_b+\frac{x_{max}}{\sqrt{2}}}^{-x_b+x_{max}} P(x,y)dxdy$$

$$= \int_{-y_b-y_{max}}^{-y_b+y_{max}} P(y)dy \int_{-x_b-x_{max}}^{-x_b+x_{max}} P(x)dx - \int_{-y_b+\frac{y_{max}}{\sqrt{2}}}^{-y_b+y_{max}} P(y)dy \int_{-x_b+\frac{x_{max}}{\sqrt{2}}}^{-x_b+x_{max}} P(x)dx$$

$$= P(x)P(y) - P_1(x)P_2(y)$$

$$(24)$$

(7) The calculation equation of $P_1(x)$ and $P_1(y)$:

$$P(x_1) = \phi(-x_b + x_{max}) - \phi(-x_b + x_{max}/\sqrt{2})$$
$$P(y_1) = \phi(-y_b + y_{max}) - \phi(-y_b + y_{max}/\sqrt{2})$$

$$(25)$$

(8) The whole collision probability:

$$P = P(z)P_d(x,y) = P(z)(P(x)P(y) - P_1(x)P_1(y))$$

$$(26)$$

4 Simulation

Assuming that there is an aircraft jsh082 and an aircraft xl021, both two are required land operations to predict whether there is a flight conflict during the landing. The aircraft jsh082 landed along the runway XL and the aircraft xl021 landed along the runway JS, the runway XL and JS met before the point SJ, and then landed along the runway CC, the specific process is shown in Fig. 2.

Assuming that the minimum safe distance between the aircraft jsh082 and the aircraft xl021 is 5 km, first, we predict the flight track of the two aircraft within the next 3 min. The aircraft jsh082 and the aircraft xl021 began to turn at 120 s, and the flight collision calculation was performed using the paielli algorithm and the improved paielli algorithm, respectively. The predicted flight collision probability is shown in the following figure.

It can be seen from Fig. 3 that the flight collision prediction calculation is conducted by the paielli algorithm. The two aircraft, the aircraft jsh082 and the aircraft xl021, before the turning, the flight collision probability prediction result is close to 0 in the corresponding 0–60 s interval. During the turning operation, the predicted rate of flight collision probability reached by 50% in the 60–100 s interval. In the safe landing process after the encounter, the flight collision probability is close to 0 in the 100–120 s interval.

Using the improved paielli algorithm, the flight collision detection probability of the aircraft jsh082 and aircraft xl021 in the interval of 0–60 s is close to 0. During the 60–100 s turning process, the flight collision detection probability is close to 0. After the landing process, corresponding to 100–120 s interval, the flight collision detection result at this stage is also close to 0.

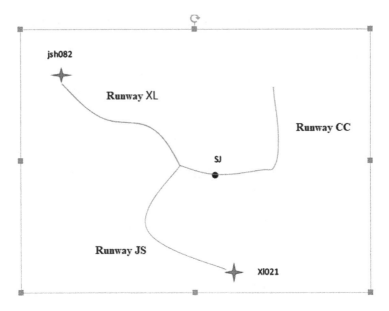

Fig. 2. Aircraft landing diagram

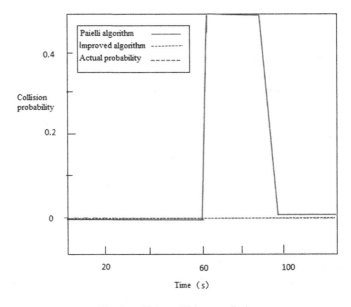

Fig. 3. Flight collision prediction

From the landing diagram of the aircraft in Fig. 3, it can be seen that there is no flight collision between the aircraft jsh082 and the aircraft xl021 during the turning. However, using the paielli algorithm to calculate the flight collision probability, the calculated value is up to 50%, which indicates that the improved paielli algorithm can effectively reduce the false alarm rate.

5 Conclusion

This paper analyzes the existing flight collision detection algorithm 'paielli algorithm' and conducts error analysis. Based on the original algorithm, by eliminating non-conflicting pairs and coordinate processing, the error rate calculated by the improved algorithm is smaller. In the process of collision probability analysis, after the two rotations, the direction of the relative speed of the aircraft in consistent with the direction of the x-axis in the coordinate system. At this time, the aircraft protection area as a reference point is in the shape of a cylinder parallel to the direction of the x-axis. However, because the rotation along the y-axis direction is performed, the interface shape formed by the y-axis and the z-axis is no longer a standard rectangle, but the projection interface in the protection zone is approximately treated as a rectangle for the simplification of the specific calculation process. The error caused by such approximation processing is worthy of in-depth analysis.

References

1. Kuchar, J., Yang, L.: A review of conflict detection and resolution. IEEE Trans. Intell. Transp. Syst. **1**(4), 179–189 (2000)
2. Hoekstra, J.M., van Gent, R., Ruigrok, R.: Designing for safety: the free flight air traffic management concept. Reliab. Eng. Syst. Saf. **75**(2), 215–232 (2002)
3. Paielli, R.A., Erzberger, H.: Conflict probability and estimation for free flight. AIAA J. Guidance Control Dyn. **20**(3), 588–596 (1997)
4. Paielli, R.A., Erzberger, H.: Conflict probability estimation generalized to non-level flight. Air Traffic Control Quart. **7**(3), 1–12 (1999)
5. Prandini, M., Hu, J., Lygeros, J., Sastry, S.: A probabilistic approach to aircraft conflict detection. IEEE Trans. Intell. Trans. Syst. Spec. Issue Air Traffic Control-Part I **1**(4), 199–220 (2000)
6. Prandini, M., Watkins, O.J.: Probabilistic aircraft conflict detection. HYBRDGE Project IST-2001-32460, Work Package WP3, Deliverable D3.2 (2005)
7. Stateczny, A., Bodus-Olkowska, I.: Sensor data fusion techniques for environment modelling. In: 16th International Radar Symposium (IRS), pp. 2155–2174. IEEE, Dresden (2015)
8. Gao, Z., Cecati, C., Ding, S.X.: A survey of fault diagnosis and fault-tolerant techniques—Part I: fault diagnosis with model-based and signal-based approaches. IEEE Trans. Industr. Electron. **62**(6), 3757–3767 (2015)
9. Li, M.: Development of foreign aircraft carrier combat system. Ship Electron. Eng. **33**(5), 6–9 (2013)

10. Yueh, S.H., Kong, J.A., Jao, J.K., Shin, R.T., Novak, L.M.: K-distribution and polarimetric terrain radar clutter. J. Electromagn. Waves Appl. **3**(8), 747–768 (2012)
11. Mou, Q., Feng, X., Xiang, S.: Design and realization of conflict detection alarm system for airport surface. J. Sichuan Univ. (Eng. Sci. Ed.) **47**(4), 104–110 (2015)
12. Wang, J., et al.: A method of OpenFlow-based real-time conflict detection and resolution for SDN access control policies. Chin. J. Comput. **38**(4), 872–883 (2015)

Cubic Spline Smoothing Algorithm for General Aviation Track

Wenyuan Xu[1], Ke Lv[1], Haoyang Yu[3(✉)], Chunyu Li[2],
and Weifeng Xu[3]

[1] Systems Engineering Research Institute, Beijing 100094, China
[2] Naval Radar and Sonar Repair Plant, Qingdao 266001, China
[3] North China Electric Power University, Baoding 071000, China
yuhaoyang_hd@163.com

Abstract. At present, in the field of general aviation track processing in China, only discrete track points can be monitored, but smooth track curves cannot be obtained. Based on the existing cubic spline curve theory, this paper proposes a navigation path smoothing algorithm based on cubic spline curve. The algorithm sequentially constructs a cubic curve equation between two adjacent points of the track. According to the curve equation, the interpolation between the two points is obtained, and the track points and the interpolation points are sequentially connected by a straight line according to the time. When the track point and the interpolation point reach a certain density, the track becomes smooth. After experimental comparison, the curve processed by the track smoothing algorithm is closer to the actual aircraft track.

Keywords: General aviation · Track processing · Cubic spline curve

1 Introduction

Since general aviation aircraft do not have advanced communication facilities, General Aviation cannot monitor real track curves like civil aviation. In practice, the navigation track monitoring can only detect discrete points of the navigation path of the navigable aircraft. This is very unfavorable for the recording, research and planning of the navigation path of the navigable aircraft, and there are huge security risks [6]. This paper expects to use the cubic spline method to process the discrete points of the track and make it a smooth track curve, so as to realize the monitoring and processing of the general aviation aircraft track.

2 Navigation Path Smoothing Algorithm Based on Cubic Spline Curve

2.1 Algorithm Principle

The basic idea of the algorithm is to construct a cubic curve equation between two adjacent points of the unsmooth track, and obtain the interpolation between the two

© Springer Nature Switzerland AG 2019
Y. Tang et al. (Eds.): HCC 2018, LNCS 11354, pp. 481–487, 2019.
https://doi.org/10.1007/978-3-030-15127-0_48

points according to the curve equation. Then, the track points and the interpolation points are sequentially connected by a straight line according to the time. When the density of the track point and the interpolation point reaches a certain level, the track can reach a smooth state [2, 5].

First, a cubic spline function is generated between two points, and N interpolation points are determined according to the obtained cubic spline function. The main function of the algorithm in this process is to process the flight data of the aircraft we are monitoring, including the geographic coordinate points of the aircraft flight and the heading angle of the aircraft at that point. Then through the coordinates of the two discrete tracks, we can find the cubic spline function between these two points.

To simplify the calculation, this derivation will use the coordinate vector method [3]. Let two points P_1 and P_2 be known, as shown in Fig. 1. Since the cubic spline function is a cubic equation, we can set the three-parameter vector equation of the curve over P_1 and P_2 to the following form.

$$P(t) = B_0 + B_1 t + B_2 t^2 + B_3 t^3 \quad (0 \leq t \leq t_1) \tag{1}$$

The meaning of each unknown is shown in Fig. 1.

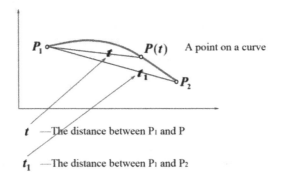

Fig. 1. Description of unknown meaning

When t = 0, the curve passes through point P_1, and the tangent vector of the curve segment at that point is P_1'. When t = t_1, the point P_2 is passed, and the tangent vector of the curve segment at this point is P_2', as shown in Fig. 2.

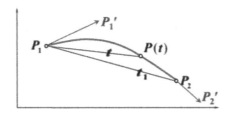

Fig. 2. Vector description of discrete track points

Derivation of Eq. (1) once.

$$P'(t) = B_1 + 2B_2 t + 3B_3 t^2 \tag{2}$$

When t = 0, the following expression can be derived from the Eqs. (1) and (2).

$$P(0) = B_0 = P_1 \tag{3}$$

$$P'(0) = B_1 = P'_1 \tag{4}$$

When t = t_1.

$$P(t_1) = B_0 + B_1 t_1 + B_2 t_1^2 + B_3 t_1^3 = P_2 \tag{5}$$

$$P'(t_1) = B_1 + 2B_2 t_1 + 3B_3 t_1^2 = P'_2 \tag{6}$$

By combining the formulas (3), (4), (5) and (6), B_0, B_1, B_2, and B_3 can be solved.

$$B_0 = P_1$$

$$B_1 = P'_1$$

$$B_2 = \frac{3}{t_1^2}(P_2 - P_1) - \frac{1}{t_1}(P'_2 + 2P'_1)$$

$$B_3 = -\frac{2}{t_1^3}(P_2 - P_1) + \frac{1}{t_1^2}(P'_2 + P'_1)$$

The coefficients B_0, B_1, B_2, and B_3 are substituted into the vector Eq. (1) and sorted.

$$P(t) = \left[1 - 3\left(\frac{t}{t_1}\right)^2 + 2\left(\frac{t}{t_1}\right)^3\right]P_1 + \left[3\left(\frac{t}{t_1}\right)^2 - 2\left(\frac{t}{t_1}\right)^3\right]P_2$$
$$+ \left[\left(\frac{t}{t_1}\right) - 2\left(\frac{t}{t_1}\right)^2 + \left(\frac{t}{t_1}\right)^3\right]P'_1 + \left[-\left(\frac{t}{t_1}\right)^2 + \left(\frac{t}{t_1}\right)^3\right]P'_2 \tag{7}$$

Equation (7) is the vector form of the cubic polynomial parametric equation passing through two points. Its component form can be expressed as follows.

$$x(t) = \left[1 - 3\left(\frac{t}{t_1}\right)^2 + 2\left(\frac{t}{t_1}\right)^3\right]x_1 + \left[3\left(\frac{t}{t_1}\right)^2 - 2\left(\frac{t}{t_1}\right)^3\right]x_2$$
$$+ \left[\left(\frac{t}{t_1}\right) - 2\left(\frac{t}{t_1}\right)^2 + \left(\frac{t}{t_1}\right)^3\right]x'_1 + \left[-\left(\frac{t}{t_1}\right)^2 + \left(\frac{t}{t_1}\right)^3\right]x'_2 \tag{8}$$

$$y(t) = \left[1 - 3\left(\frac{t}{t_1}\right)^2 + 2\left(\frac{t}{t_1}\right)^3\right]y_1 + \left[3\left(\frac{t}{t_1}\right)^2 - 2\left(\frac{t}{t_1}\right)^3\right]y_2$$
$$+ \left[\left(\frac{t}{t_1}\right) - 2\left(\frac{t}{t_1}\right)^2 + \left(\frac{t}{t_1}\right)^3\right]y_1' + \left[-\left(\frac{t}{t_1}\right)^2 + \left(\frac{t}{t_1}\right)^3\right]y_2' \tag{9}$$

Equations (8) and (9) are the result of processing the two adjacent discrete track points by the general aviation track smoothing algorithm, that is, the component representation of the cubic spline curve between two points. However, the first derivatives x_1', x_2', y_1' and y_2' at points P_1 and P_2 in Eqs. (8) and (9) are still unknown. In order to determine these four unknown first-order derivatives, the algorithm needs to process the monitored heading angle to obtain the first derivative of each track point.

According to the nature of the first derivative, the course angle is the inclination of the tangential line of the spline function at that point. Set the heading angle to a_i, then $y_i'/x_i' = \tan(a_i)$. Let $x_i' = 1$, then $y_i' = \tan(a_i)$. When $a_i = 90°/180°/360°$, let $a_{i+1} = 91°/181°/361°$. Similarly $y_{i+1}'/x_{i+1}' = \tan(a_{i+1})$, let $x_{i+1}' = 1$, then $y_{i+1}' = \cot(a_{i+1})$. When $a_{i+1} = 90°/180°/360°$, let $a_{i+1} = 91°/181°/359°$. By substituting the calculated x_1', x_2', y_1' and y_2' into the Eqs. (8) and (9), a complete cubic spline interpolation function can be obtained [4].

After obtaining the cubic spline interpolation function between two points, the next step is to use the cubic spline interpolation function to determine the interpolation between the two discrete track points [7]. The specific method is to divide the interval $[0, t_i]$ of the parameter t into k equal parts, each aliquot dt $= t_i/k$, taking t $=$ dt, 2dt, 3dt…jdt, … in order. The coordinates of the corresponding points are calculated using the cubic spline interpolation function equation, and these points are used as interpolation points. And then save the calculated interpolation point coordinates to the specified table. It is worth noting that the larger the value of K is, the more interpolation points are obtained, and the smoother the resulting track curve is. But this also increases the amount of calculations and reduces the processing speed of the algorithm. So we need to choose a moderate value when choosing K [8].

The above is the processing of the two discrete points by the algorithm. However, in practice, we have monitored N discrete track points, so we need to perform segmentation fitting on the processing results of each segment. We use the chase method to fit the N track points in turn. The specific method is as follows.

Set three track points P_1, P_2, P_3. First monitor the two points P_1 and P_2, and curve the two points. When the third point P_3 appears, the two points P_2 and P_3 are fitted. Repeat the above steps until the new track point no longer appears.

The course angle of the aircraft at these three points is constant. Ensure that the first derivative of point P_2 is the same value when smoothing the two segments P_1P_2 and P_2P_3 respectively. This ensures that the second derivative at point b is also the same, that is, the cubic spline function is continuous at point b. It can be known from the nature of the cubic spline curve that the cubic spline curve is continuous and smooth at this point, which ensures the smooth completion of the smooth connection processing of the adjacent two segments [1].

3   Algorithm Test

In the above, we propose a general aviation track smoothing curve processing algorithm based on cubic spline curve, and theoretically demonstrate the effectiveness and practicality of the algorithm. But is it characterized by high accuracy and low complexity in practical applications? We need to verify by experiment.

The source of the experimental data is the track monitoring system of General Aviation. After obtaining the discrete track data, the obtained discrete track points are processed by the algorithm, and finally a smooth track path close to the real track is obtained.

3.1   Algorithm Coding

The main way to implement the algorithm is to implement the smooth curve functions derived in Sect. 2, namely the Eqs. (8) and (9). Encapsulate the core algorithm code as a class when writing code, and generate an interpolation each time it runs. This algorithm class is called cyclically, and finally K interpolations are generated. The pseudo code is as follows.

```
public static double pi = 3.1415;
public static void Cspline(double x1, double x2, double
y1, double y2, double t, double r1,    double r2, out
double x, out double y)
        {
                double t1 = x2 - x1;
                double dy1, dy2;
                double m, m1, m2, m3, m4;
                m = t / t1;
                m1 = 1 - 3 * m * m + 2 * m * m * m;
                m2 = 3 * m * m - 2 * m * m * m;
                m3 = m - 2 * m * m + m * m * m;
                m4 = m * m * m - m * m;
                dy1 = System.Math.Tan(r1 * pi / 180);
                dy2 = System.Math.Tan(r2 * pi / 180);
                x = m1 * x1 + m2 * x2 + m3 + m4;
                y = m1 * y1 + m2 * y2 + m3 * dy1 + m4 * dy2;
        }
```

3.2   Experimental Test

The program core algorithm uses the method described above, and the final implemented test interface is shown in Fig. 4.

Click on the Get Data function, the program reads the data from the file and stores it in a temporary array. In order to compare the processed smooth track with the original discrete track, we use the display original track function. Click on the original track function and the result is shown in Fig. 3. The green polyline is the original track. The track generated in the figure is a polyline instead of a smooth curve.

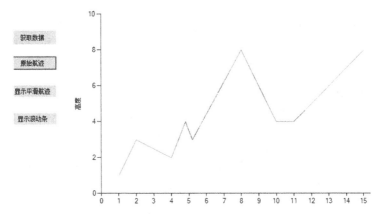

Fig. 3. Original track display (Color figure online)

Click Show Smooth Tracks to use the algorithm to process the read data and display a smooth track curve. The result is shown in Fig. 4. The red curve in the figure is the final smooth track curve. Compared with the original track, it can be found that the curve is obviously smoothed at the inflection point, which is very suitable for the actual track curve. This also proves the feasibility of this algorithm from experimental data.

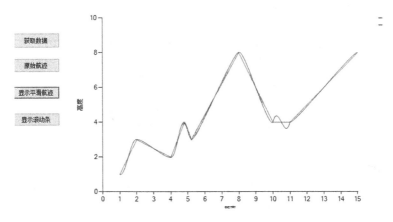

Fig. 4. Smooth track map (Color figure online)

4   Conclusion

This paper proposes a general aviation track smoothing algorithm based on cubic spline curve to solve the problem that only the discrete points of the aircraft track can be detected in the actual general aviation track monitoring. After theoretical verification and experimental testing of the algorithm, the results are as follows.

a. Based on the theory of the existing cubic spline curve, a vector representation of the cubic spline function between two discrete track points is established. This method greatly reduces the complexity and computational complexity of the algorithm.
b. Introducing the idea of the cubic spline interpolation processing curve into the general aviation track processing can achieve the effect of smoothing the track. The curve generated by the experimental results is smooth and more suitable for the actual heading curve of the aircraft.

References

1. Zhang, H.: Optimized smoothing algorithm for plane parameter cubic spline curve. J. Eng. Graph, (02), 105–108 (2009)
2. Wu, J.: Research on Multi-UAV Collaborative Track Planning and Effectiveness Evaluation Method, 05, pp. 45–47. Nanchang Aeronautical University (2012)
3. Li, Y:. Design and Implementation of Coordinate Transformation System, 03, pp. 14–18. China University of Geosciences, Beijing (2010)
4. Liu, Z.: Analysis and Control of the Singular Inflection Point of Parametric Curve, 01, pp. 10–12. Nanchang Aeronautical University (2012)
5. Dai, S.: Research on adaptive cubic spline interpolation approximation algorithm, 03, pp. 22–25. Dalian University of Technology (2008)
6. Kang, Y., Zhou, J.: General aviation development status, trends and countermeasures analysis. Mod. Navig. (05), 360–367 (2012)
7. Bai, Q.: Geometric continuity of general cubic parameter spline curve and its interpolation method, 02, pp. 14–16. Northeast Normal University (2006)
8. Fang, Z., He, Q., Xiang, B., Xiao, H., He, K., Du, Y.: A finite element cable model and its applications based on the cubicspline curve. Chin. Ocean Eng. 27(5), 683–692 (2013)
9. Wu, W.-C., Wang, T.-H., Chiu, C.-T.: Edge curve scaling and smoothing with cubic spline interpolation for image up-scaling. J. Signal Process. Syst 78(1), 95–113 (2015)
10. Habib, Z., Sakai, M., Sarfraz, M.: Interactive shape control with rational cubic splines. Comput. Aided Des. Appl. 1(1–4), 709–717 (2004)

Probability Matrix Factorization Algorithm for Course Recommendation System Fusing the Influence of Nearest Neighbor Users Based on Cloud Model

Jianguo Li, Chao Chang, Zuoxi Yang, Hailin Fu, and Yong Tang[(✉)]

South China Normal University, Guangzhou 510631, China
ytang@m.scnu.edu.cn

Abstract. With the explosion of data on the online course website, getting the required course information quickly and accurately becomes more and more difficult. In this paper, probability matrix factorization algorithm for course recommendation system fusing the influence of nearest neighbor users based on cloud model is proposed. The proposed algorithm uses the cloud model to compute user similarity and integrates social information into the course recommendation. The experimental results show that the algorithm can improve the accuracy of course recommendation effectively.

Keywords: Course recommendation · Probability matrix factorization · Social network · Cloud-model · SCHOLAT

1 Introduction

With the rapid development of information technology, more and more online course platforms are emerging one after another and anyone can choose courses on the online course platform to start learning. Therefore, the learning resources show an explosive growth, which makes it impossible for users to obtain the required information quickly and accurately in the massive course information and resources [1]. As a web-based course website for scholars, SCHOLAT course platform also encounters such problems. Recommendation system can effectively alleviate the problem of information overload by studying user data, making personalized recommendation and recommending courses that meet the user's interest.

Collaborative filtering recommendation algorithm is one of the most widely used recommendation algorithms, mainly divided into memory-based collaborative filtering and model-based collaborative filtering [2]. David Goldberg et al. proposed the recommendation system Typestry based on collaborative filtering [3]. Paul Resnick et al. proposed GroupLens, an automated collaborative filtering recommendation system based on user ratings [4]. Therefore, collaborative filtering recommendation system is gradually applied to e-commerce and social networking sites. In real world scenarios, it is impossible for users to rate every item The online course website also has the problem of data sparsity.

© Springer Nature Switzerland AG 2019
Y. Tang et al. (Eds.): HCC 2018, LNCS 11354, pp. 488–496, 2019.
https://doi.org/10.1007/978-3-030-15127-0_49

In the absence of sufficient statistics, it is difficult for the memory-based approach to provide accurate predictions. The model-based method can effectively alleviate the cold start problem by training the recommendation model with existing user information [5]. The matrix decomposition recommendation algorithm is widely used as a model-based method [6, 7, 8] because of its simplicity and high accuracy. Probability matrix decomposition [9] has achieved good results in various application areas since it is proposed.

The use of social information in social networking sites for recommendation, that is, social recommendation, has become a hot topic in the field of recommendation systems. Researcher Li [10] introduced co-author information as the user's associated information to improve the accuracy of recommendation. Researcher [11] use social tagging as a basis to solve the problem of recommendation. The course platform of SCHOLAT is different from the traditional course website. In addition to the elective function, it also has rich user social information and academic information. However, there is little research on using social information and academic information to recommend courses. In this paper, we will analyze the similarity of social information and the influence of users in social networks, and use the probability matrix decomposition method to recommend more accurate course for users.

2   Related Work

2.1   Similarity Computation of Nearest Neighbor Users

The methods of measuring user similarity mainly include cosine similarity, Pearson correlation coefficient and Euclidean distance. All three methods are based on the similarity computation of vectors, which is a strict match between object attributes.

(1) Cosine similarity: User ratings are used as vectors in n-dimensional space, and user similarity is measured by the angle between the cosines of the two vectors. Given two vectors of attributes, the scores of user i and user j in n-dimensional space are vector $\vec{i}$ and $\vec{j}$ respectively, the cosine similarity between user i and user j is:

$$\cos(\vec{i}, \vec{j}) = \frac{\vec{i} \cdot \vec{j}}{\|\vec{i}\| \cdot \|\vec{j}\|} \tag{1}$$

(2) Pearson correlation coefficient: Set X_i be the actual score of user X in item i, Y_i be the actual score of user Y in item i, $\overline{X}$ be the average score of user X, and $\overline{Y}$ be the average score of user Y, then user X and user Y's Pearson correlation coefficient is:

$$r = \frac{\sum_{i=1}^{n}(X_i - \overline{X})(Y_i - \overline{Y})}{\sqrt{\sum_{i=1}^{n}(X_i - \overline{X})^2}\sqrt{\sum_{i=1}^{n}(Y_i - \overline{Y})^2}} \tag{2}$$

(3) Euclidean distance: Set two users X and Y, X_i is the actual score of user X in item i, and Y_i is the actual score of user Y in item i, then the Euclidean distance of the two users is:

$$d(X, Y) = \sqrt{\sum_{i=1}^{n} (X_i - Y_i)^2} \tag{3}$$

2.2 Matrix Factorization

Because of its relatively low time and space complexity and high prediction accuracy, matrix factorization is widely used in recommender systems. The goal of matrix decomposition is to decompose the user-course elective matrix R into the user's hidden information feature matrix $U \in R^{M \times K}$ and the hidden information and the hidden information feature matrix of the course $V \in R^{N \times K}(K \ll \min(M, N))$, K is the dimension of the hidden information feature vectors. U and V are used to predict missing items in R, and the predicted values are compared for recommendation, that is:

$$R \approx UV^T = \widehat{R} \tag{4}$$

In collaborative filtering recommendation algorithm, probabilistic matrix decomposition assumes the potential distribution of data, which makes up for the over-fitting problem of traditional matrix decomposition method and performs well in collaborative filtering algorithm.

3 Probability Matrix Factorization Algorithm for Course Recommendation System Fusing the Influence of Nearest Neighbor Users Based on Cloud Model

In order to reflect the influence of neighboring users on target users, this paper proposes a course recommendation system that integrates the influence of neighboring users and the decomposition of probability matrix.

3.1 Computation of User Similarity Based on Cloud Model

In this paper, a qualitative and quantitative transformation model named cloud model proposed by academician Deyi Li is used to compute the similarity between two users [12]. The cloud model uses E_x (expectation), E_n (entropy), He(superentropy) three parameters to characterize a fuzzy concept, so that the fuzziness of things and randomness organically combined. The model is defined as follows:

Definition: Set U be a fuzzy set and C is a qualitative concept on U. If the numerical a∈U is a random implementation on C, and A is a stable random with a tendency for the determinacy y:Y → [0,1] ∀a∈Y a → y(x). The distribution of all a in the fuzzy domain U can be called a cloud, and each numerical a in the cloud can be called a cloud droplet.

Where E_x denotes the expectation that a distributes on the fuzzy field U, E_n denotes the measurability of C, and He denotes the uncertainty of entropy, which is determined by the randomness and fuzziness of entropy.

If the cloud-model between users is similar, the three-dimensional vector formed by controlling the parameter expectation, entropy, and super-entropy generated by the cloud droplet should also be similar. In this paper, the above three parameters are obtained by the reverse cloud algorithm, and the user's three-dimensional vector is formed. The user's similarity is computed by the user's three-dimensional vector.

Let $r_1, r_2, r_3 \ldots r_{n-2}, r_{n-1}, r_n$ be cloud droplet (rating). The computation steps of the three parameters of E_x, E_n and He for controlling cloud droplet generation are as follows:

Step1: Compute the sample average of the cloud droplet: $\bar{r} = \frac{1}{n}\sum_{i=1}^{n} r_i$; Compute the first order absolute central moment of cloud droplet: $\frac{1}{n}\sum_{i=1}^{n} |r_i - \bar{r}|$; Compute the variance of cloud droplet: $S^2 = \frac{1}{n-1}\sum_{i=1}^{n} (r_i - \bar{r})^2$.
Step2: Estimated value of $E\hat{x} = \bar{r}$;
Step3: Estimated value of $E\hat{n} = \sqrt{\frac{\pi}{2}} \times \frac{1}{n}\sum_{i=1}^{n} |r_i - E\hat{x}|$;
Step4: Estimated value of $H\hat{e} = \sqrt{S^2 - \frac{1}{3}E\hat{n}}$.

Using the reverse cloud algorithm, the user preference is represented by three parameters of the cloud computed from the user's original rating data. The cloud of three parameters of the user rating is called user characteristic vector and recorded as $\vec{V} = (Ex, En, He)$, where Ex reflects the user's comprehensive performance of each score. En reflects the user's concentration in the various ratings, that is, the measures of dispersion. And He be the stability of the entropy. User similarity based on cloud model is defined as follows:

Definition: Set the characteristic vector of two users be $\vec{V_i}$ and $\vec{V_j}$, $\vec{V_i} = (Ex_i, En_i, He_i)$, $\vec{V_j} = (Ex_j, En_j, He_j)$. Their cosine angles are called the similarity of cloud i and cloud j.

$$\cos\left(\vec{V_i}, \vec{V_j}\right) = \frac{\vec{V_i} \cdot \vec{V_j}}{\left\|\vec{V_i}\right\| \cdot \left\|\vec{V_j}\right\|} \tag{5}$$

3.2 Recommendation System

PMF is a probabilistic linear model. The addition records of N course sets $(I = \{I_j | j = 1, 2, 3, \ldots N\})$ by M user sets $(H = \{H_i | i = 1, 2, 3, \ldots M\})$ constitute the observed scoring matrix R. R_{ij} denotes user i's selection of course j, the missing item or 0 item in R means that this course is not added, and 1 means that the course has been added. The PMF model We assume that the conditional distribution of obey the Gaussian distribution, and the gradient descent method is used to correct the variables. To prevent overfitting, the regularization term is added and iteratively continues until the algorithm finally converges (Fig. 1).

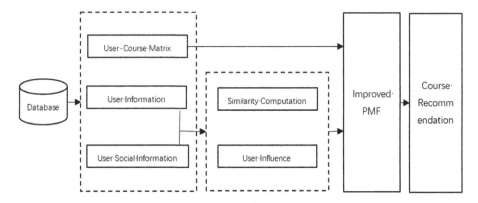

Fig. 1. Structure of proposed recommendation model

We suppose that the hidden information matrix of students-courses obeys Gauss prior distribution:

$$P(U|\delta_U^2) = \prod_{i=1}^{M} N(U_i|0, \delta_U^2 I) \tag{6}$$

$$P(V|\delta_V^2) = \prod_{j=1}^{N} N(U_j|0, \delta_V^2 I) \tag{7}$$

We suppose that the acquired student-elective scoring matrix of conditional probability also obeys the Gaussian priori distribution:

$$P(R|U, V, \delta_R^2) = \prod_{i=1}^{M} \prod_{j=1}^{N} \left[N(R_{ij}|g\left(U_i V_j^T\right), \delta_R^2) \right]^{I_{ij}^R} \tag{8}$$

Where I is the unit matrix, δ_U^2, δ_V^2 are the variances of U and V distributions respectively. I_{ij}^R is the indicator function, and if the user U_i does not join the course V_j, then its value is 0, and if it is joined, its value is 1. g(x) maps the value of $U_i V_j^T$ to the interval [0,1]. In this paper, we use $g(x) = \frac{x - x_{min}}{x_{max} - x_{min}}$, where x is the original data, x_{max} is the maximum of the original data set, x_{min} is the minimum of the original data set.

The posterior distribution function of the feature matrix can be found by Bayesian theorem:

$$P(U, V|R, \delta_R^2, \delta_U^2, \delta_V^2) \propto P(R|U, V, \delta_R^2) \times P(U|\delta_U^2) \times P(V|\delta_V^2) \tag{9}$$

Logarithmic processing of the above formula:

$$\begin{aligned}
\ln p(U, V|R, \delta_R^2, \delta_U^2, \delta_V^2) = \quad &-\frac{1}{2\delta_R^2} \sum_{i=1}^{M} \sum_{j=1}^{N} I_{ij}^R \left(R_{ij} - g\left(U_i V_j^T\right) \right)^2 \\
&-\frac{1}{2\delta_U^2} \sum_{i=1}^{M} U_i U_i^T - \frac{1}{2\delta_V^2} \sum_{j=1}^{N} V_i V_j^T \\
&-\frac{1}{2}\left(\left(\sum_{i=1}^{M} \sum_{j=1}^{N} I_{ij} \right) \ln\delta_R^2 + MK\ln\delta_U^2 + NK\ln\delta_V^2 \right) + C
\end{aligned} \tag{10}$$

Maximizing the logarithm of a posteriori is equivalent to minimizing its following objective function:

$$f = \frac{1}{2\delta_R^2} \sum_{i=1}^{M} \sum_{j=1}^{N} I_{ij} \left(R_{ij} - g\left(U_i V_j^T \right) \right)^2 + \frac{\lambda_U}{2} \sum_{i=1}^{M} U_i U_i^T + \frac{\lambda_V}{2} \sum_{j=1}^{N} V_j V_j^T \quad (11)$$

In the above formula, $\lambda_U = \frac{\delta_R^2}{\delta_U^2}$, $\lambda_V = \frac{\delta_R^2}{\delta_V^2}$.

The cloud model is used to compute the similarity between users, and a collection of neighboring users of each user is obtained. The influence of users is computed according to their academic information, academic degree, number of papers, number of projects, number of patents, number of works, etc., and the influence of nearby users is filled into the influence matrix D of nearby users. $D_i \in [1, 2]$ be the rating of the user U_i in the influence matrix. If the user U_j is the user U_i neighbor user, D_{ij} equals the user's influence score, when the user U_j is not the user U_i neighbor user, D_{ij} equals 1. The rating contributions of different users were determined according to the influence of these users on target users. The more influential the users, the greater the proportion of the rating prediction of target users.

We use gradient descent method, using U_i and V_j as parameters. To reduce the computational complexity, $\lambda_U = \lambda_V = \lambda$. U_i, V_j iterate over each iteration. $U_i \leftarrow \left(U_i - \gamma \cdot \frac{\partial f}{\partial U_i} \right) D_i$. Weighted control of each iteration by multiplying a similar user influence matrix. Where γ be the threshold. We get the hidden information characteristic matrix U and V, and then we can predict the reconstructed score matrix $\widehat{R}$ based on the eigenvector.

4 Experimental Setup and Analysis

4.1 Data Set

In this paper, a comprehensive teaching and scientific research collaboration platform developed by SCHOLAT research is used as an experiment. We use the data set provided by the SCHOLAT until March 16, 2018. A total of 43905 users and 1698 courses were recorded. User information included number of papers published, number of applications, landing score, dynamic score, academic score, academic degree, patents, number of teams joined, number of courses joined, number of friends joined, number of praise points, etc.

4.2 Evaluation Measures

In this paper, the root-mean-square error (RMSE) is used as the measure standard. The smaller the value of the RMSE, the higher the accuracy of the algorithm.

$$RMSE = \sqrt{\frac{1}{n} \sum_{i \in m, j \in n} \left(r_{ij} - \widehat{r}_{ij} \right)^2} \quad (12)$$

Where r_{ij} is real relationship between the user and the course selection, $\widehat{r}_{ij}$ is the predictive relationship between the user and the course in the recommendation system.

4.3   Experimental Results and Analysis

We compare the proposed recommendation model with the traditional probability matrix decomposition method, the nearest neighbor user influence decomposition method using cosine similarity, the nearest neighbor user influence decomposition method using Pearson similarity, and the RMSE value of the nearest neighbor user influence decomposition method using Euclidean distance.

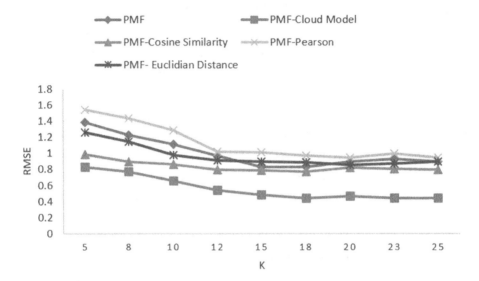

Fig. 2. The RMSE value polygon diagram of different algorithms when K takes different values

As can be seen from the above Fig. 2, the number of neighbors and λ of the parameters are both the best parameters. As the dimension K of hidden feature vector increases, the accuracy of all algorithm recommendations is improved. On the other hand, the problem of overfitting and increasing computational complexity may arise. In each dimension, the proposed algorithm is superior to the traditional PMF algorithm and other similar competition algorithms.

As the number of nearby users increases, the recommendation results gradually improve. RMSE values tend to stabilize after the number of nearby users exceeds 6. As the number of neighbor users increases, the algorithm will consume more time. Therefore, we choose neighbor users as 6 (Fig. 3).

From the above Fig. 4, we can see that the RMSE value is the smallest when $\lambda = 0.01$, and the precision of the algorithm begins to decrease as lambda continues to increase, so select $\lambda = 0.01$ as the lambda parameter value of this recommendation system.

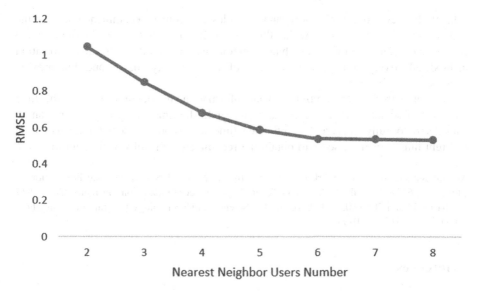

Fig. 3. The RMSE value of the algorithm in different neighborhood user numbers

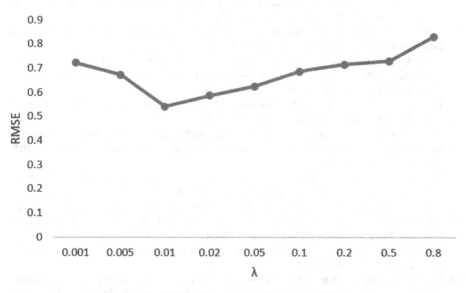

Fig. 4. The RMSE value of the algorithm in different λ

5 Conclusion

Aiming at the problem of information overload on the online course website, this paper proposed a course recommendation system that integrates the influence of neighboring users and the decomposition of probability matrix by studying the SCHOLAT course data of the new course platform, and provides the course recommendation for the system users. Experiments show that this method can provide better and more accurate course recommendation for users than the traditional probability matrix decomposition method, effectively improve the efficiency of users' inquiry course, and thus provide better services for users.

Although this paper combines social information, it does not consider the time sequence of social information. In future research, the time stamp mechanism can be considered to fully analyze the influence time and scope of social information in different time sequence, so as to obtain the recommended results of time information.

Acknowledgements. This work is supported by the National Natural Science Foundation of China (No. 61272067), the Science and Technology Project of Guangdong Province (Nos. 2017 A040405057 and 2016A030303058), and the Science and Technology Program of Guangzhou, China (No. 201604046017).

References

1. Xia, Z., Song, A., Fang, D., et al.: A collaborative filtering recommendation mechanism for cloud computing. J. Comput. Res. Dev. **51**(10), 2255–2269 (2014)
2. Chen, L., Chen, G., Wang, F.: Recommender systems based on user reviews: the state of the art. User Mod. User-Adap. Inter. **25**(2), 99–154 (2015)
3. Goldberg, D., Nichols, D., Oki, B.M., et al.: Using collaborative filtering to weave an information tapestry. Commun. ACM **35**(12), 61–70 (1992)
4. Resnick, P., Iacovou, N., Suchak, M., et al.: GroupLens: an open architecture for collaborative filtering of netnews. In: ACM Conference on Computer Supported Cooperative Work 1994, pp. 175–186. ACM, Chapel Hill (1994)
5. Pagare, R.A., Patil, S.: Study of collaborative filtering recommendation algorithm scalability issue. Int. J. Comput. Appl. **67**(25), 10–15 (2014)
6. Pham, T.A.N., Li, X., Cong, G., et al.: A general graph-based model for recommendation in event-based social networks. In: International Conference on Data Engineering 2015, pp. 567–578. IEEE (2015)
7. Bokde, D., Girase, S., Mukhopadhyay, D.: Matrix factorization model in collaborative filtering algorithms: a survey. Procedia Comput. Sci. **49**, 136–146 (2015). Icac
8. Song, Y., Zhuang, Z., Zhao, Q., et al.: Real-time automatic tag recommendation. In: International ACM SIGIR Conference on Research and Development in Information Retrieval 2008, vol. 6, pp. 515–522. ACM (2008)
9. Mnih, A., Salakhutdinov, R.R.: Probabilistic matrix factorization. Adv. Neural. Inf. Process. Syst. **20**(2), 1257–1264 (2007)
10. Li J, Xia F, Wang W, et al.: ACRec: a co-authorship based random walk model for academic collaboration recommendation. International Conference on World Wide Web 2014, pp. 1209–1214. ACM (2014)
11. Rendle, S., Schmidt-Thieme, L.: Pairwise interaction tensor factorization for personalized tag recommendation. In: ACM International Conference on Web Search and Data Mining 2010, pp. 81–90. ACM (2010)
12. Zhang, G.-W., Li, D.-Y., Li, P., et al.: A collaborative filtering recommendation algorithm based on cloud model. J. Softw. **18**(10), 2403–2411 (2007)

Cold Chain Logistics Service Quality on the Willingness of Online Shopping Fresh Products Based on Logistic Regression Model

Ting Wu, Shijun Tang$^{(\boxtimes)}$, and Zhe Yuan

City College of WUST, Wuhan 430083, China
Wut0208@163.com, Shijuntang@126.com, 779423510@qq.com

Abstract. Based on the background that fresh-chain e-commerce logistics is rapidly developed recently in China, this paper attempts to reveal factors that may affect the consumers' willingness to purchase fresh products online and further obtain the corresponding data combining consumer behavior theory and cold-chain service elements. Logistic regression model was used to quantitatively analyze the influence of different cold chain service elements on consumers' willingness to purchase fresh products online. Combining the empirical results, the corresponding improvement suggestions for fresh product cold chain logistics service quality were proposed.

Keywords: Logistic regression model · Cold chain logistics · Online shopping willingness

1 Introduction

The rapid development of e-commerce has brought innovation and development opportunities to the traditional fresh products sales model. As fresh product e-commerce is rising, it has greatly promoted the development of agriculture. Although the "Internet + Agriculture" model has been highly valued by the government and enterprises [1], the penetration rate of fresh product e-commerce in the whole fresh market is low. And the online shopping amount is not as great as expected. The reason to these phenomena is attributed that the online purchase willingness is weak. Only 1% companies can make profits in fresh product e-commerce in 4,000 companies. In ref. [2], Lansink found that information exchange and producer expectation of continuity of the relationship positively affect performance in the seller-buyer transaction. In the customer-oriented market, it is crucial to know the consuming behavior and how to meet consumer demand. This paper will study the impact on consumers' online shopping willingness from the perspective of cold chain logistics service quality.

With fresh products online shopping is developing under the background of Internet of things, the cold chain logistics distribution of fresh products online shopping presents following characteristics [3]: (1) ensuring product quality is of paramount importance; (2) logistics cost and timeliness always determines business profit; (3) development of the cold chain logistics distribution proposes high technical and

Y. Tang et al. (Eds.): HCC 2018, LNCS 11354, pp. 497–502, 2019.
https://doi.org/10.1007/978-3-030-15127-0_50

equipment requirements; (4) distributing location is geographically dispersed. Since China can provide a cold chain market of 150 billion yuan for a consumer group of more than 1.3 billion people, the large capacity and high demand of fresh market are the advantages for developing China's fresh industry. Meanwhile, the laws promulgated by the government on logistics also show great importance to cold chain logistics. However, the problem of merchant loss and poor quality by improper delivery have not only affected the customer's shopping experience, but also brought a serious loss to the fresh e-commerce. The reason to the delayed distribution is due to the customer's dispersion, underdevelopment of fresh product technology and imperfect cold chain system at the present stage. Based on the literature review, the authors divide the factors that affect consumers' willingness to purchase fresh products online into three aspects: (1) personal factors, which include consumers' age, gender, occupational nature, income and personal perception; (2) fresh product factors that include price and quality of fresh product; (3) quality factors of cold chain logistics services.

2 Investigation and Statistical Analysis of Fresh Product Online Shopping Willingness

The quality of cold chain logistics service means that, on the basis of the quality of general logistics services, special cold chain equipment and technology should be used in all aspects of product circulation to keep the products in a cold or frozen state. The corresponding quality indicators of cold chain logistics service are classified as shown in Table 1.

Table 1. Cold chain logistics service quality indicators

Primary indicator	Secondary indicator
Timeliness	E-commerce delivery speed
	Order processing and delivery speed
	After-sales feedback speed
Flexibility	Pick-up way and pick-up time
	Payment method
Reliability	Cold chain logistics equipment and packaging
	Product delivery integrity
	Quantity accuracy of goods
	Attitude of delivery staff
Economy	Logistics cost
	Return cost
Information	Accuracy of logistics information

In this empirical study, a total of 486 online questionnaires were distributed through the social network, and 450 valid questionnaires were selected. The effective recovery rate of the questionnaire was 92.6%. From the questionnaire survey results, it showed that consumers who purchased fresh products online accounted for 58.8% of the sample, and 41.2% of those who did not have online purchases expressed their willingness to order the fresh product online. There are many factors that might affect their online purchasing willingness, which include product quality, variety richness, price, and logistics services. The detailed statistical data are shown as Fig. 1 depicted.

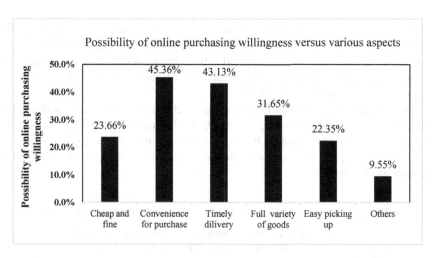

Fig. 1. Factors analysis of fresh product online purchase willingness

As it is shown in the figure, for people who have not yet experienced online shopping, about 45.36% of them considered that the purchase convenience is the most important factor that decide their fresh product online purchase willingness. The possibility for the factor of timely delivery is almost the same as the possibility for convenience, which is 43.13%. Factors such as full variety of goods, cheap and fine, and easy picking up also influence on the online shopping willingness, whose possibility are 31.65%, 23.66% and 22.35% respectively. All these factors form the main content of logistics services quality. It can be seen that for consumers who do not have experiences of online purchasing fresh products, it is also worthy paying attention to improve the quality of cold chain logistics services.

3   Impact Analysis of Cold Chain Service Elements Based on Logistic Regression Model

In this research, the logistic regression model is utilized for impact analysis of cold chain service elements. When analyzing the survey data, parameter Y in the regression model, represents the customers' online shopping willingness, and X1 represents the e-commerce delivery speed. X2 denotes order processing and delivery speed. X3 represents after-sales feedback speed. X4 is pickup mode and time. X5 and X6 are payment method and cold chain logistics equipment respectively. X7, X8, X9, X10,

X11 and X12 refer to product delivery integrity, cargo quantity accuracy, delivery personnel attitude, logistics cost, return cost, and logistics information accuracy correspondingly. The established logistic regression equation is shown as formula (1),

$$\ln\left(\frac{p}{1-p}\right) = \beta_0 + \beta_1 x_1 + \beta_2 x_2 + \ldots + \beta_{12} x_{12} \tag{1}$$

$P(Y = 1, Y = 0)$ is a dependent variable, indicating the probability to show whether the consumer is willing to purchase fresh products online. 0 means no willingness to buy, while 1 means willingness to buy. β_0 is the constant term of regression equation. β_i denotes regression coefficient corresponding to various influence factors. n refers to the number of regression factors [4, 5].

Among the collected 450 valid questionnaires, 265 consumers have experienced the online shopping of fresh products, and evaluated the quality of cold chain logistics services accordingly. In the paper, the Enter method is used in the logistic regression model for analyzing data of 265 questionnaires. Analysis results of correlation test for regression model are shown in Table 2.

Table 2. Regression analysis of various cold chain logistics service factors

	Regression coefficient B	Standard error S.E.	Chi-square value	P value sig	Dominance ratio
E-commerce delivery speed	1.236	0.285	19.812	0.001	3.412
Order processing and delivery speed	0.207	4.844	0.179	0.372	1.230
After-sales feedback speed	0.025	0.234	0.437	0.378	1.011
Pick-up way and pick-up time	0.679	0.233	8.494	0.004	1.969
Payment method	0.275	0.292	1.652	0.182	1.455
Cold chain logistics equipment and packaging	0.103	1.443	0.449	0.326	1.109
Product delivery integrity	0.567	0.242	5.505	0.019	1.763
Quantity accuracy of goods	0.244	0.214	1.279	0.165	1.289
Attitude of delivery staff	0.355	0.334	1.257	0.044	1.132
Logistics cost	−2.622	1.631	7.780	0.022	0.072
Return cost	−0.878	0.375	5.435	0.220	0.415
Accuracy of logistics information	0.005	0.246	0.478	0.384	1.005
Constant	2.038	0.285	51.667	0.000	7.675

In above analysis, we mainly analyze the two terms of P value and regression coefficient. The P value is the probability, which reflects the occurrence probability of an event. In the regression analysis, if $P < 0.05$, it is generally assumed to have an effect, and when $P < 0.01$, it is considered to have a significant effect. The larger the absolute value of the regression coefficient, the greater the influence of the corresponding factor on Y. If the independent variable is negative, it has a negative influence on Y. The chi-square value is used to test the regression coefficient to determine whether the regression coefficient is zero. If it is zero, the corresponding dominance ratio, Exp(B), is 1, indicating that the independent variable has no effect on Y. Therefore, when the chi-square value is larger, it makes greater impact on Y [6, 7].

From above results, the P value of e-commerce delivery speed as well as pick-up way and pick-up time is smaller than 0.01, which means that these two elements impact significantly on the cold chain logistics service quality. With regard to elements of logistics cost, return cost, product delivery integrity and attitude of delivery staff, the regression coefficient values are −2.622, −0.878, 0.567 and 0.355, which are much greater than that of other elements. These elements are also deemed to be the main influencing factors.

The pursuit of fresh products is its freshness. Therefore, the faster the fresh product is distributed, the better service quality the consumers will experience, and the stronger the customer's willingness to shop online. The impact of pick-up way and pick-up time on customers' online shopping fresh products is second only to the speed of delivery, and the more flexible way of picking up goods and time provided by cold chain logistics companies will make greater impact on customer's willingness. The quality of the goods will directly affect the consumer's satisfaction with their enjoyment, which in turn affects customers' online shopping willingness. The degree of integrity of the goods after delivery has a significant impact on consumers' online shopping willingness. The attitude of logistics service personnel also greatly influence customers' online shopping intentions. If the attitude is friendly, and the requirements and questions of the customers can be solved smoothly, it can provide consumers a great sense of trust and further increase online shopping intentions. Regarding the logistics cost and return cost, both of them have negative influences on consumers' online shopping willingness. The lower the cold chain logistics cost and return cost, the stronger the willingness of consumers to purchase fresh products online.

Therefore, in order to improve the quality of China's cold chain logistics services and promote the willingness of fresh online shopping, the following suggestions can be made. The first suggestion is to improve the timeliness of delivery. Increasing the timeliness of distribution requires the company to have a fast and efficient system process [8]. Through this way, it ensures that the fresh products are delivered to customers in the fastest speed and in the shortest time. The second measurement is to improve the flexibility of service quality. Before the delivery, the service provider can get in touch with the customer in advance, which can satisfy the customer's needs and facilitate the logistics personnel to arrange the distribution plan reasonably. The third is to improve management and continually update information. Additionally, enhancing the standardization level of services and strengthening the training of logistics personnel can also promote the development of fresh cold chain logistics.

4   Conclusion

Based on the research background of fresh cold chain logistics, this paper proposed to study the impact of cold chain logistics service quality on the willingness of online shopping fresh products. By determining cold chain logistics service quality indicators, a questionnaire survey is conducted. Among the 450 valid questionnaires, 58.8% of this crowd have experiences in online shopping, while others do not have any experiences.

For people who do not have online shopping experiences, statistical analysis of fresh product online shopping willingness is conducted, which indicated that purchase convenience is the most important factor that decide their fresh product online purchase willingness. Regarding people who have the online shopping experiences, logistic regression model is introduced to analyze the relationship between the quality of cold chain logistics service and the fresh willingness of online shopping. As the P value of e-commerce delivery speed is calculated to be 0.001, E-commerce delivery speed is deemed to be the key factor that may impact cold chain logistics service quality.

References

1. Chunhua, X.I.E.: The influence of perceived risk on purchase intention—a case study of taobao online shopping of fresh fruit. Asian Agric. Res. 9(5), 30–35 (2017)
2. Lansink, A.O., Bijman, J., Omta, O., et al.: Relationship characteristics and performance in fresh produce supply chains: the case of the Mexican avocado industry. J. Chain Netw. Sci. 10 (1), 1–15 (2010)
3. Huang, L., Feng, J., et al.: Study on the perceived risk about the online shopping for fresh agricultural commodities and customer acquisition. Asian Agric. Res. 6(6), 1–7 (2014)
4. Hasan, B.: Perceived irritation in online shopping: the impact of website design characteristics. Comput. Hum. Behav. 54(35), 224–230 (2016)
5. Well, J.D., Valacich, J.S., Hess, T.J.: What signals are you sending? how website quality influences perceptions of product quality and purchase intentions. MIS Qual. 35(2), 373–396 (2011)
6. Defraeye, T., Opara, U.L.: Towards integrated performance evaluation of future packaging for fresh produce in the cold chain. Trends Food Sci. Technol. 44(2), 203–205 (2015)
7. Kirezieva, K., Luning, P.A.L.: Factors affecting the status of food safety management systems in the global fresh produce chain. Food Control 52, 85 (2015)
8. Nguyen, H.T.N.: Achieving customer satisfaction through product-service systems. Eur. J. Oper. Res. 247, 179–190 (2015)

Depression Identification from Gait Spectrum Features Based on Hilbert-Huang Transform

YaHui Yuan[1], Baobin Li[1(✉)], Ning Wang[2], Qing Ye[1,3], Yan Liu[1], and Tingshao Zhu[4]

[1] University of Chinese Academy of Sciences, Beijing, China
libb@ucas.ac.cn
[2] Beijing Institute of Electronics Technology and Application, Beijing, China
[3] China Center for Modernization Research, Chinese Academy of Sciences, Beijing, China
[4] Institute of Psychology Chinese Academy of Sciences, Beijing, China
tszhu@psych.ac.cn

Abstract. Depression is a common mental illness, which is extremely harmful to individuals and society. Timely and effective diagnosis is very important for patients' treatments and depression preventions. In this paper, we treat the trajectory of gait as signal, proposing a new direction to detect the depression with gait frequency features based on Hilbert-Huang transform (HHT). Two groups of participants are recruited in this experiment, including 47 healthy people and 54 depressed patients, respectively. We process the gait data with HHT and build the classification models which verification method is leave-one-out. The best result of our work is 91.09% when the model I is adopted and the classifier is SVM. The corresponding specificity and sensitivity are 87.23% and 94.44% respectively. It verifies that the gait frequency of patients with depression is significantly different from that of healthy people, and the frequency domain features of gait are helpful for the diagnosis of depression.

Keywords: Depression · Gait · Hilbert-Huang transform · Kinect

1 Introduction

Depression is a disease characterized by persistent grief, losing interests in things that are usually enjoyed, and the inability to engage in daily activities. It will make patients feel bad and lose their living ability. Patients with severe symptoms even have the idea of self-mutilation or suicide. About 322 million people in various ages suffer from depression all over the world. The number of patients with depression has increased by 18% from 2005 to 2015 worldwide [9]. However, less than half of patients with depression (in many countries, fewer than 10%)

Baobin Li—This work was supported by NSFC under Grant U1536104, 11301504.

receive effective diagnosis and psychological treatments [9]. The most important reasons are a lack of resources and trained health-care providers. In many countries, it is difficult to resolve in the short term. In addition, social discrimination also make depressed people feel stressed. When do the scale evaluation or have a face to face communication with a doctor, they may deliberately conceal some facts and that will leads to misdiagnosis. Based on the facts above, a objective, non-invasive and easy-to-use assisted method to help the diagnosis of depression is necessary. In this article, we will try to use gait data for assisted diagnosis of depression.

People with depression are significant different from healthy people in gait. Lemke et al. have proved that compared with healthy people, the gait speed of depressed patients is significantly reduced and the stride length is shortened [6]. When one walks, he will repeats the actions of lifting and landing the foot. In fact, all the joints in the body will move in a certain pattern. Collecting the joints' motion trajectory during one's walking with Kinect, we can get a set of discrete motion signals. Previous research mainly focused on time domain features such as step size, speed, etc. [6], while in our study, we will use the signal processing method, Hilbert-Huang Transform (HHT) [5], to study the frequency domain characteristics of gait and extract the frequency domain features for the classification of patients with depression and healthy people. HHT is a time-frequency analysis method in which the signal is subjected to empirical mode decomposition to get a set of IMFs, and then each IMF is subjected to Hilbert transform to obtain time-frequency properties of the signal. We will introduce it in detail in Sect. 3.

The remainder of this paper is organized as follows. First, we introduce the related work and description about HHT in detail in Sects. 2 and 3. Section 4 shows the entire experimental process and how to extract gait spectrum features by HHT. Based on these feature, we establish identifying models for depression in Sect. 5, and experimental results show that these spectrum features are effective in characterizing and recognizing depression. Finally, Sect. 6 will provide a brief summary and discussion of future work.

2   Related Work

Psychomotor retardation is a prominent clinical feature of major depression. Changes in psychomotor behavior manifest in various motor domains including speech, facial expression, and locomotion. Therefore, many researchers are trying to find objective and effective motor features to assist in the diagnosis of depression.

Studies for depression detection with speech are popular. Researchers have proved that depressed patients consistently demonstrate prosodic speech abnormalities, such as reduced variation in loudness, repetitious pitch inflections and stress patterns, and monotonous pitch and loudness [2]. Nik Wahidah extract acoustic features from speech and predict the HAMD scores of MDD patients with it approximately [4]. Kuan Ee Brian et al. extracted four acoustic features, including prosodic, glottal, Teager's energy operator, and put forward a multichannel method for clinical depression of adolescents. Their method provided an accuracy of 73% [8]. Facial expression is also one of the focus of depression auxiliary diagnosis. This seems to be natural, because patients with depression usually stay unhappy for a long time and rarely smile. Girard et al. analyzed a series of video recordings of visiting depressed patients by coding the FACS motor unit [3]. They found that when symptom severity was high, participants showed more contempt expressions and smiled less. Anastasia et al. proposed a method for facial feature selection to help depression recognition. The experiments are based on two depression databases, AVEC2013 and AVEC2014. But their results showed that just 50% of the samples are correctly classified, which is a little better than random guessing [11]. He further improved the results in [12], reaching 88% for manual FACS, 79% for AAM, and 79% for vocal prosody. Zhou et al. proposed a multi-region DepressNet, training deep regression models for different face regions and fusing them on the same database [18]. The MAE of the two database was 6.20 and 6.21 respectively, but they did not put the precision or F-score in their result.

Changes in motor are one of the clinical manifestations of depression, too. Sloman et al. found that depressed patients walk with a lifting motion of the leg, while control subjects walk forward [14]. Compared to controls, depressed patients showed significantly lower gait velocity, reduced stride length, double limb support and cycle duration [6,15]. The above researches suggest that gait abnormalities may be an important manifestation of depression. Although the scholars have done a lot on the effectiveness of gait features, no one has established a system to distinguish depression patients from the healthy people with gait. Moreover, these studies focused on time domain features, such as speed, step size, etc., but no one paid attention to the frequency domain characteristics of gait. Continuous motion trajectories can be considered as signals. By performing frequency analysis method on them, we can obtain the frequency features of gait. In this paper, we automatically track 25 joints of the human body and sample them at 30 Hz with Kinect to obtain the discrete motion signals. Subsequently, we process the discrete signals with HHT and extract the effective features. Eventually we establish a system for categorizing the depression and healthy controls. The contributions of this paper are as follows:

- Process the gait signal with HHT to extract effective gait features;
- Establish a classification system with the extracted gait characteristics to classify the depression and the healthy controls.

3 Hilbert-Huang Transform

Hilbert-Huang transform has been proposed by Norden E. Huang in 1998, which is designed for analyzing non-linear and non-stationary signals [5]. Up to now, it has been widely studied and discussed in fields of signal processing. It can be divided into two parts: the empirical mode decomposition (EMD) and the Hilbert spectral analysis. Doctor Huang puts forward that only instantaneous frequency of single component signal has obvious physical meaning. However, natural signals are basically multi-component, and the corresponding instantaneous frequency has no obvious physical meaning. To solve this problem, Huang et al. proposed the concept of Intrinsic Mode Function (IMF) and the empirical mode decomposition method, and believed that any signal would be decomposed into several IMFs. The IMFs need to meet the following conditions:

- The number of extreme points and zero points are equal to or at most one difference from each other;
- The mean value of the envelope formed by local maxima and the envelope formed by local minima is zero.

EMD is referred as a "filtering" process. Suppose the original signal is represented as $y(t)$, then the EMD process is as follows:

Algorithm 1. The Empirical Mode Decomposition

1: Initialize $h^j(t) = y(t), i = 1, j = 0$;
2: **repeat**
3:   **repeat**
4:     $j = j + 1$;
5:     $x_i^j(t) = h^j(t)$;
6:     Find all the local maximum points of the signal $x_i^j(t)$ and interpolate them with cubic spline to obtain the upper envelope $LMx(t)$;
7:     Find all the local minimum points of the signal $x_i^j(t)$ and interpolate them with cubic spline to obtain the lower envelope $LMi(t)$;
8:     $m^j(t) = (LMx(t) + LMi(t))/2$;
9:     $h^j(t) = x_i^j(t) - m^j(t)$;
10:   **until** $h^j(t)$ meets the conditions listed above
11:   $imf_i(t) = h^j(t)$;
12:   $r_i(t) = x_i^j(t) - imf_i(t)$;
13:   $i = i + 1$;
14:   $x_i^j(t) = r_i(t)$;
15:   Reset $j = 0$;
16: **until** r_i is monotonic

EMD decomposes the multi-component signals into a series of IMFs, and each IMF is linear and stable. Figure 1 shows every component of the signal $y(t) = 10sin(2\pi \cdot 10t + 10sin(2\pi \cdot 30t)$, in which the top figure is original signal, the 2–4 are IMFs while the bottom is the residual element. Because the signal is decomposed on the basis of itself, this method is intuitive and adaptive compared to other methods like short-time Fourier transform, wavelet decomposition, etc.

In the second part, for each IMF, we perform Hilbert transform as

$$\widehat{imf_i}(t) = H\left[imf_i(t)\right] = \frac{1}{\pi} \int_{-\infty}^{\infty} \frac{imf_i(\tau)}{t - \tau} d\tau,$$

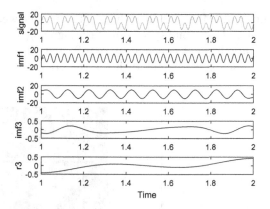

Fig. 1. The original signal and a series of IMFs obtained by EMD.

and construct an analytic function

$$z_i(t) = imf_i(t) + j \cdot \widehat{imf_i}(t) = a_i(t)e^{j\varphi_i(t)}, \tag{1}$$

while the amplitude function and phase function are defined as

$$a_i(t) = \sqrt{imf_i(t)^2 + \widehat{imf_i}(t)^2}, \tag{2}$$

$$\varphi_i(t) = arctan(\frac{\widehat{imf_i}(t)}{imf_i(t)}). \tag{3}$$

We can get the instantaneous frequency from the derivative of phase function

$$w_i(t) = \frac{1}{2\pi}\frac{d\varphi_i(t)}{dt}. \tag{4}$$

the Hilbert marginal spectrum is the amplitude distribution in frequency domain:

$$h_i(w) = \int_0^T H_i(w, t)dt, \tag{5}$$

where $H_i(w, t)$ is the Hilbert spectrum, and defined as the real part of $a_i(t)e^{j\int w(t)dt}$. That is, an signal marginal spectrum can be computing as integrating the Hilbert spectrum over time. Moreover, Adding all $h_i(w)$, we can get the total Hilbert marginal spectrum

$$h(w) = \sum_{i=1}^{k} h_i(w), \tag{6}$$

where k is the number of IMFs, and T is the length of sequence. Figure 2 shows the Hilbert spectrum and marginal spectrum of the signal $y(t) = 10sin(2\pi \cdot 10t + 10sin(2\pi \cdot 30t)$. The amplitude-frequency curve reflects that the main frequency is 10 Hz and 30 Hz, which is consistent with the the original signal $y(t)$.

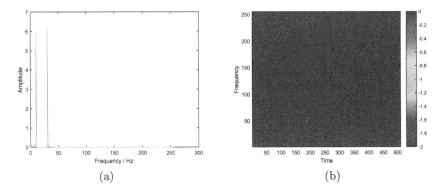

Fig. 2. The Hilbert spectrum (a) and marginal spectrum (b) of the signal $y(t) = 10sin(2\pi \cdot 10t + 10sin(2\pi \cdot 30t)$.

HHT is widely used in various types of anomaly detection. The most common applications are various types of anomaly detection, such as the depiction of seismic reflection signals [1], machine faults [13], and heart rate variability [7]. Similarly, spectral features will help signal classification in cases where the spectral discrimination of different signal classes is obviously. Yang extracted spectral features of Chinese strokes for font detection [17], and Zong et al. studied four kinds of physiological signals for emotional classification with HHT [19]. In [10], Oweis analyzed EEG signals with HHT and tried to distinguish healthy people from patients with epilepsy. In this paper, we will use HHT to extract spectrum features from gait signals and propose two methods to distinguish depressed patients and healthy people.

4   Experiments

In this section, we will give the details about the gait data collection and feature extraction. We take the Microsoft Kinect as the gait acquisition equipment for its portable and non-intrusive properties. It will track a person's movements and provide locations of 25 body joints in 3-D space, as shown in Fig. 3. In the coordinate system of Kinect, the origin is the camera lens, Z-axis is along the direction of shooting, X-axis is horizontal and perpendicular to Z-axis, and Y-axis is straight up.

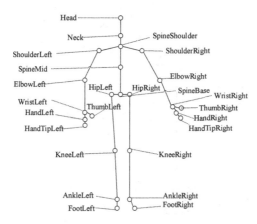

Fig. 3. The 25 makers on human body catched by Kinect through its built-in depth sensors and software.

4.1 Gait Data Collection with Kinect

Two groups of participants have been recruited in this experiment. They are 47 healthy participants and 54 patients diagnosed with depression by Beijing Anding Hospital, China.

In order to ensure that the participants stay within the sight of the Kinect during the experiment, we design an experimental walking path as Fig. 4. The rectangular footpath is 3 m wide and 6 m long, and two Kinects are placed on the two vertices on the diagonal of the rectangle. Participants will walk on the footpath anti-clockwise naturally for two minute. When the participant walks on the longer side, the Kinects record the walking data. When the participant walks on the short side, he will leave the Kinects' collection range and the data record is temporarily interrupted. The sampling rate of Kinect is about 30 Hz. In this way, we get a set of gait sequences with breaks, where each frame contains the three-dimensional spatial position of 25 markers with time stamp.

4.2 Data Preprocessing

Raw gait data is noisy, complicated and inconsistent. We will perform coordinate translation, filtering and clipping before extracting features.

In the original coordinate system, Kinect itself is the origin of coordinates. This is not conducive to the analysis of data, translating the origin of the coordinate to the person's body will make it more convenient to observe the movement of people. In order to reduce the influence of body movement, we chose spine as the new origin of coordinate, because it changes little during movement and thus can serve as a good reference. We also smooth the data with mean filter to remove the noise generated by discontinuous sampling.

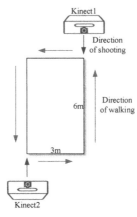

Fig. 4. The footpath used in the experiments. Participants will walk counterclockwise along the footpath. When they walk within the shooting range (the long side of rectangular), the Kinects will record their trajectory.

In the collection process, participants will get out of the sight when they walk to the short side, which results in discontinuous of gait data. We divide the collected data according to the natural segment, and cut off the unstable data caused by turning around the corner. In order to maintain data consistency, we adjust the starting position of all natural segments to right foot. If the length of the resulting signal clip is shorter than 64, this natural segment will be abandoned. After preprocessing, there are 4 to 13 clips of gait data for each people, which is a normal distribution. In order to keep everyone's data consistent and make maximum use of the data, we adjust the number of segments for all people to 8 segments. If the number of segments is more than 8, the first eight segments will be used, and if the number of segments is insufficient, then it will be supplemented by segment copies from itself. At last, we get 8 gait segments of each participant.

4.3   Feature Extraction

We show the Hilbert marginal spectrum of several gait sequence obtained by HHT in Fig. 5, which consists of healthy people (blue) and depressive patients (red) for comparison. To reduce the irrelevant component, we delete the IMFs whose correlation coefficient with original signal is less than 0.1. According to the sampling theorem, we only take the marginal spectrum corresponding to the effective part of the frequency (half of the sampling frequency, 30 Hz). Comparing the healthy to the depression, we found that the Hilbert marginal spectrum of healthy people is slightly higher than that of depressed patients, and their

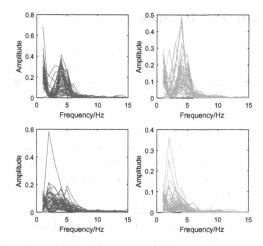

Fig. 5. The Hilbert marginal spectrum of three markers in one direction, the left (blue) is healthy, and the right (red) is depression (Color figure online).

distributions are also quite different from each other. Based on the facts above, we select three statistic features from marginal spectrum:

- Average. The average amplitude of the signal over given frequency band reflects the intensity of the signal in frequency domain, which is defined as

$$\overline{h} := \frac{\sum_{w=1}^{N} h_w(w)}{N}. \tag{7}$$

- Variance. The variance of $h(w)$ can reflect the frequency stability of the signal, and its definition is

$$s^2 = \frac{\sum_{w=1}^{N}(h(w) - \overline{h})^2}{N}. \tag{8}$$

- Maximum. The maximum frequency can reflect the dominant frequency of the signal, which is an important feature. It is calculated as

$$Max_{h(w)} = max(h(1), h(2)...h(N)). \tag{9}$$

Except for the above three features, we also take the Hilbert marginal spectrum itself (the $h(w)$) as a feature. At last, there are four kinds of features used for classification.

5   Classification and Results

In this chapter, we will introduce two methods to establish detecting models. Both methods are designed to make full use of the redundant gait data and improve the robustness of the system. We adopt three classifiers, Linear kernel Support Vector Machines (SVM), Simple Decision Tree (DTree) and K-Nearest Neighbors (KNN) [16].

In method I, we treat segments as partially independent unit. The verification method is leave-one-out. In training procession, we treat each segment of each individual as an independent training sample whose class label is same as the class of the people to which it belongs, that is to say, we have 100×8 samples in each training pass. After training, we test each segment of the people we have chosen as the test and vote to determine the final category of him(her). Due to the small amount of experimental data, the original result may be subject to significant overfitting tendency. Treating the gait clips as an independent sample in training process will significantly expand the data volume and increase the data credibility. The use of voting can also improve accuracy and reduce the impact of instability.

We take each participant as independent unit in method II. Each Participant has 8 clips of gait data, to eliminate the instability of the gait data, we take the mean of the 8 segments which belong to the same person to make full use of the redundant data. This method will significantly reduce the amount of data and make the data more streamlined. We also use the leave-one-out method for classifier verification. Compared to Method 1, this method is easier, and the reduction in data volume does not impact the credibility of the classification.

The results of the two methods are shown in Table 1. Among the metrics used in the tables, accuracy(acc) is the test accuracy of all participants, sensitivity(sen) is the test accuracy of depressed people, and specificity(spe) is the test accuracy of healthy people. They are defined as follows

$$\begin{cases} acc & = \frac{C1+C2}{101}, \\ sen & = \frac{C1}{54}, \\ spe & = \frac{C2}{47}, \end{cases}$$

where C1 is the number of healthy people which are correctly classified, and C2 is the number of the patients which are correctly classified. We can see that the accuracy of all features in three classifiers in Method I is above 75%. We get the best result of 91.09% when the feature is Hilbert marginal spectrum and the classifier is SVM. In the vast majority of cases, the index of sensitivity is higher than the specificity. In generally, the classification results of the SVM are the best, the decision tree is the second, and the result of KNN is worst of three. Among all features, Hilbert marginal spectrum performs better than any other features. In Method II, SVM and KNN get the best accuracy 90.10%, while the worst one is obtained for decision tree with variance features.

Table 1. The accuracy, sensitivity and specificity of the classification results using different features and classifiers by method 1.

Features	DTree	Method 1			Method 2		
			KNN	SVM	DTree	KNN	SVM
Hilbert marginal spectrum	acc	0.8713	0.8713	0.9109	0.8515	0.9010	0.9010
	spe	0.8511	0.7872	0.8723	0.8511	0.8723	0.8936
	sen	0.8889	0.9444	0.9444	0.8519	0.9259	0.9074
Average	acc	0.8218	0.8020	0.8515	0.7723	0.7426	0.7723
	spe	0.8085	0.8085	0.8936	0.7872	0.7660	0.8085
	sen	0.8333	0.7963	0.8148	0.7593	0.7222	0.7407
Maximum	acc	0.8119	0.7822	0.8812	0.8317	0.8416	0.8812
	spe	0.8085	0.7234	0.8723	0.8723	0.8298	0.8511
	sen	0.8148	0.8333	0.8889	0.7963	0.8519	0.9074
Variance	acc	0.8713	0.8218	0.8515	0.6931	0.8020	0.8713
	spe	0.8511	0.7872	0.7660	0.5532	0.7660	0.8298
	sen	0.8889	0.8519	0.9259	0.8148	0.8333	0.9074

In conclusion, both methods get the best performance with Hilbert marginal spectral accompanied by SVM. Except for very few cases (variance accompanied by DTree), the accuracy in all other cases is higher than 75%. The overall performance of Method I is better than Method II including the highest accuracy, but since Method II is simpler, we still think it is meaningful. Experiments have shown that our method is helpful for the detection of depression.

6 Conclusion

Gait information is one of expressive modalities for revealing depression. This paper has mainly discussed how to automatically detect depression from healthy people with gait frequency features. By the aid of Kinect, we have collected gait data of healthy people and patients with depression, and to extract spectrum features with HHT, then we established two detecting models for depression detection. Experimental results show that the detecting models work well, and among these spectrum features, the Hilbert marginal spectrum performs better than its statistical features. The amplitude of Hilbert marginal spectrum of the healthy is significantly higher than that of depressed patients. At present, our gait data only includes healthy people and unipolar depression patients, and excludes other related factors such as other physical or mental illnesses. In the future, we will try to use gait to identify more complex branches of depression, such as distinguishing unipolar depression from bipolar depression, and study the differences between the two and healthy people.

References

1. Battista, B.M., Knapp, C., McGee, T., Goebel, V.: Application of the empirical mode decomposition and hilbert-huang transform to seismic reflection data. Geophysics **72**(2), H29–H37 (2007)
2. Darby, J.K., Simmons, N., Berger, P.A.: Speech and voice parameters of depression: a pilot study. J. Commun. Disord. **17**(2), 75–85 (1984)
3. Girard, J.M., Cohn, J.F., Mahoor, M.H., Mavadati, S., Rosenwald, D.P.: Social risk and depression: evidence from manual and automatic facial expression analysis. In: 2013 10th IEEE International Conference and Workshops on Automatic Face and Gesture Recognition (FG), pp. 1–8. IEEE (2013)
4. Hashim, N.W., Wilkes, M., Salomon, R., Meggs, J., France, D.J.: Evaluation of voice acoustics as predictors of clinical depression scores. J. Voice **31**(2), 256–e1 (2017)
5. Huang, N.E., Shen, Z., Long, S.R., Wu, M.C., Shih, H.H., Zheng, Q., Yen, N.C., Tung, C.C., Liu, H.H.: The empirical mode decomposition and the hilbert spectrum for nonlinear and non-stationary time series analysis. In: Proceedings of the Royal Society of London A: Mathematical, Physical and Engineering Sciences, vol. 454, pp. 903–995. The Royal Society (1998)
6. Lemke, M.R., Wendorff, T., Mieth, B., Buhl, K., Linnemann, M.: Spatiotemporal gait patterns during over ground locomotion in major depression compared with healthy controls. J. Psychiatr. Res. **34**(4), 277–283 (2000)
7. Li, H., Kwong, S., Yang, L., Huang, D., Xiao, D.: Hilbert-huang transform for analysis of heart rate variability in cardiac health. IEEE/ACM Trans. Computat. Biol. Bioinf. **8**(6), 1557–1567 (2011)
8. Ooi, K.E.B., Lech, M., Allen, N.B.: Multichannel weighted speech classification system for prediction of major depression in adolescents. IEEE Trans. Biomed. Eng. **60**(2), 497–506 (2013)
9. World Health Organization: Depression and other common mental disorders: global health estimates (2017)
10. Oweis, R.J., Abdulhay, E.W.: Seizure classification in eeg signals utilizing hilbert-huang transform. Biomed. Eng. Online **10**(1), 1–15 (2011)
11. Pampouchidou, A., Marias, K., Tsiknakis, M., Simos, P., Yang, F., Meriaudeau, F.: Designing a framework for assisting depression severity assessment from facial image analysis. In: 2015 IEEE International Conference on Signal and Image Processing Applications (ICSIPA), pp. 578–583. IEEE (2015)
12. Pampouchidou, A., Simantiraki, O., Vazakopoulou, C.M., Chatzaki, C., Pediaditis, M., Maridaki, A., Marias, K., Simos, P., Yang, F., Meriaudeau, F., et al.: Facial geometry and speech analysis for depression detection. In: 2017 39th Annual International Conference of the IEEE Engineering in Medicine and Biology Society (EMBC), pp. 1433–1436. IEEE (2017)
13. Rai, V., Mohanty, A.: Bearing fault diagnosis using fft of intrinsic mode functions in hilbert-huang transform. Mechan. Syst. Signal Process. **21**(6), 2607–2615 (2007)
14. Sloman, L., Berridge, M., Homatidis, S., Hunter, D., Duck, T.: Gait patterns of depressed patients and normal subjects. Am. J. Psychiatry **139**, 94–97 (1982)
15. Stoll, A.L., Hausdorff, J.M., Peng, C.K., Goldberger, A.L.: Gait unsteadiness and fall risk in two affective disorders: A preliminary study (2004)
16. Tribbey, W., Press,W.H., Teukolsky, S.A., Vetterling, W.T., Flannery, B.P.: Numerical Recipes 3rd edition: The Art of Scientific Computing. Cambridge University Press, Cambridge (2007). Hardback, 1235 p. ACM (2010). ISBN 978-0-521-88068-8

17. Yang, Z., Yang, L., Qi, D., Suen, C.Y.: An emd-based recognition method for chinese fonts and styles. Pattern Recogn. Lett. **27**(14), 1692–1701 (2006)
18. Zhou, X., Jin, K., Shang, Y., Guo, G.: Visually interpretable representation learning for depression recognition from facial images. IEEE Trans. Affect. Comput. (2018)
19. Zong, C., Chetouani, M.: Hilbert-Huang transform based physiological signals analysis for emotion recognition. In: 2009 IEEE International Symposium on Signal Processing and Information Technology (ISSPIT), pp. 334–339. IEEE (2009)

A Study on Depression Detection Using Eye Tracking

Shuai Zeng, Junhong Niu, Jing Zhu[✉], and Xiaowei Li

School of Information Science and Engineering, Lanzhou University,
Lanzhou, China
{zengsh16,niujh16,zhujing,lixwei}@lzu.edu.cn

Abstract. Depression has become one of the most common mental illnesses in the past decade, affecting millions of patients and their families. However the methods of diagnosing depression almost exclusively rely on questionnaire-based interviews and clinical judgments of symptom severity, which are highly dependent on doctors' experience and makes it a labor-intensive work. Our study aims to develop an objective and convenient method to assist depression detection using the eye tracking technology. Eye movement data was collected from over 50 subjects using an emotional faces free viewing task paradigm. After data preprocessing, the highest accuracy of 76.04% was achieved by the Support Vector Machine (SVM) classifier. Results indicate that with the improvement of the classification accuracy, eye movement features hold the potential to form a feasible method for depression detection.

Keywords: Depression detection · Eye tracking · Classification

1 Introduction

Depression is a common mental disorder that already affects more than 350 million people worldwide [1]. It will not only make a bad influence on the patients but also on their families. The World Health Organization said that depression will become the second leading cause of illness by the year 2020 [2]. However, the assessment methods of diagnosing depression almost exclusively rely on the patient-reported and clinical judgments of the symptom severity [3]. Current diagnostic techniques of depression have obvious disadvantages, which are associated with the patient denial, poor sensitivity, subjective biases and inaccuracy [4]. Finding an objective, accurate and practical method for depression detection still remains a challenge.

Some diagnostic techniques of depression can achieve over 90% accuracy in depression detection, such as the EEG (Electroencephalo-graph) and sMRI (Structural Magnetic Resonance Imaging). However, they all need the professional and extremely expensive apparatuses. By contrast, eye tracking is much more accessible (even the Kinect V2 can support eye tracking and it only cost less than 200$), so it is growing in popularity amongst the researchers from different disciplines. Usability analysts, sports scientists, cognitive psychologists, reading researchers, neurophysiologists, electrical engineers, and other have a vested interest in eye tracking for different reasons [5]. Compared with EEG and sMRI the convenience and accessibility make eye tracking a

© Springer Nature Switzerland AG 2019
Y. Tang et al. (Eds.): HCC 2018, LNCS 11354, pp. 516–523, 2019.
https://doi.org/10.1007/978-3-030-15127-0_52

more practicable approach for depression detection in mass usage. In the future, with the development of the camera and sensor in the mobile phone we can even make an app for depression detection with eye tracking (now iPhone X can support eye tracking for some functions such as face ID). Although the classification accuracy of eye tracking for depression detection is inferior to EEG and sMRI, the generalizability and accessibility are irreplaceable advantages of eye tracking.

1.1 Related Work

Dating back to the spread of psychology's "cognitive revolution" to psychotherapy, attentional biases for emotional stimuli have been a key mechanism in theoretical accounts of affective disorders [6]. Therefore, most of the previous studies mainly focus on finding out the difference of attention bias between depressed patients and healthy controls. Peckham suggested an attention bias for dysphoric stimuli and possible neglecting of positive stimuli in depression [6] and Ms. Li also found an attention bias using a free view task paradigm [7]. Besides, as far as we know, only a few of studies pay attention to the depression detection using eye movement. Li used a free view task paradigm and KNN to got 81% accuracy for mild depression detection [8]. Sharifa used an audio-video experimental paradigm and SVM to got 75% accuracy for major depression detection [9]. The recent evidence in the mental health assessment have demonstrated that the facial appearance could be highly indicative of depressive disorder [10–12]. Over all, in this paper we used an emotional faces free viewing task paradigm to detect major depression.

1.2 Subjects

For the experimental validation, we used the real-world data, collected in a study at Lanzhou University Second Hospital in Lanzhou, China, a hospital granted with the title of Class A Grade 3 Hospital. The subjects in this study included patients who have been diagnosed with depression as well as healthy controls. Data was collected from over 28 depressed subjects who had been diagnosed with depression, and over 24 healthy controls. Data was acquired after obtaining the informed consent from the participants in accordance with the approval from the local institutional ethics committee. In this paper, to balance the gender, age, and education level, we chose 18 subjects undergoing depression and 18 healthy controls (shown in Table 1). Although we have to admit that the amount of data is small, it is a common problem in similar studies. As we continue to collect more data, the future study will be able to report on a large dataset. All participants gave informed consent before enrolment into the study, which was approved by Lanzhou University Second Hospital's ethics committee. Each participant received a reimbursement of approximately USD $16 for the participation after experiment.

1.3 Apparatus

The eye movement data was collected by an EyeLink 1000 Eye Tracker (SR Research Ltd., Mississauga, Ontario, Canada) with a sampling rate of 250 Hz. We only recorded

Table 1. Basic information for experiment subjects

Variables (Mean ± S.D.)	Depressed patients (n = 18)	Healthy controls (n = 18)	p
Gender (Males: Females)	9:9	10:8	0.747
Age (Years)	31.56 ± 8.60	31.33 ± 9.26	0.941
Education level (Years)	15.06 ± 3.04	15.50 ± 3.24	0.674
PHQ9	18.17 ± 4.00	3.06 ± 4.06	0.000*

P = p-value

the eye movement from the left eye of the subjects, because two eyes of people without eye diseases have the same movement patterns. The experimental stimuli were shown on a 17 inch liquid crystal display monitor at a resolution of 1024×768. The participants' eyes were kept at a distance of approximately 60 cm from the monitor and 60 cm from the eye tracker a fixed chin rest was used to keep the participants' heads steady.

1.4   Stimuli and Procedure

The stimuli we used consisted a set of images including 24 happy faces (12 males, 12 females), 24 sad faces (12 males, 12 females), 24 angry faces (12 males, 12 females), 24 astounded faces (12 males, 12 females), and 64 neutral faces (32 males, 32 females). All the pictures were selected from the NimStim Set of Facial Expressions image library [13]. All images were processed by Photoshop software and the size, gradation and resolution were made consistent (i.e. image size 190×250 pixels; 6.5×5 cm). Subjects would see four pictures in one trail and all the stimuli could be divided into four kinds (as shown in Fig. 1). The first kind included 1 emotional face expression and 3 neutral face expressions, the second kind included 2 emotional face expressions and 2 neutral face expressions, the third kind included 3 emotional face expressions and 1 neural face expression, and the fourth kind included 4 emotional face expressions. The emotion of stimuli could be happy, sad, astounded, and angry. There would not be the same emotion in one stimulus except neutral face expression because the neutral face expression was used as comparison in the whole experiment. The positions of the emotional images were counter balanced.

Firstly, a calibration would be performed to ensure that our eye tracker can catch the pupil and the eye movement data were recorded accurately (an error of below 0.5° had been achieved). Secondly, there would be 4 practice trails to make sure that subjects were familiar with the experimental procedures and reduce mistakes, the practice trails were totally the same as real trails. Each trial began with a white cross on a gray background. This cross was presented as a prompt for 1 s in the center of the screen to inform the participant that the emotional picture is to come. Participants were asked to stare at the fixation cross. Afterwards, an image with four faces was displayed for 3 s and participants were instructed to view them freely. A rest time of 2 s preceded the next trail. There were a total of 40 trails for each subject.

In one stimulus, each face will be regarded as an interest area (every stimulus has four faces, therefore there will be four interest areas) and the fixation time on each

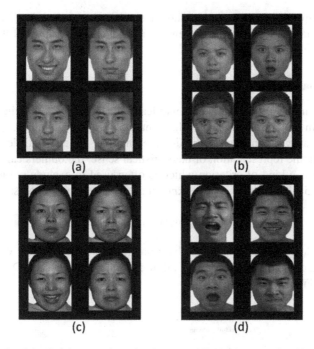

Fig. 1. Example of the facial expression stimulus materials: (a) one emotion (the emotion can be happy, sad, astounded, angry) faces and 3 neutral faces, (b) 2 emotion faces and 2 neutral faces, (c) 3 emotion faces and 1 neutral faces, (d) 4 emotion faces.

interest area was designated the total fixation time independently. Each stimulus would produce four data records, one subject would have 160 data records, for instance in Fig. 1(a) would produce one data record for happy face and three data records for neutral faces.

2 Data Processing

2.1 Data Cleaning

Because the experiment used a free view task paradigm, which would produce data record for each face in one stimulus, it might produce empty values in some specific situations. For example, if the participants blinked for a long time or blinked too frequently the eye tracker may lose the pupil of the subject and this situation will surely produce a lot of empty values in one trail, another situation is that the participants who did not focus on the interest areas can also lead to missing data.

There are different strategies to deal with tuples with empty values. If a tuple contains many attributes (more than 50% amount of attributes) with missing values, this tuple will be deserted. If a tuple just contains a few attributes with missing values then we use a measure of central tendency for the attribute (e.g., the mean or median) to fill in the missing value. For normal (symmetric) data distributions, the mean can be used, while skewed data distribution should employ the median [14]. In this paper because our data distribution was a skewed distribution, we employed the median to filling the missing value.

2.2 Data Quality Control

Outliers mixed in the experimental data will lead to inaccuracy of experimental results. Hence to obtain the correct results, outliers must be properly eliminated. The outliers are not necessarily wrong but only unrepresentative on the statistical properties. Therefore, from the representation of the statistical sense, the outliers can be removed. Pauta criterion is a kind of criterion which is often used to exclude outliers in experimental data [15]. Pauta criterion, also known as 3σ criterion, is expressed as follows,

$$|x_i - \bar{x}| > 3\sigma, (i = 1, 2, \ldots, m) \tag{1}$$

The x_i would be considered as an abnormal value and be rejected, where σ is the standard deviation, and $\bar{x}$ is the mean of all measurement values [16]. Detailed process is depicted in Fig. 2.

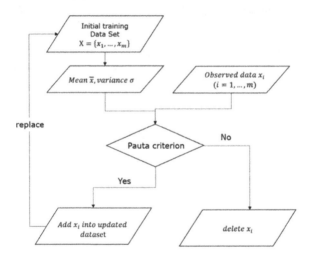

Fig. 2. The flow chart of data quality control

2.3 Data Standardization

The measurement unit can have effect on the data analysis. In general, expressing an attribute in smaller units will lead to a larger range for that attribute, and thus tend to give such an attribute greater effect or "weight". To help avoid dependence on the choice of measurement units, the data should be normalized or standardized. This involves transforming the data to fall within a smaller or common range such as [−1,1] or [0.0, 1.0]. (The terms standardize and normalize are used interchangeably in data preprocessing, although in statistics, the latter term also has other connotations.) [14].

In this paper, we used method of z-score normalization for analysis. Here we give the mathematical expression of the method. Let A be a numeric attribute with n observed values, $v_1, v_2, \ldots, v_m$. In z-score normalization (or zero-mean normalization), the values for an attribute, A, are normalized based on the mean (i.e., average) and standard deviation of A. v_i, is normalized to v_i' by computing:

$$v_i' = (v_i - \bar{A})/\sigma, (i = 1, 2, \ldots, m) \qquad (2)$$

Where $\bar{A}$ and σ_A are the mean and standard deviation, respectively, of attribute A.

2.4 Feature Used

The raw data used in this paper was generated by the EyeLink Data Viewer (SR Research Ltd., Mississauga, Ontario, Canada), which is used to display, filter, and create output reports from EyeLink 1000 EDF data files. In this study, we selected 87 attributes of eye movement data, which could directly and continuously record visual attention such as average_fixation_duration, fixation_count, fixation_duration_max, fixation_dura-tion_max_time, fixation_duration_min, fixation_duration_min _time, pupil _size_mean, pupil_size_max, pupil_size_max_time, pupil_size_max_X, pupil_size_max_Y, pupil_-size_min, pupil_size_min_time, pupil_size_min_X, pupil_size_min_Y and etc.

3 Results and Discussion

A 10-fold cross validation was used in this paper, training data and testing data were strictly separated. Data samples from one subject could not be divided into both training data and test data, lest it achieve a falsely high classification accuracy. From a large number of classifiers we chose: k-Nearest Neighbor (KNN), Support Vector Machine (SVM), Naïve Bayes, Bayes Net, Random Forest, J48, and Logistic Regression, which are often used in other cognitive researches. We also compared the results that used the feature selection algorithm and the results that directly used all features. The idea of the feature selection is to choose a subset of features that improve the performance of the classifier especially when we deal with high dimension data. In this paper, we adopted Correlation-based Feature Selection (CFS) together with the search method Best First or Greedy Stepwise, on account of selecting a subset of features which are highly correlated with the class while having low inter-correlation.

Table 2 demonstrated that the SVM classifier got the highest accuracy of 76.04% ± 3.83%, this result is in accordance with the relative researches mentioned previously. As seen in Table 2, the Random Forest, SVM and KNN classifiers are better than other classifiers, these three algorithms achieve the best classification accuracies and this phenomenon is same as the relative studies both in the eye tracking and EEG [8, 9, 15–17], in their studies the SVM, KNN and Random Forest classifiers also outperformed than other classifiers. From Table 2 we also can see that the feature selection algorithms didn't improve the accuracies remarkably, in some classifiers the results of using feature selection algorithm even worse than using all features. The statistics results of our study showed no statistically significant differences and according to our results it was difficult to demonstrate attentional bias in depression in a statistical way but this result was also same to many researchers' results [18].

Table 2. Classification results of several classifiers

Classifiers	Best first	Greedy stepwise	All features
BayesNet	69.24% ± 7.00%	71.10% ± 7.73%	70.40% ± 7.15%
Logistic	59.21% ± 8.40%	63.60% ± 7.05%	63.27% ± 7.91%
RandomForest	73.07% ± 7.75%	72.37% ± 6.25%	74.29% ± 7.68%
J48	49.09% ± 9.68%	54.23% ± 8.56%	58.55% ± 8.58%
NaiveBayes	59.02% ± 8.48%	61.10% ± 6.27%	64.13% ± 5.62%
SVM	73.53% ± 3.99%	74.01% ± 2.66%	76.04% ± 3.83%
KNN	72.34% ± 6.30%	73.99% ± 6.92%	74.55% ± 6.07%

The accuracy of the eye tracking is not as high as EEG in depression detection area. However the eye tracking possess several irreplaceable advantages. Firstly, the price of equipment used in eye tracking studies are much lower than EEG equipment. Secondly, in recent years, the eye tracking is becoming more and more easily to access than ever (even the Kinect V2 can support eye tracking and it only cost less than $200). In the future, with the development of the camera and sensor in mobile phone we can even make an app for depression detection using eye tracking. Finally, the raw data of eye tracking contains few artifacts to remove, it even can be used directly and give a diagnosis very fast. For now depression detection researches using the eye tracking still in its infancy, we believe this situation will be improved with the development of eye tracking equipment and data mining methods. With a higher accuracy, eye tracking may become an ideal technology for depression detection.

4   Future Work

In the future, we will continually focus on depression detection algorithm improving and using a different experiment paradigm to collect eye movement data to get a more reliable classification accuracy for real-time applications, which will make eye tracking a more convenient, economical, and a popular method for depression detection.

Acknowledgement. This work was supported by the National Basic Research Program of China (973 Program) [No. 2014CB744600]; the National Natural Science Foundation of China [Nos. 61632014, 61210010]; the International Cooperation Project of Ministry of Science and Technology [No. 2013DFA11140]; and the Program of Beijing Municipal Science & Technology Commission [No. Z171100000117005].

References

1. Marcus, M., Yasamy, M.T., van Ommeren, M., Chisholm, D., Saxena, S.: Depression: a global public health concern. WHO Department of Mental Health and Substance Abuse 1, 6–8 (2012)
2. Brundtland, G.H.: Mental health: new understanding, new hope. JAMA **286**, 2391 (2001)
3. Mundt, J.C., Snyder, P.J., Cannizzaro, M.S., Chappie, K., Geralts, D.S.: Voice acoustic measures of depression severity and treatment response collected via interactive voice response (IVR) technology. J. Neurolinguistics **20**, 50–64 (2007)
4. Sung, M., Marci, C., Pentland, A.: Objective physiological and behavioral measures for identifying and tracking depression state in clinically depressed patients. Massachusetts Institute of Technology Media Laboratory, Cambridge, MA, Tech. Rep. TR 595 (2005)
5. Holmqvist, K., Nyström, M., Andersson, R., Dewhurst, R., Jarodzka, H., Van de Weijer, J.: Eye tracking: a comprehensive guide to methods and measures. OUP Oxford (2011)
6. Peckham, A.D., McHugh, R.K., Otto, M.W.: A meta-analysis of the magnitude of biased attention in depression. Depression Anxiety **27**, 1135–1142 (2010)
7. Lu, S.: Attentional bias scores in patients with depression and effects of age: a controlled, eye-tracking study. J. Int. Med. Res. **45**, 1518–1527 (2017)
8. Li, X., Cao, T., Sun, S., Hu, B., Ratcliffe, M.: Classification study on eye movement data: towards a new approach in depression detection. In: 2016 IEEE Congress on Evolutionary Computation (CEC), pp. 1227–1232. IEEE (2016)
9. Alghowinem, S., Goecke, R., Wagner, M., Parker, G., Breakspear, M.: Eye movement analysis for depression detection. In: 2013 20th IEEE International Conference on Image Processing (ICIP), pp. 4220–4224. IEEE (2013)
10. Alghowinem, S., et al.: Multimodal depression detection: fusion analysis of paralinguistic, head pose and eye gaze behaviors. IEEE Trans. Affect. Comput., 1 (2017)
11. Zhu, Y., Shang, Y., Shao, Z., Guo, G.: Automated depression diagnosis based on deep networks to encode facial appearance and dynamics. IEEE Trans. Affect. Comput., 1 (2017)
12. Girard, J.M., Cohn, J.F., Mahoor, M.H., Mavadati, S.M., Hammal, Z., Rosenwald, D.P.: Nonverbal social withdrawal in depression: evidence from manual and automatic analysis. Image Vis. Comput. **32**, 641–647 (2014)
13. Tottenham, N., et al.: The NimStim set of facial expressions: judgments from untrained research participants. Psychiatry Res. **168**, 242–249 (2009)
14. Han, J., Pei, J., Kamber, M.: Data Mining: Concepts and Techniques. Elsevier, New York (2011)
15. Zhang, X.-L., Liu, P.: A new delay jitter smoothing algorithm based on Pareto distribution in cyber-physical systems. Wirel. Netw. **21**, 1913–1923 (2015)
16. Zhao, H., Min, F., Zhu, W.: Cost-sensitive feature selection of numeric data with measurement errors. J. Appl. Math. **2013**, 1–13 (2013)
17. Cai, H., Zhang, X., Zhang, Y., Wang, Z., Hu, B.: A case-based reasoning model for depression based on three-electrode EEG data. IEEE Trans. Affect. Comput. 1 (2018)
18. Mogg, K., Bradley, B.P.: Attentional bias in generalized anxiety disorder versus depressive disorder. Cogn. Ther. Res. **29**, 29–45 (2005)

Research on the Interaction Design of AR Picture Books via Usability Test

Rui Cao$^{(\boxtimes)}$ and Wenjun Hou

School of Digital Media and Design Arts,
Beijing University of Posts and Telecommunications, No 10, Xitucheng Road,
Haidian District, Beijing 100876, People's Republic of China
1170695087@qq.com

Abstract. In children's books, picture books account for seventy percent. With the development of information technology, picture book with Augmented Reality (AR) is an emerging application under the influences of digital technologies, which is a new kind of picture book with the auxiliary of games. The AR picture book bases on the children's cognitive level and the operation of the mobile devices, satisfying their needs of exploring and learning new things. Under the literature review and competitive analysis, we exploratively put forward the prototypes of interactive AR picture books, especially for the kids aged from 5 to 8. Furthermore, we designed the usability test on one of our prototypes, Rocket Dream, to better understand the process of interaction design. According to the results, we testify the feasibility of AR picture books as well as figure out some issues in the interaction processes, such as the weak guidance in the interactive operation, incomplete essential information in the interface, as well as the fact that children's preference of visible interaction and touch operations and they are more sensitive to large objects. Also, we propose potential solutions to these issues in the following parts, which could be useful for the future of AR picture books.

Keywords: AR picture books · Human-computer interaction ·
Interaction design · Child education · Usability test

1 Introduction

With the popularity of mobile intelligent terminal and the development of mobile communication network technology, the development of mobile applications start booming. More and more children are beginning to contact digital mobile platform. Digitalization is changing the children's reading, learning, education methods [1]. Gradually the standard of people's life is improving, and the investment to children's education also gradually increase. And the development of children-related applications become popular. And with the emergence of a growing number of mobile applications, the forms of human-computer interaction are becoming more and more diverse, from the initial simple clicks and drags, to today's more attractive interactions, like the multi-touch and the virtual space. People continue to pursue a more rich, three-dimensional and more vivid interactive experience.

© Springer Nature Switzerland AG 2019
Y. Tang et al. (Eds.): HCC 2018, LNCS 11354, pp. 524–534, 2019.
https://doi.org/10.1007/978-3-030-15127-0_53

The picture book is an important medium for children to read and study, which is described as "the presentation of early childhood culture" and is also known as the golden key to unlocking the treasure house of life [2]. And the single reading model in the traditional kind already can't satisfy the needs of children exploring the world. Children in growth keep looking for more interesting and more operational models of learning. The diversity and technicality in children's publications directly impact on the experience gained by the children in reading [3]. Picture book combining with Augmented Reality (AR) emerges under this background, with the help of AR interactions, to combine the real space and virtual 3D models. Also, it could maximally give a virtual scene, containing complete and convenient experience, which can not only meet the paper reading experience, but offer a magical effect from plain text to 3D motions, to bring brand-new "more than one paper screen" reading experience.

According to the 2010 Children's User Scale and Internet Behavior Survey Report [4], by 2010, the number of children netizens in China had reached 89.582 million, and the penetration rate of children using the Internet had reached 51%, of which 67.7% of children think that the Internet is their favorite form of media. 5–8 years old child is in a critical period of the development of education. At this stage, children have strong curiosity and strong thirst for knowledge, thinking and consciousness. Although their consciousness fails to certainly fit the objective world, they already have preliminary abstract thinking. However, AR picture books could covert the obscure knowledge, into dynamic learning contents, to satisfy children's curiosity, attract their attentions, enhance the multi-dimensional knowledge transition, improve the experience of children's education, and also increase the communications between parents and children.

Interaction activity is the biggest difference between AR children picture books and traditional paper picture books. A good interaction design can attract children to read, enhance children's understanding of the content, and improve the immersive reading. It allows the transformation from the previously "fragmented" browsing reading mode into a state of in-depth reading in a visual, sensible space [5]. The AR children's picture books develop rapidly, while many products are not particularly mature. The main issues are listed following: the age positioning is not clear; the performance of the user experience is poor; the interactive design is not in conformity with the children's characteristics of operation and so on. This paper aims to focus on the improvement of interaction design in AR picture book, via the usability test on current prototype of it.

2 Background in AR Picture Books

Virtual Reality (VR) is a kind of computer simulation system, which can create and experience the virtual world. It utilizes computers to generate a simulation environment in both vision and physic, which combines multi-source information and interactive 3D, from the visual, auditory level to provide virtual scenarios for users.

Augmented Reality (AR) is a growing technology on basis of VR. Compared with VR, AR has real and simple scenario models. Augmented Reality technology provides the possibility to realize a variety of information presentation under the combination of virtual and real. No matter what kind of interface mode [6], it will rely on AR technology for visual presentation to achieve the purpose of natural interaction. The report of Battelle Memorial Institute showed top 10 strategic technics in 2020, where AR ranked 10th place [7].

The picture book is not only a tool for children to read, but also a means for children to participate in interaction and exchange of ideas, hence the interaction of children's picture books is particularly important [8]. At present, in the entire AR book market, the proportion of books in early childhood education is as high as 90%, while the AR publications in the field of children's education and publishing mainly focus on children's knowledge learning publications, children's science publications and text-books [9]. The common format of AR books is "books + mobile devices + app".

At the beginning, AR picture books allow users scanning 2D figures to display 3D models with pre-set gestures. After a period of time, the trend of AR picture books evolves from "look" to "play". "Paint & Fun" is a new interaction of AR game, which has white paper and wireframes for painting. After scanning with the mobile devices, users could see the 3D model of their paints, also they could make some interactions with the models, such as moving, zooming, rotating, etc. Today's AR picture books possess multimodal interaction patterns, and pay great attention to the optimal presentation mode of the combination of virtual and real interface and 3D model [10]. As shown in Fig. 1, taking The Adventure of Trilobite Pangpang AR book as an example, readers could download apps on their mobile phones or tablet. Then they could use the App scanning on the specified page in the book, to see dinosaurs. Through clicks and drags by single finger, and scaling by two fingers, they can interact with the dinosaurs, like rotation and scaling, accompanying with the sounds of roar.

Fig. 1. The Adventure of Trilobite Pangpang Interaction diagram

3   Rocket Dream

Rocket Dream uses the combination of multiple comics and cross-page with a total of 19 large pages, of which 10 pages are storytellings and 9 pages are interactive games, where the interactive games are interspersed in the story. The story draws forth the knowledge point explanation of the interactive game, while the interactive game promotes the narrative process of the story. The screenshot of the picture book is shown in Fig. 2.

The application interface and functions are shown in Figs. 3, 4 and 5.

Fig. 2. The finished book and picture book of *Rocket Dream*

Fig. 3. Virtual and real interaction - space-aware rocket structure assembly game

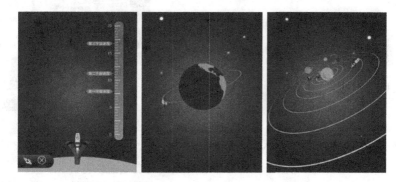

Fig. 4. Virtual and real interaction - rocket launch game

Fig. 5. Virtual and real interaction - AR coloring interactive gameplay

The interactive games are divided into the following 9 categories. The specific types and screenshots are shown in Table 1:

Table 1. *Rocket Dream* interactive game classification

Content	Interactivity Type	Screenshot
History Introduction	2D Interactive Animation	
Aircraft	Paper Painting	
Cosmic velocity	Entertainment Game	
Spacecraft	Paper Painting	
Rocket parts	Virtual Assemble	
Launch process	Video Display	
Solar system	Virtual 3D Model	
Space station	2D Plane Assemble	
Rocket return cabin	Video Display	

4 Usability Test Design

This experiment aims to evaluate the usability of the AR interactions in *Rocket Dream* AR. According to the interview, we collect the data of their feedbacks. Then, we suggest potential improvements based on the analysis of the data.

4.1 Sample Space

Details of the sample space are shown in the Table 2.

Table 2. Sample space of the volunteers

Volunteer	Gender	Age	Grade	School	Parents' occupation	Ever tried AR products?	Ever read AR picture books?
A	M	5	Preschool	Houxiaohe Primary School	Staff	No	No
B	M	6	1st	Houxiaohe Primary School	Middle school teacher	No	No
C	F	6	1st	Beijing World Youth Academy	Staff	No	No
D	F	7	1st	Beijing World Youth Academy	Freelancer	No	No
E	F	7	1st	Beijing World Youth Academy	Freelancer	No	No
F	F	8	2nd	Beijing World Youth Academy	College teacher	Yes	Yes
G	M	8	2nd	Houxiaohe Primary School	Middle school teacher	No	No
H	F	8	2nd	Houxiaohe Primary School	Doctor	Yes	Yes

4.2 Tasks

Task1: read the instruction, open the app and use the scan function to scan one page
Task2: turn to page 10, launch a rocket with the second cosmic velocity

Task3: turn to page 17, find the suspending fairing, complete assembling and launch the rocket

Task4: turn to page 24, find the orbit period of Venus

4.3   Process

Test is performed in the usability laboratory, with independent test room and observe room. There are two video cameras in the testing room: one is for facial expressions and the other is for the interactions between volunteer and mobile device. And tester takes the responsibility of observing and note taking. Detailed process is shown in the Fig. 6.

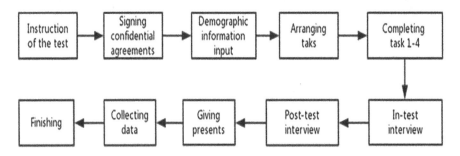

Fig. 6. The flowchart of testing process

5   Results

The metrics of this test are error rate, efficiency and user satisfaction.

Error is divided into two categories: technical error (system collapse, power off, etc.) and non-technical error.

Success: finish the test with/without hint

Hint: finish the test after hint

Help: finish the test after help

Independent Success (Ind. Success): finish the test without hint

Error rate is shown in the Table 3.

Efficiency was counted based on the time spent on each task, which is timed between the start word "start" and end word "finish" by the facilitator. Detailed data is shown in the Table 4.

User satisfaction was collected by the in-test interview. Volunteers were asked to grade each task from four perspectives, difficulty, expectation, fault tolerance, and learnability. The grade was scaled from 1 to 5. And volunteers were asked to explain their grades if the grade was below 2.

Table 3. Error rate

Success		Volunteer									
		A	B	C	D	E	F	G	H	Success rate	Error rate
Success	Task1	Yes	Yes	Yes	Yes	Yes	Yes	Yes	Yes	100%	0%
	Task2	Yes	No	Yes	Yes	No	Yes	Yes	Yes	75%	25%
	Task3	Yes	Yes	Yes	Yes	Yes	Yes	Yes	Yes	100%	0%
	Task4	No	Yes	No	No	No	Yes	Yes	Yes	50%	50%
Hint	Task1	Yes	Yes	No	No	No	No	No	No	75%	25%
	Task2	Yes	Yes	Yes	Yes	Yes	Yes	Yes	Yes	100%	0%
	Task3	Yes	Yes	Yes	yes	Yes	Yes	Yes	Yes	100%	0%
	Task4	Yes	Yes	Yes	Yes	Yes	Yes	Yes	No	87.5%	12.5%
Help	Task1	No	No	No	No	No	No	Yes	No	12.5%	87.5%
	Task2	Yes	Yes	Yes	Yes	No	Yes	No	Yes	75%	25%
	Task3	Yes	Yes	Yes	No	No	No	Yes	No	50%	50%
	Task4	Yes	Yes	Yes	Yes	Yes	Yes	Yes	Yes	100%	0%
Ind. Success	Task1	No	No	No	No	Yes	No	No	Yes	25%	75%
	Task2	No	No	No	No	No	No	Yes	No	12.5%	87.5%
	Task3	No	No	No	No	No	No	No	Yes	12.5%	87.5%
	Task4	No	No	No	No	No	No	No	No	0%	100%

Table 4. Time spent on each task per volunteer

Task	Volunteer								
	A	B	C	D	E	F	G	H	Avg.
Task1	45 s	58 s	65 s	48 s	39 s	37 s	51 s	42 s	48 s
Task2	106 s	81 s	72 s	60 s	54 s	45 s	26 s	27 s	59 s
Task3	90 s	71 s	152 s	93 s	72 s	55 s	56 s	64 s	82 s
Task4	170 s	111 s	80 s	53 s	34 s	25 s	56 s	47 s	72 s

Q1: What is the difficulty of this task? 1 refers to hard, and 5 refers to easy.
Q2: Does the task result meet your expectation? 1 refers to no, and 5 refers to yes.
Q3: Do you think this task is fault tolerable? 1 refers to no, and 5 refers to yes.
Q4: How do you grade the learnability of this task? 1 refers to hard, and 5 refers to easy.

The grades of user satisfaction are shown in the Table 5.

Table 5. Results of user satisfaction

Task		Volunteer								
		A	B	C	D	E	F	G	H	Avg.
Task1	Q1	4	4	5	5	3	3	5	5	4.3
	Q2	2	4	2	4	1	1	1	2	2.1
	Q3	2	1	1	2	1	3	2	1	1.6
	Q4	4	5	4	5	4	4	5	5	4.5
Task2	Q1	4	3	2	2	3	3	3	3	2.9
	Q2	2	3	2	5	2	3	3	2	2.8
	Q3	3	4	2	4	3	4	4	3	3.4
	Q4	3	4	3	3	2	3	5	5	3.5
Task3	Q1	4	4	3	4	3	3	5	5	3.9
	Q2	2	4	5	5	5	2	5	4	4.0
	Q3	3	4	4	4	5	3	5	5	4.1
	Q4	3	4	3	3	2	4	5	4	3.5
Task4	Q1	2	1	4	4	1	3	2	1	2.3
	Q2	4	4	5	4	4	4	2	5	4.0
	Q3	1	1	2	5	5	5	2	2	2.9
	Q4	4	4	1	3	1	4	5	5	3.4

6   Discussion

Through the analysis of the behavior of children's using mobile devices, we find that compared with adults, children's acceptance to complex gestures is low. Children tend to click on the simple gestures. Their curiosity is strong. They may click on any elements on the screen to discovery the new contents, and carries on the repetitive operations. In addition, children aged from 5 to 8 are in the developing stage of their listening, speaking, reading and writing abilities. Comparing with the abilities of speaking, reading and writing, their abilities of voice recognition are relatively high. The voice interaction not only decreases the difficulty of the use of applications for children but also attracts their attentions [11]. There are three main practical issues in our test as following.

6.1   The Lack of Guidance

This issue appears in two occasions. Firstly, there is no feedback when the users give inputs or interactions, thus users will get confused. Secondly, there is no correct guidance or feedback when the users give wrong actions.

For our users, low-age kids, real-time feedback is required. Especially, AR is a brand-new interaction for them. It is necessary to let them know clearly where they are and what the next is. We could have a hint, like animation, for the next step if users do not give correct inputs in 5 s.

6.2 The Lack of Essential Information

Young-age users have uncertainties about the location of clickable areas. There are always invalid clicks. The design of icons directly effects the identification and efficiency of children while using, and simple icons with the necessary text can directly convey the operation instruction [12]. We would better distinct where clickable areas are by using special icons, shades, animations, etc. These areas should be good-looking and easy to understand. The size of the buttons and icons should also be appropriate, and the convenience of child operation should be satisfied. Moreover, interface feedback and tips should be in a way that children can easily understand and remember. In addition, the interaction design should be more visual, and should minimize the use of text as much as possible; the interactive interface should show the current state, so that children do not be confused and misunderstood [13].

6.3 Preference of Children for Visible Interaction and Touch Operations

Compared to clickable interactions, children are more sensitive to large objects and colorful interfaces. Children prefer to interactive operation by their hands. They love to walk around in the room or outsides best. The designers should bring the sensory experience into the design, where they could implement the improvements of visual, auditory, tactile and other sensory experience functions [14]. According to the facts, the design of AR interactions needs to meet the child's psychological and physiological cognition. Thus, we may have long-term games or tasks for kids. For example, kids need to find parts and pieces by walking around to assemble the rocket. At the same time, the interface will refresh per 10 min and those parts and pieces will appear, which could be set in the system backend. Also, the refreshing time may be reduced by giving correct answers to the pre-set questions.

7 Conclusion

The popularity of mobile Internet and smart phones promotes the development of AR picture books. And new technology gradually infiltrates into the field of younger users. The study by Saidin (2015) confirms that in learning and education, if new techniques cannot promote critical thinking, meaning generation, or metacognition, technology will create a process of passive learning [15]. Therefore, in order to impress consumers, designers need to focus on not only the contents, but also the interactive operation, like AR. And the user-centric design for kids is a key to the success of AR products [16]. Many AR children picture books on the current market does not match the cognitive ability of school-age children. They do not conduct comprehensive analysis and researches on the operation abilities and the characteristics of mobile devices for children. Also, the interactions are very simple, mostly in the form of 3D model and animations after scanning, which are poorly made. While now published AR educational picture books based on mobile devices, to some extent hinder the communication between parents and children. Parent-child reading can cultivate children's reading interest and ability, also can help with meeting the attachment psychology of children, and promote the social development of them [17].

Research work of this article mainly divides into four parts, introduction of product background, product status analysis, user research, and summary of user demand analysis. We prototyped an AR picture book, Rocket Dream, for kids aged 5 to 8, and conducted usability test. Then, we concluded the feasibility of improving parent-child communication and increasing child learning interest with the platform of AR picture book, by providing excellent interaction experience.

The deficiency and subsequent recommendations of this article: this article belongs to the exploratory research. The research on 5 to 8-year-old children's cognitive characteristics, behavior and psychological characteristics still needs further study. And the practice research of AR picture books is unrepresentative. Due to restrictions of time and resources, research results of this article still need further improvement and demonstration. In general, we believe there will be promising future of the AR applications, especially in children education and entertainment.

References

1. Luo, H.T.: Research on the interaction design of children picture books. Jiangnan University (2014)
2. Su, M.Y.: AR interaction design for children's parent-child picture book education
3. Zhao, Y.Q.: The application of AR and VR technology in children's publishing industry. News Dissemination (12), 108–110 (2016)
4. Liang, Y.H.: Research on the usability of mobile products for children and interaction design
5. Zhang, J., Yu, S.: The Change of AR Book Reading Mode in the Perspective of Cognitive Theory
6. Sonnenwald, D.H., et al.: Experimental comparison of 2D and 3D technology mediated paramedic physician collaboration in remote emergency medical situations
7. Qi, J.M., Wang, Y.: The application of AR picture books in early education. Technol. Inf. **13** (26), 158–159 (2015)
8. Hao, Q.: Interactive Design Study of Illustrations in Children's Picture Books
9. Yi, Q.: Discussion on the Application Status and Prospects of AR Technology in Book Publishing
10. Smith, C., Karayiannidis, Y., Nalpantidis, L., et al.: Dual arm manipulation—a survey
11. Jiu, Z., Huang, X.Y.: Predicament and countermeasure of the publication of digit children picture book on mobile devices. Univ. Publishing (2), 47–49 (2017)
12. Zheng, J.: Research on interactive product design of education for preschool children. Xi'an University of Technology (2014)
13. Vatavu, R.D., Cramariuc, G., Schipor, D.M.: Touch interaction for children aged 3 to 6 years: experimental findings and relationship to motor skills
14. Ye, J.Q.: Research on the development of AR+ children's picture books in the media environment. 2017.19.039
15. Saidin, N.F.: A review of research on augmented reality in education: advantages and applications. Int. Educ. Stu.
16. Xiong, H.: Aesthetic taste and the editing and publishing of domestic children's picture
17. Zhang, Y.: Application and design strategy of AR technology in children's science books

System Design of Driving Behavior Recognition Based on Semi-supervised Learning

Chaonan Xu[1(✉)], Yong Zhang[1(✉)], Da Guo[1], Wei Wang[2], and Baoling Liu[1]

[1] School of Electronic Engineering, Beijing University of Posts and Telecommunications, Beijing 100876, China
{xuchaonan, yongzhang, guoda, blliu}@bupt.edu.cn
[2] School of Chemical Engineering and Technology, Tianjin University, Tianjin 300072, China
wwangg@tju.edu.cn

Abstract. Driving Behavior Recognition (DBR) has always been a key problem in the field of vehicle driving and traffic safety. However, there are some problems in previous studies that only the coarse-grained features are extracted and a large number of unlabeled samples consume human and material resources to label them. In this paper, we propose a driving behavior recognition system based on semi-supervised learning to solve these problems. First, wavelet decomposition is used to process the sensor signal, giving a decomposition of sensor signal into a set of approximate and detailed coefficients. Then, we extract 300 features from the decomposed signal to capture the patterns of driving behaviors, which contains fine-grained features especially. And 34 features are selected by a feature selection algorithm based on random forest algorithm (RF). Finally, an improved semi-supervised algorithm is proposed to optimize the strategy of selecting the unlabeled samples. And the experiment results show that fine-grained features are also effective and important relatively. And the improved algorithm has better classification performance compared to the previous algorithm using the dataset with different labeling rate.

Keywords: Fine-grained · Wavelet decomposition · Semi-supervised

1 Introduction

The WHO (World Health Organization)'s road safety report shows that traffic accidents have become one of the important causes of leading death [1]. In addition, with the rapid development of intelligent transportation and autonomous driving, a lot of researches are focusing on driving behavior recognition technology in order to reduce traffic accidents and improve driving safety.

There have been works on driving behavior recognition technology based on cameras [2, 3] and Inertial Measurement Unit (IMU) sensors of smartphone for data collection [4–7] which incur the high cost and unstable. The threshold analysis method [4], Hidden Markov Model (HMM) [3], k-Nearest Neighbor (k-NN) [5] and Dynamic

© Springer Nature Switzerland AG 2019
Y. Tang et al. (Eds.): HCC 2018, LNCS 11354, pp. 535–546, 2019.
https://doi.org/10.1007/978-3-030-15127-0_54

Time Warping algorithms [6] based on distance are used for classification and recognition. Since thresholds may be affected by car type and sensors' sensitivity, they cannot be used in the most driving scene. And Hidden Markov Model usually needs road conditions and status information of the driver for classification which relies on many kinds of data sources. As for methods based on distance, they require high computational complexity for analyzing data.

The methods above are based on raw data. In general, machine learning algorithms use a series of features as input rather than raw data. Therefore, it is more effective to extract features from the data for analysis and recognition. The authors extracted the features in the time domain of raw data based on the 3-axis accelerometer of smartphone sensors for distinguishing normal and abnormal driving behaviors [7, 8]. However, the traditional method of extracting features only in the time domain cannot dig out more important features of sensor data, which is coarse-grained. As a powerful tool for time-frequency domain analysis [9], wavelet analysis was used to extract time-frequency features in brainwave signal research, which has good effects [10, 11]. Since the brainwave signal and the sensor data signal are both time series, multi-resolution features in time-frequency domain can be extracted: approximate features and detail features of the sensor data signal using the method of wavelet analysis.

However, the existing methods of driving behavior recognition based on machine learning are almost supervised algorithms. Unfortunately, in the process of driving behavior data collection, there are only a few labeled samples, a large number of unlabeled samples are unutilized unless using manual labeling which is time-consuming and expensive. Semi-supervised learning [12] can make the classifier learn from a small number of labeled samples to label unlabeled samples, and improve the performance of supervised learning at the same time. In the field of semi-supervised learning, Zhou has done a lot of advanced research. Zhou proposed a Co-Forest algorithm based on ensemble learning [13] which has a better performance than Co-Training algorithm [14] and Tri-Training algorithm [15]. However, the Co-Forest algorithm simply selects the unlabeled samples to be added to the labeled set only by the high confidence of the other component classifiers, while the low confidence samples are wasted. Active learning can improve this shortcoming [16]. Considering that the low confidence samples may have more helpful information, the label of low confidence samples can be provided by medical experts, such as manual labeling [17] in Query by Committee [18], an active learning mechanism. However, this is not feasible because some unlabeled samples cannot be recognized through the naked eye of people.

Based on the above analysis, the main contributions of this paper can be summarized as follows:

(1) Instead of only coarse-grained features being extracted, a fine-grained feature extraction method based on wavelet decomposition is proposed to excavate the approximation and detail of the sensor signal. This method aims to observe signals in multiple resolutions (also called multi-scale) and dig out more relatively important features of the sensor signal.

(2) An improved semi-supervised algorithm is proposed to optimize the strategy of selecting the unlabeled samples, which can make full use of unlabeled samples in

the process of collecting driving behavior data. This algorithm implements an automated labeling process instead of manual labeling to save human and material resources in the process of driving behavior data collection.

The remainder of this paper is organized as follows: Sect. 2 presents the parts of the system model and gives detail algorithms. The experiment results are presented in Sect. 3 and Sect. 4 concludes the paper.

2 System Design

2.1 Overview

In previous studies of the field of Driving Behavior Recognition (DBR), only the coarse-grained features are extracted and a large number of unlabeled samples consume human and material resources to label them. In this section, we present the architecture of our proposed system to solve these problems. In this paper, we collect driving behaviors samples using MPU-6050 according to readings from 3-axis acceleration sensor $A = [a_x, a_y, a_z]$ and the 3-axis angular velocity sensor $G = [w_x, w_y, w_z]$. And the whole data set contains a labeled set and unlabeled set. The whole system of Driving Behavior Recognition is separated into three part, data processing, feature extraction and selection, training and classification. It is as shown in Fig. 1.

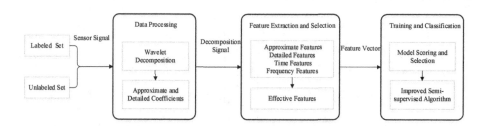

Fig. 1. System architecture

In the process of data processing and the process of feature extraction and selection, different from the traditional method of extracting features only in the time domain, this paper mainly uses wavelet decomposition to extract more detailed and fine-grained features of signals. In the process of training and classification, based on the effective features chosen by the selected model, this system used an improved semi-supervised algorithm to optimize the strategy of selecting the unlabeled samples.

2.2 Data Processing

The Mallat algorithm [19] is a fast algorithm for discrete wavelet transform, which can decompose and observe signals from coarse-grained to fine-grained in multiple resolutions. The wavelet transform which is based on the Daubechies mother wavelet gives a decomposition of sensor signal into a set of approximate (A_i) and detailed (D_i)

coefficients of the level i (i = 1, ...n, n ∈ Z). Each coefficient can correspond to a decomposition signal of the raw signal in the different frequency range. In this paper, every dimension of the sensor signal is decomposed into four levels, which obtains D_1, D_2, D_3, D_4 and A_4 five coefficients. The frequency range and length in the time domain of D_1, D_2, D_3, D_4 and A_4 are shown in Table 1. In a word, we obtain multiple resolution sensor signals by Mallat.

Table 1. Multi-resolution in the time-frequency domain

Wavelet decomposition coefficients	Frequency range (Hz)	Signal length (point)
A_4	$0 \sim f/16$	M/16
D_4	$f/16 \sim f/8$	M/16
D_3	$f/8 \sim f/4$	M/8
D_2	$f/4 \sim f/2$	M/4
D_1	$f/2 \sim f$	M/2
Raw	$0 \sim f$	M

Moreover, the A_4 coefficient can be regarded as the outline of the raw sensor signal, so we can distinguish the various driving behaviors from the A_4 coefficient in a visual and simple way at first. Figure 2 illustrates the raw signal and an A_4 signal of the 3-axis accelerometer (a) (b) and 3-axis angular velocity signal (c) (d).

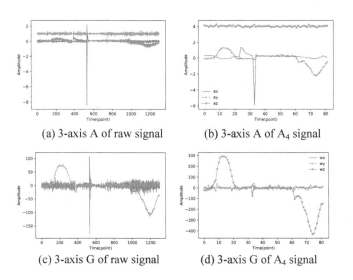

(a) 3-axis A of raw signal   (b) 3-axis A of A_4 signal

(c) 3-axis G of raw signal   (d) 3-axis G of A_4 signal

Fig. 2. Decomposition signal

2.3   Extracting and Selecting Effective Features

Feature Extraction

Before extracting features, each sample has six dimensions (acceleration $A = [a_x, a_y, a_z]$, angular velocity $G = [w_x, w_y, w_z]$). Each dimension is a time series. In order to make machine learning learn more features, it is necessary to extract features from raw data. In this paper, we extract 35 fine-grained features of D_1, D_2, D_3, D_4 and A_4 five coefficients from each dimension of raw data as wavelet features. The wavelet features are divided into 42 approximate features and 168 detailed features. We also extract 7 coarse-grained features in the time domain and 8 coarse-grained features in the frequency domain from each dimension of raw data. Finally, we obtain 300 features for each sample, including 210 newly extracted fine-grained features and 90 coarse-grained features. In this paper, we will more focus on 210 fine-grained features to dig out more relatively important features of the sensor signal.

Feature Selection

In order to select the more effective features and obtain the relatively important features, a feature selection algorithm is used in this paper which based on random forest algorithm (RF). The algorithm is described as follows.

(1) Calculate the importance score of features (*FI*).

The Gini coefficient (*GI*) is used as an indicator of feature importance evaluation [20]. For each tree in the random forest, the importance score of the feature X_j (*FI*) at the node p, that is, the change of the Gini coefficient before and after the node p splits into the left node and right node can be described by:

$$FI_{jp}^{(Gini)} = GI_p - GI_l - GI_r \tag{1}$$

As for a forest including n trees, the importance score of the feature X_j (*FI*) can be described by:

$$FI_j^{(Gini)} = \sum_{i=1}^{n} FI_{jp}^{(Gini)} \tag{2}$$

(2) Sorting feature variables in random forests in descending order of *FI*.
(3) Determine the selection ratio of 50%, select the top 50% important features from the current 300 features, and obtain a new feature set.
(4) Iterate 10 times, take the intersection of these 10 sets.

2.4   Improved Semi-supervised Algorithm

PAC with Noisy Examples

According to Angluin [21], the size of training data m, the hypothesis worst-case error rate ε and the noise rate η satisfy the following relationship:

$$m = \frac{c}{\varepsilon^2(1 - 2\eta)^2} \tag{3}$$

Where c is a constant. By reforming (3), the utility function can be described by:

$$u = \frac{c}{\varepsilon^2} = m(1 - 2\eta)^2 \tag{4}$$

According to (4), the Co-Forests algorithm and improved algorithm in this paper ensure that the classification error rate ε can be reduced iteratively, that is the utility function u can be increased iteratively in each iteration of adding new labeled samples into the training set.

The Co-Training Process of Co-Forest

Co-Forest is an ensemble algorithm contains N classifiers (Classification and Regression Tree, CART), which is donated by H^*. The H^* is divided into a component classifier h_i and other component classifiers in H^* except h_i, which is the concomitant ensemble of h_i, denoted by H_i. Let L and U denote the labeled set and unlabeled set, in each learning iteration of Co-Forest, H_i predicts each sample s in U. If the voting rate, that is confidence value exceeds a threshold θ, the sample will be copied into labeled set L, the new labeled set is donated by L'.

In the Co-Forest algorithm, the confidence value for an unlabeled example is simply estimated by the number of classifiers that agree on the label assigned by H_i. There are two kinds of noisy samples, the noisy samples in L and the samples in L_i that are misclassified by H_i. The noisy rate is donated by η_0 and e. e is obtained by the out-of-bag error estimation.

Improved Co-Forest Algorithm

In the Co-Forest algorithm, only high confidence samples are taken into account, the low confidence samples are wasted. However, Query by Committee (QBC) algorithm focuses on highly divergent samples. To a certain extent, it can be understood as low confidence sample in the Co-Forest algorithm. Learning from the idea of QBC, the active learning mechanism, this paper makes full use of all samples including low confidence samples.

The main contributions of the improved Co-Forest algorithm can be summarized as follows:

According to confidence with upper-threshold θ_H and lower-threshold θ_L, unlabeled samples U are divided into whiteness (high confidence) samples, grayness samples, and blackness (low confidence) samples, donated by U_w, U_g, and U_b. In each iteration, grayness samples will be divided into three sets U_w, U_g, and U_b by updated ensemble classifier H^*.

As for blackness samples, which are highly divergent samples, the ensemble classifier H^* can't predict the blackness samples accurately. Considering the clustering assumptions, this paper uses k-Nearest Neighbor (k-NN) to label blackness samples as experts labeling in QBC. This method can make full use of highly divergent samples which have more helpful information and features for ensemble classifier H^*.

Therefore, the noise rate η in each iteration can be expressed as:

$$\eta = \frac{m_{noisy}}{m_{all}} = \frac{\eta_0 * L + e_w * U_w + e_b * U_b}{L \cup U_w \cup U_b} \tag{5}$$

Where e_w and e_b are the noisy rate (error rate) of whiteness samples labeled by H^* and blackness samples labeled by the k-Nearest Neighbor algorithm, which are obtained by the out-of-bag error estimation.

The utility of H^* in each iteration can be expressed as:

$$u = m_{all}(1 - 2\eta)^2 = (L \cup U_w \cup U_b)(1 - \frac{\eta_0 * L + e_w * U_w + e_b * U_b}{L \cup U_w \cup U_b})^2 \tag{6}$$

In summary, the algorithm is described as follows:

Algorithm: Improved Co-Forest algorithm

Input: the labeled set L, the unlabeled set U, the noisy rate η_0 of L, the number of classifiers N, the upper-threshold θ_H and lower-threshold θ_L of confidence value.

Output: New-classifier H^*, new labeled set L'

1.   Construct a random forest H^* consisting N trees(CART) using L.
2.   **While**:
3.   Calculate confidence value : $conf(s)$, $s \in U$
4. $\{U_w, U_g, U_b\} \xleftarrow{\theta_H, \theta_L} U$
5.   Update L, $L \leftarrow L + U_w$
6. $e_w \leftarrow Outofbag Error(H^*, L)$
7.   Label s by finding k-nearest neighbors' labels of s, $s \in U_b$
8.   Update L, $L \leftarrow L + U_b$
9. $e_b \leftarrow Outofbag Error(H^*, L)$
10.   Update U, $U \leftarrow U_g$
11.   Calculate η_t and u_t of the t-th iteration
12.   **If** $(U = \Phi$ or $u_t < u_{t-1})$
13.   **Break**
14.   Update $H^*(L)$

---

3   Experiments and Evaluations

3.1   Data Set

In this paper, the experimental vehicle is a wireless remote control car (HSP 94111) which is equipped with the MPU6050, the SD card storage module and the WIFI wireless transmission module to realize the real-time data acquisition. The data is collected at 100 Hz. Initially, the data is stored in hexadecimal data and then converted to decimal data by parsing and calculation.

Finally, a total of 1200 samples were collected containing seven kinds of behaviors, acceleration, collision, constant speed, left turn, right turn and accelerated left turn and accelerated right turn. And each sample has acceleration $A = [a_x, a_y, a_z]$, angular velocity $G = [w_x, w_y, w_z]$ six dimensions.

3.2   Metrics

To evaluate the performance of the model, we adopt the Accuracy (ACC) and F1-Score (F1) as evaluation metrics based on the True Positive (TP), True Negative (TN), False Positive (FP) and False Negative (FN) which are described as Table 2.

Table 2. Description of TP, FP, TN, FN

The ground truth	Prediction	
	0	1
0	TP	FN
1	FP	TN

$$Accuracy = \frac{TP + TN}{TP + FN + FP + TN} \tag{7}$$

$$Precision = \frac{TP}{TP + FP} \tag{8}$$

$$Recall = \frac{TP}{TP + FN} \tag{9}$$

$$F1 = \frac{2 * Precision * Recall}{Precision + Recall} \tag{10}$$

3.3   Experimental Result

The original 1200 samples are time series of varying lengths. Each sample can be changed to a feature vector with the shape of 1*300 according to Sect. 2.3.

After the feature selection method is adjusted, relatively effective and important features are obtained finally. Instead of 300 dimensions, a feature vector with 34 dimensions is selected for each sample.

We can conclude that a_x, a_y, w_z are the main axes which distinguish driving behaviors. And instead of only coarse-grained features being extracted, the fine-grained features which are extracted in this paper are also effective and important relatively (see Table 3).

Table 3. Effective features of six axes

	a_x	a_y	a_z	w_x	w_y	w_z	Sum
Fine-grained features	2	5	0	1	0	7	15
Coarse-grained features	7	3	1	0	1	7	19
Total	9	8	1	1	1	14	34

Model Scoring and Selection

In this process, we compare several algorithms in order to select the model which has the best performance. We put our experiment on a computer with 4G memory and use machine learning packages which written in Python.

Six machine learning algorithms including Support Vector Machine (SVM), k-Nearest Neighbors (k-NN, k = 5), Naive Bayes (NB), Decision Tree (DT), Random Forests (RF) and Xgboost are applied for comparison. The input of these algorithms is a feature matrix with the shape of 1200 * 34. The original 1201 samples are divided into training set and test set. The percentage of the test set is 20% of the total data set.

As shown in Table 4, SVM, NB, and KNN have better performance with 34 features after feature selection than 300 features. And those algorithm based CART already has good performance with 300 features. Finally, KNN and those algorithm based CART have highest accuracy and F1-score with 34 features. In order to reduce the computational complexity of the model and reduce the over-fitting phenomenon, we finally choose the Random Forest (RF) model as the final model used in this paper. The accuracy of different behaviors with the RF Model is shown in Fig. 3.

Table 4. Accuracy and F1-score of different algorithms

Data sets	Metrics	SVM	NB	KNN	DT	Xgboost	RF
300 features	Acc	0.668	0.934	0.950	0.969	0.967	**0.971**
	F1	0.575	0.938	0.950	0.967	0.966	**0.970**
34 features	Acc	0.876	0.950	0.971	0.971	0.968	**0.971**
	F1	0.859	0.950	0.970	0.971	0.969	**0.971**

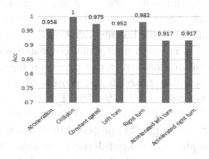

Fig. 3. The accuracy of different behaviors with RF Model

Results of the Improved Algorithm

Based on the RF model and 34 features chosen in this paper, the Weka interface, a machine learning tool written in Java is used to realize the improved semi-supervised algorithm. And a method of constructing virtual samples is used to make more samples to verify the performance of the algorithm. This method is similar to SMOTE (Synthetic Minority Oversampling Technique) algorithm [22]. The idea is to interpolate a new sample between two similar samples. And the method of this paper is that adding six noisy sequences to the six axes of a new sample, which have the same distribution as six axes.

There are new 2100 samples, that is, 300 samples for each type of driving behaviors. Finally, there are 3300 samples used in this part. It should be noted that the test set here also serves as a test set for supervised learning. Except for 20% samples in 1200 samples, the rest of the data set with 3300 samples is as the training set in this part.

In order to evaluate the performance of the improved algorithm, the training set is divided into a labeled set and unlabeled set with the different labeling rate. Moreover, the number of cooperative classifiers is 15 (N = 15), upper-threshold θ_H is 0.85 and the lower-threshold θ_L is 0.47. In each iteration with labeling rate of 1%, 2%, 5%, the change of whiteness samples (a), grayness samples (b), and blackness samples (c) are shown in Fig. 4.

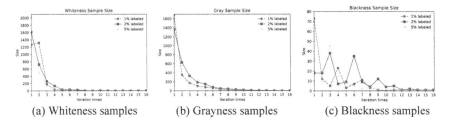

(a) Whiteness samples   (b) Grayness samples   (c) Blackness samples

Fig. 4. The change of unlabeled samples in each iteration

The results of the improved algorithm with the labeling rate of 1%, 2%, 5% is shown in Fig. 5. The accuracy and F1 are significantly improved in the first 2–3 iteration times with the large whiteness samples and darkness samples adding to the initial labeled set in Fig. 4. And at the eighth and ninth iteration, the accuracy and F1 reaches highest. This result shows that although the number of the labeled data set is always increasing, the samples in the last few iteration times has a little positive effect on the results. In addition, it is worth noting that compared with the Co-Forest algorithm, the improved algorithm proposed in this paper has higher accuracy. The accuracy of the improved algorithm raised 6.7%, 4.6% and 3.0% than the Co-Forest algorithm with the labeling rate of 1%, 2%, 5%. The F1 of the improved algorithm raised 4.9%, 5.4% and 2.1% than the Co-Forest algorithm with the labeling rate of 1%, 2%, 5%. In short, the improved algorithm proposed in this paper optimizes the strategy of selecting unlabeled samples and makes full use of all unlabeled samples. Moreover, the overall prediction accuracy rate reached 98.0% with labeling rate of 5% on test data set in Table 5 and shows good robustness.

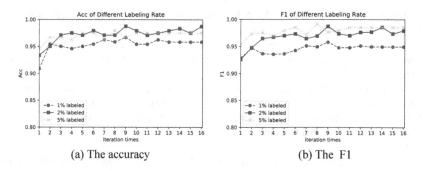

(a) The accuracy   (b) The F1

Fig. 5. The results of the improved algorithm in each iteration

Table 5. The comparison of the improved algorithm and Co-forest algorithm

Labeling rate	Metrics	Co-forest algorithm	Improved algorithm	Improved
1%	Acc	0.892	**0.959**	**6.7%**
	F1	0.900	**0.949**	**4.9%**
2%	Acc	0.929	**0.975**	**4.6%**
	F1	0.924	**0.979**	**5.4%**
5%	Acc	0.950	**0.980**	**3.0%**
	F1	0.964	**0.985**	**2.1%**

4 Conclusion

In the system proposed in this paper, we have presented a fine-grained feature extraction method based on wavelet decomposition and an improved semi-supervised algorithm based on RF. The results show that the fine-grained features are also effective and important relatively. This idea means that it is effective to observe signals from coarse-grained to fine-grained in multiple resolutions. Moreover, the improved semi-supervised algorithm can perform better using the 3300 samples with different labeling rate than Co-Forests, which optimizes the strategy of selecting the unlabeled samples. In a word, our proposed system has important application value for the optimization of the classification model in the field of Driving Behavior Recognition.

Acknowledgments. The authors would like to thank the reviewers for their detailed reviews and constructive comments, which have helped improve the quality of this paper. This work is supported by the National Natural Science Foundation of China under (Grant No. 61372117).

References

1. World Health Organization: Global status report on road safety 2013: supporting a Decade of Action. Inj. Prev. **15**(4), 286 (2015)
2. Lee, B.G., Chung, W.Y.: Driver alertness monitoring using fusion of facial features and bio-signals. Sens. J. **12**(7), 2416–2422 (2012)

3. Gadepally, V., Krishnamurthy, A.: A framework for estimating long term driver behavior. J. Adv. Transp. **2017**(3), 1–11 (2017)
4. Sharma, H., Reddy, R.K., Karthik, A.: S-CarCrash: real-time crash detection analysis and emergency alert using smartphone. In: International Conference on Connected Vehicles and Expo, pp. 36–42 (2016)
5. Ali, A.H., Atia, A., Mostafa, M.S.M.: Recognizing driving behavior and road anomaly using smartphone sensors. Int. J. Ambient Comput. Intell. **8**(3), 22–37 (2017)
6. Johnson, D.A., Trivedi, M.M.: Driving style recognition using a smartphone as a sensor platform. In: International Conference on Intelligent Transportation Systems, pp. 1609–1615 (2011)
7. Chen, Z., Yu, J., Zhu, Y.: D3: abnormal driving behaviors detection and identification using smartphone sensors. In: International Conference on Sensing, Communication, and Networking, pp. 524–532 (2015)
8. Vaitkus, V., Lengvenis, P., Žylius, G.: Driving style classification using long-term accelerometer information. In: International Conference on Methods and Models in Automation and Robotics, pp. 641–644 (2014)
9. Haddad, S.A.P., Serdijn, W.A.: Ultra Low-Power Biomedical Signal Processing: An Analog Wavelet Filter Approach for Pacemakers, pp. 25–35. Springer, Dordrecht (2009). https://doi.org/10.1007/978-1-4020-9073-8
10. Hendel, M., Benyettou, A., Hendel, F.: Automatic heartbeats classification based on discrete wavelet transform and on a fusion of probabilistic neural networks. J. Appl. Sci. **10**(15), 1554–1562 (2010)
11. Yang, S., Shen, H.: Heartbeat classification using discrete wavelet transform and kernel principal component analysis. In: Tencon Spring Conference, pp. 34–38 (2013)
12. Chapelle, O., Schölkopf, B., Zien, A.: Semi-supervised learning. In: Pacific-Asia Conference on Advances in Knowledge Discovery and Data Mining, pp. 588–595. Springer, Heidelberg (2009)
13. Li, M., Zhou, Z.H.: Improve Computer-Aided Diagnosis With Machine Learning Techniques Using Undiagnosed Samples. IEEE Press, Piscataway (2007)
14. Blum, A.: Combining labeled and unlabeled data with co-training. In: Conference on Computational Learning Theory, pp. 92–100 (1998)
15. Zhou, Z.H., Li, M.: Tri-Training: exploiting unlabeled data using three classifiers. IEEE Trans. Knowl. Data Eng. **17**(11), 1529–1541 (2015)
16. Liu, J., Yu, H., Yang, W., Sun, C.: Combining active learning and semi-supervised learning based on extreme learning machine for multi-class image classification. In: He, X., et al. (eds.) IScIDE 2015. LNCS, vol. 9242, pp. 163–175. Springer, Cham (2015). https://doi.org/10.1007/978-3-319-23989-7_18
17. Yang, J., Yang, P., Jin, X.: Multi-classification for malicious URL based on improved semi-supervised algorithm. In: International Conference on Computational Science and Engineering, pp. 143–150 (2017)
18. Seung, H.S., Opper, M., Sompolinsky, H.: Query by committee. In: Proceedings of the Workshop on Computational Learning Theory, pp. 287–294 (1992)
19. Mallat, S.G.: A theory of multiresolution signal decomposition: the wavelet representation. Pattern Anal. Mach. Intell. **11**(7), 674–693 (1989)
20. Genuer, R., Poggi, J.M., Tuleau-Malot, C.: Variable Selection Using Random Forests. Elsevier Science Inc., New York (2010)
21. Angluin, D., Laird, P.: Learning From Noisy Examples. Kluwer Academic Publishers, Boston (1988)
22. Chawla, N.V., Bowyer, K.W., Hall, L.O.: SMOTE: synthetic minority over-sampling technique. J. Artif. Intell. Res. **16**(1), 321–357 (2002)

Video Anomaly Detection Based on Hierarchical Clustering

Chunyue Zhao, Beichen Li, Qing Wang, and Zhipeng Wang[✉]

School of Electrical Automation and Information Engineering,
Tianjin University, Tianjin 300072, China
zpwang@tju.edu.cn

Abstract. Recently the video surveillance market has developed rapidly, but judging whether there is abnormal behavior in the video relying on manpower is too expensive. Therefore, a method is needed to identify the abnormal behavior automatically. Scholars at home and abroad have done in-depth research on video abnormal events detection in different scenarios. However, the current detection technology still needs improvement in the speed of the algorithm. From this point of view, this paper proposes a video abnormal event detection method based on hierarchical clustering.

In order to construct the sparse coefficient matrix more accurately and quickly, the hierarchical clustering is introduced into sparse coding in this paper. And the structure information of the sparse coefficient matrix is used as the clustering criteria, which improves the standard group sparse coding method. In addition, the BK-SVD algorithm is used to train the dictionary so that we can further improve the speed of the algorithm through dictionary division. In the experimental part, we prove that the proposed algorithm has great performance in frame level and pixel level in MATLAB environment.

Keywords: Video abnormal event detection · Dictionary learning · Sparse coding

1 Introduction

Nowadays, video anomaly detection technology attracts more and more attention because video surveillance has a wide range of applications in various fields such as production and life. It is too expensive to judge whether there is abnormal behavior in the surveillance video relying solely on manpower. Therefore, we urgently need a technology that can automatically recognize abnormal behavior.

The difficulty of detecting anomalies is diverse in different scenes. And the research on dense scenes is the most challenging but also the most practical work. This paper proposes a video anomaly detection algorithm based on hierarchical clustering for video anomaly detection in dense scenes. The algorithm can automatically detect abnormal behaviors and meet the needs of practical applications.

In the work of abnormal event detection, there are mainly three parts that need to be solved, namely feature extraction, dictionary learning and sparse coding. At present, previous researches have focus on feature extraction methods. However, dictionary

© Springer Nature Switzerland AG 2019
Y. Tang et al. (Eds.): HCC 2018, LNCS 11354, pp. 547–559, 2019.
https://doi.org/10.1007/978-3-030-15127-0_55

learning and sparse coding algorithms still need more in-depth research. Therefore, the content of this paper mainly focuses on these two parts.

In the field of dictionary learning and sparse coding, a relatively new and effective idea is to consider the structural characteristics of the dictionary or the sparse matrix, and add some constraints during dictionary learning and coefficient matrix construction to make the trained dictionary reconstruct the original information more accurately. However, there is always a general deficiency in this kind of ideas. This is because the artificially defined constraints always have certain limitations, which leads to some deviations from the restoration of the original features. Therefore, this paper supposes to improve the way in which artificial constraints are formulated. In addition, the idea of dictionary division also plays a dominant role in dictionary learning, because the division of sub-dictionary is beneficial to reduce the computation time and improve the robustness of the algorithm. Therefore, this paper uses the idea of dictionary partitioning and hierarchical clustering to automatically fuse atoms in the dictionary, which greatly improves the accuracy and speed of abnormal event detection.

The contributions of this paper: (1) On the basis of making full use of the structure of the sparse coefficient matrix, hierarchical clustering is used in the sparse representation. (2) BK-SVD is used in the dictionary learning process. Blocking the dictionary atom helps to improve the speed of dictionary training. (3) Deformation of the reconstruction error by the least squares method makes it easier to judge the abnormality in experiments.

2 Related Work

2.1 Trajectory-Based Video Anomaly Detection

In the sparse scene, the complexity of the feature is not high (this complexity generally refers to the dimension of the feature) because there is no occlusion. Therefore, detecting anomalies by analyzing motion trajectories is more robust. The process of obtaining the trajectory is actually the process of dividing the input video into video blocks of a specified size. Since the video block cannot be directly analyzed quantitatively, the trajectory need to be described by the model. And then the algorithm is used to analyze the obtained trajectory.

Scholars of Northwestern University proposed a dynamic hierarchical clustering anomaly detection method based on trajectory [1]. The author first described the target trajectory with HMM model. The distance between different trajectories calculated by the Bayesian information criterion is used as the criterion for the similarity. The cluster of the target is then obtained by the 2-depth greedy algorithm. After each clustering, all remaining trajectories are retrained and classified to achieve the effect of dynamic clustering. The algorithm has a good effect on the accuracy of detection and the speed of the algorithm, so it is of great significance to promote the development of video anomaly detection technology.

2.2 Population-Based Video Anomaly Detection

In dense scenes, it is not easy to analyze the video directly because each frame contains numerous complex information. We often selectively extract features of each frame in the video, and then describe them by the dictionary and sparse coefficient matrix. In different algorithms, the way to train the dictionary is diverse. But all algorithms are for strong generalization ability of the dictionary, and the trained dictionary can meet the speed requirements of practical application.

Scholars have proposed many excellent algorithms for the crowd-based video anomaly detection. These algorithms have achieved good detection results in practical applications.

The scholars from Aalborg University [2] proposed an unsupervised dictionary learning method. The authors paid attention to the distribution problem of sparse coefficient matrix. They not only used reconstruction error as the criterion for judging anomalies, but also believed that features corresponding to dense non-zero coefficients are basically normal. Therefore, dictionary corresponding to normal behavior was trained and then linked into a complete dictionary corresponding to all normal behaviors. The dictionary was supplemented with atoms corresponding to dense non-zero coefficients in order to obtain a higher recovery dictionary. However, this algorithm does not focus on the speed problem, so it is not suitable for scenarios with lots of operations.

Scholars from the Chinese Academy of Sciences have proposed a structural dictionary learning method for complex scenes [3]. The authors believe that the sparse coefficient matrix corresponding to normal features are similar in many ways, so the normal features will have resemble sparse representations. And the authors also believe that features that are close to the normal feature space are basically normal, so the sparse matrices of features with smaller spatial distances are similar. The authors use these two spatial structure features as constraints for dictionary learning and coefficient matrix construction, which obtain relatively high precision in the detection of abnormal events.

Professors of Tianjin Normal University proposed a method for detecting abnormal events by compact and low rank sparse learning (CLSR) [4]. The authors believe that the characteristics corresponding to normal behavior are similar, so the dictionary trained has low rank. The author also believes that the sparse matrices corresponding to normal behavior are similar, so the tightness rule is proposed to constrain the construction of sparse coefficient matrices. Compared with many algorithms that ignore structural information, the experimental part of the paper verifies that the method of dictionary learning and sparse coefficient construction under the low rank rule and the tightness rule improves the detection accuracy.

Scholars from the Chinese University of Hong Kong have proposed a fast dictionary learning method [5], which can achieve a detection rate of 140–150 frames per second in MATLAB software. This paper automatically combines the dictionary atoms (columns in the dictionary) but it need promise the sufficient number of combinations and minimum reconstruction error. Finally, a target dictionary composed of many sub-dictionaries is formed. Since the fusion process of the sub-dictionary is automatic and the dictionary is sequentially updated by sub-dictionary, the speed of the algorithm is guaranteed. At present, the idea of dictionary division plays an important role in dictionary learning, because it may has a positive effect on algorithm speed, algorithm robustness or algorithm accuracy.

Zelnik-Manor scholars proposed a method of dictionary optimization using block sparse representation [6]. The author proposes a BK-SVD (block K-singular value decomposition) dictionary training method, which is a breakthrough in the research of dictionary learning. We are more familiar with the K-SVD dictionary construction method, in which the dictionary atom is updated column by column. Comparatively, the block updating strategy in the BK-SVD algorithm can help to improve the update speed. This paper also takes the speed of updating into consideration, so the BK-SVD algorithm is adopted.

Based on references [7–13], we can understand that there are two main types of algorithms with good performance in the field of sparse coding representation, namely greedy algorithm and L1 norm approximation, respectively. Greedy algorithms include MP, OMP, StOMP and there are mainly BP and FOCUSS in L1 norm approximation algorithms. According to the literature, we can discover that the greedy algorithm is faster and more suitable for larger computing needs. Since research in this paper is aimed at dense scenes with crowds which means a lot of operations, we adopt the greedy algorithm in the sparse coding part. In addition, this paper takes advantage of the structural characteristics of the sparse coefficient matrix itself and utilizes it as the clustering criterion for hierarchical clustering.

This paper studies crowded and complex scenarios, so we detect the anomalies based on the crowd.

In summary, it can be clearly recognized that there are many excellent algorithms in dictionary learning and sparse matrix construction, but basically no algorithm can meet the requirements of practical application for both calculation speed and accuracy. It means that there is still a long way to go. We not only need to construct a sparse coefficient matrix and a dictionary which can describe the training samples, but also require the dictionary to have generalization ability. Otherwise, it will increase the missed detection and misjudgment of abnormal events.

The paper is organized as follows: Section 1 presents an overview of the background of the topic. Section 2 introduces the related work. Section 3 discusses the sparse coding algorithm for video anomaly detection. In Sect. 4, this paper refers to the K-SVD algorithm and the BK-SVD algorithm in details, followed by comparison of experimental results and conclusion in Sect. 5.

3   Sparse Coding Based on Hierarchical Clustering

3.1   Several Basic Clustering Methods

Clustering is to divide a data set into several disjoint subsets according to the specified criteria. There are great similarities among data in the same subset and little similarities between different subsets.

Prototype Clustering. The K-means algorithm [14] is a typical prototype clustering algorithm. The clustering process is the process of minimizing the square error. At this time, each object closely surrounds the mean of all objects within the cluster. The process of clustering is not easy, because there is no prior knowledge about this dataset and there is not any tag information. Therefore, it needs to find the cluster corresponding to the least square error which is actually an NP-hard problem. Applying the greedy algorithm to solve such clustering problems and find the optimal clustering results by continuous iterative optimization. The learning vector quantization algorithm [15, 16] is also a type of prototype clustering, but the difference is that it assumes that the data samples have category markers.

Hierarchical Clustering. The [17–20] is divided into two categories: "top-down" and "bottom-up" [21]. The "top-down" algorithm, also known as the divisive method, treats the collection to be classified as a large class. And after each iteration, smaller classes will be generated. The "bottom-up" algorithm can also be known as the agglomerative method. In this algorithm, each element in the set to be classified is regarded as a class, and the number of classes decreases every iteration.

The amount of data in this paper is relatively large, and the final number of clusters cannot be determined in advance. Therefore, from the perspective of computational complexity and algorithm speed, hierarchical clustering is more suitable for sparse coding. In this paper, the similarity of sparsity degree is taken as the criterion of clustering, and the maximum size of each clustering set is defined. It is hoped that the objects with the closest sparsity degree are gathered together. In summary, hierarchical agglomerative clustering is more suitable for sparse coding in this paper.

Applying hierarchical clustering to sparse coding can also achieve the purpose of dictionary partitioning. This is because the dictionary atoms corresponding to the sparse coefficient sub-matrix are automatically aggregated into one piece, and finally the purpose of dictionary division is achieved.

3.2 Group Sparse Dictionary Learning

Group sparse representation is also an important research direction in machine learning, and its application is very extensive. The combination of group sparse coding and dictionary learning has been widely used in the research of image processing. Applying group sparse and graph rules to medical image denoising and medical image fusion [22] greatly reduces the distortion probability of the image. This paper also introduces group sparse idea into the dictionary encoding.

If all the pixels are used to redisplay an image, the calculation speed cannot meet the requirements of the actual application. This is because there are too many pixels in each frame. Therefore, a part of the pixels must be selected to construct a dictionary. And the original image is obtained by multiplying the dictionary with a sparse coefficient matrix. Assuming that the extracted training features are $Y = [y_1, ..., y_n]$, the corresponding dictionary $D = [d_1, ..., d_n] = [D_1, ..., D_g]$, $d_i(i = 1, ..., n)$ is the column vector of dictionary D, $D_i(i = 1, ..., g)$ is the sub-dictionary, and the sparse coefficient matrix $X = [X_1^T, ..., X_L^T]^T$, then Y can be represented by the dictionary D and the sparse coefficient matrix X in the formula (1).

$$Y = DX = \left(D_1, \ldots, D_g\right)\left(X_1^T, \ldots, X_g^T\right)^T \tag{1}$$

Among them, $X_i(i = 1 \ldots g)$ can be solved by the optimization problem shown by Eq. (2), and its physical meaning is that each sub-matrix in the sparse coefficient matrix is desirably as sparse as possible.

$$\min_X \|X\|_{0,i} \quad \text{subject to} \quad Y = DX, \; i = 1, \ldots g \tag{2}$$

In addition, the L0 norm's calculation is difficult, so the L0 norm will be relaxed to the L1 norm. And the solution problem of X_i will be solved by the Eq. (3). Where k is an arbitrary small positive real, and the smaller the value of k, the smaller the reconstruction error.

$$\min_X \|X\|_{1,g} \quad \text{subject to} \quad \|Y - DX\|_2^2 \leq k \tag{3}$$

3.3 Sparse Coding Based on Hierarchical Clustering

In the sparse coding process, this paper uses the BOMP algorithm to obtain the sparse coefficient matrix. The BOMP algorithm [23, 24] is a variant of the OMP algorithm [25, 26], which integrates the grouping idea into the OMP algorithm. In the BOMP algorithm, the value of a sparse coefficient sub-matrix is updated, which improve the algorithm speed.

Different from the general block sparse dictionary learning, this paper considers the structural information of the sparse coefficient matrix itself when dividing the block. We know that normal behavior is common and the distribution is relatively regular, so their corresponding sparse coefficient matrices are similar. We can consider the row vector with "short distance" as the clustering object in the sparse coefficient matrix. This paper stipulates that the maximum number of matrix rows is s after clustering, so that the s rows with the most similar sparsity are gathered together.

Since there is no prior knowledge about the sparse coefficient matrix, we first need to initialize the clustering block and get the initial sparse coefficient matrix by OMP algorithm. Then, under the condition that the sparse coefficient sub-matrix is required to be as sparse as possible, sub-matrixes just obtained are clustered again to obtain a new block partitioning result. Finally, the obtained block is used to perform iterative updating by using the BOMP algorithm to obtain a sparse coefficient matrix.

Suppose the training feature is $Y = [y1, \ldots, yn]$, the corresponding dictionary $D = [d1, \ldots, dn] = [D1, \ldots, DL]$, di ($i = 1, \ldots, n$) is the column vector of dictionary D, Di($i = 1, \ldots, L$) is the sub-dictionary after dividing the block, and L is the number of blocks in the dictionary. The sparse coefficient matrix $X = [X_1^T, \ldots, X_L^T]$T, Pj($j = 1, \ldots, L$) is the number of row vectors included in each sparse coefficient sub-matrix, where L is the number of blocks of the sparse coefficient matrix X. 1 is the sparsity of each sparse coefficient sub-matrix, which means that the non-zero element's number of each column in the sparse coefficient sub-matrix does not exceed l. Therefore, the construction of the dictionary D and the sparse coefficient matrix X in this paper needs to satisfy the formula (4).

$$\min_{D,X,L} \|Y - DX\|_2^2 \quad \text{subject to} \quad \|X_m\|_{0,n} \leq l, P_j \leq s \tag{4}$$

Where m = 1, ..., G (G is the number of columns in the sparse coefficient sub-matrix Xn (n = 1, ..., L)), $\|X\|_{0,n} \leq l$ means that the sum of the number of non-zero elements in each column in each sparse coefficient sub-matrix need to be less than l which is specified in this paper. In the clustering process, each sparse coefficient sub-matrix should not include more than s rows and the non-zero element's number of each column does not exceed l.

The following article describes in detail how the clustering process is implemented in the algorithm. Assume that the set of column indices of non-zero element positions in each row is Ri (i = 1, ..., L), where L is the number of blocks of the sparse coefficient matrix X. In this paper, we follow Eq. (5) to find out the two rows (or two block) with the closest sparsity.

$$[u, v] = \arg \max_{u \neq v} (R_u \cap R_v) \quad \text{subject to} \quad P^* \leq s \tag{5}$$

Where u, v are the two blocks or two rows with the closest sparsity, and P* is the number of row vectors included in the new set to be found. If two blocks or two rows found satisfy the requirements of Eqs. (3–5), the two blocks or two rows are combined into one block and the corresponding row index will be removed from the sparse coefficient matrix.

In this paper, the original intention of clustering the sparse coefficient matrix lies in the division of the corresponding dictionary. The rows in the sparse coefficient matrix are divided into blocks as cluster objects, so that the columns in the corresponding dictionary can automatically generate sub-dictionaries. The clustering process of the sparse coefficient matrix is shown in Fig. 1.

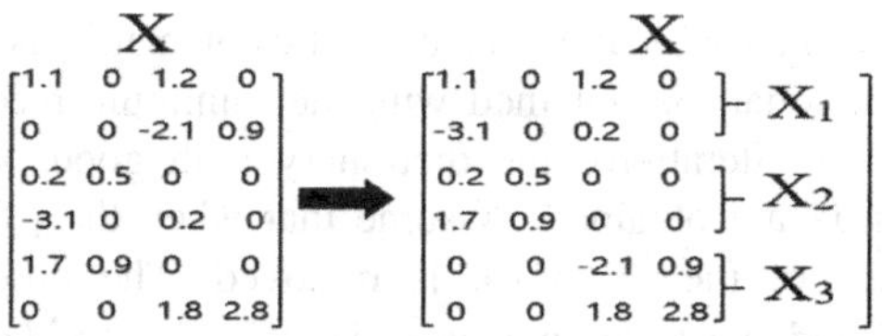

Fig. 1. Clustering process of sparse coefficient matrix

In fact, the difference from the general group sparse algorithm is that this paper not only considers the structural characteristics of the sparse coefficient matrix, but also limits the sparsity of each column in the sparse coefficient sub-matrix in the clustering process.

4 Dictionary Learning Based on Block K-Singular Value Decomposition

4.1 BK-SVD Algorithm

The BK-SVD algorithm (block K-singular value decomposition algorithm) is a variant of the K-SVD algorithm. This algorithm updates the dictionary atom block by block. The hierarchical clustering has divided the sparse coefficient matrix into blocks, which corresponds to the sub-block of the dictionary. So the sub-dictionary is also obtained. When updating each block, the remaining blocks are guaranteed to be fixed, and the singular value analysis is performed on the current error term.

4.2 The K-SVD and BK-SVD Algorithm

The update of the sparse coefficient matrix and the dictionary is actually performed crosswise, which means that after the dictionary initialization, the BOMP algorithm is used to find the sparse coefficient matrix at the time when keeping D unchanged. After that, we keep the updated sparse coefficient matrix unchanged and update the dictionary D with the BK-SVD algorithm. It is assumed that the nth dictionary learning is performed, and the sparse coefficient matrix is the one obtained by updating the n-1th step. After using the sparse coefficient matrix obtained, the dictionary in the iterative process can be obtained.

$$X^{(n)} = \arg\min_x \left\| Y - D^{(n-1)}X \right\|_2^2 \quad \text{subject to} \quad \|X_i\|_{0,j} \leq l, P^* \leq s \quad (6)$$

Where i = 1, ..., G (the number of columns of the sparse coefficient matrix), j = 1, ..., L (the number of blocks in the dictionary). This update step needs to be done with the help of the BOMP algorithm.

$$D^{(n)} = \arg\min_D \left\| Y - DX^{(n)} \right\|_2^2 \quad (7)$$

Equation (7) shows that after the sparse coefficient matrix is obtained in the nth iteration, the target dictionary is obtained with the minimum reconstruction error. In this experiment, after 50 iterations, the dictionary with good performance and the sparse coefficient matrix are obtained. Assume that when the j (j ε [1, L]) block is updated, the influence of the jth block is removed. The difference between the reconstructed feature and the training feature is $W_j = Y - \sum_{i \neq j} D_i X_i^T$, and the recon-struction error of the training feature is $E = \left\| W_j - D_j X_j^T \right\|_2^2$. The singular value decomposition is performed on the reconstruction error, then $W_j = U\Delta V'$. At this time, the general form of the dictionary D and the sparse coefficient matrix is obtained, as shown in Eq. (8).

$$D_j = [U_1, \ldots, U_t]$$
$$X_j^T = [\Delta_1^1 V_1, \ldots, \Delta_t^t V_t]^T \tag{8}$$

Equations (4–3) is the result of applying the BK-SVD algorithm to dictionary learning, where t is the number of columns corresponding to the current block in the dictionary. The singular value decomposition of the corresponding error of each block is better than the singular value decomposition of the corresponding error of a single column, because the final result in former case is more likely to achieve global optimality. So this is part of the reason why the BK-SVD algorithm is better than the K-SVD algorithm.

4.3   Judgment of Abnormal Events

In general, the method of judging anomalies is mainly through reconstruction error. If the reconstruction error is bigger than a given threshold, the test feature is judged to be abnormal; otherwise, the test feature is judged to be normal. The reconstruction error E is represented by Eq. (9), which is basically the most general representation of the reconstruction error. But different algorithms will deform it to more easily and accurately detect abnormal behavior.

$$E = \|Y - DX\|_2^2 \tag{9}$$

Since it is desirable to minimize the reconstruction error(even close to zero), the corresponding sparse coefficient matrix can be found by the least squares method. The general representation of the sparse coefficient matrix solved by the least squares method is given by Eq. (10).

$$X = (D^T D)^{-1} D^T Y \tag{10}$$

After the Eq. (10) is obtained, the reconstruction error E can be rewritten into the form of the Eq. (11). We can see that the reconstruction error at this time is only related to the dictionary D and the test feature Y. Although this kind of solution brings some errors to some extent, it provides great convenience for the anomaly detection in this paper. From the perspective of the simplicity of the algorithm, this paper chooses this method.

$$E = \|Y - DX\|_2^2 = \left\| \left(D(D^T D)^{-1} D^T - I \right) Y \right\|_2^2 \tag{11}$$

Where, I is the identity matrix. We define an auxiliary variable F to facilitate the operation.

$$F = \left(D(D^T D)^{-1} D^T - I \right) \tag{12}$$

Therefore, when the reconstruction error $E = \|FY\|_2^2$ is bigger than a given threshold, it is judged as an abnormal event; otherwise, it is judged as a normal event.

The above is the main idea of this algorithm. This algorithm has made some improvements to the more advanced algorithms to some extent. The following paper will prove the advancement of this algorithm in video anomaly detection through experiments.

5 Experiments

5.1 Feature Extraction

The data set used in this paper is the Avenue data set [5], which contains 16 training videos and 21 test videos. Each frame has 120*160 pixels. For each video, this paper first divides it into 10*10*5 pixel blocks, which means that each frame of 5 consecutive frames consists of 10*10 small pixel blocks. It need to be judged whether there is motion behavior in each pixel block. If there is motion behavior, it means that this part of the feature is relatively significant, and the feature will be extracted. The process of extracting features is actually a compression process as for each frame in the original video. In this process, this paper hopes that the finally obtained features can ensure the accurate restoration of the original video as much as possible. The features extracted in this way have a high advantage for the retention of information. However, the number of such features is still a considerable amount. So after the preliminary feature extraction, 500*16000 training features can be obtained. The PCA algorithm is used to reduce the dimension. Finally 10*16000 training features are obtained, which will be used to train the dictionary in this paper.

5.2 Parameter Settings

In the end, this paper hopes to build a 10*96 dictionary. Although the size of the dictionary will have certain deficiencies as for the accuracy of the restoration training features, it is very beneficial to the speed of the algorithm. This article sets the value of s to 3, that means each block in the dictionary does not include more than three columns. The setting of this value is affected by the size of the dictionary.

In this paper, the number of iterations of the dictionary is set to 50. In each iteration process, this paper runs sparse coding based on hierarchical clustering and dictionary learning algorithm based on BK-SVD. In the experiment, it was found that even if iterating 50 times, the speed of the algorithm is still relatively fast. After 50 iterations, we can get a dictionary with better performance.

In this experiment, the value of l is set to 2, which means that the number of non-zeros of each column in each sparse coefficient sub-matrix does not exceed 2. The size of this value is actually affected by the size of the s value. In this experiment, the value of s is 3, which means that each block in the sparse coefficient matrix has only 3 rows, so the value of l must be less than the value of s. When the value of l is 1, the requirement for the sparse condition is too high, because it may cause a large loss of the original information, so the value of l is set to 2. It will ensure that the coefficient matrix is as sparse as possible and the original information is not lost as much as possible.

5.3   Frame-Level and Pixel-Level ROC Curves

At the frame level, the abnormal frame is defined as follows: If one of the pixels in the frame is abnormal, the frame is abnormal. The frame-level ROC curve in this paper is shown in Fig. 2. The AUC value corresponding to this ROC curve is 0.75218.

At the pixel level, the criterion for determining a frame as anomalies is as follows: When the frame in Groundtruth is abnormal, if more than 40% of the pixels in the testing frame are abnormal, then the frame is judged as Abnormal; when the frame in Groundtruth is normal, if one of the pixels of the testing frame is abnormal, the frame is judged to be abnormal. The pixel-level ROC curve in this paper is shown in Fig. 3. The AUC value corresponding to this ROC curve is 0.5701.

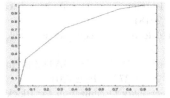

Fig. 2. Frame-level ROC curve   **Fig. 3.** Pixel-level ROC curve

Table 1. Frame-level and pixel-level AUC values

Algorithm	Frame-level AUC(%)	Pixel-level AUC(%)
Lu's algorithm [5]	81.75	63.3
Ours	75.2	57.0

The AUC values of the present algorithm and Lu's algorithm at the frame level as well as at the pixel level are listed in Table 1.

It can be seen from Table 1 that although the detection accuracy of the algorithm at the frame level and the pixel level is lower than that of the Lu's algorithm, the

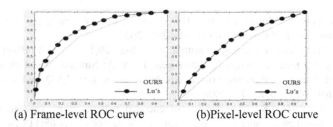

(a) Frame-level ROC curve   (b)Pixel-level ROC curve

Fig. 4. ROC curve of two algorithms at the frame level and pixel level

algorithm still has a satisfactory detection result. In the future, we can further improve the detection accuracy of the algorithm by other methods. Figure 4 shows the ROC curves of the two algorithms at the frame and pixel levels.

Acknowledgments. This work was supported by National Natural Science Foundation of China under Grant 61731003, the projects of International Cooperation and Exchanges NFSC under Grant 61520106002.

References

1. Jiang, F., Wu, Y., Katsaggelos, A.K.: A dynamic hierarchical clustering method for trajectory-based unusual video event detection. IEEE Trans. Image Process. **18**(4), 907–13 (2009). Publication of the IEEE Signal Processing Society
2. Ren, H., Liu, W., Olsen, S.I., et al.: Unsupervised behavior-specific dictionary learning for abnormal event detection. IEEE Trans. Signal Process. **17**(2), 99–111 (2015)
3. Yuan, Y., Feng, Y., Lu, X.: Structured dictionary learning for abnormal event detection in crowded scenes. Pattern Recogn. **62**(4), 129–138 (2018)
4. Zhang, Z., Mei, X., Xiao, B.: Abnormal event detection via compact low-rank sparse learning. IEEE Intell. Syst. **31**(2), 29–36 (2016)
5. Lu, C., Shi, J., Jia, J.: Abnormal event detection at 150 FPS in MATLAB. In: IEEE International Conference on Computer Vision, pp. 2720–2727. IEEE (2014)
6. Zelnik-Manor, L., Rosenblum, K., Eldar, Y.C.: Dictionary Optimization for Block-Sparse Representations, pp. 34–47. IEEE Press, Piscataway (2012)
7. Ren, H., Pan, H., Olsen, S.I., et al.: An in-depth study of sparse codes on abnormality detection. IEEE International Conference on Advanced Video and Signal Based Surveillance, pp. 66–72. IEEE Computer Society (2016)
8. Aharon, M., Elad, M., Bruckstein, A.: The K-SVD: an algorithm for designing overcomplete dictionaries for sparse representation. IEEE Trans. Signal Process. **54**(11), 4311–4322 (2006)
9. Li, S., Yin, H., Fang, L.: Group-sparse representation with dictionary learning for medical image denoising and fusion. IEEE Trans. Biomed. Eng. **45**(23), 531–541 (2015)
10. Olshausen, B.A., Field, D.J.: Natural image statistics and efficient coding. Network: Comput. Neural Syst. **2**(7), 333–339 (1996)
11. Lesage, S., Gribonval, R., Bimbot, F., et al.: Learning unions of orthonormal bases with thresholded singular value decomposition. In: Proceedings of the IEEE International Conference on Acoustics, Speech, and Signal Processing, pp. 293–296. IEEE (2005)
12. Duarte-Carvajalino, J.M., Sapiro, G.: Learning to sense sparse signals: simultaneous sensing matrix and sparsifying dictionary optimization. IEEE Trans. Image Process. **18**(7), 1395 (2009)
13. Kreutz-Delgado, K., Murray, J.F., Rao, B.D., et al.: Dictionary learning algorithms for sparse representation. Neural Comput. **15**(2), 349–396 (2003)
14. Hartigan, J.A.: Means clustering algorithm. Appl. Stat. **28**(1), 100–108 (1979)
15. Asikainen, A., Kolehmainen, M., Ruuskanen, J., et al.: Structure-based classification of active and inactive estrogenic compounds by decision tree, LVQ and kNN methods. Chemosphere **62**(4), 658–673 (2006)
16. Ganesh Murthy, C.N.S.: Classification of encoded patterns using constructive learning algorithms based on learning vector quantization (LVQ). Tech. Rep. **65**(11), 245–257 (1996)

17. Liu, Y., Xu, H., Yi, H., et al.: Network anomaly detection based on dynamic hierarchical clustering of cross domain data. In: IEEE International Conference on Software Quality, Reliability and Security Companion, pp. 200–204. IEEE (2017)
18. Ren, W.W., Liang, H., Zhao, K., et al.: An efficient parallel anomaly detection algorithm based on hierarchical clustering. J. Networks **8**(3), 672–679 (2013)
19. Hu, L., Ren, W.W, Ren, F.: An adaptive anomaly detection based on hierarchical clustering. In: International Conference on Information Science & Engineering, pp. 1626–1629. IEEE (2009)
20. Chen, W., Liu, X., Li, T., et al.: A negative selection algorithm based on hierarchical clustering of self set and its application in anomaly detection. Int. J. Comput. Intell. Syst. **4** (4), 410–419 (2011)
21. 周爱武, 潘勇, 崔丹丹等.: AGNES算法在K-means算法中的应用. 微型机与应用 **30**(23), 79–81 (2011)
22. Li, S., Yin, H., Fang, L.: Group-sparse representation with dictionary learning for medical image denoising and fusion. IEEE Trans. Biomed. Eng. **59**(12), 3450–3459 (2012)
23. Eldar, Y.C., Kuppinger, P., Bölcskei, H.: Block-sparse signals: uncertainty relations and efficient recovery. IEEE Trans. Signal Process. **58**(6), 3042–3054 (2010)
24. Eldar, Y.C., Bolcskei, H.: Block-sparsity: coherence and efficient recovery. In: IEEE International Conference on Acoustics, Speech and Signal Processing, pp. 2885–2888. IEEE (2009)
25. Goklani, H.S.: A review on image reconstruction using compressed sensing algorithms: OMP, CoSaMP and NIHT. Int. J. Image Graphics Sig. Process. **9**(8), 30–41 (2017)
26. Xu, Y., Sun, G., Geng, T., et al.: An improved method for OMP-based algorithm; using fusing strategy. In: IEEE, International Colloquium on Signal Processing & ITS Applications, pp. 202–207. IEEE (2017)

Research on the Design Method
of the Command Simulation Training System

Dongmei Zhao[1(✉)], Hao Li[1], Wenyuan Xu[1], Shengxiao Zhang[1],
Li Guo[1], and Yuxiao Yang[2(✉)]

[1] CSSC Systems Engineering Research Institute, Beijing 100094, China
[2] North China Electric Power University, Baoding 071003, China
yangyuxiao_ncepu@outlook.com

Abstract. Considering the factors of command decision-making, tactical warfare method, weapon equipment, enemy confrontation and natural environment in the process of military operation, the basic environment of distributed command and countermeasure simulation training is constructed by using the simulation and deduction of attack and defense against the red and blue two sides in complex battlefield environment. The battle plan and operational action inspection and evaluation will provide technical support for the rapid upgrading of the information level of military actual combat training. In this paper, the functional requirements of the system are obtained from the analysis of the operational requirements, and the system design is carried out. The system architecture of software and hardware and the typical application process of the system are put forward. On this basis, some key problems which are urgently needed to be solved in the current realization are discussed.

Keywords: Simulation training · Attack-defense confrontation · Inspecting evaluation

1 Introduction

As an important part of military training, operational command training is also a difficult point in military training. It is mainly embodied in the following aspects: first, it is difficult to fight against the actual combat, the special characteristics of the combat task, the combat object and the combat equipment make the organization close to the practical and antagonistic comprehensive command and training. Two, it is difficult to test the results of the command operation, because the command can not be actually executed, which leads to the execution of the operational command. The effect is difficult to quantify. At present, the traditional training methods, such as artificial situation induction and situation decision, have many problems, such as the subjectivity of the battlefield situation verdict too strong, the situation guidance too arbitrary, and the qualitative extensive training evaluation, which leads to the inadequacy of the elements of the command and training coverage, and the lack of actual combat and antagonism, which restricts the ability of the training and support of command confrontation.

With the continuous development and change of military equipment, operational task, combat target and battlefield environment under the condition of information

Y. Tang et al. (Eds.): HCC 2018, LNCS 11354, pp. 560–570, 2019.
https://doi.org/10.1007/978-3-030-15127-0_56

warfare, the gap between command training support ability and command training level is becoming more and more gap. It is necessary to take into consideration all kinds of elements of tactical warfare, command decision, weapon equipment, enemy confrontation and natural environment in the process of military operation, and make full use of modern simulation and simulation training technology to simulate the battlefield environment, enemy and US operations and antagonism effect. A set of command simulation training system is built to realize operational planning based on actual operational environment, actual operational purpose and actual operational capability, combat plan formulation, operational action execution, aircraft command decision-making and operational action deduction inspection evaluation.

2   System Function Design

Based on the design concept of building an equivalent force of "virtual army", this paper fully combs the content and process of military operations, analyzes and designs the functions of the command simulation training system, and its specific functions are as follows.

1. Simulation function of combat entity. It includes red and blue entity models and battlefield environment models. By means of component-based and parameterized modeling ideas, the basic component models of different types of weapons are built by setting component parameters, and then the entity model is constructed through the assembly of different component models.
2. Action simulation function. The operational action of the Red action is mainly the operation process of the operational entities, according to logical process, constraint rules and ability level, such as the command post, the main combat unit and the support unit, respectively in the stage of operational preparation, the stage of operation and the end of the operation. Blue operations are mainly for command and countermeasure simulation training to provide confrontation input, such as reconnaissance and surveillance, precision strike, anti-missile interception and so on.
3. Command function of combat scheme. The system can decompose the imported job results and the initial situation into standardized and formatted instructions, and give the instructions to the corresponding simulation entities to drive the running of the entity.
4. Man machine interactive function: First, the scenario editing function is provided to support the initial situation setting, operational rules setting, and battlefield environment setting. Secondly, the comprehensive guidance function is provided, which can monitor the battlefield situation, provide situation guidance, simulation process control and so on, and support the guidance personnel to guide the red and blue action against the action from the global. Third, command control of red side and virtual soldier control of blue side are supported. Finally, the operational evaluation function is provided to assist the leader in adjudication of the result of confrontation, and to evaluate and analyze the effect of command and decision.

3 System Architecture

3.1 Software Architecture

On the basis of fully analyzing the simulation requirements of military operations, the system is designed under the guidance of the principle of separation of business and function, separation of model and data, separation of model and platform. Then, a hierarchical structure is formed, which is divided into basic layer, resource layer, supporting layer and application layer from bottom to top. The system architecture is shown in Fig. 1.

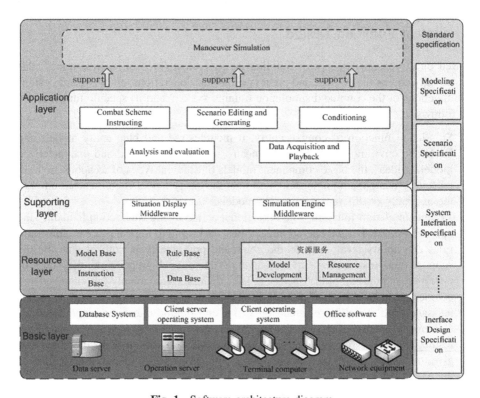

Fig. 1. Software architecture diagram

Application layer: In order to prepare, implement and summarize the different requirements of the manoeuver simulation, the functions of command, edit and generation, integrated guidance, data collection and replay, and analysis and evaluation are provided. It supports the application of military operations simulation through flexible combination and configuration of function modules, models and data in resource layer, support layer and application layer.

Supporting layer: It mainly provides functions such as model scheduling management, simulation situation displays and so on.

Resource layer: It mainly provides the model and data support for the upper layer, including the functions of model base, database and resource service.

Basic layer: It mainly provides the hardware and software support environment for command simulation training system, including network equipment, server, client, operating system and database, etc.

Standard specification: It is the guidance of the system. which mainly regulates the modeling, thinking, and system integration of the system construction process to improve the efficiency and quality of the system development, and improve the extensibility and flexibility of the system.

3.2 Hardware Architecture

Under the principle of deployment design separated from model computation and business operation, the hardware system of the command simulation training system is composed of high performance simulation computing server, high speed network switching equipment, interactive client of simulation system and mass storage system. The hardware structure is shown in Fig. 2.

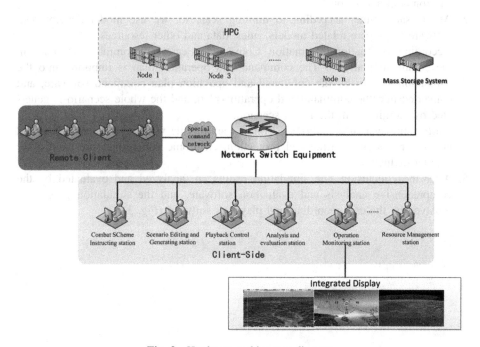

Fig. 2. Hardware architecture diagram

The blade server is adopted in the simulation computation server, which has good expansibility. The system can be configured flexibly according to different simulation scales, It effectively improves the running speed of the system simulation.

The interactive client uses the commercial computer to realize the human-computer interaction, and completes the operation instruction input, the editing, the operation process control, the operation process situation and the result display.

A commercial high-performance router is used between the simulation computing server and the interactive client to realize the high-speed data exchange between the client and the server.

4   System Application Process

The application process of the system is mainly designed for different stages of training preparation, training implementation, summary and evaluation, such as different activities, business logic sequence and so on. The main application process is as follows:

1. According to the operation plan and the operational plan, the initial situation of the initial situation is to be set up by using the initial situation, such as the force arrangement, the force deployment, the enemy threat behavior, the battlefield environment and so on.
2. At the same time, according to training scenario, we use model development software to prepare related models, rules, data and other resources.
3. According to situation information, Commanders carry out command operation in the command platform. The command and operation plan is imported into the simulation system by the command and operation plan command software, and converted into the command and operation plan, and the whole scenario is generated by merging with the initial scenario.
4. Starting the system operation, to conduct antimissile warfare, reconnaissance and anti reconnaissance, strike and protection, electronic countermeasures and other counter deductions.
5. After the conclusion, the simulation results are analyzed and evaluated by the comprehensive analysis and evaluation software, and the simulation process is replayed by the data recording and playback software (Fig. 3).

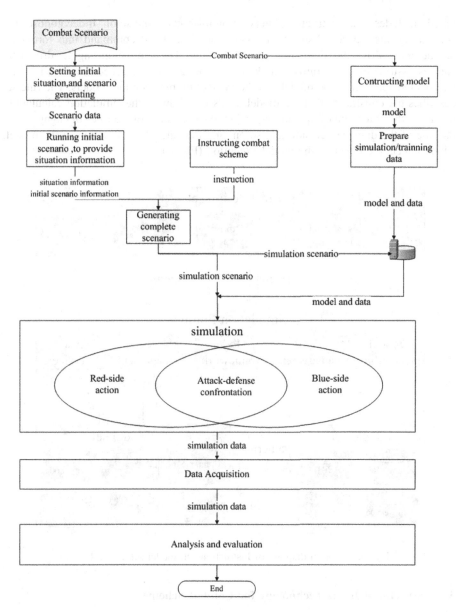

Fig. 3. System application flow

5   Key Design of System

5.1   Three Set Core Function Design

The relationship between instruction set, model set and rule set is shown in the following figure. The instruction set is the formatted external input of the system, which mainly includes basic instruction, intervention instruction and so on. It decomposes the combat scheme, special disposition scheme and command command into formatted instruction to drive the simulation engine running. The simulation engine, driven by various instructions, centralizes all kinds of models in the scheduling model, and realizes the solution of the model and the interaction between the models. The rule set describes the constraints that the model needs to follow in the simulation calculation, including the constraint of operational action, the constraint of the rules of engagement, the rule of the damage decision and so on, and the calculation function of the model, which affects the effect of the model execution (Fig. 4).

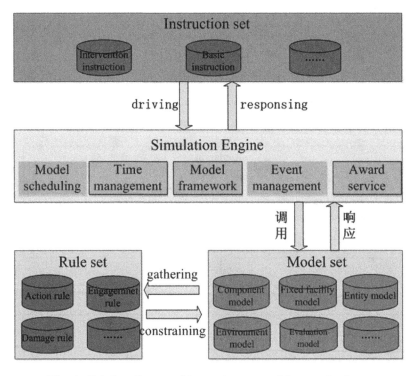

Fig. 4. Relation diagram of Instruction set, model set, and rule set

5.2   Structured Input Technology for Combat Scheme

At present, the command of the command information system is usually issued in the form of documents. The main description is the natural language of the commanders, and the standardized and formatted results are not completely formed. The computer can not be identified automatically. The interaction between the command information system and the command simulation training system needs to analyze, organize and abstract the commands of the command and training process, transform and describe the commands in the formatted form, realize the information access simulation system of command and instruction in the training process, and realize the interconnection and intercommunication between the command information system and the simulation system.

In this paper, we refer to the EATI template of the US Army to extract the fighting entity, the action, the task, the interaction between the entities, and the interaction between the entities from the operational process. The entity metamodel, action metamodel, task metamodel and interaction metamodel are established respectively. Taking the vehicle maneuver as an example, its metamodel is constructed as follows:

entity metamodel. EN::=<N,TE,AT,AC>

N(Entity name)——Task execution unit ;

TE(Task of entity execution)——From the starting point to the destination ;

AT(A collection of entity attributes)——AT::={Number, model, vehicle length and width, maximum speed, maximum acceleration, minimum turning radius, fuel capacity, minimum highway grade requirements, target characteristics, vulnerability, etc. } ;

AC(Action set of entity)——CT::={Motor operation, damage repair, etc. }.

action metamodel. AC::=<N,EN,SC,IC,EC>

N (Action name)——maneuvering operation ;

EN (Executive entity)——Task execution unit ;

SC (Starting conditions)——Receive the maneuvering command and satisfy the maneuverability requirement ;

IC (Interruption conditions)——Be destroyed, etc. ;

EC (Terminate conditions)—— {Receive the covert orders, arrive at camping sites, arrive at destinations, etc.}

task metamodel. TS::=<TN, EN, XN, AX, AR, EC>

TN (Task name)——From the starting point to the destination ;

EN (Executive entity)——Task execution unit ;

XN (Related entities)——Command post, support detachment, etc.

AX (Action set)——{ Motor operation, damage repair, etc. } ;

AR(Action execution order and scheduling rules) ——Maneuver according to plan, and perform other actions according to the rules of action ;

EC(Task termination condition) ——reach destination ;

interaction metamodel. IA::=<IN, FN, JN, JR>

IN (Interactive name)——Maneuver command ;

FN (Sending entity)——Command post ;

JN (Receiving entity)——Task execution unit ;

JR (Interactive content)——{ Maneuver command }。

Taking maneuvering instructions for example, part of its structured input template is as follows (Fig. 5).

Instruction Term	Instruction Content
Instruction Name	
Instruction Number	
Function Specification	
Sending Entity	
Starting Time	
Ending Time	
Starting Site	
Ending Site	
Passing Site	
Returning Location	
Execution Condition	
Execution Mode	

Fig. 5. Schematic diagram of structured input template

5.3 Cross-Platform Model Adaptation Technology

The model is the core asset of the command simulation training system. In order to improve the generality, extensibility and portability of the model, these paper separations the model function from the model call, analyzes the model properties and behavior, and provides the development template Through the configuration of the template parameters, the framework code of the corresponding model is generated to realize the unified description of the similar model. Based on this addition of the business model code, the model dynamic link library is generated. At the same time, in order to improve the applicability and convenience of the model, this paper provides a standardized model invocation interface to realize the unified invocation of the model DLL for improving the applicability and convenience of the model (Fig. 6).

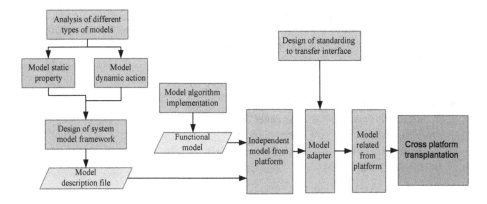

Fig. 6. Cross-platform model adaptation technology

5.4 High Efficiency Simulation Engine Technology

The calculation amount of command simulation training system in the process of simulation is closely related to the complexity of combat plan and the precision of calculation model. In the case of a large number of entity types and numbers, complex entity interaction, fine mathematical model, and frequent command and scheduling, the computation is large. In order to ensure the stable or even super real time deduction of the calculated frame rate (speed up the deduction rate), this paper uses the high efficiency simulation engine technology to realize the management and scheduling of the simulation entity (Fig. 7).

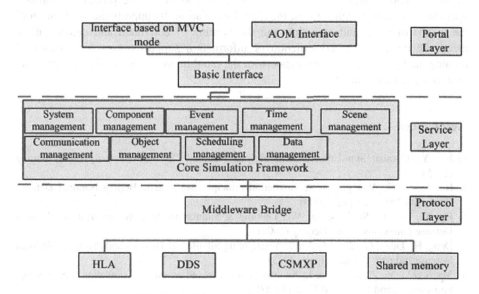

Fig. 7. Architecture of simulation engine

The functions of this system's high efficiency simulation engine can be summarized as communication transparency, integration component and programming objectification.

First, communication transparency. In a simulation step, the state data of the remote simulation object is transferred to the process that needs access to the object, and the simulation engine automatically maintains the life cycle of the simulation object and the data consistency between the distributed nodes. And this process is completely transparent to the application.

Second, integrated component-based, the developed simulation model is encapsulated in the form of independent components. When the model needs to participate in the simulation system, only the components of the model need to be deployed to the simulation system. The simulation engine can automatically load the simulation component and complete the model assembly.

Finally, Programming objectification, the interface provided by the simulation engine is the object oriented interface. When the application needs to be called, it only needs to define the object of the simulation engine to realize the object oriented interface call.

6 Summary

The command simulation training system provides a support platform for tactical training in complex battlefield environment. In the virtual battlefield environment, the simulation and deduction of the whole elements and the whole process of military operations are carried out through the digital simulation. It supports the verification of the results of command operation and the training of command and confrontation, meets the needs of the construction of information support means for operational training, and provides technical support for the rapid improvement of the training level of military actual combat.

References

1. Ma, Y.: Combat Simulation System. Beijing National Defense University Press, Beijing (2005)
2. Hu, X., Si, G.: War gaming & simulation principle and system. Beijing National Defense University Press, Beijing (2009)
3. Hu, X., Luo, P., Si, G., et al.: War Gaming & Simulation Principle and System. National Defense University Press, Beijing (2005)
4. Deng, H., Deng, G., Zhao, Q., et al.: Warfighting Simulation Theories and Practices. National Defense Industry Press, Beijing (2013)
5. Ma, S., Ma, Y., Zhu, M., Liu, J.: Analyzing studies based on warfare simulation commanding. Fire Control and Command Control (2008)
6. Huang, X., Ling, X., Wen, W.: Research on developing platform for battle simulation system. Fire Control Command Control (2014)
7. Cheng, L.: Design and implementation of operational scheme deduction system. Ship Electron. Eng. (2014)
8. Guo, Q., Tang, Z., Luo, X., et al.: Military Equipment Warfighting Simulation. National Defense Industry Press, Beijing (2013)

Study on Double Piezoelectric Layers Driven Pump Based on MEMS and 3D Printing

Liang Huang, Gang Tang[✉], Min Hu, Zhibiao Li, Xiaoxiao Yan,
Bin Xu, and Xiaozhen Deng

Jiangxi Province Key Laboratory of Precision Drive and Control,
Nanchang Institute of Technology, Nanchang 330099, China
tanggangnit@163.com

Abstract. A piezoelectric pump of small size, compact structure, light weight and large flow rate has been developed in this paper. The pump consists of two circular piezoelectric vibrators, two valve covers and three check valves. Furthermore, two circular piezoelectric vibrators made by MEMS technology are proposed as the dual transducer of the pump body. The simulation analysis consequences show that beryllium bronze is optimal metal material as substrate of piezoelectric vibrator, compared with other metal materials including alloy steel, titanium alloy and aluminum alloy. A prototype of the piezoelectric pump, with the external dimensions of 40 mm × 40 mm × –20 mm, was made by 3D printing technology that can replace conventional techniques and greatly shorten the processing cycle. The experimental results show that water rised 5 mm through the inlet pipe, when double piezoelectric vibrators were driven by AC voltages.

Keywords: MEMS technology · Double vibrators · 3D printing

1 Introduction

Piezoelectric pumps have been developed in recent years and their typical operate mode is driven by piezoelectric vibrator to produce positive fluid flow. The piezoelectric pump has the advantage of small volume, compact structure, light weight and no electromagnetic interference. It is easy to be processed and produced. Therefore, It is widely used in the fields of cooling of electronic devices, fuel cells, medical research, aerospace and lubrication of mechanical equipments. Especially, piezoelectric pumps are widely used in the cooling of electronic equipment because of low energy consumption and low operating noise. With the development of diversification, complication and integration of electronic devices, heat will be gradually increased. Because air-cooling dissipation has many limitations in the application of electronic products. So the research on compact, lightweight and large-flow liquid micropumps is of great significance. For large-flow piezoelectric pumps, many scholars at home and abroad have done similar thorough research. For example, Los Angeles scholars, the University of California, used piezoelectric stacks as drivers of piezoelectric pump to achieve a flow rate of 3 L/min. But the pump was finished with the dimensions of 38 mm × 140 mm (length × width). The advantage of the piezoelectric stack drive is

© Springer Nature Switzerland AG 2019
Y. Tang et al. (Eds.): HCC 2018, LNCS 11354, pp. 571–579, 2019.
https://doi.org/10.1007/978-3-030-15127-0_57

that the output force is large. But the cost of the pump is high, and it is not conducive to miniaturization [1–7]. A two-chamber series piezoelectric pump, driven by double piezoelectric vibrator, is low in cost, and the overall volume of the pump is reduced to 40 mm × 40 mm. So the structure is simple. The structure of the single vibrator piezoelectric pump is easy to control the thickness of the pump body with only one piezoelectric vibrator working, which is suitable for micro-pumps with small precision driving flow requirements. Thus, it's working performance has certain limitation [8, 9]. The PZT thick film in the preparation of piezoelectric vibrators usually adopts the traditional process of screen printing. The main disadvantage is that the sintering temperature is high and the prepared thick film has high porosity and low density [10]. In this paper, the piezoelectric vibrators based on MEMS technology are used as the double driver of the pump. Furthermore, the structure of the pump is fabricated based on the 3D printing technology. The experimental results demonstrate that the performance of the piezoelectric pump is better than others. Firstly, the displacement of piezoelectric vibrator reaches 45 μm.

2   Structure and Principle

The piezoelectric pump mainly consists of two covers, two O-rings, two piezoelectric vibrators, a pump body and three check valves. As shown in Fig. 1, The two piezoelectric vibrators are fixed on pump body through the O-rings. The second check valve usually in the lower chamber separates the two chambers, which is both the outlet valve of the upper chamber and the inlet valve of the lower chamber.

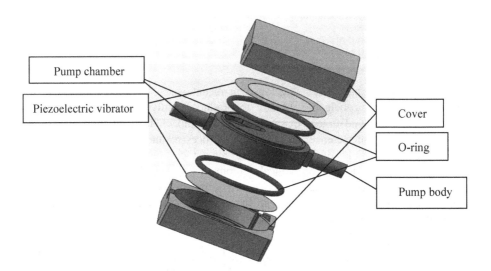

Fig. 1. Exploded view of the piezoelectric pump

Figure 2(a), (b) and (c) show the pumping mechanism. These two processes are called suction process and discharge process, respectively. The voltage driven oscillation of the PZT device creates a cyclic displacement. Firstly, two piezoelectric vibrators move upward, the volume of the first pump chamber becomes larger by the decreasing pressure. The first check valve opens with the fluid entering the upper pump chamber. It is called the suction process I as shown in Fig. 2(a). As two piezoelectric vibrators move downward, the volume of the upper pump chamber is decreased, and the pressures in the chambers are increased. At the same time, the volume of the lower pump chamber is increased, and the pressures in the chambers are decreased. Responding to the change in pressure, the first check valve closes, and the second

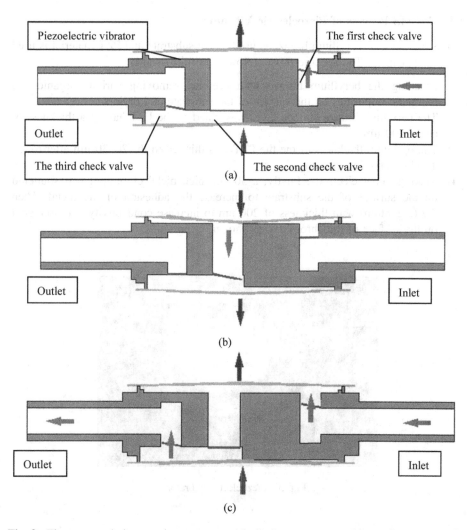

Fig. 2. Flow state during suction process and discharge process. (a) suction process I, (b) suction process II, (c) discharge and suction process

check valve opens. Meanwhile, the liquid flows into the second chamber. It is called the suction process II as shown in Fig. 2(b). Finally, the discharge process is shown in Fig. 2(c). As the lower piezoelectric vibrator move upward, the volume of the lower pump chamber is decreased, and the pressures in the chambers are increased. The second check valve closes, and the third check valve opens. So liquid in the lower chamber moves into the outlet. The fluid is aspirated into the upper pump chamber simultaneously. Repeat the above process, liquid moves continuously from the inlet to the outlet.

3   Processing Technology

3.1   MEMS Process of Piezoelectric Vibrator

In this experiment, beryllium bronze is used as the substrate and PZT material is used as the upper layer. The specific process flow is as follows [11]:

(1) Cleaning the beryllium bronze with acetone, removing various organic and inorganic impurities on the surface of the substrate and drying;
(2) The beryllium bronze surface layer was glued to the PZT having a thickness of about 400 µm.
(3) The optimum thickness of the the PZT was thinned to the 20–30 µm after rough grinding and fine grinding.
(4) Making upper electrode: Firstly, a 30 nm thick metal chromium was sputtered on the surface of the substrate to increase the adhesion of the metal. Then Ag (argentum) with thickness of 200 nm to increase conductivity was sputtered on the surface of the chromium, as shown in Fig. 3.

Fig. 3. Piezoelectric vibrator

3.2   3D Printing Process

The 3D printing type which is similar to the method of squeezing toothpaste has the advantages of low price, small size and relatively being prone to generating operations. The piezoelectric pump body can be processed by 3D printing rapid prototyping technology. The main operation steps are to design the pump model in solidworks 3D software as shown in Fig. 4. Then it's saved the file as stl format and imported into the Creality 3D software. After converted to G code, the printer starts to run and then melts the nozzle so that the plastic material PLA was melted and squeezed out from the nozzle, which was deposited at the specified position. The piezoelectric pump body measured 40 mm × 40 mm × 20 mm (length × width × height). The pump chamber (Ø32 × 0.4 mm) is linked to the inlet and outlet pipes (Ø6 × 10 mm). The check valves (Ø10 × 0.1) consisted of PET materials is incised by a laser cut. The O-ring (Ø36 mm × 1) is integrated to pump chamber. Finally, the upper pump is tightly glued to lower pump by AB glue. The model and material object are demonstrated in Fig. 4 (a), (b), (c) and (d), respectively.

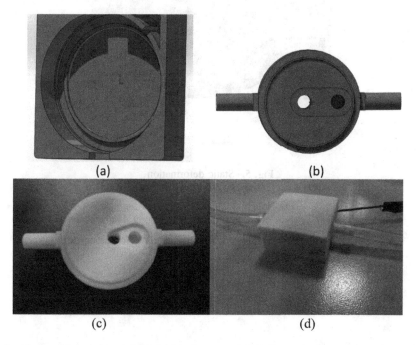

(a)   (b)

(c)   (d)

Fig. 4. The model and material object (a) Cover model (b) Pump body model (c) Pump body material object by 3D printing (d) Fabricated piezoelectric pump

4 Simulation Analysis

The mechanical deformation of the piezoelectric vibrator made by MEMS technology changes the volume and pressure of the pu(a)mp chamber and it's crucial to the performance of the piezoelectric pump. The circular piezoelectric vibrator consisted of metal substrate and PZT.

The deformation amount of vibration of different types of substrate materials will have a large difference, which will affect the performance of the piezoelectric vibrator and directly affect the working effect of the pump. In the simulation analysis, the selected substrate has diameter of 35 mm and thickness of 0.4 mm; the piezoelectric ceramic has diameter of 25 mm and thickness of 0.2 mm; the applied AC voltage is 100 V; the final static deformation of the piezoelectric vibrator is shown in Fig. 5. The center displacement of the vibrator is shown in Table 1. Choosing the appropriate metal substrate is the basis for the design and testing of the piezoelectric pump.

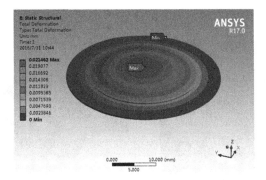

Fig. 5. Static deformation

Table 1. Piezoelectric transducer center point displacement

Substrate material	Piezoelectric vibrator center deformation (μm)
Alloy steel	21.5 μm
Bronze bronze	29.3 μm
Titanium alloy	31.9 μm
Aluminum alloy	38.8 μm

From Table 1, it can be determined that: different metal materials are used as the substrate, and the center displacement of the piezoelectric vibrator is different. The aluminum alloy as the substrate produces the largest displacement, followed by the titanium alloy, the beryllium bronze and the alloy steel center. In addition to considering the influence of the size of the center displacement, it is also necessary to comprehensively consider the conductivity, processing technology, weldability and blocking properties of the substrate. Since the aluminum alloy having a small hardness, which is susceptible to deformation in vibration and affects the effect of the piezo-electric vibrator. Furthermore, the titanium alloy is not suitable for use because of its high electrical resistivity. In conclusion, beryllium bronze is finally used as the substrate material.

The analysis of the dynamic characteristics of a circular piezoelectric vibrator mainly including the analysis of the natural frequency and mode shape, which are only related to the structure of the piezoelectric vibrator itself. For a passive valve piezo-electric pump, if the frequency of the piezoelectric vibrator is close to the natural frequency, the system will resonate and the output maximal. The four modes of the vibration mode simulated by the previously selected piezoelectric vibrator are shown in Fig. 6 and the corresponding natural frequencies are shown in Table 2. It can be seen from the vibration pattern that the first mode of the circular piezoelectric vibrator is arched. This mode is similar to the vibrator in static analysis; the second mode is a convex-concave symmetric surface; the third mode is concave at both ends. Symmetrical surface; the fourth mode is a convex surface with two convexities and two concaves.

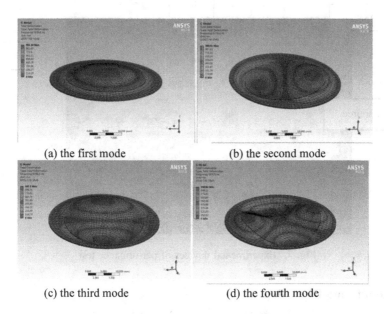

(a) the first mode   (b) the second mode

(c) the third mode   (d) the fourth mode

Fig. 6. The four modes of the piezoelectric vibrator

Table 2. Natural frequency of the first four orders of piezoelectric vibrators

Order	The first level	The second order	The third order	The fourth order
Natural frequency (HZ)	3156	6776	6780	10915

It can be seen from the vibration pattern that only the deflection of the first mode is gradually increasing from the edge of the vibrator to the middle position and has good symmetry. In this vibration mode, the combination of the vibrator and the pump body can produce the largest pump chamber volume. Therefore, the operating frequency of the piezoelectric pump power supply should be close to the first-order natural frequency of the piezoelectric vibrator.

5   Experiment and Result Analysis

5.1   Experimental Test

To test the output performance of the piezoelectric pump, an experimental system was set up and it is shown in Fig. 7. The experimental temperature is of 25 °C and water was utilized as the working medium.

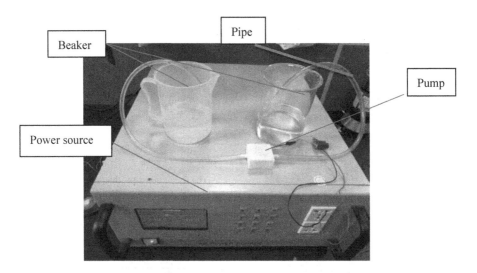

Fig. 7. Experimental devices of performance test

5.2   Results and Discussion

When the driving voltage is 200 V at a frequency of 150 Hz, water only rises 5 mm through the inlet pipe. But the result isn't optimal, the main reasons are as follows:

(1) Poor sealing performance between piezoelectric vibrator and O-ring.
(2) Displacement of the piezoelectric vibrator driven by AC voltages is not enough.

According to the above reasons, We will strengthen sealing performance between piezoelectric vibrator and O-ring and optimize design of piezoelectric vibrator in later stages.

6 Conclusions

A driven piezoelectric pump with prominent working features is presented in this paper, which is fabricated by the MEMS techniques. The piezoelectric pump body is fabricated by 3D printing technology. Compared with others, The piezoelectric pump with dimension of 40 mm × 40 mm × 20 mm have advantage of small volume, compact structure, light weight. The simulation analysis consequences show that beryllium bronze is optimal metal material as substrate of piezoelectric vibrator. Based on test results and reasons, we will optimize performance of piezoelectric pump in the near future.

Acknowledgments. This work was supported by Jiangxi Science and Technology Support Project (20151BBE50044).

References

1. Valdovinos, J., Williams, R.J., Levi, D.S., et al.: Evaluating piezoelectric hydraulic pumps as drivers for pulsatile pediatric ventricular assist devices. J. Intell. Mater. Syst. Struct. **25**(10), 1276–1285 (2014)
2. Zhang, Z., Wang, W., Chen, X.: Study on the performance of piezoelectric micro pump for insulin injection. Chin. J. Med. Instrum. **39**(1), 64–67 (2015)
3. Research on water cooling system of computer chip whose power source is pressure electric pump. Jilin university (2005)
4. Ludwigs, A.: Micropumps-past, progress and future prospects. Sens. Actuators **105**(1), 28–38 (2005)
5. Liu, J., Sun, X., Liu, G., et al.: Experiment on dual-chamber parallel piezoelectric stack pump for electrorheological fluids. In: International Conference on Mechatronics and Automation. IEEE (2011)
6. Doll, A., Heinrichs, M., Goldschmidtboeing, F., et al.: A high performance bidirectional micropump for a novel artificial sphincter system. Sens. Actuators, A **130**(2), 445–453 (2006)
7. Peng, T.: Investigation on piezo-pump used in water-cooling-system for computer CPU Chip. J. Refrig. **30**(3), 30–34 (2009)
8. Shou, L., Yao, L.Q., Wang, Y.H., et al.: A kind of concentric piezoelectric membrane micropump for medical use. Piezoelectrics Acoustooptics **6**, 021 (2005)
9. Kim, J.H., Kang, C.J., Kim, Y.S.: A disposable polydimethylsiloxane-based diffuser micropump actuated by piezoelectric-disc. Microelectron. Eng. **71**(2), 119–124 (2004)
10. Ullmann, A.: The piezoelectric valve-less pump - performance enhancement analysis. Sens. Actuators, A **69**(1), 97–105 (1998)
11. Nong, N.V., Samson, A.J., Pryds, N., et al.: Microstructure and thermoelectric properties of screen-printed thick films of misfit-layered cobalt oxides with Ag addition. J. Electron. Mater. **41**(6), 1280–1285 (2012)
12. Tang, G., Yang, B., Hou, C., Liu, J., et al.: A piezoelectric micro generator worked at low frequency and high acceleration based on PZT and phosphor bronze bonding. Sci. Rep. **6**, 38798 (2016)

The Design and Practice of Data-Driven Teaching Evaluation Model in Colleges and Universities

LingYun Sun[✉]

Shandong University of Finance and Economics, Jinan, China
sunly123@hotmail.com

Abstract. With the rapid development of information technology in the 21st century, we have entered the era of big data. Based on information technology and big data, there are new teaching concepts, teaching models and teaching methods. In the last two years of teaching, we have tried to design and practice a data-driven college teaching evaluation model. Build an intelligent auxiliary teaching platform to provide teachers, students and teaching management personnel with intelligent assistance; take the "diversification-value process-testing ability" as the guiding ideology to carry out curriculum assessment reform, stimulate students' enthusiasm for learning, and promote students' all-round development; On the basis of constructing the classroom teaching index system and collecting a large amount of course data, the paper constructs a data-driven classroom teaching evaluation model: establishing a curriculum real-time evaluation teaching goal, constructing a classroom evaluation index system, and designing data-driven classroom teaching.

Keywords: Data driven · Incentive mechanism · Intelligent learning

1 Introduction

At present, there are two major problems in Chinese universities: first, the classroom attendance is low, participation is low, students do not study at ordinary times, before the test, only pay attention to test results, do not pay attention to the learning process, the study and examination of the significance of cognitive bias; the second is the college students' employment rate is not ideal, college students professional knowledge, There is a deviation between the ability of learning and the ability of innovation and the demand of employing units.

The reform of the course examination mode is an important breakthrough to solve these two problems. The basic aim of curriculum assessment is to transform the curriculum knowledge into students' professional cognitive ability and application ability, and to convert them into lifelong learning ability and creative ability. Whether the course examination method is reasonable not only affects the effect of the assessment directly, but also relates to the effect of classroom teaching and students' learning effect.

© Springer Nature Switzerland AG 2019
Y. Tang et al. (Eds.): HCC 2018, LNCS 11354, pp. 580–590, 2019.
https://doi.org/10.1007/978-3-030-15127-0_58

We propose to take "pluralistic-value process-examination ability" as the guiding ideology, it aims at stimulating students' learning enthusiasm, improving learning effect, promoting students' all-round development, exploring the comprehensive assessment method based on the combination of formative evaluation and summative evaluation, and forming a system of multiple assessment.

The new scheme will inevitably encounter the habits of teachers and students in the school conflict. First of all, teachers have been accustomed to the current assessment methods, especially the old teachers who have been teaching the same course for many years, using the original syllabus and examination mode to be familiar, and the implementation of the new teaching model may arouse objections. Because the curriculum examination way endows the teacher with the bigger autonomy, if the teacher does not support the new examination way, still maintains the original way, then the curriculum examination way reform will be fictitious. Secondly, because the teachers should be given greater autonomy, if the performance evaluation can not transcend the bondage of human society, reform effectiveness greatly reduced. Again, students are accustomed to getting a decent grade with the busy week before the exam, in the new assessment mode, each class, every homework and examination should be taken seriously, each paper can not use the network crash, if the students can not really understand the significance of the new assessment methods and timely adjustment of mentality and learning habits, Their scaled grades are likely to decrease. Therefore, we should fully arouse the teacher's guiding role and break the habit constraint which hinders the reform of the course examination mode.

Larry Cuban (cuban 2013) once asked: "Why is there a lot of structural changes in the school, and the change in teaching practice is still very small?" in the case of great changes in school management, curriculum, culture, teacher-student relationship–new textbooks, modern technology and new teachers are entering the classroom The basic teaching of teachers shows continuity and stability, such as back answers, paper and pencil tests are always retained by teachers [1]. However, with the 21st century's ability or core literacy becoming a growing number of countries' educational goals, teachers must be able to design and implement innovative teaching methods (innovative Pedagogies) to provide students with the conditions to achieve higher levels of learning outcomes, Thus systematically developing students' abilities or accomplishments in the 21st century, rather than achieving these results by chance.

Teaching has duality. Teaching and learning are interrelated and mutually reinforcing. It is a dynamic interaction between teaching and learning knowledge and practice. This requires motivating teachers to guide reform and implement new curriculum assessment programs. Encourage teachers to promote teaching, increase classroom teaching, course papers, group subject guidance and training, cultivate students' professional thinking and professionalism, improve students' communication skills and teamwork ability in professional knowledge. To this end, we completed the exploratory construction of the teaching evaluation system of "teaching evaluation" in colleges and universities, and proposed to use the students' learning results to evaluate the quality of classroom teaching, realize teaching interaction, and finally achieve the goal of precision teaching (Fig. 1).

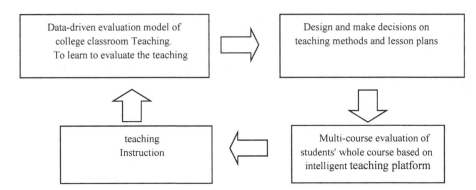

Fig. 1. Design framework of intelligent assistant teaching platform

2 The Design and Practice of Students' Multi-course Examination Mode Based on Intelligent Teaching Platform

The reform of curriculum assessment is guided by the principle of "pluralism-value process-examination ability"; To promote students' learning and to promote teaching as the assessment goal, the aim is to arouse the enthusiasm of student study, to improve the effect of learning, to promote the all-round development of students, and to make teachers adjust teaching specific goals, update teaching contents and forms in time, Improve the pertinence and effectiveness of teaching activities.

2.1 Explore the Course Examination Plan

This paper explores the comprehensive assessment method based on the combination of formative evaluation and summative evaluation, forms a system of multiple assessment, and designs and implements the reform plan of curriculum assessment according to different types of courses, and constructs the question bank with updating mechanism.

Based on the combination of formative evaluation and summative evaluation, the whole process, comprehensive, diversified and personalized assessment methods are designed.

Learning is a complex continuous process, beginning and end, from the knowledge learning, there is a start, expand, deepen and complete the process of construction, formative assessment and summative evaluation of the learning process and results of the diagnosis and evaluation. Summative evaluation is at the end of the semester, the use of the final examination methods to assess the students knowledge and ability, and the examination results of the analysis and value judgments, can reflect the curriculum knowledge and theory of memory, understanding and application ability. Formative evaluation is in the semester, the continuous use of a variety of ways to continuously assess student learning activities, and real-time, dynamic, multiple assessment results of the analysis and value judgments, this is a process of evaluation mechanism, is conducive to grasp the overall students learning process [2].

In order to carry out the diagnosis and evaluation of the whole process of students' learning, the new comprehensive system of curriculum assessment is based on the principle of the combination of formative evaluation and summative evaluation, using multiple assessment methods, highlighting the requirements of competency assessment, designing the whole process, comprehensive, pluralistic and personalized assessment.

The whole process of assessment: In the student learning process, non-stop to carry out multiple forms, multiple content assessment, to achieve the whole process of student learning activities tracking, diagnosis and feedback. In the whole process assessment, timely diagnosis and feedback of student learning effects, at any time based on the students' learning progress and real situation of students' knowledge and ability to carry out targeted teaching and guidance, improve student learning and teacher teaching; To motivate and maintain the continuous motivation of students to focus on learning every day; The result of the whole process assessment is to grasp the change and development process of students' learning dynamics, to reduce the whole process of student learning and to give the process evaluation.

Comprehensive Assessment: Learning is a comprehensive system of multi factor interaction and common progress, based on the comprehensive evaluation of students' learning activities, the construction of knowledge, ability and quality as the main content of the Comprehensive Assessment Index system. The reform of curriculum assessment should be assigned to complex and comprehensive indexes, to determine the proportion of the factors of knowledge, ability and quality in the assessment of students, in general, the curriculum assessment reform should be based on knowledge assessment, highlighting the requirements of competency assessment, taking into account the quality assessment, to guide the all-round development of students' knowledge, ability and quality.

Multi-Assessment: Diversified assessment means that many subjects of teaching activities participate widely and use a variety of assessment methods to realize the whole process of student learning and the development of students comprehensive quality assessment. In the course of teaching, the continuous release of a variety of learning tasks and real-time evaluation, the formation of various forms, rich content of multiple assessment. The system consists of the following components: Classroom performance, daily work, practical performance, Report papers and test tests, including test before class, unit test, mid-term examination, final exam.

Personalized assessment: Through the whole process of student learning to track and feedback, master the individual knowledge, ability and quality of the comprehensive situation of students to judge the characteristics of knowledge learning and quality development, the formation of individual learning development process diagnosis, Individualized learning guidance can be provided for students' special learning situations and individualized learning needs.

Using the whole process, comprehensive, pluralistic and individualized examination method to realize the whole process assessment and comprehensive assessment can impel students to pay attention to the whole process of learning, develop comprehensive ability and accomplishment, keep learning initiative and enthusiasm, improve learning efficiency, and urge teachers to adjust teaching objectives, update teaching contents and forms in time, Improve the pertinence and effectiveness of teaching activities.

2.2 Using the Intelligent Teaching Platform

Intelligent teaching Platform The platform currently contains a check-in system, job management system, examination system, question and answer system, Curriculum Knowledge Atlas system, evaluation management, teaching resources management functions. From this platform you can get student check-in, classroom online questioning, classroom online practice, classroom online Review, each chapter online test, Learning Resources browsing situation, case analysis and so on data analysis obtains the student's examination result, in addition to the achievement, but also takes the Knowledge Atlas form to display the examination result, may have the individual curriculum knowledge Atlas, may also be the class knowledge Atlas, the calendar year curriculum Knowledge Atlas and so on form comprehensively objective demonstration Course examination result.

Based on the information technology to construct the Intelligent Assistant teaching platform, can collect and analyze the students' learning process data, so it needs the function of teaching management, online learning, evaluation management, teaching resource management, and provides the intelligent assistant for teachers, students and teaching administrators. As shown in the Fig. 2:

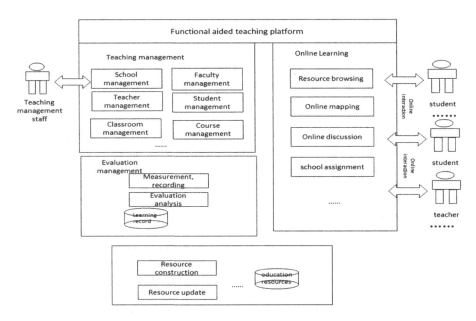

Fig. 2. The function of intelligent assistant teaching platform

2.3 Implementation Exploration

Since 2016, the intelligent teaching platform has been developed, based on the project team members, the process assessment and finalization of the management Information System course are analyzed, the goal of the course teaching is decomposed into the individual goal of the students, and a new model of curriculum evaluation is summed up according to the results of many years' practical teaching. Comprehensive evaluation of students' comprehensive achievements is made from four aspects, such as knowledge mastery, curriculum participation, personal contribution and team cooperation ability. Based on the reform guiding ideology of Formative curriculum assessment, the new assessment system pays attention to the process and development evaluation, emphasizes the improvement of students' ability and the growth of students, increases the content of students' comprehensive accomplishment and innovation ability, and enhances the weight of formative assessment.

At the same time, in the design of the weight system, the project team for the different curriculum assessment objectives and curriculum characteristics of the corresponding weight distribution. Taking into account the applicability and operability of performance evaluation tools, the use of simple empirical estimation method to determine its weight ratio, and follow the following principles: the corresponding weight of knowledge mastery should be more than 70% of its total, "Management Information System" Student performance evaluation Index System.

Knowledge Mastery (70%) consists of normal test scores (15%), Case assignments (5%) and final test scores (50%), including peacetime tests, case studies and final tests, the whole process of multiple, knowledge and competency assessment combined.

Curriculum participation (20%) consists of classroom performance, answering questions (5%), online discussion (5%), class attendance (5%) and Knowledge point Browsing (5%), focusing on classroom participation and motivating students to participate actively.

Personal contributions (5%) from the search for relevant information, theoretical preparation (2%), point of view, method unique, innovative (3%), mainly in the group members of peer review, on the platform for scoring.

Team collaboration Capability (5%) includes further research capability, knowledge development ability (1%), active participation, communication ability, collaboration ability (2%) and task completion, problem solving (2%), and the evaluation of results by teachers according to case discussion.

The effect of multi-evaluation method system: (1) constantly provide a variety of assessment methods for students to learn the whole process of tracking assessment and real-time feedback. (2) comprehensive diagnosis and evaluation of the comprehensive development of students' knowledge, ability and quality by using a variety of assessment methods. Different forms of assessment methods focus on different assessment content, such as various tests focused on the effect of knowledge learning, the paper focused on the ability to study, including the use of knowledge theory, the ability to deal with integrated information and logic, dialectical and critical thinking ability, the group task focused on innovative ability and team cooperation ability.

In the two-year "Management Information System" course teaching practice, has received the positive effect, has raised the student's study enthusiasm and the curriculum whole process participation degree, to the student this discipline comprehensive accomplishment and the ability Questionnaire show, has the obvious enhancement.

3   Data-Driven Evaluation Model Design of College Classroom Teaching

3.1   Construction of Classroom Teaching Index System

In view of the traditional teaching mode, the aim of teaching evaluation is single, one-sided evaluation, subjective problem of evaluation standard, complete the exploratory construction of teaching evaluation system of "learning evaluation teaching" in college class, evaluate the quality of classroom teaching with the result of students, realize teaching interaction, and finally achieve the goal of precision teaching.

The basic idea or assumption of the design of Teaching evaluation Index system of "learning to teach" is to evaluate the effect and quality of teachers' teaching by the performance and state of learning behavior caused and promoted by teaching. On the basis of theoretical analysis, this paper constructs the evaluation Index system of classroom teaching based on the teaching assessment. From the perspective of system theory, summarize all aspects of the classroom teaching process, and then select key elements and behavioral characteristics from the perspective of effective teaching.

3.2   Effective Use of the Teaching Aid Platform

In view of the limitation of time and space under the traditional classroom teaching mode, the incomplete evaluation of data collection, the weak analysis ability and the not timely feedback of the evaluation results, this paper constructs an intelligent assistant teaching platform to explore how to collect student data more conveniently and in time to realize the objective evaluation of teaching.

Based on the requirement of acquiring students' learning situation in real time, this platform realizes the functions of teaching management, online learning, evaluation management, teaching resource management and so on, and provides intelligent assistant for teachers, students and teaching administrators.

3.3   Constructing Data-Driven Classroom Teaching Evaluation Model

On the basis of completing the construction of classroom teaching index system and collecting a large amount of course data, the paper builds a data-driven classroom teaching evaluation model: establishing a curriculum real-time evaluation teaching goal, constructing a classroom evaluation index system, and designing data-driven classroom teaching.

(1) Analyze the correspondence between the classroom teaching index system and the specific observation points of the student classroom data.

The basic idea of the design of teaching evaluation index system based on "teaching evaluation" is to evaluate the effect and quality of teacher teaching by the performance and state of learning behavior caused and promoted by the teaching. On the basis of theoretical analysis, we construct the evaluation Index system of classroom teaching based on the teaching assessment. From the perspective of system theory, this paper sums up all the links in the course of classroom teaching, then selects the key elements and behavioral characteristics from the perspective of effective teaching, and determines whether the students' classroom data can be collected by constructing the intelligent education aid platform. At present, the index system is designed by teaching effect, teaching benefit, teaching efficiency, background variable, four evaluation items, 13 evaluation points.

There are four aspects of teaching effectiveness: (1) the comprehensiveness of learning outcomes, the achievement of students' academic achievement, mastery of knowledge and skills, and the values of emotional attitudes; (2) sustainability of learning outcomes, interest in learning, self-confidence Heart, learning ability, self-efficacy and learning habits; (3) the level of learning outcomes, whether the results of different courses taught by different teachers are different; (4) the comprehensiveness of learning outcomes, whether teaching can promote the whole Student learning and development.

There are three aspects of teaching efficiency: (1) Learn to apply, whether students can apply the knowledge gained in the course to practice; (2) Learn to study, whether the students in the process of positive thinking, active questioning, there is no internal understanding, active construction; (3) Learn to behave, Whether to actively pay attention to and guide students in teaching activities of various moral performance and moral development.

The teaching efficiency has two aspects: (1) The teaching time, whether the classroom teaching time is used in the teaching and learning activities which point to the goal; (2) The time utilization, whether the teaching activity points to the value maximization teaching content.

The background variables consider four aspects, (1) teaching objectives, whether there are stratified design goals, content, schedule, and job requirements; whether students of different levels or characteristics have different teaching objectives; (2) Whether the content of the teaching grasps the characteristics of the subject; whether the interpretation of the content is accurate (reasonable determination of the focus of teaching); whether it meets the teaching objectives; (3) Academic situation, the curriculum design is adapted to the student's experience foundation, matches the student's knowledge base, and adapts to the students' thinking ability. The curriculum design meets the students' learning needs; (4) Teaching conditions, whether the teaching behavior of this class has corresponding time, space and material conditions.

On this basis, it is the key to analyze and establish the correspondence between the Classroom Teaching index system and the students' classroom data, and construct the data-driven evaluation model by using the platform data.

(2) Multi-evaluation management and integration among modules in teaching aid platform.

Construct a developmental and diversified evaluation system to promote multi-agent and multi-dimensional interaction. Constructing a multi-evaluation system: On the subject of evaluation, teachers evaluate each other among students, students, teachers, students, students, in the evaluation content, each subject can evaluate the teaching content, teaching resources, student work content and form.

Based on the need to obtain student learning in real time, the use of information technology to build an intelligent auxiliary teaching platform, the platform has realized the functions of teaching management, online learning, teaching resource management, providing intelligent assistance for teachers, students and teaching management personnel. On this basis, improve evaluation management and its integration with other modules.

(3) build a data-driven evaluation of teaching and learning classroom teaching evaluation model.

Big data-assisted teaching platform to keep as many students and students as possible in the data information on the platform, including the teaching platform for students of all activities, such as learning knowledge map, do quizzes, upload assignments, participate in the discussion, interactive evaluation, click Resource Links. With the continuous development of students' learning, new real-time data are produced, and the data will be greatly increased in both volume and surface.

(1) to analyze a large amount of data and excavate valuable teaching information. The large data-assisted platform displays a large amount of information collected and stored in a graph, and provides teachers with data analysis tools. On this basis, teachers use classification or combination analysis, individual analysis and other methods to have a large number of real-time data analysis, accurate grasp of the individual or all students learning progress and learning needs, ideological trends and ideas, and so on to obtain a large number of valuable teaching information.

(2) using valuable information obtained from large data analysis to guide teaching activities. In the course of teaching objectives and program formulation, teachers adjust teaching objectives and programs according to the results of data analysis, and can also put forward individualized or group teaching scheme for individual or some groups of different characteristics. In the course of classroom teaching, it is fully based on the students' learning ability, learning needs and interests, thoughts and viewpoints, setting appropriate teaching contents and choosing appropriate teaching methods. In the teaching feedback and the evaluation link, synthesizes the Student classroom study situation and the unceasingly updated platform study data to the continuous and the comprehensive appraisal, consummates and enriches the appraisal system.

(3) How the big data-assisted teaching platform plays an important role in promoting students' learning. The first is to adapt to the students' networked, digitized and fragmented learning mode, so that students can easily access various learning resources on the teaching aid platform, and use the fragmented time to learn at any time to extend the learning time and space. Second, the teaching aid platform gives students full learning autonomy. Students decide their own learning and expanding content, freely participate in discussions and evaluations, and enable students to consciously internalize correct ideas. The third is to make students' online and offline learning complement and deepen each other.

4   Conclusions

To make the change happen and sustainable, the key is to motivate teachers, in large data environment, using a variety of advanced platform design teaching, teachers as a common designer (co-designer), not only the implementation of design activities. The implementation process of design based research generally includes: researchers and participants collaborate to analyze the problem-using existing theories, methods, techniques to build theoretical prototypes-iterative practice in real situations, revising the scheme, summarizing the experiences, methods and principles of design and implementation (Baoha 2015) [3]. It should be realized that in the process of change, teachers are also learners. Therefore, training for teachers should conform to effective scientific principles of learning, while common design implies learning principles such as collaborative learning, real problem solving, and inquiry learning.

In addition, the main place where teacher teaching occurs is that the classroom is nested in schools, school districts, society, countries and other systems. Therefore, the change of teaching evaluation model depends on the changes at all levels. Nancy Law links change with learning and builds a multi-level, multi-scope learning model that demonstrates how to support innovation in teaching (Fig. 3). First, changes at all levels should be understood as a concept of learning, which means that at different levels, the factors that influence student learning should also be defined as learning outcomes at that level (see figure below).

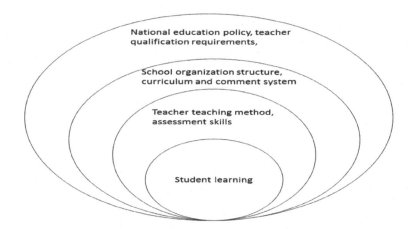

Fig. 3. Levels that affect student learning factors

Therefore, if we take a simple analysis of the problem perspective, we will think that the teaching change can be divided into a number of isolated and static parts, which can be solved by using algorithms and flow chart techniques, and once the initial problem is solved, the result is reproducible; The perspective of analysis will regard teaching change as an environment consisting of multiple factors that are interdependent and constantly changing, so the teaching change is nonlinear and uncertain. In fact, the data-driven college teaching evaluation model reflects systemicity and complexity. The new evaluation model is not realized in a vacuum and isolated environment. The macro background of students, teachers, schools, society, countries and even the world will be It has an impact on this. Therefore, in solving the problem of teaching change, it is important to have great patience and keen perception, and to be sensitive to the iterative feedback results appearing at various levels, so as to continue to advance.

Acknowledgments. This work was financially supported by the Teaching Reform Research Project of Undergraduate Colleges and Universities of Shandong Province (Z2016Z036) and the Teaching Reform Research Project of Shandong University of Finance and Economics (jy201806, jy201830, jy201810).

References

1. Jin, Q., DingXu, ShengQun: How to change the teaching–OECD the "5C" framework of the creative pedagogy. Open Educ. Res. (04) (2018)
2. HeKan: On the evaluation model of pluralistic teaching in colleges and universities. Educ. Manag. Coll. Univ. **5**(01) (2007)
3. Baoha, L.Y.: The characteristics of learning analysis under the background of large data. J. Beijing RTVU (01) (2015)

Research and Application of RFID System Security Authentication Protocol Based on ECC Algorithm

Qiaohong Zu[(⊠)] and Kui Cai

School of Logistics Engineering, Wuhan University of Technology,
Wuhan 430063, People's Republic of China
zuqiaohong@foxmail.com, 1548855492@qq.com

Abstract. The security of most existing RFID protocols, which is difficult to guarantee, is based on simple XOR or Hash operations. However, the more secure traditional encryption algorithm is difficult to apply to the RFID system directly, due to the limited computing resources of RFID tags. In order to solve the security risks of RFID in user data security and privacy, this paper studies the lightweight of elliptic curve cryptography (ECC) algorithm, which makes it more suitable for RFID tags, and then it designs a new RFID two-way authentication protocol combined the advantages and disadvantages of existing protocol based on ECC algorithms. Finally, in view of the frequent transfer of electronic tag ownership in the logistics supply chain environment, the designed RFID authentication protocol has been extended to make it more suitable for the logistics supply chain.

Keywords: RFID · Elliptic curve cryptography (ECC) ·
Algorithm lightweight · Authentication protocol · Protocol extension

1 Introduction

Currently, in the large-scale application of RFID, most of the RFID data acquisition systems are unencrypted. All data exchanges are exposed to the open wireless network environment, and the external data can be easily obtained by some means. Therefore, it is necessary to encrypt the relevant data by means of encryption [1, 2]. Although the current ECC algorithms all have high encryption intensity, which can meet the intensity requirements of RFID data encryption, it requires certain computing capacity of tags, and improving the computing capacity of tags will inevitably increase the production cost of tags, which will inevitably lead to increased difficulty in large-scale application. Therefore, in order to apply the ECC algorithm to the RFID system, it is necessary to carry out the lightweight processing to ensure that the hardware cost enables to be controlled while ensuring certain encryption strength. In view of above, the lightweight research of ECC algorithm and the application of security protocol based on ECC algorithm are effective methods to improve the security of existing RFID systems.

© Springer Nature Switzerland AG 2019
Y. Tang et al. (Eds.): HCC 2018, LNCS 11354, pp. 591–600, 2019.
https://doi.org/10.1007/978-3-030-15127-0_59

2 Lightweight Study of Elliptic Curve Cryptography

Although elliptic curve encryption can achieve high encryption intensity and secure key management, it is not feasible to apply traditional ECC encryption directly to resource-constrained RFID environment. In the ECC encryption operation, the two operations that are carried out most frequently and consume the most computing resources are scalar multiplication and point addition. Therefore, in this paper, the point addition operation and scalar multiplication operation are optimized to realize the lightweight of elliptic encryption algorithm.

2.1 Fast Point Add Operation

Point add operation is a calculation frequently used in elliptic curve encryption, and there are many researches on the optimization of point add operation. Nicolas [3] provided a method for fast point add operation with the same Z coordinate under Jacobian coordinates, and the formula is as follows:

$$
\begin{cases}
X_3 = (Y_2 - Y_1)^2 - (X_2 - X_1)^3 - 2X_1(X_2 - X_1)^2 \\
Y_3 = (Y_2 - Y_1)(X_1(X_2 - X_1)^2 - X_3) - Y_1(X_2 - X_1)^3 \\
Z_3 = Z(X_2 - X_1)
\end{cases}
\tag{1}
$$

In this case, the point add operation only requires 5 times of modulus multiplication and 2 times of modulus square operation, which is greatly improved compared with the calculation amount of 8 times of modulus multiplication and 3 times of modulus square in the fastest jacobian-affine coordinate. Venelli [4] used this formula to realize the fast algorithm of $dP + Q$. In order to improve the scalar multiplication algorithm later, this paper only used it to realize the fast point add operation of $2P + Q$. Above all, $P + Q = (X_3, Y_3, Z_3)$ in jacobian-affine coordinates can be calculated by these following equations:

$$
\begin{cases}
X_3 = 4(Z_1^3 Y_2 - Y_1)^2 - 4(Z_1^2 X_2 - X_1)^3 - 8X_1(Z_1^2 X_2 - X_1)^2 \\
Y_3 = 2(Z_1^3 Y_2 - Y_1)(4X_1(Z_1^2 X_2 - X_1)^2 - X_3) - 8Y_1(Z_1^2 X_2 - X_1)^3 \\
Z_3 = 2Z_1(Z_1^2 X_2 - X_1)
\end{cases}
\tag{2}
$$

Through the idea of transforming modular multiplication into squares, $Z_3 = 2Z_1(Z_1^2 X_2 - X_1)$ could be converted into:

$$
Z_3 = (Z_1 + Z_1^2 X_2 - X_1)^2 - Z_1^2 - (Z_1^2 X_2 - X_1)^2
\tag{3}
$$

However, the calculation of $2P + Q$ cannot directly use Nicolas's formula, because the z coordinates of $(P+Q)$ and P in $2P + Q = (P + Q) + P$ are different, but the following transformation can be performed on point P to make their Z coordinate equal.

$$
(X_1, Y_1, Z_1) = (4X_1(Z_1^2 X_2 - X_1)^2, 8Y_1(Z_1^2 X_2 - X_1)^3, 2Z_1^2(Z_1^2 X_2 - X_1))
\tag{4}
$$

The following substitutions can also be made to further reduce the amount of computation of $(2P + Q)$:

Let $\alpha = Z_1^3 Y_2 - Y_1$, $\beta = Z_1^2 X_2 - X_1$ and substitute them into the Eq. (2) and gives:

$$X_3 = 4\alpha^2 - 4\beta^3 - 8X_1\beta^2, Y_3 = 2\alpha(4X_1\beta^2 - X_3) - 8Y_1\beta^3 \tag{5}$$

$$X_1^{(1)} = 4X_1\beta^2, Y_1^{(1)} = 8Y_1\beta^3, Z_1^{(1)} = (Z_1 + \beta)^2 - Z_1^2 - \beta^2 \tag{6}$$

Let these following equations be established.

$$\theta = X_3 - X_1^{(1)} = 4\alpha^2 - 4\beta^3 - 8X_1\beta^2 - 4X_1\beta^2 = 4(\alpha^2 - \beta^3 - 3X_1\beta^2) \tag{7}$$

$$\omega = Y_3 - Y_1^{(1)} = -2\alpha\theta - 16Y_1\beta^3 = \alpha^2 + \theta^2 - (\alpha + \theta)^2 - 16Y_1\beta^3 \tag{8}$$

Equations (7) and (8) can eliminate the intermediate calculation of X_3, Y_3 and Z_3, and avoid a modular multiplication operation. At this time, compared with the calculation amount of $(8M + 3S) + (8M + 3S) = 16M + 6S$ calculated by jacobian-affine coordinate directly, the calculation quantity of $2P + Q$ is $(6M + 5S) + (5M + 2S) = 11M + 7S$, which is much lower.

2.2 NAF Improvements and Sliding Window NAF Improvements

(1) Improvement ideas

The RFID tag is not only hardware with limited computing resource, but also hardware with limited storage resource. Therefore, when designing the algorithm used in RFID system, we should take into account the efficiency of the algorithm and the amount of memory. This paper presents two versions of the improvement. One is based on the improvement of NAF, which improves the efficiency of the NAF algorithm. The other is based on the improvement of the sliding window NAF, which sacrifices a small amount of computational efficiency, but it reduces the amount of pre-calculation by half.

(2) Improvement of NAF scalar multiplication algorithm

Inspired by the $2P + Q$ algorithm implemented in Sect 2.1, the $2P + Q$ algorithm can reduce the computation amount of point add operation, apply it to the NAF algorithm, and further improve the NAF algorithm. By replacing $P + Q$ with $2P + Q$ as the basic computing unit of NAF, the calculation amount of point addition operation can be reduced, and the hamming weight of m can be reduced indirectly. The improved algorithm is as follows.

Algorithm 1: improvement of NAF scalar multiplication algorithm

Input: where the integers m and P∈E are the point on the elliptic curve group

Output: $Q = mP$

    A. If $m = 0$, output O and return Q;

    B. If $m < 0$, then set $P \leftarrow (-p), m \leftarrow (-m)$;

    C. Find the binary representation $k = (k_l k_{l-1} \dots k_1 k_0)_2$ of k;

    D. Find the binary representation $h = (h_l h_{l-1} \dots h_1 h_0)_2$ of 3k;

    E. $Q \leftarrow Q$;

    F. *For i from l − 1 downto 1 do*

        a) $Q \leftarrow 2Q$;

        b) If $k_i = 0$ and $h_i = 1$, calculate $Q \leftarrow 2Q + P$;

        c) If $k_i = 1$ and $h_i = 0$, calculate $Q \leftarrow 2Q - P$;

    Output Q

(3) Improvement of NAF algorithm based on sliding window

Since the sliding window NAF algorithm needs to pre-calculate $P_i = iP$, (*i* is an odd number), so you can pre-calculate a value every other item (Pre-calculation is done when i%4=1, and fast calculation is used when i%4=3). The calculation is performed to reduce the pre-storage of half of the sliding window NAF scalar multiplication algorithm, while using the other half of the pre-storage to ensure the efficiency of the algorithm. The improved low-memory NAF scalar multiplication algorithm is as follows.

Algorithm 2: the improvement based on sliding window NAF algorithm

Input: where the integers $m, P \in E$ are the point on the elliptic curve group, and the window width is w

Output: $Q = mP$

    A. Pre-calculate $P_i = iP (i = 1,3,5, \dots, \{2\{[-(-1)^w]\}/3 - 1)$ every other item;

    B. If $m = 0$, output O and return Q;

    C. If m<0, then set $P \leftarrow (-P), m \leftarrow (-m)$;

    D. Find the NAF expression $m = (k_l k_{l-1} \dots k_1 k_0)_2$;

    E. $Q \leftarrow P$;

    F. For *i* from $l − 1$ downto 1 do

        a) if $k_i = 0, t = 1, u = 0$

            else find t, let $t \leq w, u = (k_i k_{i-1} \dots k_{i-t+1})\%$ and $u\%2 = 1$;

        b) $Q \leftarrow 2tQ$;

            if $u > 0$,

            case: $u\%4 = 1, Q \leftarrow Q + uP$;

            case: $u\%4 = 3, Q \leftarrow Q+(u-2)P+2P$;

        c) if $u < 0$,

            case: $u\%4 = 1, Q \leftarrow Q- uP$;

            case: $u\%4 = 3, Q \leftarrow Q-(u-2)P+2P$;

        d) $i \leftarrow i - t$

    Output Q

3   Design of RFID System Security Authentication Protocol Based on Lightweight ECC Algorithm

This paper combines the advantages and disadvantages of current mainstream RFID protocols, and proposes an ECC-based RFID Authentication Protocol (hereinafter referred to as ERAP protocol) based on the second chapter ECC lightweight algorithm. The protocol uses the status flag to determine the status of the tag ownership transfer, enabling extended applications in the logistics supply chain environment, which can greatly reduce tag usage costs. At the same time, in order to prevent the leakage of ID after the transfer of ownership, MetalID replaces the ID in the RFID tag of this protocol [5, 6].

3.1   Protocol Initialization

The initialization stage mainly involves the generation of some parameters and the writing of data.

- Select the finite field F_D and the encrypted elliptic curve E.
- Define the elliptic curve on the finite field F_D, select the proper basis point G and calculate the order n of G.
- Select k_r as the private key of the reader, and calculate the public key K_r
- The server database saves (ID, MID, k_r, G)
- The tag is written to $(MetalID, K_r, G, status)$ where the initial $MetalID = hash(ID)$, abbreviated as MID, and status is set to 0.

3.2   Specific Process of the Protocol

This protocol is based on the RFID encryption protocol with improved ECC algorithm. The protocol process is executed in the figure. The protocol consists of three main components: tags, servers, and readers. Since the reader initially issues a query or transfer command to the tag, the first thing to do is to authenticate the reader and then the tag. The protocol is mainly divided into two processes: the reader authentication process and the tag authentication process. The detailed process of the protocol is shown in Fig. 1.

(1) Reader authentication process
  (1) The pseudo random number generator (PRNG) of the reader R generates a random number r_1, calculates $T_1 = r_1 G$, and sends T_1 as a query command to the tag.
  (2) The tag randomly selects the integer r_2, calculates $T_2 = r_2 G, T_3(x_3, y_3) = r_2(T_1 + K_r)$ and sends T_2 to the reader;
  (3) The reader performs the following calculation: $T_3'(x_3', y_3') = (r_1 + k_r)T_2$, $u = x_3' \oplus y_3'$, sending u to the tag;
  (4) After receiving the message from the reader, the tag is verified as follows: if the equation $u = x_3 \oplus y_3$ is established, the reader authentication will passed; otherwise, the protocol will be terminated.

(2) Tag authentication process

(5) The tag continues to be calculated as follows: $T_4 = MID + h(T_1 \parallel T_2 \parallel T_3)$, $T_5 = h(T_2 \parallel T_3 \parallel MID)$, sending T_4, T_5 to reader R.

(6) After receiving the response of the tag, the reader forwards T_2, T_4, T_5, r_1 and T_1 to the server together, and the server calculates $T_3' = T_3(r_1 + k_r)T_2$, $MID = T_4 - h(T_1 \parallel T_2 \parallel T_3)$, $T_5' = h(T_2 \parallel T_3 \parallel MID)$, and then retrieves the database table by using the MID as an index. If the database table exist $MID = MID'$ and $T_5' = T_5$, the tag is verified, and the background server sends the relevant information of the tag to the reader.

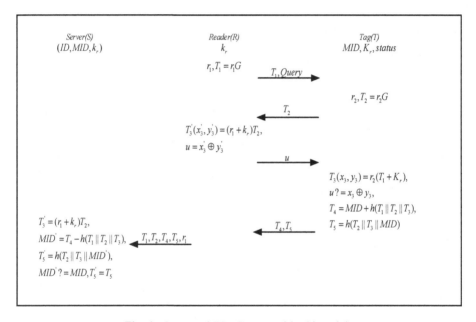

Fig. 1. Improved ERAP protocol in this article

4   Extended Application of ERAP Protocol in Logistics Supply Chain Environment

In the application of RFID system in the field of logistics supply chain, due to the circulation of goods, the label will often change the ownership of users, and its security requirements are different from the normal environment. In the context of the supply chain, the protocol should not only take into account security issues outside the RFID system, that is, various kinds of external attacks by lawbreakers, but also security issues within the system, that is, ownership transfer parties are not necessarily reliable. Therefore, in order to ensure the security of RFID application in logistics supply chain, the extension protocol not only needs to meet the security requirements of general RFID systems, but also needs to meet the privacy security requirements of the original owner of the tag and the privacy security requirements of the new owner of the tag [7]. This chapter expands the ownership transfer of the ERAP protocol designed above, applies it to the environment of logistics supply chain, and verifies the validity of security protocol through simulation.

4.1 Ownership Transfer Extension Scheme of ERAP Protocol

The transfer of ownership can be abstracted into two main bodies: the old owner (OO), that is, the supplier and the new owner (NO) of the tag, that is, the retailer. OO will trade some goods with RFID tags to NO, NO will check the goods after receiving the goods. The goods that have passed the inspection will be accepted, and the unqualified goods will be returned to OO. In such an application environment, the above protocol is extended as follows. The extended solution process is shown in Fig. 2, where K'_r is the reader's public key for the new owner of the tag.

The specific description is as follows:

(1) NO sends its reader's public key K'_r to OO through the secure channel, OO also sends the data table (MID, ID) and K_r to NO through the secure channel, and NO saves it in the RFID system server database.

(2) After completing the reader and tag authentication, OO sends the public key K'_r of NO to the tag and sends a change instruction. The tag calculates $K = K'_r \oplus K_r$ (modulo operation in the prime domain) and sets the status to 1 (transfer status).

(3) After receiving the goods, the NO sends a change command and K_r to the tag. The tag is received as the following calculation $K'_r = K \oplus K_r$, and the above label and reader authentication process is performed.

(4) If the authentication is successful, the server calculates $T_6 = h(MID \parallel T_1 \parallel T_2 \parallel T_3)$, and updates the database $MID = h(MID \parallel T_3)$, sends T_6 to the reader, and sends a confirmation message to the OO server. If the authentication fails, it returns a random binary string ER (ER and T_6 are equal in length) to the reader. The role of the ER is mainly to interfere with the attacker.

(5) The reader receives the message T_6 from the server and forwards it to the tag. After the tag receives T_6, it verifies $T_6 = h(MID \parallel T_1 \parallel T_2 \parallel T_3)$. If the verification is passed, $MID = h(MID \parallel T_3)$, status = 0, $K = K'_r$ are updated, and then the ownership transfer is completed. If the authentication fails, no update operation will be performed.

The tag will check the value of its status before the reader's authentication every time. If it is 1, it needs to perform an exclusive XOR operation to obtain the public key K_r or K'_r of the authentication reader. At the end of the authentication, if the value of status is 1, a Hash operation is performed to update the MetalID inside the tag, and the public key is updated with a value of 0.

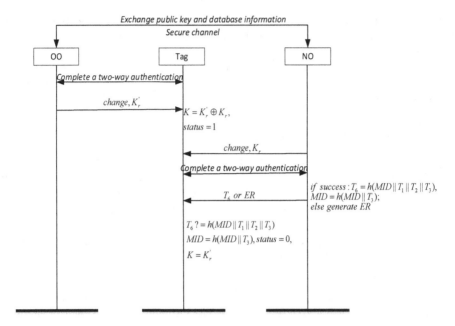

Fig. 2. Extended application of protocol in logistics supply chain environment

4.2   Simulation Implementation of the Protocol

In the supply chain environment, in order to verify the validity of the protocol, this paper uses the software to simulate the protocol according to process of the protocol. The simulation development environment is Visual Studio 2013, the simulation language is C++, the curve $E : y^2 = x^3 + ax + b$ used in the simulation is based on the elliptic curve on the prime domain, and the simulation algorithm calls the Tommath large number library to realize fast point add, scalar multiplication, double point operation and NAF coding. The mutual authentication between reader and tag is mainly based on logical simulation and calls the improved algorithm to implement it.

The specific simulation process is as follows:

(1) Protocol initialization

The protocol initialization process is mainly to generate the relevant parameters required by the protocol elliptic curve algorithm. The main parameters are the parameters p, a, b of the elliptic curve used by the determination algorithm, the generator G of the elliptic curve, and public key Q or the private key k of the reader required for the authentication process.

(2) Reader authentication process

The reader authentication process is challenged by the reader initially. The reader sends T_1 to the tag, and the tag responds T_2. After the reader receives T_2, the reader calculates $T_3'(x', y') = (r_1 + k)T_2$, and sends $u' = x' * y'(mod p)$ to the tag. After the tag receives u', it calculates $T_3(x, y) = r_2(T_1 + K)$, $u = x * y(mod p)$. If $u = u'$, the reader is

Fig. 3. Reader authentication simulation

authenticated; If not, the reader authentication fails. The simulation result of the reader authentication process is shown in Fig. 3.

(3) Tag authentication process

Tag authentication mainly relies on the verification of MID and T. There is a correspondence between MID and T. The essence of verification T is the verification of the reader public key. The result of the tag authentication simulation is shown in Fig. 4.

From the above figures, it can be found that both the reader authentication and the tag authentication are passed during the simulation authentication process of the newly designed protocol, which verifies the validity of the new protocol.

5 Conclusion

In this paper, first and forest, the underlying operation of ECC algorithm, point-add operation, is studied. Moreover, the $2P + Q$ fast algorithm is introduced and applied to the most commonly used scalar multiplication of ECC encryption based on the research of Nicolas, which improves the effectiveness of NAF scalar multiplication algorithm. Meanwhile, the NAF algorithm that is based on sliding window, has been improved to reduce its pre-calculation by half, which is more suitable for RFID tags. Moreover, on the basis of the improved algorithm, a new RFID authentication protocol based on ECC, ERAP, is proposed, which combines the advantages and disadvantages of existing protocols based on ECC. After that, this paper extends the ERAP protocol and introduces the status parameters so that it can facilitate the transfer of label ownership in the logistics supply chain environment. In the end, in order to verify the validity of the protocol, the protocol is simulated.

References

1. Mujahid, U., Najam-Ul-Islam, M., Shami, M.A.: RCIA: a new ultralightweight RFID authentication protocol using recursive hash. Int. J. Distrib. Sens. Netw. **2015**(3), 1–8 (2015)
2. Guo, C., Zhang, Z.J., Zhu, L.H., et al.: A novel secure group RFID authentication protocol. J. China Univ. Posts Telecommun. **21**(1), 94–103 (2014)
3. Nicolas, M.: Fast and Secure Elliptic Curve Scalar Multiplication Over Prime Fields Using Special Addition Chains. IACR Cryptology Eprint Archive (2006)
4. Venelli, A., Dassance, F.: Fast scalar multiplication for elliptic curve cryptosystems over prime fields: US, US 8369517 B2 (2013)
5. Liao, Y.P., Hsiao, C.M.: A secure ECC-based RFID authentication scheme integrated with ID-verifier transfer protocol. Ad Hoc Netw. **18**(7), 133–146 (2014)
6. Lee, Y.C., Hsieh, Y.C., You, P.S., et al.: An Improvement on RFID Authentication Protocol with Privacy Protection. IEEE Computer Society (2008)
7. Xie, R., Jian, B.Y., Liu, D.W.: An improved ownership transfer for RFID protocol. Int. J. Netw. Secur. **20**(1), 149–156 (2018)

A Systematic Framework for Malicious Traffic Detection Based on Feature Repository

Shuai Liu, Yong Zhang$^{(\boxtimes)}$ (iD), Lei Jin, Xiaojuan Wang, Mei Song, and Da Guo

Beijing University of Posts and Telecommunications, Beijing, China
yongzhang@bupt.edu.cn

Abstract. Machine learning becomes an effective method to detect malicious traffic. With the proliferation of network traffic, malicious traffic categories are greatly increased, which puts forward higher requirements for the computation time and detection accuracy of machine learning. A feature selection framework is proposed to balance the computation time and detection accuracy. First, we construct a feature repository of traffic information with high dimensions. In order to reduce the computation time and minimize the loss of accuracy, we investigate the feature selection algorithms. The algorithm based on chi-square test and xgboost algorithm are adopted to evaluate the proposal. The experiments on CTU dataset show that the proposal can reduce the computation time while ensuring the accuracy.

Keywords: Malicious traffic detection · Machine learning ·
Feature repository · Feature selection

1 Introduction

Due to the wide application of computers and smart phones in various fields, the Internet has gradually become the main carrier of various businesses, and various network applications have also emerged in an endless stream. In the huge network traffic, there are inevitably many malicious traffic. The malicious traffic not only affects the network bandwidth and service quality, but also poses a security risk to personal privacy and even national security to a certain extent [1].

The traditional method for identifying malicious traffic mainly relies on deep packet inspection technology and the deep packet inspection technology mainly identifies the malicious traffic by matching the application layer data of the data packets [2]. This feature-based matching technology can only work when all packet content is accessible. However, with the use of encryption technology, deep packet inspection technology can play a smaller role [3].

In recent years, the application of machine learning in computer vision, speech recognition, and natural language processing has achieved remarkable results, demonstrating the powerful ability of machine learning in solving classification and prediction [4–6]. Therefore, using machine learning technology to

© Springer Nature Switzerland AG 2019
Y. Tang et al. (Eds.): HCC 2018, LNCS 11354, pp. 601–612, 2019.
https://doi.org/10.1007/978-3-030-15127-0_60

solve the problem of cyberspace has attracted the attention of many researchers [7–9]. However, machine learning based algorithm requires huge datasets to train the classifier, and how to extract valid features from the original traffic data has also become a major challenge. In addition, it is well known that the traffic data is very large, and how to improve the detection speed while ensuring the detection accuracy is also a problem which is difficult to handle.

In order to solve the above problem, we have done the following work:

1. We proposed a new method to extract as much information as possible from the raw traffic data by building a feature repository.
2. We proposed a feature selection method to select features from the feature repository to ensure the speed and performance of the detection.
3. We proposed a systematic malicious traffic identification process.

The rest of this paper is organized as follows: The second section introduces related work. The third section describes in detail the approach we have proposed. The fourth section shows some experiments. The fifth section is conclusion and future work.

2   Related Work

In the existing literature, there are many researches on traffic classification, which are mainly divided into three categories: 1. Port-based. 2. Based on deep packet inspection. 3. Based on machine learning.

The port-based traffic classification method mainly performs port number detection on the TCP/IP packet header [10], and identifies a certain traffic type according to a well-known port number. This method was very simple and convenient during the early traffic classification times. However, with the development of network applications and the emergence of technologies such as port confusion, the scope of adaptation is getting smaller and smaller [11,12].

The deep packet based detection is a method with high traffic classification accuracy. The so-called "deep" is compared with the normal packet analysis method. The normal packet detection only analyzes the content below the 4 layers of the OSI model. This includes source address, destination address, source port, destination port, and protocol type [13,14]. In addition to the previous hierarchy analysis, DPI also adds application layer analysis to identify various applications and their content. Xu [15] and others made a simple survey on deep packet inspection technology, and systematically summarized the application, algorithm and hardware deployment platform of deep packet inspection technology. Rodrigues [16] and others set up a honeypot system in which abnormal flow was detected by deep packet inspection of network traffic payloads. However, this system has nothing to do with encrypted traffic.

The network traffic classification method based on machine learning mainly calculates the statistics of network traffic packets, and then sends these data as features to the machine learning model for training, thereby completing the classification of different network traffic. The methods in machine learning mainly include supervised learning and unsupervised learning. Unsupervised learning does not require labeled data, and clusters directly based on the similarity of samples. In the study of using unsupervised learning methods to classify traffic, Erman [17] and others based on the K-means algorithm to classify the unidirectional traffic of the network and found the server-to-client direction in the TCP connection. The experiments show that the unidirectional flow statistics are classified with higher accuracy. Finamore et al. [18] proposed a clustering algorithm that automatically determines the number of clusters based on the enhanced K-means algorithm, achieving 0.95 accuracy on actual UDP traffic. However, in the unsupervised classification, since the actual flow type cannot be determined, the constructed model is difficult to apply to the actual scene. Supervised learning may solve this problem. Xu [19] et al. proposed the Incremental KNN-SVM method for malicious traffic detection. The KDD CUP99 dataset shows that this method has great advantages in dealing with huge traffic and multi-classification problems. Yang [20] proposed a SVM hybrid model based on particle swarm optimization. Yang extracted a few features such as port number and packets from the original data, and the extracted features are entropy calculated before sent to the SVM classifier. But only the binary classification of normal and abnormal are made. Ali [21] et al. proposed a fast learning network PSO-FLN based on particle swarm optimization algorithm, and compared with Extreme Learning Machine (ELM) [22], PSO-FLN performed better than other methods on KDD CUP99 dataset, and the author found that the accuracy of recognition can be significantly improved by increasing the number of network neurons, but the author did not do extra work on other datasets. Umer [23] et al. proposed a network intrusion detection system for next generation networks. In terms of feature extraction, the flow records in all datasets are in NetFlow v5 format and they used 9-tuple flow records in the experiment. Alshammari [24] et al. did a lot of meaningful work in feature extraction in order to identify encrypted VoIP traffic. They mainly calculate the statistical information of the network flow to extract the features, and test the extracted features on a variety of classification algorithms to verify the validity of these features. Table 1 gives the features used in related research.

Table 1. Feature used in related research.

Solution	Dataset	Dimensions	Feature type
[19]	KDD CUP99	41	TCP connections information
[20]	1998 DARPA dataset	6	Five tuples and packet information
[21]	KDD CUP99	41	TCP connection information
[23]	Open dataset [27]	9	Netflow records
[24]	Collected by author	29	Network traffic connections
[25]	Collected by author	12	Statistics feature
[26]	Provided by third party	19	Statistics feature

3 Proposed Approach

3.1 Raw PCAP File

In this paper, the data we use is the original captured PCAP file. This is a common data packet storage format. The PCAP file consists of a PCAP file header and multiple PCAP data packets. The PCAP packet is composed of the packet header and the packet content. Obviously, the classifier cannot directly learn the PCAP file. The content of the PCAP packet is the TCP/IP protocol cluster data that we will analyze and restore. Therefore, before sending it to the classifier, we use pyshark and python scripts to turn it into a vector that can be recognized by the classifier.

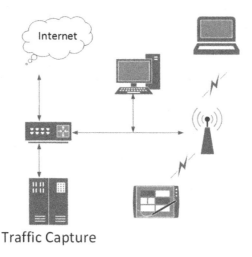

Fig. 1. The traffic capture scheme framework.

3.2 Framework

Figure 1 gives the traffic capture scheme we designed. In this scheme, all the traffic passing through the switch is copied from the source port to the destination port by port mirroring of the switch, and then the traffic is captured by the server deployed by us at the destination port. Figure 2 gives the entire process flow of our proposed method. The main steps shown in Fig. 2 are as follows:

Step 1: Capture the original PCAP traffic packet.

Step 2: Parse the captured traffic packet and perform feature extraction to build a feature repository.

Step 3: Executive feature selection algorithm.

Step 4: Divide the selected features into training set and test set, and send them to the classifier for training and testing.

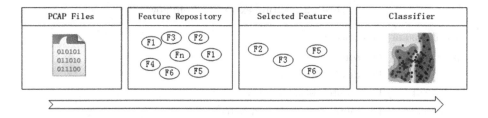

Fig. 2. The traffic capture scheme framework.

Feature Repository. As we mentioned in Sect. 2. In the known references, there are many methods of feature extraction. Although the feature extraction methods proposed by each paper are different, they all have good performance. In the field of machine learning, it's all known that it's the key point to extract the right feature to build a perfect classifier. However, it is not an easy task to propose complete and effective features in a short period of time. In addition, the features extracted may only have a good effect on a particular dataset, thus failing to guarantee the universality of the proposed method. In order to solve this problem, we propose a novel method to identify different malicious traffic types, that is, first establish the feature repository, then select features from feature repository that have greate influence on the performance of classifier, we should pay attention to extracting as many features as possible from the original traffic data, expanding the repository as much as possible, and then selecting the features in the designed feature selection phase. In this paper, we process network traffic as single direction flow, the direction of the flow is from attacking host to the attacked host. The extraction of all features is based on the flow in this direction.

Table 2 gives the details of our feature repository. The byte_dis is a 256-dimension vector, and the value of each dimension represents the number of related bytes in the stream. For example, byte_dis[99] is equal to 25, which means that 0x63 (99) appears 25 times in all packets of the flow. And the total dimension in our feature repository is 473.

Feature Selection. As we mentioned above, in order to extract as many features as possible from the original PCAP file, we propose the concept of the feature repository and build a huge feature repository containing a large number of original features, of course its dimensions is particularly high. However, in this step, we reduce the dimension by removing redundant and uncorrelated features from the features of the high-dimensional space by feature selection. The reason we do this is that with a limited number of samples, using a large number of features to design the classifier is too computationally expensive and does not necessarily guarantee better classifier performance. Therefore, in the step of feature selection, our main job is to filter out well-characterized features from the feature repository. And here are some basic selection principles for the

Table 2. Feature extracted from flow

Feature name	Description
sp	The source port
dp	The destination port
inbyte_cnt	The number of inbound bytes
outbyte_cnt	The number of outbound bytes
inpkt_cnt	The number of inbound packets
outpkt_cnt	The number of outbound packets
pkt_length	Long sequence of packets of the first hundred packets in a stream
pkt_time	First hundred packets of arrival time interval sequence in a stream
byte_dis	The number of each byte in a flow
byte_persec	The number of bytes peer second
max_pkt	The number of the maximum packet
min_pkt	The number of the minimum packet
mean_pkt	The average length of the packet
std_pkt	The variance of the length of the packet
max_bd	The max value in in byte_dis
min_bd	The minimum value in byte_dis
entropy_bd	The entropy of byte distribution
std_bd	The variance of byte distribution
duration	Duration of a flow

feature selection work, for example, obtain the smallest possible feature subset; does not significantly reduce the classification accuracy; does not affect the class distribution and the feature subset should have stable adaptability and so on. There are several common feature selection algorithms:

(1) Information gain.

The information gain comes from the field of communication, Shannon's theory of information entropy. The definition of information entropy is as follows:

$$H(X) = -\sum_{i}^{n} P_i \bullet \log_2 P_i \tag{1}$$

Its essence is to measure the magnitude of the uncertainty of an event. The information gain is relative to a specific feature. For example, for a certain feature X, the corresponding value has n kinds $(x_1, x_2, x_3 ... x_n)$ respectively, and the feature X is calculated. Information entropy at $x_1, x_2, x_3 ... x_n$. The information gain can be expressed as the original information entropy-conditional entropy. As follows:

$$G(T) = H(C) - H(C|X) \tag{2}$$

$$H(C|X) = \sum_{i=1}^{n} P_i H(C|X = x_i) \tag{3}$$

(2) Information gain rate.

The number of feature values has a large interference to the information gain. In order to avoid the number of values of some features, some features have fewer values, which makes the information gain cannot be a fair measure of a feature. Thus, the information gain rate is normalized to the information gain based on the original information gain. Its expression is:

$$GR = \frac{G(T)}{H(C)} \tag{4}$$

(3) Correlation coefficient.

The correlation coefficient is to judge the correlation between the two variables. The more commonly used is the person correlation coefficient. The larger the absolute value of the correlation coefficient, the higher the correlation between X and Y. The formula is as follows:

$$\rho_{XY} = \frac{E((X - EX)(Y - EY))}{\sqrt{D(X)D(Y)}} \tag{5}$$

(4) Chi-square test.

The chi-square test is to test whether the two variables are independent of each other. The application in feature selection is to check whether the feature is independent from the target variable. It is assumed here that the feature and the target variable are independent of each other. The deviation between the value of the characteristic and the expected value is statistically calculated. Since the deviation cannot take a negative value, the square of the deviation between the value of the characteristic and the expected value is calculated and summed. The final result serves as a criterion for judging whether the null hypothesis is true. Its formula is as follows:

$$\chi^2 = \sum_{i}^{n} \frac{(x_i - E)^2}{E} \tag{6}$$

4   Experiments and Analysis

4.1   Dataset

In order to generate the dataset from original captured PCAP file that can be used for machine learning algorithm, we mainly use wireshark and python scripts to clean and reorganized the data. The specific steps are described in detail in Sect. 3. The dataset used in the experiments was collected by Malware Capture Facility Project which was born in the CTU University of Prague in

Czech Republic. There are more than three hundred captured files in the list of their dataset. In their dataset, each folder has a corresponding specific virus known. In a LAN environment, a known host is infected with a known virus, and then traffic data for a certain period of time is captured at the gateway. Due to the IP of the infected host and the virus we already know, it is very easy to extract the malicious data stream from the original data and label them.

During the experiment, we used the specific virus species name directly in relation to the Arabic numerals. For example, '0' stands for the virus Simda, and '1' stands for the virus CoinMiner and so on. Table 3 gives the details of the dataset we used.

Table 3. Dataset details information.

Label	Name	Number
0	Artemis	10473
1	CoinMinerXMRig	17896
2	TrickBot	8211
3	None	20376
4	Trickster	45462
5	WebCompanion	59303
6	Dridex	27998
7	CoinMiner	50225

In machine learning tasks, we often encounter this kind of trouble: data imbalance. Data imbalance problems mainly exist in supervised machine learning tasks. When encountering unbalanced data, the traditional classification algorithm with the overall classification accuracy rate as the learning target will pay too much attention to the majority class, thus reducing the classification performance of a few class samples. Most common machine learning algorithms do not work well for unbalanced dataset. There are many ways to solve such problems. For example, downsample or upsample the sample and resample with different ratios and more. Here we use the method of upsampling. Experiments show that the accuracy after upsampling increases by 3% points.

4.2   Performance Evaluation

The performance of the model is measured by the following metrics:

True Positive (TP) is the number of positive sample that is predicted by the model to be positive;

True Negative (TN) is the number of negative sample that is predicted by the model to be negative;

False Positive (FP) is the number of negative sample that is predicted by the model to be positive;

False Negative (FN) is the number of positive sample that is predicted by the model to be negative.

Precision (Pre): Reflects the proportion of the true positive sample in the positive case determined by the classifier.

$$Precision = \frac{TP}{TP + FP} \quad (7)$$

Recall: Reflects the proportion of positive cases that are correctly judged as the total positive examples.

$$Recall = \frac{TP}{TP + FN} \quad (8)$$

Accuracy (Acc): Reflects the proportion of all forecast pairs in the total number of samples.

$$Accuracy = \frac{TP + TN}{TP + FN + FP + TN} \quad (9)$$

F1-Score (F1): Harmonic average of accuracy and recall.

$$F1 - Score = \frac{2 * Recall * \mathrm{Pr}\,ecision}{Recall + \mathrm{Pr}\,ecision} \quad (10)$$

4.3 Results

The experimental platform configuration is as follows: Xeon E2610 CPU, ubuntu16.04 OS, 32G RAM. Figure 3 shows the performance of our proposed method on four metrics. During the training, we randomly scrambled all the data after resampling, and selected 80% as the training set and the remaining 20% as the test set. On the 8 classification task, the original features were used to train the XGBoost classifier, and the final accuracy rate reached 99.8%. With the selected features, the accuracy rate still reaches 98.2%, but due to the reduction of the number of features, the time spent on training is greatly reduced. Table 4 give the training time for the feature repository and feature selection on XGBoost. We believe that in the case of high real-time requirements, by sacrificing a small amount of accuracy in exchange for a lot of time cost is acceptable.

In order to verify the validity of our proposed method, we performed experiments based on the feature extraction method proposed by Alshammari [24] and compared them with the dataset we used. Figure 4 gives the results of our proposed method and Alshammari's method on the four performance indicator. As shown in Fig. 5a, b, it can be seen from the ROC curve that in the ROC curve of the proposed method, the AUC (area under the ROC curve, AUC) for each type of malicious traffic is very close to 1. This shows that the performance of our proposed method is still relatively good.

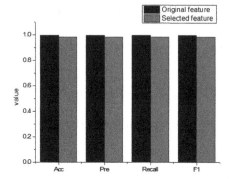

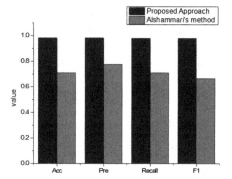

Fig. 3. Original feature and selected feature performance on XGBoost classifier.

Fig. 4. Comparison between our proposed method and Alshammari's method.

Table 4. Time spent comparison.

Method	Original feature	Feature selection
Time spent/s	180.8	**34.4**

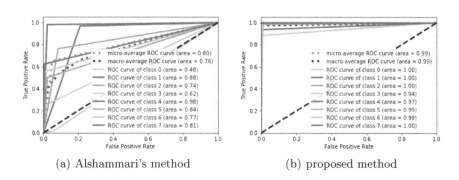

(a) Alshammari's method

(b) proposed method

Fig. 5. The ROC curve of the compared method and proposed method

5   Conclusion and Future Work

In this paper, we briefly introduce the current research methods in the field of traffic classification. Then we introduce the method based on feature repository and feature selection. The establishment of the feature repository ensures the full mining of the original data information, but the high dimensions of feature will greatly increase the computation time cost, and the feature selection method solves the problem well. Experiments have shown the proposed method met our expectations: it takes less time to train a well-behaved classifier. In the future work, we will focus on the selected features and study what the selected features represent specifically.

Acknowledgements. The research is funded by the National Natural Science Foundation of China (Grant No. 61372117).

References

1. Yang, X.N., Wang, W., Xu, X.F., Pang, G.R., Zhang, C.L.: Research on the construction of a novel cyberspace security ecosystem. Engineering **4**(1), 47–52 (2018)
2. Kumar, S., Dharmapurikar, S., Yu, F., Crowley, P., Turner, J.: Algorithms to accelerate multiple regular expressions matching for deep packet inspection, vol. 36, pp. 339–350. ACM (2006)
3. Bujlow, T., Carela-Español, V., Barlet-Ros, P.: Independent comparison of popular dpi tools for traffic classification. Comput. Netw. **76**, 75–89 (2015)
4. Khan, S., Rahmani, H., Shah, S.A.A., Bennamoun, M.: A guide to convolutional neural networks for computer vision. Synth. Lect. Comput. Vis. **8**(1), 1–207 (2018)
5. Abdelaziz, A.H.: Comparing fusion models for dnn-based audiovisual continuous speech recognition. IEEE/ACM Trans. Audio Speech Lang. Process. **26**(3), 475–484 (2018)
6. Young, T., Hazarika, D., Poria, S., Cambria, E.: Recent trends in deep learning based natural language processing (review article). IEEE Comput. Intelli. Mag. **13**(3), 55–75 (2018)
7. Buczak, A.L., Guven, E.: A survey of data mining and machine learning methods for cyber security intrusion detection. IEEE Commun. Surv. Tutor. **18**(2), 1153–1176 (2017)
8. Nishani, L., Biba, M.: Machine learning for intrusion detection in manet: a state-of-the-art survey. J. Intell. Inf. Syst. **46**(2), 391–407 (2016)
9. Wang, M., Cui, Y., Wang, X., Xiao, S., Jiang, J.: Machine learning for networking: workflow, advances and opportunities. IEEE Network, PP(99), 1–8 (2017)
10. Wang, Y., Xiang, Y., Yu, S.Z.: An automatic application signature construction system for unknown traffic. Concurr. Comput. Pract. Exp. **22**(13), 1927–1944 (2010)
11. Moore, A.W., Papagiannaki, K.: Toward the accurate identification of network applications. In: Dovrolis, C. (ed.) PAM 2005. LNCS, vol. 3431, pp. 41–54. Springer, Heidelberg (2005). https://doi.org/10.1007/978-3-540-31966-5_4
12. Sen, S., Spatscheck, O., Wang, D.: Accurate, scalable in-network identification of P2P traffic using application signatures. In: Proceedings of ACM WWW (2004)
13. Tongaonkar, A., Torres, R., Iliofotou, M., Keralapura, R., Nucci, A.: Towards self adaptive network traffic classification. Comput. Commun. **56**, 35–46 (2015)
14. Gomes, J.V., Pereira, M., Monteiro, P.P.: Detection and classification of peer-to-peer traffic:a survey. ACM Comput. Surv. **45**(3), 1–40 (2013)
15. Xu, C., Chen, S., Su, J., Yiu, S.M., Hui, L.C.K.: A survey on regular expression matching for deep packet inspection: applications, algorithms, and hardware platforms. IEEE Commun. Surv. Tutor. **18**(4), 2991–3029 (2016)
16. Rodrigues, G.P., et al.: Cybersecurity and network forensics: analysis of malicious traffic towards a honeynet with deep packet inspection. Appl. Sci. **7**(10), 1082 (2017)
17. Finamore, A., Mellia, M., Meo, M.: Mining unclassified traffic using automatic clustering techniques. In: Domingo-Pascual, J., Shavitt, Y., Uhlig, S. (eds.) TMA 2011. LNCS, vol. 6613, pp. 150–163. Springer, Heidelberg (2011). https://doi.org/10.1007/978-3-642-20305-3_13

18. Erman, J., Mahanti, A., Arlitt, M., Williamson, C.: Identifying and discriminating between web and peer-to-peer traffic in the network core. In: International Conference on World Wide Web, pp. 883–892. ACM (2007)
19. Xu, B., Chen, S., Zhang, H., Wu, T.: Incremental k-NN SVM method in intrusion detection. In: IEEE International Conference on Software Engineering and Service Science, pp. 712–717. IEEE (2017)
20. Yang, L.: Network anomaly traffic detection algorithm based on SVM. In: International Conference on Robots and Intelligent System, pp. 217–220. IEEE Computer Society (2017)
21. Ali, M.H., Mohammed, B.A.D.A., Ismail, M.A.B., Zolkipli, M.F.: A new intrusion detection system based on fast learning network and particle swarm optimization. IEEE Access, PP(99), 1–1 (2018)
22. Huang, G.B., Zhu, Q.Y., Siew, C.K.: Extreme learning machine: a new learning scheme of feedforward neural networks. In: Proceedings of IEEE International Joint Conference on Neural Networks, 2004, vol. 2, pp. 985–990. IEEE (2005)
23. Umer, M.F., Sher, M., Bi, Y.: A two-stage flow-based intrusion detection model for next-generation networks. Plos One 13(1), e0180945 (2018)
24. Alshammari, R., Zincir-Heywood, A.N.: Identification of voip encrypted traffic using a machine learning approach. J. King Saud Univ. Comput. Inf. Sci. 27(1), 77–92 (2015)
25. Liu, Y., Chen, J., Chang, P., Yun, X.: A novel algorithm for encrypted traffic classification based on sliding window of flow's first N packets. In: IEEE International Conference on Computational Intelligence and Applications, pp. 463–470. IEEE (2017)
26. Aceto, G., Ciuonzo, D., Montieri, A., Pescapé, A.: Multi-classification approaches for classifying mobile app traffic. J. Netw. Comput. Appl. 103, 131–145 (2017)
27. Sperotto, A., Sadre, R., van Vliet, F., Pras, A.: A labeled data set for flow-based intrusion detection. In: Nunzi, G., Scoglio, C., Li, X. (eds.) IPOM 2009. LNCS, vol. 5843, pp. 39–50. Springer, Heidelberg (2009). https://doi.org/10.1007/978-3-642-04968-2_4

Indoor Location and Tracking System Using Computer Vision

Adrián J. Ramírez-Díaz$^{(\boxtimes)}$, José Rodríguez-García, Sonia Mendoza, and Amilcar Meneses Viveros

Department of Computer Science, Centro de Investigación y de Estudios Avanzados del IPN, Mexico City, Mexico
jramirez@compuntacion.cs.cinvestav.mx,
{rodriguez,smendoza,ameneses}@cs.cinvestav.mx

Abstract. In ubiquitous computing systems, determining the location of objects in the environment can provide basic information about the context of such objects. In closed environments an Interior Positioning System (IPS) helps to determine the location of people or robots through the use of a point-based reference system placed in the environment. Several mechanisms can be used to locate references, for example: light sensing, radio frequencies, sound, or images. In this paper, it is presented an image-based IPS that finds the location of a robot in a zone and provides functions to generate paths for the robot. The zones are identified through reference markers, which are analyzed in a server using image processing and Cloud Robotics, in order to minimize processing load in the robot. Once the marker is analyzed, a route is sent to the robot.

Keywords: Indoor location · Tracking · Computer vision

1 Introduction

Since the ubiquitous computing came into existence, the objective has been the creation of computer systems capable of manage the environment, where the system interacts with users through intelligent devices as robots and electronic components embedded in the environment, among others [20]. These elements provide information and services, at the same time that they are aware of the changes occurred in the environment. In other words, they are aware of the context. This allows decision-making under different scenarios in which the user can be found, or even, adjust the behavior of the system according to the preferences of the users.

Context awareness attempts to select the information that determines the actions of the ubiquitous system components. This context can be extracted from different sources, including: the users actions, the characteristics of the environment, the system status and its components, and even monitoring the user's health.

© Springer Nature Switzerland AG 2019
Y. Tang et al. (Eds.): HCC 2018, LNCS 11354, pp. 613–624, 2019.
https://doi.org/10.1007/978-3-030-15127-0_61

Within these context sources, elements and users location in the ubiquitous environment become one of the most analyzed sources, since this allows to establish a relationship between the different zones and the available services.

Therefore, it is necessary to create tools that allow to establish, according to a reference system, the position of the different objects managed by the system.

An IPS is a mechanism that provides the location of objects, such as robots or people, in an environment through referenced magnitudes [2]. The location of these objects can be expressed as a symbolic reference, "kitchen", for example, or as a coordinate-based reference [22].

Regardless of the way in which the location of an object is expressed, an IPS determines the location through a physical characteristic, provided by reference points (also called nodes or anchors) with a known location. This characteristic must be proportional to the distance between the reference point and the object [15].

Reference points make use of different technologies to provide contextual information, which is necessary to determine the location of objects in the environment. Therefore, it is possible to classify an IPS according tho the technology used to provide a reference magnitude [21], such as: (1) light-based systems (LIDAR) [17], which use the reflective capacities of different materials to measure the time in which the light is reflected, similar to the operation of a radar; (2) camera-based systems, which extract location and movement information from images taken by a camera [9]; (3) radio-based systems that are the most used ones, since they determine the position through intensity analysis of radio waves, such as WiFi, RFID or bluetooth [12]; and (4) audio-based systems, which are based on the propagation of sound waves [8].

All these methods to build reference points present a set of positive and negative characteristics, for example: accuracy, interference, economic cost, computational cost, or security. However, due to the characteristics and requirements presented by the applications that make use of an IPS, it is not possible to say that one system is better than another.

One of the main uses of an IPS is robot positioning within an environment, with the aim of driving a robot through different areas, where its presence is required. Beyond needing a robot to be located in coordinates x, y, it is needed that the robot moves from zone A to zone B, tracing a path between both zones. This paper presents a navigation and tracking indoor system for robots, through the use of computer vision techniques.

One of the main drawbacks of using computer vision techniques is the complexity of image analysis, which leads to a high computational cost [16]. This complexity is due to the amount of information obtained from an image, causing difficulties such as: multiple variations in natural environments, algorithmic complexity, and privacy problems [11]. However, if the information about the objects in the image is known, it simplifies the analysis of the image.

For this reason, the system described in this paper makes use of visual reference points (markers), from which the position of a robot is determined, by analyzing the information contained in marker and establishing a route that allows the robot to move through the environment. This paper is organized as follows: in Sect. 2, it is presented a general description of this proposal. Section 3 gives details of the tracking system. In Sect. 4, it is explained how the robot displaces between markers. Finally, a conclusion and ideas for future work are presented in Sect. 5.

2 General Description of the VINS System

The system presented in this paper, named Visual Indoor Navigation System (VINS), allows robot navigation in a specific environment, through the tracking of visual markers, based on the approach known as Cloud Robotics. This approach allows us to establish an efficient cooperation between robots, intelligent environments, and humans [1], through the communication of the robot with the system that manages the environment. This allows us to distribute data processing between the robot and the system.

VINS consists of two main components: the *robot displacement manager* and the *route generator*. The purpose of the *robot displacement manager* is to guide the robot to a reference point, meaning that it is in charge of managing the mobility to guide the robot towards the markers in the environment. The *route generator* establishes the set of markers that the robot must visit to move from one area to another. In addition, for each marker visited by the robot (control point) the *route generator* obtains the marker identifier, in order to validate that the robot is on the correct route. Otherwise, it generates a new route based on the current position.

It is pertinent to point out that the tasks performed by the *route generator*, such as the marker identification or the selection of a route, are processes that require a greater computing capacity, in comparison with the tasks performed by the *displacement manager*. Therefore, it is convenient to use Cloud Robotics, since it allows processing of tasks with greater requirements in a computer with greater computing capacity, reducing processing time and energy consumption in the robot.

The tasks performed by the robot are those corresponding to obtaining and filtering images, the classification of objects contained in filtered images, the displacement of the robot according to the position of the object in the image, and the sending of images to the server. On the other hand, the server is responsible for the identification of the marker near the robot, and therefore the identification of the area where it is located. The server is also responsible for generating the appropriate path for the robot to move from one area to another, indicating the direction it must follow each time the robot reaches a control point. In Fig. 1, the mentioned tasks are presented as well as the inputs and outputs of each task.

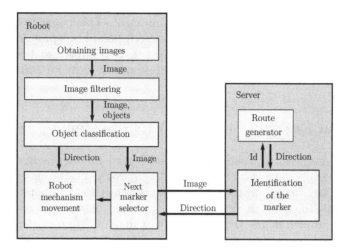

Fig. 1. Components of the VINS system

3 Markers Design Processing

3.1 Marker Design

Due to the large amount of information that can be extracted from the environment through images, the marker, used as a reference point, has the necessary characteristics to make it stand out in the environment, with the aim of facilitating filtering and extraction of objects, by placing the marker within the images captured by the robot.

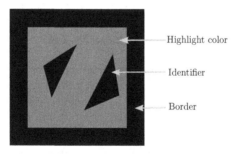

Fig. 2. Proposed design for the position marker. (Color figure online)

Figure 2 shows an example about the design of the position markers used by VINS. This design has three main characteristics: highlight color, border, and identifier. The highlight color facilitates the discovery of the marker, since this color is different from the objects belonging to the navigation environment, decreasing, in this way, the number of objects extracted from the image.

The border limits the space of the marker and provides a second information filtering method, since it allows the classification of objects by their form. The identifier allows VINS to recognize, in a unique way, the analyzed marker, providing at same time, a method for identifying areas in the environment.

3.2 Filtering the Images, from RGB to HSV

The RGB model allows us to describe a color based on the intensity of three essential colors (channels): red, green and blue, by setting the intensity of each color in a range between 0 and 100 % [10]. To perform the extraction of objects under this schema, it is necessary that the color used as characteristic has distinctive levels of saturation in contrast to the colors in the environment.

The primary colors under the RGB schema are easy to identify, since they present high saturation in only one of the three primary colors (channels); unfortunately, these basic colors are widely present in the environment. This results in the selection of multiple objects, complicating in this way the task of classification. Also, the secondary colors (magenta, cyan, and yellow) have high saturation in two of the three primary channels and low saturation in the third one. This allows the creation of selection rules using saturation levels. Likewise, the secondary colors have the advantage of not being common in the environment, decreasing the number of objects obtained.

In the RGB schema, the received image is broken down into three grayscale images. Each one of these images represents the saturation level of each primary color. The resulting images are passed through a filter that selects the pixels with high or low saturations, according to the characteristics described by the color selected. The pixels chosen by those filters are considered as objects indicating the color of the marker.

However, the correct functioning of the filtering process lies in the choice of the right ranges to determine the saturation of each pixel. A first filter was implemented using the RGB schema, and the saturation level was established with a 20% of acceptance, that is, pixels with saturation levels higher than 200 were selected in high filters, and pixels with saturation levels lower than 50 in low filters were also selected for images with saturations between 0 and 255.

This filter works correctly on images with sharp colors. However, in images with a dark or clear composition, the filter does not select markers correctly. This generates the need to change the model used to represent colors in a model that allows us to use tonalities insensitive to light.

The HSV schema is formed by three parameters: hue, saturation, and value. This schema is not based on intensities of primary colors. Hue is the term that indicates the color used and is represented in degrees, where a degree indicates the color used. Saturation refers to the tones of a color. The value represents the intensity of light in a color [19].

To move from the RGB schema to the HSV schema, it is necessary to identify the greater and lesser values of RGB components, denominated max and min respectively; then, the hue is obtained using Eq. 1.

$$H = \begin{cases} 0°, & \Delta = 0 \\ 60° \left(\frac{G-B}{\Delta}\right), & max = R \\ 60° \left(\frac{B-R}{\Delta} + 2\right), & max = G \\ 60° \left(\frac{R-G}{\Delta} + 4\right), & max = B \end{cases} \tag{1}$$

where Δ represents the difference between max and min. The *saturation* and *value* are calculated using Eqs. 2 and 3, respectively.

$$S = \begin{cases} 0, & max = 0 \\ \frac{\Delta}{max}, & max \neq 0 \end{cases} \tag{2}$$

$$V = max \tag{3}$$

Once the pixels have been converted to the HSV schema, it is necessary to establish a range for the values of the hue where the desired color is found, avoiding in this way the problems associated with luminosity in the pixels. It is also suggested to eliminate the pixels with colors close to black and white, by setting a lower limit for *valueandsaturation*.

3.3   Classification of Objects

After the objects have been extracted from the image, the transformations that an object may have due to the position of the robot's camera with respect to the marker, must have been taken into account. Therefore, it is necessary to use characteristics for automatic classification, which are invariant to transformations that an object could have, such as scale, translation, and rotation [3].

The invariant Hu's moments are one of the most popular techniques in two dimensional pattern recognition. A non-linear combination of the regular moments is used to generate a set of moments that have the capability to remain invariant to transformations in scale, translation, and rotation [18]. For this reason Hu's moments are the classification characteristics selected to separate markers.

Once the characteristics are determined, it is necessary to select and implement a classification algorithm, which allows us to analyze the characteristics of objects and determine whether an object belongs to the interest class. One of the most used algorithms for this task is *K-Nearest Neighbor (KNN)*, which allows us to classify vectors in different classes. This classification is done through the evaluation of distances between the vector to be classified and the characteristic vectors of each class. The last vectors are selected in the training stage and represent the most significant values of the classes or the statistical measurements of them [14].

Next, a test bench is generated where the Hu's moments correspond to the markers, and other objects selected by the filter are included. This test has been conducted with the objective of measuring the efficiency of the implemented classifiers.

When using KNN as first classifier, it is necessary to calculate the statistical mean of the training data to use them as characteristic vectors. However, the results prove that the use of statistical means as characteristic vectors does not represent, in a significant way, each one of the classes. Therefore, when evaluating the classifier with the evaluation data, some classification errors are generated.

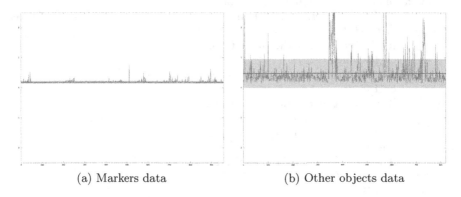

(a) Markers data   (b) Other objects data

Fig. 3. First Hu's moment graphics

When performing the statistical analysis of data, it is observed that the Hu's moments obtained from markers are lower than the data obtained from other objects. For example, in Fig. 3, it is presented the data corresponding to the first moment, where it is possible to observe the dispersion of the object's data. This allows us to establish a limit to decide whether a new element may or may not be selected as a marker. This limit is defined with the help of a genetic algorithm, establishing as function to optimize the classification error count. A classification error is the selection of an object that is not a marker or the rejection of a real marker. The parameters established for the construction of the genetic algorithm are: real representation for individuals, 100 elements as population size, total arithmetic cross with 0.9 of probability, mutation with 0.01 of probability, and selection by binary tournament. This set of parameters gives the best results for our proposal.

4   Displacing Between Markers

As mentioned, the navigation process of VINS is based on capturing images by the camera's robot, the movements of the robot are used to center the images of markers into the camera's lens. Therefore, it is necessary to describe the process through which the robot moves from a starting point to a particular marker.

Algorithm 1 describes how the robot moves from an initial position to a control point. Four initial parameters are required: *targetMarker*, *range*, *timelapse*, and *maxArea*. Given that, in some cases, more than one marker can appear in the image, the *targetMarker* allows us to establish which of those

---

Algorithm 1. Robot displacement algorithm

Require: *targetMarker, range, timelapse, maxArea*
  lost = true, lastTurn = right, stop = false
  repeat
    Get *image*
    Filter *image*
    for all *marker* in the filtered *image* **do**
      if *marker = targetMarker* **then**   ▷ Made by the server
        lost = false
        if *marker* is below the *range* **then**
          lastTurn = left
          Robot turns left for a *timelapse*
        end if
        if *marker* is above the *range* **then**
          lastTurn = right
          Robot turns left for a *timelapse*
        end if
        if *marker.area = maxArea* **then**
          stop = true
        end if
      end if
    end for
    if *lost = true* **then**
      Robot turns contrary to the *lastTurn* for a *timelapse*
    end if
    Robot goes straight ahead
    lost = true
  until *stop = true*

---

markers is the robot's target. The *range* defines the segment of the image where the marker can be considered centered, indicating that the robot does not need to rotate. The parameter *timelapse* represents the spin that the robot will have in case the marker is not centered in the image; *timelapse* can also be seen as the parameter that controls the correction of the robot's route. Finally, *maxArea* represents the area that a marker must reach within an image to consider that the robot has reached the target.

4.1   Identification of Markers and Route Management

The identification of markers is done by analyzing his information and the characteristics contained in the identifier. This information can be represented through objects, symbols, or by more complex encoding patterns, such as Quick Response (QR) codes [6]. Some other proposals are a fiducial marker, this is an automatic detection and identification system that allows the calibration of the camera, at the same time obtaining its position and orientation (*pose*) [13].

    The type of identifier created for the marker depends on the requirements given by the ubiquitous computing system, because if the system requires a high number of possible markers, it makes necessary to increase the complexity of the marker, and consequently, to manage a greater number of areas in the environment.

In the work by Cruz and de la Fraga [4], it is presented the design of a marker that combines the identification and fiducial markers. This marker reduces the errors produced by transformations such as rotation, translation, and perspective generated by the position of the robot's camera. However, using more complex markers requires a deepest analysis to identify them, for this reason, in the tests carried out for this system, geometric figures are used as identifiers, using Hu's moments to classify these objects.

Once the identifiers have been established, it is necessary to define the paths between them, in order to establish routes through which the robot can move. When using visual markers, if the robot can locate the position of the marker B placed in the marker A, and if it can establish a line of sight to the marker B, it is said that it is possible to establish a path AB. This is possible only if the characteristics of terrain together with the movements of the robot allow this.

Once the markers are located and the connections between them are defined, the navigation map of the robot can be represented as a directed graph, which allows the implementation of algorithms to find the shortest route, such as the Dijkstra algorithm [5].

5   Implementation and Testing

The robot used for the implementation of VINS performs the image processing using a *Raspberry Pi Zero W*, "a low cost, credit-card sized computer," as it is defined by the provider [7]. This computer allows the connection with the server through WIFI and with the movement mechanism of the robot through the USB port. The robot has four wheels with independent traction each one; every wheel can have three states: advance, back, and stop. These states allow us to control the movements of the robot: forward, backward, turn to the right, or turn to the left.

The test scenario consists of placing the robot in an environment that uses markers with one, two and three figures with the aim of analyze in first instance, that the robot moves towards the marker, in second instance, that the robot correctly selects the identifier present in the marker. Each test starts by placing the robot in the direction of a marker, this robot moves to the marker and reaches the target. Next, the robot sends the images of the marker to the server and the server determines the corresponding identifier, indicating that the robot turns to the direction indicated by the server. Finally, the robot turns to identify a new marker and begins the process again.

Under this scenario, it was found that the algorithm described above works correctly, since it was able to move the robot towards markers in the different tests. On the other hand, the calculation of the identifier in the marker is a process that requires special attention because it allows establishing a relationship between the position of the robot and the area in which the robot is located.

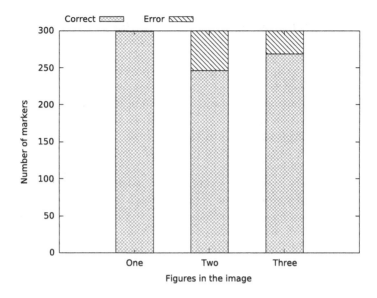

Fig. 4. Evaluation of marker identification process

Figure 4 presents results obtained from the evaluation of this procedure, it shows that markers with one figure, the evaluation was performed correctly, however, in markers with more figures, the number of errors increased. For this reason, it was necessary to analyze the images that the robot sends, this leads to identify that errors occur in those cases in which the image received contains an incomplete marker. Due to this problem, it is necessary that the robot verifies that the image is centered on the marker before sending it to the server, otherwise, the robot makes the corresponding movements to center the marker, that is, turn to the right or to the left until the image is centered.

6 Conclusions and Future Work

The choice of a location and navigation system for robots in interiors is closely linked to the requirements of the system, giving rise to multiple proposals found in the literature. The solution presented in this paper focuses on systems where the cost of implementation for the navigation system must be kept low, or where the installation of electronic reference points in the environment is not possible given the capabilities of the robot's processors.

This work also shows that it is possible to use computer vision techniques in robots with low processing capacity when using a Cloud Computing approach, which in turn has the advantage of the interaction capability of the navigation system with others systems that manage the environment.

The effectiveness of the HSV colors representation scheme is corroborated for filtering objects by their color compared against the RGB scheme. Similarly, an object classification method based on the analysis of Hu's moments.

The use of markers to determine the area in which the robot is located gives rise to the possibility of adding more precision to the location of the robots. Since it is possible to identify relevant points within the marker and compare them with a previously stored model. With this, a more precise location of the robot can be generated with respect to the marker, for example establishing distances from the data stored in the model, identifying even the angle of the robot with respect to the marker.

References

1. Bonaccorsi, M., Fiorini, L., Cavallo, F., Esposito, R., Dario, P.: Design of cloud robotic services for senior citizens to improve independent living and personal health management. In: Andó, B., Siciliano, P., Marletta, V., Monteriú, A. (eds.) Ambient Assisted Living. BIOSYSROB, vol. 11, pp. 465–475. Springer, Cham (2015). https://doi.org/10.1007/978-3-319-18374-9_43
2. Caron, C., Chamberland-Tremblay, D., Lapierre, C., Hadaya, P., Roche, S., Saada, M.: Indoor Positioning, pp. 1011–1019. Springer, Cham (2017). https://doi.org/10.1007/978-3-319-17885-1
3. Chen, Q., Petriu, E., Yang, X.: A comparative study of fourier descriptors and hu's seven moment invariants for image recognition. In: Canadian Conference on Electrical and Computer Engineering 2004 (IEEE Cat. No.04CH37513), vol. 1, pp. 103–106, May 2004
4. Cruz-Hernández, H., de la Fraga, L.G.: A fiducial tag invariant to rotation, translation, and perspective transformations. Pattern Recognit. **81**, 213–223 (2018). http://www.sciencedirect.com/science/article/pii/S0031320318301146
5. Dijkstra, E.W.: A note on two problems in connexion with graphs. Numer. Math. **1**(1), 269–271 (1959)
6. Focardi, R., Luccio, F.L., Wahsheh, H.A.M.: Usable cryptographic QR codes. In: 2018 IEEE International Conference on Industrial Technology (ICIT), pp. 1664–1669, February 2018
7. Foundation, R.P.: What is a raspberry pi (2018). https://www.raspberrypi.org/help/what-%20is-a-raspberry-pi/
8. Ifukube, T., Sasaki, T., Peng, C.: A blind mobility aid modeled after echolocation of bats. IEEE Trans. Biomed. Eng. **38**(5), 461–465 (1991)
9. La Delfa, G.C., Catania, V., Monteleone, S., De Paz, J.F., Bajo, J.: Computer vision based indoor navigation: a visual markers evaluation. In: Mohamed, A., Novais, P., Pereira, A., Villarrubia González, G., Fernández-Caballero, A. (eds.) Ambient Intelligence - Software and Applications. AISC, vol. 376, pp. 165–173. Springer, Cham (2015). https://doi.org/10.1007/978-3-319-19695-4_17
10. Li, D. (ed.): RGB, pp. 1791–1791. Springer, Boston (2008). https://doi.org/10.1007/978-0-387-48998-8
11. Rashidi, P., Mihailidis, A.: A survey on ambient-assisted living tools for older adults. IEEE J. Biomed. Health Inform. **17**(3), 579–590 (2013)
12. Schwiegelshohn, F., Wehner, P., Werner, F., Gohringer, D., Hubner, M.: Enabling indoor object localization through bluetooth beacons on the radio robot platform. In: 2016 International Conference on Embedded Computer Systems: Architectures, Modeling and Simulation (SAMOS), pp. 328–333, July 2016

13. Shabalina, K., Sagitov, A., Li, H., Magid, E.: Comparing fiducial marker systems occlusion resilience through a robot eye. In: 2017 10th International Conference on Developments in eSystems Engineering (DeSE), pp. 273–278, June 2017

14. Shang, W., Huang, H., Zhu, H., Lin, Y., Wang, Z., Qu, Y.: An improved kNN algorithm – fuzzy kNN. In: Hao, Y., et al. (eds.) CIS 2005. LNCS (LNAI), vol. 3801, pp. 741–746. Springer, Heidelberg (2005). https://doi.org/10.1007/11596448_109

15. Sharma, O., Pandey, J., Akhtar, H., Rathee, G.: Navigation in AR based on digital replicas. Vis. Comput. **34**, 925–936 (2018)

16. Song, Z., Jiang, G., Huang, C.: A survey on indoor positioning technologies. In: Zhou, Q. (ed.) ICTMF 2011. CCIS, vol. 164, pp. 198–206. Springer, Heidelberg (2011). https://doi.org/10.1007/978-3-642-24999-0_28

17. Starek, M.J.: Light Detection and Ranging (LIDAR), pp. 383–384. Springer, Dordrecht (2016). https://doi.org/10.1007/978-94-017-8801-4

18. Teh, C.H., Chin, R.T.: On image analysis by the methods of moments. IEEE Trans. Pattern Anal. Mach. Intell. **10**(4), 496–513 (1988)

19. Wang, Q., Li, Z., Wang, J., Liu, S., Li, D.: A fast processing method of foreign fiber images based on HSV color space. In: Li, D., Chen, Y. (eds.) CCTA 2012. IAICT, vol. 392, pp. 390–397. Springer, Heidelberg (2013). https://doi.org/10.1007/978-3-642-36124-1_47

20. Weiser, M.: The computer for the 21st century. SIGMOBILE Mob. Comput. Commun. Rev. **3**(3), 3–11 (1999)

21. Werner, M.: Basic Positioning Techniques, pp. 73–99. Springer, Cham (2014). https://doi.org/10.1007/978-3-319-10699-1_3

22. Youssef, M.: Indoor Localization, pp. 1004–1010. Springer, Cham (2017). https://doi.org/10.1007/978-3-319-17885-1

A Hybrid Recommendation Model Based on Time Dimension for Academic Teams

Yong Tang$^{(\boxtimes)}$, Jihong Lin, Hanlu Chu, Junyi He, and Fengjie Luo

South China Normal University, Guangzhou, Guangdong, China
{ytang,jhlin,hlchu,jyhe,lawfj}@m.scnu.edu.cn

Abstract. Recommending academic teams for users of scholars' social network systems is of great practical value for promoting communication among scholars. This paper proposed a hybrid recommendation model based on time dimension for academic teams. The model combines the three dimensions (the similarity of user and team, good friends and hot teams), and generates a list of team recommendation based on different weights given by the team's creation time. Experiments on the SCHOLAT data set show that the proposed model can effectively improve the recommendation accuracy and coverage, and solve the cold start problem to a certain extent.

Keywords: Academic team · Social networks · Hybrid model · Recommendation system

1 Introduction

Collaboration has become vital for successfully carrying out creative work. In academic/research world, this is manifested, but finding the right collaborators is not an easy problem [1]. It is significant to explore team-based issues with the increasing interests of information exploration in big scholarly data [2]. Teams typically produce more frequently cited research than individuals do [3]. A new trend in online social networks is to recommend for groups instead of individuals [4].

Information overload phenomenon occurs, when more information is presented than the ability of information seekers to process and handles the information [5]. In recent years, recommendation systems have proven to be a valuable tool for dealing with information overload issues [6].

At present, two issues [7] are crippling the academic team recommendation system, one is "how to handle new users", and the other is "how to surprise users".

The former is well-known as cold-start recommendation. When a user is new to recommender System, the system cannot recommend items that are relevant to her/him because of lack of previous information about the user. This problem is known as cold-start, which remains open because it does not have a final solution [8]. Zhao et al. [9] indicated that existing recommendation systems often display the same recommendation item over and over again, and Top-N does not consider whether the recommendation item has been shown to the user previously, that is, whether the recommendation item is fresh or old.

Y. Tang et al. (Eds.): HCC 2018, LNCS 11354, pp. 625–637, 2019.
https://doi.org/10.1007/978-3-030-15127-0_62

The latter can be called as long-tail recommendation. Long-tail items [10] are an important key to practical success for recommender systems. Since short-head items are likely to be well known to many users, the ability to recommend items outside of this band of popularity will determine if a recommender can introduce new products to users.

There are many researches on academic teams at home and abroad. Sun et al. [11] proposed a social-aware group recommendation framework that jointly utilizes both social relationships and social behaviors to not only infer a group's preference, but also model the tolerance and altruism characteristics of group members. Yuan et al. [12] proposed a multidimensional team recommendation model which combines social friendship and popular team information. Datta et al. [13] proposed social web application for team recommendation, an integrated multidisciplinary team recommendation framework for identifying potential teams. Zhang et al. [14] proposed a behavior and score similarity based association rule group recommendation algorithm, in which the rule with its scores is regarded as an expert, and the experts with same conclusion are grouped together. However, the above method only solves the cold start problem of the recommender system, and does not consider the long tail effect.

It is well-known that recommendation algorithms have biases towards popular items [11]. Zhang et al. [15] proposed a hybrid recommendation algorithm based on social relations and time series topics. The time dimension can solve the long tail effect to some extent. Therefore, the time dimension is integrated into the recommendation model of this article.

Moreover, Sharma et al. [16] showed that friend-based recommendation algorithms are not worse than network-based recommendation algorithms, even though they require less historical data. So based on the friend's recommendation algorithm, the cold-start problem can be effectively solved. Sharma et al. [17] considered people's behavior is seldom influenced by their friends. Most people's behavior is driven by their own preferences. At the same time, experiments have shown that less active users are more susceptible to friends. Therefore, this paper adopts different recommendation strategies according to different user situations.

The main contributions of this paper are:

(1) This paper proposes a hybrid recommendation model of academic team. According to the completeness of user's personal information, different recommendation strategies can solve the problem of information overload and cold start to a certain extent.
(2) The long tail effect is fully considered in this paper. So the time dimension is integrated into the recommendation model, which can improve the coverage and novelty of the recommendation system to a certain extent.

2   Core Algorithm Description

In this paper, a hybrid recommendation model based on time dimension for academic teams is proposed. As shown in Fig. 1, this model combines three dimensions: similarity of user and team information, friendship and popular team to recommend academic team for users.

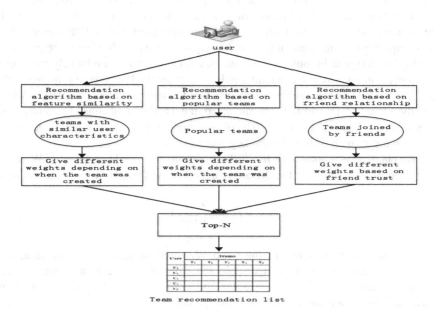

Fig. 1. A hybrid recommendation model based on time dimension for academic teams

(1) A recommendation algorithm based on similarity of features is used to find teams with similar user characteristics, and different weights are given according to the team's creation time. Finally, a team recommendation list is formed;

(2) A recommendation algorithm based on popular teams is used to find the hot teams and then the recommendation list is obtained. According to the team's creation time to give different weights, and finally form a recommendation list;

(3) A recommendation algorithm based on friend relationship, using friends who are similar to the user feature, have a relatively close relationship, and interact relatively frequently, and assigning different weights according to the trust of the friend to form a team recommendation list.

2.1   Recommendation Based on Similarity Between Users and Teams

Word Segmentation and Term Frequency. First, M team information and N user information in social network are segmented. After that, removal of stop words, including modal particles, adverbs, prepositions, conjunctions and so on. Usually stop words themselves have no clear meaning, only put it into a complete sentence will have a certain role, such as the common "the", "in". Removal of stop words can save storage space and improve the efficiency of feature processing.

Then the word segmentation result of the team information is counted to the thesaurus table $R = \{w_1, w_2, \cdots, w_q\}$, w_i represents the words of team information documents, and q represents the total number of words in the thesaurus table.

Then use TF-IDF to do word frequency statistics. TF (Term Frequency), the number or frequency of occurrences of a word in an article, if a word in an article appears many times, then the word may be more important words. IDF (inverse document frequency) is a measure of the word "weight". On the basis of word frequency, if a word is less frequent in multiple documents, it means that it is a relatively rare word, but it appears many times in a certain article, the larger the IDF value of the word, the larger its "weight" in this article. So when a word is more common, its IDF is lower. When the values of TF and IDF are calculated, the TF-IDF is obtained by multiplying the two values. The higher the TF-IDF of this word, the more important it is in this article, the more likely it is to be the key word of the article.

Team characteristic matrix $T \in \mathbf{R}^{m \times q}$, T_{ij} represents the weight of the word W_j in the team t_i information based on the thesaurus table R. The formula for T_{ij} is as follows.

$$T_{ij} = TF \cdot lg\left(\frac{m}{DF_i}\right) \qquad (1)$$

TF_{ij} denotes the frequency of the word W_j based on team T_i information; DF_i denotes the number of teams that contain the word W_j in team information. User characteristic matrix $U \in \mathbb{R}^{n \times q}$, U_{ij} represents the word frequency TF_{ij} of the word C_j in the user u_i information based on the thesaurus table R.

$$TF_{ij} = \frac{c}{q} \qquad (2)$$

Among them, C represents the number of words C_j appears in user u_i information.

PCA Feature Extraction. Principal component analysis (PCA) is one of the most widely used methods in feature extraction. This method mainly uses fewer new variables instead of more variables on the basis of the correlation among the features, and makes these fewer new variables retain as much information as possible from more variables as possible. The algorithm is as follows:

Set multi-feature sample X as q dimensional random variable $X = (x_1, x_2, \cdots, x_q)$, linear transformation is carried out, get new variables $y_i (i = 1, 2, \cdots, d)(d \leq q)$, Then there are $y_i = a_i^T X$. The purpose of PCA is to find a set of transformation vectors $a_i (i = 1, 2, \cdots, d)$, and satisfy $a_i^T a_i = 1$, $a_i^T a_{j \neq i} = 0$, The variance of y_i is maximized so that the largest variance y_i is the first principal component of sample X, the second largest variance is the second principal component, and so on. The solution result a_i

should be the eigenvector of the covariance matrix Σ_x of sample X. It is a solution to the following linear equation:

$$(\Sigma_x - \lambda_i E) \cdot a_i = 0 \tag{3}$$

The λ_i is the eigenvalue corresponding to the eigenvector a_i. Sort the d eigenvalues of the covariance matrix Σ_x by $\lambda_1 > \lambda_2 > ...> \lambda_d$, the first principal component y_1 satisfies $y_1 = a_1^T X$, a_1 is eigenvector corresponding to eigenvalue λ_1, the first d principal component y_d satisfies $y_d = a_d^T X$, a_d is eigenvector corresponding to eigenvalue λ_d. So the former d (d $\leq$ n) principal components $y_i (i = 1, 2, \cdots, d)$ are to replace the original more variable (Table 1).

Table 1. Team feature extraction process.

Input:Team characteristic matrix T=$(w_1, w_2, ..., w_q)$; Dimension of low dimensional space d.
Process: 1:Centralization of all samples:$w_1 \leftarrow w_1 - \frac{1}{m} \sum_{i=1}^q w_1$;   2: Calculating the covariance matrix of samples Σ_x;   3: Eigenvalue decomposition for covariance matrix Σ_x;   4:The characteristic vector $y_1, y_2, ..., y_d$ corresponding to the largest d eigenvalue is obtained;   Output: Projection matrix Y= $(y_1, y_2, ..., y_d)$.

So the former d (d $\leq$ n) principal components $y_i (i = 1, 2, \cdots, d)$ are to replace the original more variables, So as to achieve the purpose of dimensionality reduction.

User and Team Similarity. By calculating the cosine value of two text vectors, we can know their similarity in statistical methods.

The characteristic attributes of users and teams can be represented by user characteristic matrix and team characteristic matrix respectively, The similarity between users and teams can be measured by the cosine angle between the feature matrices:

$$\cos(T_i, U_j) = \frac{\sum (T_i \times U_j)}{\sqrt{\sum (T_i)^2 \times \sum (U_j)^2}} \tag{4}$$

U is the user feature matrix, T is the team feature matrix, U_j is the feature vector of user j, T_i is the feature vector of team i.

Recommendation Based on Feature Similarity. Considering the coverage and novelty, the recommendation system needs to recommend some newly created and highly similar academic teams to users. Therefore, a recommendation value based on

user and team characteristics similar to the fusion time dimension is proposed (Rvsf). It
is defined as:

$$Rvsf_{ij} = \alpha \cos\left(T_i, U_j\right) + \beta \frac{(t_i - t_{min})}{(t_0 - t_{min})} \tag{5}$$

Among them, t_i is the creation time of team i, t_{min} is the earliest creation time of all
teams, t_0 is the current time, α and β are derived from experiments. The K team with
the highest recommended value forms a team recommendation list (Rvsf$_{list}$) based on
user and team characteristics.

2.2   Recommendation Based on Popular Team

A popular team is a team that has received a wide range of user attention within a
certain period of time and within a certain range. Team click-through rate, the number
of team members, the dynamic number of teams, dynamic browsing, the number of
team announcements, announcement browsing these indicators to measure the hot
team.

Each index measurement unit of the multi-index evaluation system is different, in
order to be able to participate in the calculation of popularity of all indicators. It is
necessary to standardize the indicators and map their values to a certain numerical
range through function transformation. Therefore, the formula for team popularity
(TPF) is defined as follows:

$$TPF_i = \sum_{j=0}^{k} \frac{x_{ij} - x_{min}}{x_{max} - x_{min}} \tag{6}$$

Among them, TPF$_i$ is the popularity of team i. The x_{min} and x_{max} respectively
represent the smallest value and the largest value of the indicator in the team. Although
the newly created hot team is not as good as the popular team that has been created for
a long time, the newly created hot team has the trend of developing new academic
research to a certain extent. This model joins the team's creation time metric when
forming a list of recommendations for popular teams, so that the newly created hot
team can be prioritized. Therefore, the recommended value based on popular teams
(Rvht) is defined as follows:

$$Rvht_i = \alpha\, TPF_i + \beta \frac{(t_i - t_{min})}{(t_0 - t_{min})} \tag{7}$$

Among them, α and β are derived from experiments. Finally, the top K teams in
Rvht form a popular team (Rvht$_{list}$).

2.3   Recommendation Based on Friend Relationship

At present, the communication environment of social networks is relatively closed.
This closed social environment has created absolute familiarity among friends. At the

same time, people expand their relationships in social networks so that they can reach more of their favorite friends and teams. Because of the distance between close friends and the number of friends in common, these can affect the credibility of friends. Therefore, in this paper, we use similarity, frequency of interaction and the rate of common friends as three indicators when seeking the confidence of friends.

Friend Feature Similarity. Calculate the similarity between users and friends, use formula (5) to get $Rvsf_{ij}$. The $\cos(i,j)$ is the similarity between user i and user j, where t_i is the time when user i and user j establish friendship, and t_{min} is the time when user i establish friendship with the first friend.

Interaction Frequency. The strength of the relationship between the user and his friends can be expressed by the frequency of interaction between the contacts, and because different users have different ways of expressing. Therefore, the frequency of interaction between a user and a friend in this article is represented by the amount of interaction between the user and the friend divided by the total amount of interaction between the user and all friends.

The interaction frequency can be expressed as dynamic point number, number of messages in the station, number of online chat records, and so on. The formula for the interaction frequency (IF) of this article is as follows:

$$IF_{ij} = \frac{|F(i,j)|}{|N(i)|} \qquad (8)$$

The frequency of interaction between user i and friend j is IF_{ij}, and N(i) is the total amount of interaction of user i. $F(i,j)$ is a function to calculation the amount of interaction between user i and user j.

Common Friend Rate. There is a big difference in the total amount of user friends, this article uses the ratio of the number of friends and a friend's total friends to the total number of users' friends to represent the common friend rate. The formula for common friend rate (CM) is as follows:

$$CM_{ij} = \frac{|f(i,j)|}{|n(i)|} \qquad (9)$$

Where n(i) is the total number of friends of user i, and f(i, j) is the number of friends of user i and user j.

Friend's Trust. In summary, the friend's trust formula (DT) is as follows:

$$DT_{ij} = \alpha Rvsf_{ij} + \beta IF_{ij} - \gamma CM_{ij} \qquad (10)$$

The recommended value based on the friend relationship can be expressed as $Frec_i = \sum_j DT_{ij} \cdot [T_{j1}, T_{j2}, \cdots, T_{jm}]$, T_j is the team that user i's friend j joined. Then K teams with the highest recommendation value based on the friend relationship are extracted to form a recommendation list based on the friend relationship ($Frec_{list}$).

2.4   Hybrid Recommendation Model Algorithm Based on Time Dimension for Academic Teams

Combine recommendation lists based on similarity between users and teams ($Rvsf_{list}$), recommended list based on popular teams ($Rvht_{list}$), recommendation list based on friend relationship ($Frec_{list}$). The recommended values of the hybrid recommendation model based on time dimension for academic teams (HRM_{list}) are defined as:

$$HRM_{list} = \alpha Rvsf_{list} + \beta Rvht_{list} + \gamma Frec_{list} \tag{11}$$

Among them, α, β, γ are obtained by experiments. Finally, the K teams with the top HRM rankings are selected to form the recommendation list (HRM_{list}) of the hybrid recommendation model based on time dimension for academic teams.

3   Academic Team Recommendation Process

The SCHOLAT [18] has 1000+ academic teams. In this social network environment, information overload and cold start are urgent problems to be solved. How to help scholars and academic teams to establish fast and effective contact has become a key issue for the academic social network platform to improve their services. According to the completeness of user data, users who log in to the SCHOLAT can be divided into the following four situations (see Fig. 2):

(1) The user does not have personal information, and there is no friend relationship: recommend popular teams for the user.
(2) The user does not have personal information, but there is a friend relationship: the teams are recommended for the user according to the information of the two dimensions of the friend relationship and the popular team.
(3) The user has personal information, but there is no friend relationship: the teams are recommended for the user according to the information of the two dimensions of the personal information and the popular team.

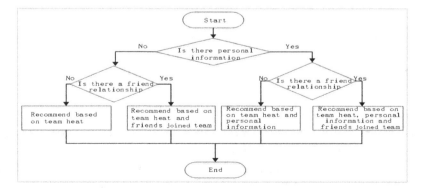

Fig. 2. Team recommendation flow chart

(4) The user has personal information and has a friend relationship: the teams are recommended for the user according to the information of the three dimensions of the personal information, the friend relationship and the popular team.

4   Experimental Result

4.1   Data Set Introduction

This paper uses the academic social network platform (SCHOLAT) [18] data set for academic team recommendation. From this data set, the personal information of 800 users on March 12, 2018, and their friend relationship and 200 teams information were extracted. The team information and personal information are shown in Fig. 3.

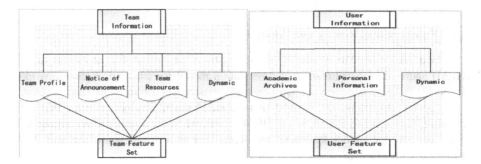

Fig. 3. Team information and personal information.

4.2   Evaluation Indicators

When recommending an academic team, this paper gives users a personalized recommendation list. This recommendation is called Top-N recommendation. The prediction accuracy recommended by Top-N generally passes Precision, recall, F-measure and coverage.

Let R(u) be the academic team recommendation list given to user u based on the completeness of the user data, and T(u) is the team list of user u itself, I is the number of teams in the data set.

The Recall of the recommendation results is defined as:

$$R = \frac{\sum_{u \in U} |R(u) \cap T(u)|}{\sum_{u \in U} |T(u)|} \tag{12}$$

The Precision of the recommendation results is defined as:

$$P = \frac{\sum_{u \in U} |R(u) \cap T(u)|}{\sum_{u \in U} |R(u)|} \tag{13}$$

The F-measure of the recommendation results is defined as:

$$F = \frac{2 \cdot \text{Precision} \cdot \text{Recall}}{\text{Precision} + \text{Recall}} \tag{14}$$

The Coverage of the recommendation results is defined as:

$$C = \frac{\left| \bigcup_{u \in U} Ru \right|}{|I|} \tag{15}$$

4.3   Experimental Results and Analysis

A hybrid recommendation model based on time dimension for academic teams (HRM) and multi-faceted team recommendation model for academic social networks (PGNMF) [12] are compared experiments; the experimental results are shown below.

It can be seen from Figs. 4 and 5 that the HRM model is superior to the PGNMF model in Precision and Recall. In the HRM model, when the number of recommended teams is four, the model Precision and Recall are improved by 36.36% and 20% compared with PGNMF.

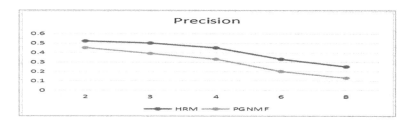

Fig. 4. Precision.

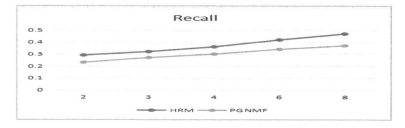

Fig. 5. Recall.

According to Fig. 6, in the F-measure, when the number of recommended teams is larger, the recommended effects of the two recommended models are deteriorated. When the number of recommended teams is 4, the F-measure of the HRM model is higher, which is 27.4% higher than that of the PGNMF model. As can be seen from Fig. 7, as the number of recommendations increases, each model will increase in coverage. The HRM comparison PGNMF has obvious advantages in the evaluation index of recommended coverage.

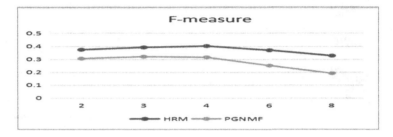

Fig. 6. F-measure.

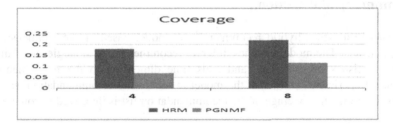

Fig. 7. Coverage.

In summary, the model proposed in this paper makes full use of the user's academic information, social relations and team information. The model is applied to the SCHOLAT for the recommendation of the academic team, which can more effectively recommend the academic team of interest to the scholar to achieve personalized recommendation effects, as shown in Fig. 8.

Fig. 8. Academic team recommendation results.

5   Summary and Outlook

This paper proposes a hybrid recommendation model based on time dimension for academic teams. The model matches different recommendation strategies according to the characteristics of user data, and adds time dimension to the recommendation algorithm. Experiments show that the model improves the recommendation accuracy. At the same time, the coverage of the recommendation list is improved to some extent.

Acknowledgement. This work is supported by the National Natural Science Foundation of China (No. 61772211), the Science and Technology Program of Guangzhou, China (Nos. 201604046017 and 201704020203), the Science and Technology Project of Guangdong Province, China (No. 2016A030303058).

References

1. Ventresque, A., Yong, J.T.T., Datta, A.: Impact of expertise, social cohesiveness and team repetition for academic team recommendation. In: Datta, A., Shulman, S., Zheng, B., Lin, S. D., Sun, A., Lim, E.P. (eds.) Social Informatics. SocInfo 2011. Lecture Notes in Computer Science, vol. 6984, pp. 296–299. Springer, Heidelberg (2011). https://doi.org/10.1007/978-3-642-24704-0_33
2. Yu, S., Xia, F., Zhang, K., Ning, Z., Zhong, J., Liu, C.: Team recognition in big scholarly data: exploring collaboration intensity. In: IEEE, International Conference on Dependable, Autonomic and Secure Computing, International Conference on Pervasive Intelligence and Computing, International Conference on Big Data Intelligence and Computing and Cyber Science and Technology Congress, pp. 925–932. IEEE (2017)

3. Wuchty, S., Jones, B.F., Uzzi, B.: The increasing dominance of teams in production of knowledge. Science **316**(5827), 1036–1039 (2007)
4. Skowron, P., Faliszewski, P., Lang, J.: Finding a collective set of items: from proportional multi representation to group recommendation ☆. Artif. Intell. **241**, 191–216 (2016)
5. Swar, B., Hameed, T., Reychav, I.: Information overload, psychological ill-being, and behavioral intention to continue online health information search. Comput. Hum. Behav. **70**, 416–425 (2016)
6. Ricci, F., Rokach, L., Shapira, B.: Recommender Systems: Introduction and Challenges. In: Ricci, F., Rokach, L., Shapira, B. (eds.) Recommender Systems Handbook, pp. 1–34. Springer, Boston (2015). https://doi.org/10.1007/978-1-4899-7637-6_1
7. Li, J., Lu, K., Huang, Z., Shen, H.T.: Two birds one stone: on both cold-start and long-tail recommendation. ACM on Multimedia Conference, pp. 898–906. ACM (2017)
8. Camacho, L.A.G., Alves-Souza, S.N.: Social network data to alleviate cold-start in recommender system: a systematic review. Inf. Process. Manage. **54**(4), 529–544 (2018)
9. Zhao, Q., Adomavicius, G., Harper, F.M., Willemsen, M., Konstan, J.A.: Toward better interactions in recommender systems: cycling and serpentining approaches for top-n item lists. In: Proceedings of the 2017 ACM Conference on Computer Supported Cooperative Work and Social Computing, pp. 1444–1453 (2017)
10. Abdollahpouri, H., Burke, R., Mobasher, B.: Value-aware item weighting for long-tail recommendation (2018)
11. Sun, L., Wang, X., Wang, Z., Zhao, H., Zhu, W.: Social-aware video recommendation for online social groups. IEEE Trans. Multimedia **19**(3), 609–618 (2017)
12. Yuan, C., Zeng, B., Tang, Y., Wang, D., Zeng, H., Computer, S.O.: Multi-faceted team recommendation model for academic social networks. J. Front. Comput. Sci. Technol. (2016)
13. Bossi, L., Braghin, S., Datta, A., Trombetta, A.: The zen of multidisciplinary team recommendation. J. Assoc. Inform. Sci. Technol. **65**(12), 2518–2533 (2014)
14. Zhang, J.L., Liang, J.Y., Pang, J.F., Wang, B.L.: Behavior and score similarity based algorithm for association rule group recommendation. Comput. Sci. **41**(3), 36–40 (2014)
15. Zhang, Y., Tu, Z., Wang, Q.: Temporec: temporal-topic based recommender for social network services. Mob. Netw. Appl. **22**(6), 1182–1191 (2017)
16. Sharma, A., Gemici, M., Dan, C.: Friends, strangers, and the value of ego networks for recommendation. ICWSM **13**, 721–724 (2013)
17. Sharma, A., Dan, C.: Distinguishing between Personal Preferences and Social Influence in Online Activity Feeds. In: ACM Conference on Computer-Supported Cooperative Work & Social Computing, pp. 1091–1103. ACM (2016)
18. SCHOLAT Homepage. http://www.scholat.com/

An Improved Method and Implementation of Indoor Position Fingerprint Matching Localization Based on WLAN

Yanling Lu[1,2(✉)], Di Zhang[1], Yao Cheng[1], Yalong Pan[1],
Lingpan Shi[1], Caiwei Liu[1], and Dongyuan Huang[3]

[1] College of Geomatics and Geoinformation, Guilin University of Technology,
Guilin 541004, China
358498163@qq.com, 985640911@qq.com
[2] Guangxi Key Laboratory of Spatial Information and Geomatics, Guilin
University of Technology, Guilin 541004, China
[3] Nanning Land Resources Center of Surveying, Mapping and Geoinformation,
Nanning 53000, China

Abstract. The position fingerprint matching methods are analyzed, this paper improves the insufficient. In the database establishment phase, an AP selection algorithm based on stability is proposed, the Gaussian filtering method is applied to the processing of WLAN signal, Thus, a more stable signal strength characteristic value is extracted. In the matching phase, an AP selection algorithm based on MAC address is presented, narrow the range of matches, combine with KNN algorithm to Fingerprint matching, to increase the speed of the match. By verify through actual data, the paper analyzes the results of the improved off-line phase, effectively improve the quality of fingerprint database and reduce the burden of database. It not only narrow the matching range, but also improve the accuracy of the localization algorithm. The 3D model is built by 3D Max which is a virtual simulation system for user interaction, Virtual reality technology is applied to indoor positioning to complete the establishment of map. The coordinate transformation algorithm is established, and the position fingerprint matching method is used to present the positioning technology to virtual reality to realize the indoor positioning human-computer interaction based on WLAN.

Keywords: Indoor positioning · WLAN · Access point select

1 Introduction

Based on the application and development of related technology of users' locating information, location-based service has become a basic service requirement of daily work and life. With the increasing demand for indoor positioning and higher precision requirements. The existing satellite positioning system cannot be directly applied to indoor positioning with complex environment because the satellite signal cannot penetrate walls, steel and other obstacles. Other indoor positioning technology came into being [1]. At present, the techniques used for indoor positioning are: ultra-broadband

© Springer Nature Switzerland AG 2019
Y. Tang et al. (Eds.): HCC 2018, LNCS 11354, pp. 638–649, 2019.
https://doi.org/10.1007/978-3-030-15127-0_63

technology, RFID (Radio Frequency Identification) Technology, ultrasonic technology, infrared technology, Bluetooth technology and WLAN (Wireless Local Area Networks) technology and so on. Compared with other positioning technologies, WLAN positioning technology has obvious advantages. Firstly, WLAN has wide coverage, easy access, easy extension, and the signal propagation based on WLAN is less affected by obstacles. Secondly, mobile phones or laptops can be used to pick up wireless signals in public places, workplaces and campuses, except for the AP (Wireless Access Point) in those places which are access to the Internet, but also to achieve their own location. Therefore, WLAN technology is more suitable for indoor positioning of buildings.

2 Indoor Location Fingerprint Matching Method Based on WLAN

WLAN is a direct arrival wave, and the distance between the test point and AP can be calculated according to the product of propagation time and propagation velocity. Then the location coordinates of the test points are calculated by using the mathematical geometry algorithm. TOA, TDOA and AOA are typical WLAN-based localization methods. In addition to these three methods, there are RSS-based signal propagation model location method and location fingerprint matching location method. Unlike the other four location methods, location fingerprint matching method does not require ranging, does not require known AP coordinates, and does not require specific equipment. Because it does not need to consider the problem of signal attenuation due to obstruction, it is also less affected by the environment, it only according to the idea of matching, software can be used to achieve positioning. To sum up, the location fingerprint matching method is the most suitable location method based on WLAN. However, due to the interference of environmental factors, the precision of the localization algorithm is not high enough, and the cost is increased or the calculation of location is increased. Therefore, how to study WLAN location fingerprint matching algorithm, and how to improve positioning accuracy and location efficiency, which have become a hot research field of indoor location.

The location of fingerprint matching method is completed in two stages. Firstly, establishing fingerprint library and then making online matching. The establishment of the location fingerprint database means that the location process is half done. In order to determine the position of the reference point, the location area should be divided into several grids. Then, the RSS signals of each reference point are collected in turn, including the number of AP received at the reference point, the signal strength of each AP signal intensity and the MAC address. The coordinate of the reference point is recorded while the RSS signal is collected. After traversing all the reference points, the RSS signal of each reference point and its actual coordinates are recorded in the database. It completes the location fingerprint matching method off-line phase of the database establishment. As shown in Fig. 1.

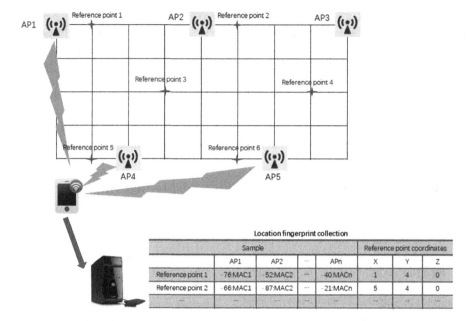

Fig. 1. Construction of indoor location fingerprint matching method

When the location database is established, the terminal at the location point sends the WLAN signal information from the current location to the server, which matches the fingerprint in the database, the matching algorithm is used to calculate and find the most matched fingerprint, then the location is successful.

3   An Improved Method of Indoor Position Fingerprint Matching Localization Based on WLAN

3.1   Extraction of RSS Signal Features Based on Gaussian Filtering Model

The location fingerprint matching method is based on the characteristics of the relationship between RSS signal and actual coordinates, and the establishment of accurate and reliable location fingerprint library are the premise and guarantee of improving positioning accuracy. However, due to the complexity of the indoor environment, radio waves are affected by various factors in the process of propagation, resulting in the acquisition of various types of noise, to a certain extent, these noises will have an uncertain effect on the location results. Therefore, in order to gain more reliable fingerprint data, the existing processing methods include filtering the collected signal by Kalman filter or Gaussian filter.

Kalman filter is a linear extrapolation filter with minimum variance. According to the data of the previous state, the data of the next state is predicted. It can effectively deal with the real problem with time-varying characteristics, but it cannot correct or eliminate the singular value of signal fluctuation, and its error increases with the increase of prediction value. Gauss filter is relatively simple compared with Kalman filter. It is a linear smoothing filter. In the process of processing signal, a Gaussian filter model is established and small probability data are eliminated by the model. The process of retaining high frequency data and finding the mean value of high frequency data.

Because many random variables are subject to or approximate to the normal distribution [2], the signal strength value received after transmission through multiparty channel is generally subject to log-normal distribution. Thus, the classical WLAN signal model can be represented as follows [3]:

$$RSS(d) = RSS(d_0) - 10nlg\left(\frac{d}{d_0}\right) - X_\sigma \tag{1}$$

RSS(d) and RSS(d_0) represent the signal strength values measured at distance d and reference point d_0 respectively. X_σ is a Gaussian random distribution variable with a mean of 0 and a standard deviation of σ (usually 4–10). It is a measurement error caused by signal refraction, reflection, people moving around and other reason.

Therefore, assuming that the RSS signal sampling value $\{RSS_1, RSS_2, RSS_3......RSS_n\}$ received from a fixed AP at a certain location obeys the normal distribution, the mean value is:

$$\mu = \frac{1}{n}\sum_{i=1}^{n} RS \tag{2}$$

The standard deviation is:

$$\sigma = \sqrt{\frac{1}{n}\sum_{i=1}^{n} (RSS_i - \mu)^2} \tag{3}$$

The probability density function is:

$$F(RSS) = \frac{1}{\sigma\sqrt{2\pi}}e^{-\frac{(RSS_i-\mu)^2}{2\sigma^2}} \tag{4}$$

In the above formulas, RSS_i represents the strength of the first signal and i indicates the total number of data collected at a certain location. Selection probability F(RSS). If the empirical range is greater than 0.6 and less than 1 is the high probability occurrence area, the reserved range is F(RSS), Greater than or equal to 0.6 and less than equal to 1.

The results of Gaussian filtering vary from amount of the sampled data, and the greater the n of the sampled data, the more accurate the Gaussian probability model is, the closer the filtered RSS data is to the actual RSS value. In the off-line data acquisition phase of location fingerprint matching method, more samples need to be collected to establish a database to ensure more accurate online matching and more accurate positioning. Therefore, it is feasible to use Gaussian filtering in off-line phase for data filtering. Figures 2 and 3 are signal intensity figs before and after Gaussian filtering, and the signal intensity distributions before filtering are compared, and the data after Gaussian filtering are more smooth.

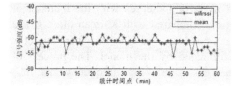

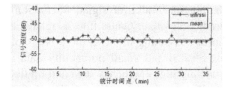

Fig. 2. Pre-filter signal fluctuation diagram

Fig. 3. Post-filtering signal fluctuation diagram

3.2 Implementation of AP Selection Algorithm

Because of the wide coverage of the wireless local area network, WLAN signals can be received and used in many large buildings for location, it is one of the advantages of choosing WLAN technology for indoor positioning [4]. The WIFI integrated analyzer software test found that not all of the AP with the location of the movement of significant changes, that is, not all of the AP is conducive to indoor positioning. If all the *RSS* signals received by the AP are recorded in the database as sample data, this will not only result in poor quality of fingerprint library and lower positioning accuracy, but it also will cause the burden of fingerprint library construction, increase the online matching time and increase the complexity of the algorithm. In the real environment [5], some of the AP signal fluctuations are relatively large, or the mapping relationship with the location is relatively vague, not all of the AP contain useful feature information for location resolution and judgment, so it is necessary to make an effective choice of access point AP, in order to ensure that sufficient AP for location or to ensure that the selected AP signal contains obvious eigenvalues [6].

3.2.1 Stability Based AP Selection Algorithm in Off-Line Phase

In order to select the signal strength of the AP with high frequency and stability in the reference point, we choose the AP after taking sample data off line and building the library. It reduces the burden on the database and ensures accuracy. The location method based on WLAN fingerprint database is based on the mapping relationship between signal intensity and position. If the fingerprint library is established, the signal strength of the AP with large fluctuation is chosen for fingerprint. Therefore, the stability of AP plays an important role in location fingerprint matching. The stability of an AP has two reference points. One is the fluctuation of the signal strength of the AP and the other is the frequency of the AP at the reference point.

Firstly, the AP received at each reference point and its signal intensity are different in the location area, and the sampling I times need to be collected continuously at each reference point. The collection of AP received at the reference point is $\{AP_1, AP_2, AP_3 \ldots \ldots AP_N\}$, The corresponding *RSS* value is $\{RSS_1, RSS_2, RSS_3 \ldots \ldots RSS_n\}$, Then we need to select m stable RSS in n APs. The I sample value for the first AP is represented a $\{RSR_1^1, RSS_1^2, RSS_1^3 \ldots\ldots RSR_1^i\}$, Then the signal fluctuation of the Kth AP can be described by its variance:

$$V(AP_k) = \frac{1}{I} \sum\nolimits_{i=1}^{I} \left(RSR_K^i - \overline{RSS}\right)^2 \tag{5}$$

In Eq. (5), $\overline{RSS}$ shows the mean value of the first sample of the Kth AP, and RSR_K^i indicates the signal strength of the first sample of the Kth AP. The deviation degree between the random variable and its mathematical expectation (mean) is generally measured by the variance of the signal intensity. The smaller the variance, the smaller signal amplitude, the better the stability of the AP [7]. The stability of an AP not only takes into account the signal intensity fluctuation of an AP, but also the frequency of the AP at the reference point. The first K AP continuous I sampling, the frequency can be expressed:

$$F(AP_K) = \frac{Sum_K}{I} \tag{6}$$

In Eq. (6), I indicates the total number of samples taken at the reference point, and Sum_K indicates the number of times the first Kth of AP appears, and the greater the value of $F(AP_K)$, indicating that the frequency of the AP is higher in the I sampling, That is, the more stable the AP is for this reference point. The Eqs. (5) and (6) know that the stability of an AP can be expressed as:

$$Sta(AP_K) = \frac{1}{\frac{1}{I} \sum\nolimits_{i=1}^{I} \left(RSR_K^i - \overline{RSS}\right)^2} * \frac{Sum_K}{I} \tag{7}$$

The larger $Sta(AP_k)$, the better the stability of the AP, from large to small in order, to take the previous maximum of the AP as the choice of the AP, and when m is 5, the positioning accuracy is the highest. In Eq. (7), if the denominator is 0, it is represented directly by Eq. (6), but this is almost impossible because signals sometimes change and there are constant changes at each moment. Therefore, the off-line phase can be completed before building by selecting stable AP and then using Gaussian filter.

3.2.2 Online Phase AP Selection Optimization Match Based on MAC

In the online phase, it just need to match the real-time signal strength and MAC address of each AP received at the site to the fingerprint library. The estimation of location can be achieved by finding the optimal match, the optimal point is that the MAC address and signal strength of the AP received through MAC verification are the same or similar to that of the AP of a fingerprint in the fingerprint library, but considering the need to match all the information with each fingerprint in the database one by one, the calculation is very large. Therefore, before implementing the conventional matching algorithm, the AP selection of fingerprint library is needed to narrow the matching range. Search the MAC address of the most powerful and stable AP that has been received many times online and contains all the fingerprints of this MAC address in the fingerprint library, filter out other fingerprints away from the site to be fixed, and then use the matching algorithm to calculate the location coordinates.

Suppose there are 10 fingerprint samples in the fingerprint library of the signal, respectively:

$$
\begin{bmatrix}
(RSS_1 : MAC_1, RSS_2 : MAC_2, RSS_4 : MAC_4, RSS_5 : MAC_5, RSS_6 : MAC_6) \\
(RSS_1 : MAC_1, RSS_3 : MAC_3, RSS_5 : MAC_5) \\
(RSS_3 : MAC_3, RSS_4 : MAC_4, RSS_5 : MAC_5) \\
(RSS_1 : MAC_1, RSS_5 : MAC_5, RSS_4 : MAC_4, RSS_7 : MAC_7) \\
(RSS_3 : MAC_3, RSS_6 : MAC_6, RSS_1 : MAC_1, RSS_7 : MAC_7) \\
(RSS_3 : MAC_3, RSS_4 : MAC_4, RSS_5 : MAC_5, RSS_6 : MAC_6) \\
(RSS_2 : MAC_2, RSS_1 : MAC_1, RSS_5 : MAC_5, RSS_8 : MAC_8, RSS_3 : MAC_3) \\
(RSS_6 : MAC_6, RSS_2 : MAC_2, RSS_5 : MAC_5, RSS_1 : MAC_1) \\
(RSS_2 : MAC_2, RSS_4 : MAC_4, RSS_5 : MAC_5, RSS_8 : MAC_8) \\
(RSS_3 : MAC_3, RSS_7 : MAC_7, RSS_2 : MAC_2, RSS_6 : MAC_6)
\end{bmatrix}
$$

The signals collected online are:

$$(RSS_1 : MAC_1, RSS_2 : MAC_2, RSS_4 : MAC_4, RSS_5 : MAC_5)$$

AP_1 is the strongest signal received, so its MAC as a search condition, in 10 fingerprints, through search screening, including some of the fingerprints left the following fingerprint:

$$
\begin{bmatrix}
(RSS_1 : MAC_1, RSS_2 : MAC_2, RSS_4 : MAC_4, RSS_5 : MAC_5, RSS_6 : MAC_6) \\
(RSS_1 : MAC_1, RSS_3 : MAC_3, RSS_5 : MAC_5) \\
(RSS_1 : MAC_1, RSS_5 : MAC_5, RSS_4 : MAC_4, RSS_7 : MAC_7) \\
(RSS_3 : MAC_3, RSS_6 : MAC_6, RSS_1 : MAC_1, RSS_7 : MAC_7) \\
(RSS_2 : MAC_2, RSS_1 : MAC_1, RSS_5 : MAC_5, RSS_8 : MAC_8, RSS_3 : MAC_3) \\
(RSS_6 : MAC_6, RSS_2 : MAC_2, RSS_5 : MAC_5, RSS_1 : MAC_1)
\end{bmatrix}
$$

Through the strongest and most stable AP MAC address, the corresponding fingerprint subset is selected, and then the Euclidean distance is calculated by KNN algorithm. The above is just a list of 10 fingerprints, but in reality, the signal propagation distance of wireless router is not long, and the location reference point in a building is far more than 10. The more advantageous it is, the better it is to adopt this method; In terms of accuracy, the AP received at the close reference point is close to the same, the signal intensity is different, and the AP received at the distance is basically different. The small Euclidean distance can be avoided, but the reference point is not in the vicinity of the site to be fixed, it can effectively improve the positioning accuracy.

4 Implementation of Fingerprint Matching Method for Indoor Position Based on Virtual Reality Technology

4.1 Construction of Virtual Reality Scene Oriented to Indoor Position

The application of virtual reality technology in indoor positioning is a virtual environment simulation based on real data. It is based on graphics and uses 3D MAX 3D modeling software to model scenes. Then using ArcGIS Server and other software as development tools to realize the roaming and interactive functions of the scene. According to the virtual reality technology, the three-dimensional dynamic results are presented to the locator, so that the locator can realize the relevant location information and location scene. Therefore, human-MAC interaction can be realized through the establishment of three-dimensional scene, the transformation of coordinates and the method of location fingerprint matching [8]. Firstly, collecting the geometric data, coordinate data and texture data needed for the virtual reality scene, and then use the data to model the real scene, including the transformation of coordinates. Finally, the real-time rendering of the virtual scene is realized by the three-dimensional engine on the output device, and the roaming of the whole scene is completed [9]. Taking the Yanshan campus of Guilin University of Technology in China as an example, a three-dimensional model was established.

(1) Preparatory: According to the above model data acquisition method to collect information, taking texture photos, as far as possible to record the route and the name of the building, and then use software to build useful images. As shown in Fig. 4.

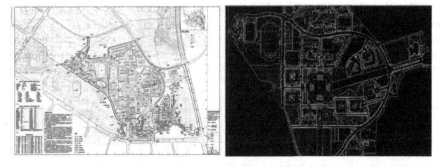

Fig. 4. Planning and 1:1000 topographic map of Guilin University of Technology

(2) Modeling: The CAD plan is introduced into 3D MAX software, according to the length and width of the plan and the height measured, stretch, extrude the column, wall and separate floors, then deal with the floor, windows and other details. Do the same for each scene inside the building, and use the sun and VR light source to make the scene lighting-deployment. As shown in Fig. 5.
(3) Post-processing: To integrate and optimize the model of each scene.

Fig. 5. The overall modeling effect of building

4.2   Implementation of Improved Method for Indoor Fingerprint Matching and Location Based on WLAN

After the building is built with the 3D model of the interior, the coordinate system of the model is only a custom coordinate system. Based on WLAN indoor location, users can estimate their location coordinates by location fingerprint matching method (x, y, z). If the positioning system shows the azimuth target directly in the virtual scene, a series of coordinate transformation is required.

First, the actual longitude and latitude coordinates are converted to the world coordinates, (which are at the pixel points on the plane view), and then the world coordinates are converted into virtual scenes. That is, the local or national coordinate system and virtual scene model coordinate system conversion. The coordinate transformation relationship between the two-dimensional map and the three-dimensional virtual map can be solved by using the coordinate of the two-dimensional plan and the three-dimensional virtual scene. Generally speaking, the control point of a building will manually select three of the ground base points of the building as the control point. The coordinates of [A D], [B E] and [C F] are assumed to be 3 control points. The coordinate of the two-dimensional plan is (Sx, Sy) and the coordinate of the virtual reality scene is (Gx, Gy). The matrix equation based on the least square method is as follows:

$$\begin{bmatrix} A & D \\ B & E \\ C & F \end{bmatrix} = \begin{bmatrix} n & \sum S_x & \sum S_y \\ \sum S_x & \sum S_x^2 & \sum S_x S_y \\ \sum S_y & \sum S_x S_y & \sum S_y^2 \end{bmatrix}^{-1} * \begin{bmatrix} \sum G_x & \sum S_y \\ \sum S_x G_x & \sum S_x S_y \\ \sum S_y G_x & \sum S_y G_y \end{bmatrix} \tag{8}$$

Equation (8) is used to solve A, B, C, D, E, F, and to establish the normal calculation formula of coordinates according to the result:

$$S_x = A * S_x + B * S_y + C \tag{9}$$

$$S_y = D * S_x + E * S_y + F \tag{10}$$

The equation of coordinate inverse calculation is obtained by recounting Eqs. (9) and (10):

$$S_x = \frac{E * G_x - B * G_y - CE + BF}{AE - BD} \tag{11}$$

$$S_y = \frac{D * G_x - A * G_y - CD + AF}{BD - AE} \tag{12}$$

The viewpoint is set up in the virtual scene to realize the 3D geometry transformation. The three-dimensional geometric transformation can be represented by the fourth order matrix T_{3D}:

$$\overline{T_{3D}} = \begin{bmatrix} a_{11} & a_{12} & a_{13} & a_{14} \\ a_{21} & a_{22} & a_{23} & a_{24} \\ a_{31} & a_{32} & a_{33} & a_{34} \\ a_{41} & a_{42} & a_{43} & a_{44} \end{bmatrix} \tag{13}$$

$\overline{T_{3D}}$ can be divided into four sub-matrices, respectively to achieve the graphics to reduce, enlarge, pan, projection transformation and the whole scene of the zoom transformation.

In a three-dimensional geometric transformation matrix, any point $P(x, y, z)$ can be converted to $P'(x', y', z')$ by the following expression:

$$\begin{bmatrix} x \\ y \\ z \\ 1 \end{bmatrix} = \overline{T_{3D}} \begin{bmatrix} x' \\ y' \\ z' \\ 1 \end{bmatrix} \tag{14}$$

Thus, after the relationship between the 3D scene model and the coordinate system transformation, the virtual scene needs to be properly expressed. That is, the design of human-computer interaction function can complete the translation, scaling, rotation and positioning of virtual scene [10]. According to the improved WLAN indoor fingerprint matching method, the location coordinates of the locator are calculated, and the corresponding position of the virtual scene is calculated by the coordinate transformation formula. At the same time, the virtual character is positioned to the corresponding position. The work-flows for online positioning can be represented in Fig. 6.

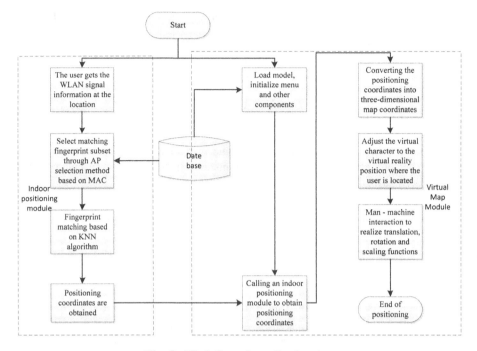

Fig. 6. Work-flows for online location

5   Conclusion

In WLAN-based indoor position method, the accuracy of location fingerprint matching depends on the stability of fingerprint library. And the indoor environment is much more complicated than the outdoor environment. Once the indoor environment changes a lot, the signal received will change a lot. Even if the location of the on-line phase acquisition signal is the same as that of the off-line phase acquisition reference point, the signal will be very different. Therefore, in the matching phase, it is very difficult to match the correct fingerprint, which makes the deviation of location estimation very big, which is the problem of fingerprint database mismatch. Aiming at the shortcomings of the location fingerprint matching method, the two processes of location are improved, and the stability-based AP selection algorithm and Gaussian filtering are established. The quality of the sample signal is improved by the purposeful selection of AP and the processing of RSS signal intensity. In the online location phase, an AP selection algorithm based on MAC address is established to narrow the matching range, and the improved online location algorithm is verified by an example. The accuracy has also been improved to a certain extent. At the same time, compared with two-dimensional maps, virtual reality technology can display more abundant location information, according to the situation of location tracking in three-dimensional form. This realizes the fusion of real space and information space.

Acknowledgements. The project was funded by Guangxi Key Laboratory of Spatial Information and Geomatics (15-140-07-14, 16-380-25-17) and the 'Ba Gui Scholars' program of the provincial government of Guangxi.

References

1. Xu, H.Y., Li, Z., Wang, J.: Application status and prospect of indoor map. Bull. Surv. Mapp. **9**(1), 119–121 (2014)
2. Xu, J.: Research and Implementation of Indoor Location Technology Based on WLAN Location Fingerprints. Beijing University of Technology, Beijing (2014)
3. Liu, Z.: Research and Implementation of Indoor Positioning Technology for Smartphones. South China University of Technology, Guangzhou (2015)
4. Luo, Y.J., Law, C.L.: Indoor positioning using UWB-IR signals in the presence of dense multipath with path overlapping. IEEE Wireless Commun. **11**(10), 3734–3743 (2012)
5. Shen, Y.: Study on the Selection Algorithm of Access Points in Wireless Indoor Location Based on Fingerprint. Zhejiang University, Hangzhou (2014)
6. Li, J., Zhang, Y., Luan, F.: Probability match room location method based on multipath fingerprints. J. Tsinghua Univ. (Sci. Technol.) 05, 514–519 (2015)
7. Zhang, J.: Research on Indoor Positioning Technology Based on WLAN. Shenyang University of Technology, Shenyang (2015)
8. Wan, Z.: Research on Indoor Positioning Technology Based on Power Cord and Location Fingerprint in Intelligent Space. Beijing University of Technology, Beijing (2014)
9. Miao, Z., Ma, J.: Foundation and Application of Virtual Reality Technology. Tsinghua University Press, Beijing (2014)
10. Li, J.: Realization of Three-Dimensional Modeling and Interaction System of Interior Architecture Based on Unity3D. China University of Mining and Industry, Jiangsu (2014)

Research on Bionic Robot Micro Adsorption Mechanism SMA Driven System

Junmei Xi[1,2(✉)], Huan Wang[1,2], Zhen Wang[2], Songhai Zhang[2], and Anbang Dai[2]

[1] Jiangxi Province Key Laboratory of Precision Drive and Control,
Nanchang, China
2331223957@qq.com
[2] School of Mechanical and Electrical Engineering,
Nanchang Institute of Technology, Nanchang, China

Abstract. In view of the shape memory alloy (SMA) actuation and sensing properties of smart materials, using of SMA output larger driving force in the process of phase transformation, a biasing SMA rigid elastic coupling drive sucker control system has been designed. Using field programmable gate array (FPGA) for generating a pulse width modulated control mode, the resistance detection feedback circuit controls the current feedback, and monitors the pressure of the adsorption mechanism, thus achieving the micro-adsorption mechanism SMA drive control. Experiments show that the system not only has better precision and a certain response speed, but also has good adaptability and self-protection ability.

Keywords: Bionic robot · Driver · PWM · Micro-suction cups

1 Introduction

Shape Memory Alloy is a new functional material with shape memory effect. The SMA driver can replace the traditional driver in special occasions, and has the characteristics of large power/mass ratio, simple structure, no noise pollution, and easy control. Therefore, it is widely used in micro-mini robot drives [1]. Adsorption mechanism sucking with negative pressure needs to simulate the characteristics of bio-muscle contraction, the current common materials can not be achieved. Through the use of smart material shape memory alloy (SMA) to imitate this characteristic, the adsorption mechanism of negative pressure adsorption is Studied, and a new design method is provided for wall-adsorption mechanism design of wall-climbing robot [2]. Because SMA springs are driven by shape memory effect, there are problems such as stress lag, phase transition and temperature, stress, and direction of the movement. SMA is affected by heating and cooling conditions in practical applications. The stress-strain curve of the shape memory alloy is nonlinear, and its mechanical properties vary greatly at different temperatures, making it difficult to drive, and the response frequency

Xi Junmei (1967-), Professor, is engaged in research on advanced mechanical design methods.

is a technical difficulty for the application of the SMA driver [3]. One-way shape memory alloys are used in bionics design, and two-way actuators are configured for the biasing elements. The temperature changes and elongations are relatively linear and easy to control and stable. One-way memory alloys have a relatively large internal resistance to Cu and are prone to heating. In practice, this type of actuator has a better application. For the bionic micro-adsorption mechanism driven by SMA, due to the space limitation of the micro-adsorption mechanism, to realize the SMA drive control is to accurately regulate the temperature of the shape memory alloy wire, the drive technology is one of the key technologies that must be solved. Based on the above reasons, a biased SMA elastically coupled and driven adsorption mechanism was designed. The SMA heating control and resistance detection feedback circuit was adopted to realize the SMA drive control of the bionic micro-adsorption mechanism.

2 SMA-Driven Sucker Structure Design

At the same length, the SMA spring drive has a large stroke, simplifying the structure. Therefore, the SMA spring driver is designed based on a one-way memory alloy and configured with an oscillating element (bias spring). The output displacement of a single-way SMA spring is linearly distributed. It is easy to control the change of elongation by the control current. The change of elongation and temperature is linear, and the biasing spring makes it possible to stop the heating of the SMA spring. It produces a corresponding restoring force. The structure of the SMA spring-driven suction cup is shown in Fig. 1. The suction cup is mainly composed of a rigid support body 1, a bias spring, a slider, an SMA spring, a guide rod, a guiding element, an elastomer thin plate 1, a rigid support body 2, and a spring thin plate. 2, cavity body. The bias spring is coaxial with the SMA spring so that a multi-cavity structure can be formed. One end of the guide rod is glued with the elastomer plate, the other end is threadedly connected with the slider through the guide member, the slider can slide in the cylinder body at the upper end of the rigid support body, and the SMA spring and the biasing spring are respectively placed on both sides of the slider. When the suction cup is in contact with the suctioned surface, the elastic body plate, the edge of the elastic body and the suctioned surface enclose a sealed air chamber. One end of the guide rod is glued with the elastomer plate, the other end is threadedly connected with the slider through the guide member, the slider can slide in the cylinder body at the upper end of the rigid support body, and the SMA spring and the biasing spring are respectively placed on the slider on both sides. When the suction cup is in contact with the suctioned surface, the elastic body plate, the edge of the elastic body and the suctioned surface enclose a sealed air chamber. The bottom of the suction cup also adopts a two-stage multi-cavity structure to improve the adaptability of the wall surface. Even if there are small gaps on the adsorption wall surface, the suction cup minimizes the influence of the air leakage on the adsorption effect of the entire suction cup.

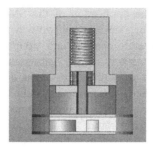

1.Rigid support body.1 2.Bias spring 3.Slider 4.SMA spring 5.Guide rod 6.Guiding element 7.Elastomer thin plate.1 8.Rigid support body.2 9.Spring thin plate.2 10.Cavity body

Fig. 1. Sucker structure diagram

3   Drive Control System Design

Aiming at the characteristics of biased elastic-coupled SMA mechanism, the SMA heating power drive control circuit, detection feedback circuit and pulse width modulation circuit are designed based on high-speed single-chip microcomputer and FPGA, and a complete control system is formed with the adsorption mechanism and host computer. The system structure is shown in Fig. 2.

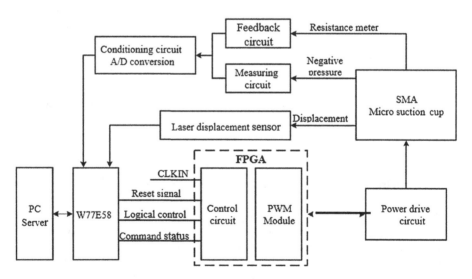

Fig. 2. Schematic diagram of the control system

The control system compares the set displacement value with the current position value, selects the feedback acquisition circuit and the power drive heating, and outputs a PWM voltage signal. When the PWM wave is at a low level, the power circuit controls the SMA heating and the resistance feedback circuit stops working; when the

PWM wave is at a high level, the power drive circuit stops working, and the feedback circuit collects the voltage change of the SMA bridge, and after the amplifying circuit, Analog/Digital conversion is sent to the MicrocontrollerW77E58, the duty cycle of the PWM is adjusted by the control algorithm, and then the power of the heating drive is controlled to complete the SMA drive control of the adsorption mechanism.

3.1 Pulse Width Modulation Circuit

Pulse width modulation circuit uses EPM7000S logic device, EPM7000S achieve pulse width modulation, logic control and other functions, system JTAG interface and EPM7000S chip connect circuit. The Verilog HDL hardware description language is used on the Quartus II development platform to perform logic design on the FPGA. The input compilation, function simulation, input optimization, placement and routing, and post-simulation are designed using the VHDL description language. After the compilation, the SRAM object file SFO and PFO programming object files are generated, the POF file is downloaded and loaded into the EPC configuration chip, and the SFO file is configured with the SRAM structure. The sof file can also be converted to JTAG's indirect configuration file jic, using JTAG to download both EPC and FPGA devices.

3.2 Sensor Detection Circuit

The adsorption mechanism seals around the insertion hole of the air pressure sensor input pipe with a sealant, and the adsorption mechanism is directly placed on the input air pipe of the sensor to measure the air pressure. In the adsorption mechanism, the SMA generates heat through the electric current, and the restoring force of the spring causes the SMA spring mandrel to slide away from the platen, creating a negative pressure within the chassis to form a negative pressure [4]. The air pressure sensor detects the negative pressure signal and passes it to the SCM after conditioning, filtering and digital/analog conversion. The least squares method is used for calibration to ensure the accuracy of signal acquisition. Due to the space limitations of the micro-adsorption mechanism, the LB-1000 series of high-precision long-distance laser displacement sensors are used. The sensors include the LB-300 sensor head and the LB-1200 controller. The sensor has a measuring distance of 300 mm, a voltage range of −5 V to +5 V, a linear relationship between displacement and voltage, and a sensitivity of 50 mV/mm. Before using the sensor, set and connect the controller and terminal [5]: (1) Select "HIGH" for the RESPONSE speed switch; (2) Select "AUTO" for the SENS switch. (3) DIP switch. The closed-loop control method is used to control the position of the SMA drive and monitor the displacement of the SMA drive. The detection feedback circuit collects the voltage signal of the SMA resistance change through the bridge, passes the filter processing, converts the analog/digital circuit into the single-chip processor, and compares the initial position parameter values. Set the FPGA to generate a variable duty cycle PWM pulse wave, quickly control the SMA drive circuit on and off to achieve the drive control requirements. The bridge detection circuit converts the small resistance change of the SMA into a voltage signal. The conditioning amplification is entered into the A/D converter in a differential manner. The SPI serial bus is then sent to the main microprocessor for data calibration and related processing to realize high-precision voltage measurement.

3.3   Power Drive Circuit

Due to the insufficient PWM pin driving capability of the FPGA, a corresponding driving circuit is designed to drive the load circuit to provide a sufficient heating current to drive the SMA so that the SMA can reach the phase change temperature. When SMA heating is stopped, the SMA is completely isolated from the heating power source and ground. When the SMA heating power supply voltage is constant, the duty cycle PWM signal determines the magnitude of the incoming SMA current. The power driver circuit is composed of isolation, voltage regulation, comparison drive and heating protection circuit, as shown in Fig. 3.

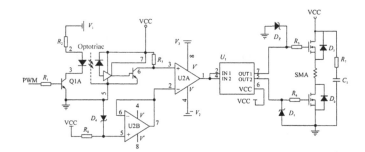

Fig. 3. Power drive circuit

The drive system uses a diode and on the SMA to release the energy stored by its inductance. The P6KE200A CG suppression diode is selected as the protection function. When the voltage exceeds its rated voltage, the interference pulse is embedded in the specified value to ensure that the power switch tube is not damaged. The performance of SMA actuators is often affected by the amount of current and predeformation, cooling rate, and drive load. The cooling rate of the SMA spring restricts its response speed. Its drive current is proportional to the recovery speed, the current needs to be controlled within a reasonable range, the current is too low to reduce its response speed, and the drive memory performance is damaged. The temperature of the SMA is greatly influenced by the environment, and it is difficult to accurately control only the control current. Therefore, current feedback control is performed based on parameters such as temperature, output displacement, and force.

4   System Test and Analysis

The SMA's heat is controlled by controlling the current input to the SMA to control the drive output displacement and force. The PC communicates with the upper computer through the serial port, controls the SCM system, and acquires the measured data, and dynamically displays various types of data and reports. Considering the space limitations of the adsorption mechanism, the control uses current pulse width modulation and natural cooling. It is easy to introduce random noise in the real-time acquisition, reduce the sensor's drift and nonlinear errors, and use digital filtering to eliminate the interference signal and obtain a good correction effect. It is easy to introduce random noise in the real-time acquisition. By reducing the sensor's drift and nonlinear errors, and using digital filtering to eliminate the interference signal, it obtain a good correction effect. The diameter of the SMA spring wire used in the experiment was 2 mm, the diameter of the spring was 12 mm, and the length was 10 mm. At different driving currents, the offset, air pressure and negative pressure response time of the SMA elastomer sheet were measured. The ambient temperature during the test was approximately 20 °C. The sucker experimental device and sucker are shown in Fig. 4.

Fig. 4. Experimental device

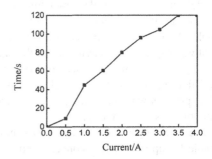

Fig. 5. Relationship between current and bias

The current and offset diagrams are shown in Fig. 5. From this figure, it can be seen that as the current value increases, the deflection generated by the bottom plate of the elastic material is larger, and from the current value of 4A, this offset has a significant increase. Explaining that the current value is 4A is an important turning point in the

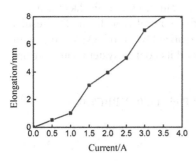

Fig. 6. Relationship between current and time

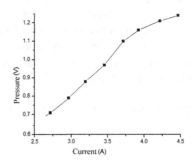

Fig. 7. Relationship between current and pressure

change of results of the sucker. The curve of current and time variation is shown in Fig. 6. It can be seen from the figure that the greater the energizing current, the shorter the required time. With the increase of the current value, the change of energizing time is basically a linear distribution.

The relation between the negative pressure generated by a cavity of the sucker and the current change is shown in Fig. 7. It can be seen from the figure that the negative pressure changes substantially linearly with the current. When the current value is 4A, a point of curvature change occurs, and the negative pressure around this current value changes rapidly, indicating that the current of 4A is the point where the deformation rate of the SMA spring is better.

From the above analysis, it can be seen that as the SMA spring drive current increases, the deflection of the elastomer thin plate becomes larger, the negative pressure generated by the suction cup increases, and the response time is shortened. The deformation of the elastomer thin plate is relatively large and the air pressure is smaller than the theoretical value. This is mainly due to the following reasons: The initial volume of the air cavity under the elastic body cannot be measured; Because the initial volume of the air cavity below the elastic body is relatively large, and the suction cup is always leaking, the deformation of the elastic body is often very large, not a small deformation, but a large deformation; The force exerted on the elastomer by SMA spring is variable in the process of elongation, while a constant force is adopted in the process of modeling.

The force applied to the elastomer is not concentrated force, and the above model is suitable for approximately concentrated force. The negative pressure produced by the sucker can reach over 12 kPa, and the sucker can absorb about 10 s when the SMA spring stops heating.

5   Conclusion

As the acting force, displacement and response time of SMA drive are often affected by the size of current and predeformation, cooling speed and load, the micro-adsorption mechanism of SMA rigid-elastic coupling drive is designed. Build heating power drive, resistance meter feedback, displacement and pressure detection circuit, using field FPGA to generate PWM heating drive method, use field effect tube as power drive switch, taking into account parameters such as temperature, output displacement and force, bionic micro-adsorption mechanism SMA drive control is realized. The results show that the passive adsorption mechanism can be realized based on SMA drive, with small overall mass and volume, and the mechanism and its control system hardware are relatively simple and the design is flexible.

Fund Project. Jiangxi Science and Technology Support Project (20123BBE50088).

References

1. Tang, S., Dong, E., Xu, W., et al.: New SMA torsional actuator design and experimental study. J. Funct. Mater. Devices **17**(4), 407–411 (2011)
2. Hu, B., Wang, L., Fu, Z., et al.: Design, modeling and experiment of bionic suction cup driven by shape memory alloy. J. Shanghai Jiaotong Univ. **43**(11), 1698–1702 (2009)
3. Zhang, Y., Dong, E., Xu, W.: Research on improving working frequency of shape memory alloy linear actuator. Mod. Manuf. Eng. **9**, 93–96 (2010)
4. Dong, Z., Zhao, Y., Fu, Z., et al.: Research on bionic passive negative pressure suction chuck based on SMA drive. Mechatronics **9**, 47–51 (2008)
5. Wang, L.: Research on control system of variable wing SMA driver based on DSP. Nanjing University of Aeronautics and Astronautics, Nanjing (2009)

The Transient Dynamics Analysis of Electric Flip Plate Roadblock Machine

Min Feng[1], Kang Geng[2(✉)], and Hongjian Zhao[3]

[1] Tianjin Research Institute for Water Transport Engineering,
M.O.T., Tianjin 301500, China
[2] School of Mechanical Engineering, Hebei University of Technology,
Tianjin 300130, China
2774792891@qq.com
[3] Beijing Oriental Tea Valley Electronics Co., Ltd., Beijing 100176, China

Abstract. The reliability under the condition of impact load from vehicle is the most important indicator for roadblock machine. Since the impact load is very large and happens in a short time, the traditional statics analysis using finite element method is not applicable, and transient dynamics analysis plays a very important role in this situation. In this paper, the transient dynamics analysis is carried out for electric flip plate roadblock machine using ANSYS software based on statics analysis for structure and modal analysis for lifting process of cover plate, these results (strain and stress) are obtained to assist and optimize the design scheme with the statics analysis and modal analysis together.

Keywords: Roadblock machine · Impact load · Transient dynamics · ANSYS

1 Introduction

As a kind of security equipment widely used in recent years, roadblock machine has been developed rapidly and the research topic of roadblock machine has gradually increased both at home and abroad. The use of finite element software is also increasingly accepted by designers, which provide assistance for mechanical design. As a kind of equipment of intercepting vehicles, every index of roadblock machine has definite requirements, such as elevation, structural strength, ability to withstand impact force, etc., which makes the design of roadblock machine very demanding. In terms of structural strength, the finite element statics analysis method is often used to analyze the stress of the structure under static state and the weak position is analyzed by the finite element cloud map.

When under working condition, roadblock machine bears the impact load from the target intercepted vehicle. Because the collision occurs in a very short time, therefore, the statics mechanical analysis of finite element analysis is no longer applicable under this condition. The transient dynamics analysis of roadblock machine is needed.

Transient dynamics analysis can be used to analyze the state of the model bearing time-varying loads. It is also called the time history analysis method, by which the time-varying index of the model under different loads can be obtained, such as deformation, stress and strain, etc.

© Springer Nature Switzerland AG 2019
Y. Tang et al. (Eds.): HCC 2018, LNCS 11354, pp. 658–665, 2019.
https://doi.org/10.1007/978-3-030-15127-0_65

The model under transient dynamics analysis can be rigid body or deformation body, even arbitrary structure, after the correct mesh division and modal analysis of the model. The response with time can be obtained by applying time-varying loads. For deformation body, the strain value and stress can be obtained by analysis.

In this paper, the transient dynamics analysis of electric roadblock machine is carried out by using ANSYS finite element analysis software, which provides a design method for roadblock machine.

2   Structure of Electric Flip Plate Roadblock Machine

As shown in Fig. 1, the electric roadblock machine is driven by electric power, which is based on plane four bar mechanism that apply first grade gear transmission to achieve motion transfer. Gears are installed on the output shaft of the motor, by which the rotating motion of the motor is transferred to the second shaft which is connected with the planar four-bar mechanism and driven as a driving force to drive the linkage mechanism. Finally, the second shaft drive the flip plate to move up and down. After the plate is completely lifted, the self-locking characteristic of four-bar mechanism is used to ensure its safety and reliability.

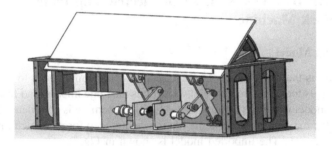

Fig. 1. Three-dimensional model of electric roadblock machine

3   The Theoretical Basis of Transient Dynamics Analysis

The general equation of transient dynamic analysis is:

$$[M]\{x''\} + [C]\{x'\} + [K]\{x\} = \{F(t)\} \tag{1}$$

$$\gamma i = \{xi\}T[M]\{D\} \tag{2}$$

$$\{xi\}T[M]\{xi\} = 1 \tag{3}$$

In the formula, [M] is the mass matrix of the body; [C] is the damping matrix of the system; [K] is the stiffness matrix of the body; {x} is the motion displacement of the body; {F (t)} is the force vector; {x'} is the velocity of the body's motion; {x''} is the acceleration of the body's motion [1].

ANSYS Workbench has two methods to solve the above equations: implicit solution method and explicit solution method [2].

3.1   Implicit Solution Method

In the implicit solution, ANSYS software adapts the open solution method, which is also known as Newmark time integral method. This method requires that the time step should not be too large. If the time step is too large, the convergence of the solution function may be affected, causing solution time is too long [3].

3.2   Explicit Solution Method

In the explicit solution method, ANSYS software uses the closed method to solve the problem. This method requires that the time step should not be too large. Compared with the implicit solution method, the explicit solution method can be used in the case of nonlinear problems. The accuracy of the solution is controlled by setting the size of the calculation step, and the solution speed is relatively fast. The displacement of the solution is obtained by solving the equation with the displacement unknown quantity [4].

4   Transient Dynamic Analysis of Electric Flip Plate Roadblock Machine

4.1   Building Model

This condition is that the electric flip plate roadblock machine opens to the maximum state to intercepts the target vehicle and withstands the huge impact load from the target vehicle. The model is built in Solidworks. After the assembly is complete, the flip plate is opened to the maximum state and the file is saved as step format which imported into ANSYS Workbench. The imported model is shown in Fig. 2.

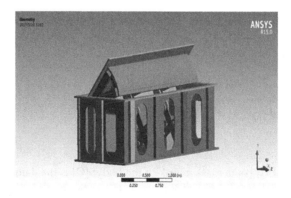

Fig. 2. Finite element model of electric flip plate roadblock machine

4.2 Adding Materials

The material of gear is 40Cr and material of shaft is 40Cr. The material parameter of 40Cr is that the density is 7900 kg/cm3, Poisson's ratio is 0.3, elastic modulus is 2e + 11 pa. The other parts are made of structural steel. The material table is shown in Fig. 3.

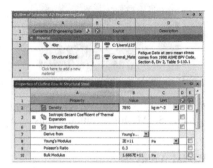

Fig. 3. Material table

4.3 Contact Establishment

Contact establishment mainly generates from the rotating pair of linkages, the contact surface between the shaft and the bearing and the contact surface between the rod needed to turn the flip plate up and down and flip plate. As is shown in Fig. 4, there are 36 contact relations in contact relations.

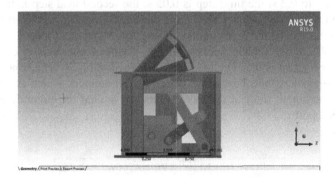

Fig. 4. Contact position diagram

4.4 Mesh Generation

When mesh is divided, the grid parameters are set to 0.1 m and 98446 meshes are obtained. In order to obtain better grid distribution, the correlation center level is the highest, which is fine. At the same time, the maximum size of the default TNDS is defined, and the value is set to 100 in the Tolerance Value in the detail column, so that when the size of the parts in the assembly is smaller than this value, the system will be generated by default as TNDS. The final mesh generation results are shown in Fig. 5.

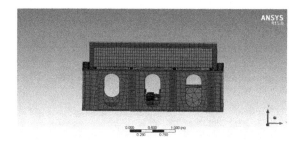

Fig. 5. Mesh generation results

4.5 Preprocessing of Modal Analysis

Before the transient dynamics analysis, the modal analysis needs to be preprocessed. Because of the limited space, the process of modal analysis is omission.

4.6 Transient Dynamics Analysis

First, the modal project needed to share is processed, the transient analysis module is found in Toolbox, and dragged to the solution item in model analysis module of the main interface, then the module layout of data sharing can be obtained.

Then the transient dynamic load is added which consists of two steps: the first load step is from 0 s to 0.2 s, the time step is 0.05 s, the second load step is from 0.2 s to 0.4 s and the time step is 0.05 s. Considering the actual work, the curved surface of the flip plate is impacted, so the load direction is negative along the Z axis, and the impact force which the target vehicle applies on the roadblock machine is 600 KN. The damping of the system is 0.5. The load table is shown in Fig. 6. The constraint of the impact force model is that the bottom plate is fixed support, which is called Fixed Support and the output shaft of the motor is rotated around the circumference that is Cylindrical Support.

	Steps	Time [s]	✔ X [N]	✔ Y [N]	✔ Z [N]
1	1	0.	= 0.	= 0.	0.
2	1	0.2	0.	0.	-6.e+005
3	2	0.4	= 0.	= 0.	0.

Fig. 6. Loads table

4.7   Nephogram of Post-processing Results and Analysis

After the preprocessing of modal analysis and the loads application are completed, it can be entered into the post-processing. In the post-treatment, the total deformation and stress distribution can be obtained by selecting two post-processing results which consist of total deformation and equivalent stress. The total deformation nephogram is shown in Fig. 7.

Fig. 7. Transient dynamics analysis of total deformation

From the relevant digital information in Fig. 7, it can be seen that the position of the largest deformation is on the front line of the flip plate. The maximum deformation is 0.006 m. So the deformation is small, which has little influence on the efficiency of roadblock machine intercepting the target vehicle. The total deformation mainly occurs in the upper part of the machine and there is almost no deformation in the lower part of the machine. It can be seen that as the foundation of the roadblock machine the lower part of the machine has very little deformation and the stability is very satisfactory. For the upper part of the machine which mainly bears the impact load, deformation is not more than 0.006 m, the deformation of the arc surface of the flip plate is 0.005 m and the maximum deformation of the link mechanism is 0.001 m, which is within the acceptable range.

Figure 8 shows the equivalent stress cloud map. In order to better observe the stress distribution, the parts with small partial stress are hidden. It can be seen that the maximum equivalent stress is placed on the flip plate from the Fig. 8. In the position where the rotating center of the flip plate is in contact with rod of the center of rotation, the stress is the greatest, which is 1019 Mpa. Due to the ultimate stress of Q235 steel is 235 Mpa, apparently, the position is no longer working. At the same time, the position of the connection between the flip plate and the link mechanism also produces the stress value which exceeds the ultimate stress of the structure steel. The minimum stress occurs on the rod at the rotating center of the flip plate, which is 78.5 pa. As one of the important parts to ensure roadblock machine can work normally, the equivalent

Fig. 8. Equivalent stress cloud map of transient dynamics analysis

stress of the link mechanism is less than the ultimate stress, which is about 0.5 times of
the ultimate stress.

Figures 9 and 10 are curves of maximum Z-axis displacement with varying time
and maximum equivalent stress with varying time. It can be seen that the displacement
and equivalent stress reach the maximum point at 0.2 s, the maximum displacement is
0.006 m and the maximum stress is 1019 MPa. In other words, when the roadblock
machine is subjected to impact load, the deformation and stress of the machine will
change in a very short time.

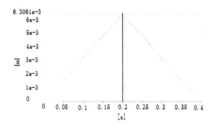

Fig. 9. Z axis displacement-time relationship curve

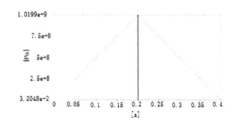

Fig. 10. Equivalent stress-time relationship curve

5   Conclusion

In this paper, on the basis of modal analysis of electric flip plate roadblock machine, the
transient dynamics analysis about vehicle under impact condition is carried out. The
main conclusions are as follows:

5.1   The Reasonable Design Method

Besides statics analysis and modal analysis of the structure, transient dynamics analysis
of the electric flip plate roadblock machine about vehicle under impact condition is also
carried out. The design process is more in line with the actual working conditions and
requirements, the design method and analysis means are more reasonable and comprehensive and the design results are more reliable and safe.

5.2   Complex Research Conditions

In this paper, compared with the three working conditions of the electric flip plate roadblock machine which are closed state, the lifting process and the open state, the impact condition of vehicle and the force under this condition are more complicated. At the same time, the working condition is also the most important. The transient dynamics analysis of this condition is more necessary.

References

1. Yu, G.Z.: The finite element analysis and research on topology optimization technology vehicle frame based on ANSYS workbench. Graduate School of Harbin University of Technology, Shenzhen (2015)
2. Yu, Y.J.: Current situation and analysis of ANSYS workbench. Mech. Electr. Eng. Technol. (03), 70–76 (2014)
3. Zhu, X.P., Zhang, J.P., Chen, L.J., et al.: Research on lightweight design and optimization of a light duty truck frame based on ANSYS workbench. J. Qingdao Univ. (Eng. Technol. Ed.) (03), 70–76 (2014)
4. Han, X.D., Wang, G.P., Yang, C., et al.: Optimal design of transformer tank based on ANSYS workbench. High Volt. Appar. (08), 110–114 (2014)

Structural Design of Precise Flow Valve Driven by Giant Magnetostrictive Material

Zhibiao Li$^{(\boxtimes)}$, Bin Xu, Hongmin Wen, and Qinghua Cao

Nanchang Institute of Technology, Nanchang, China
123284289@qq.com

Abstract. In this paper is proposed and designed a kind of precise flow valve which adopts giant magnetostrictive material (GMM) as its actuator. The valve drives the spool movement through the GMM rod's scalability and deformation under the magnetic field and effectively improves the displacement output by adopting a hydraulic proportional magnifying structure. The opening of the flow valve mouth is then controlled, realizing a precise control on the flow. After design and calculation, where the structure parameters of the valve are determined, a prototype is manufactured and an experimental platform is built as well. Tests prove that the flow control precision of the GMM precise valve designed in this paper reaches 18 mL/min.

Keywords: Precise flow valve · GMM · Optimized design

1 Introduction

With the development of technologies in the fields such as fluid control, precise machinery and control, fine chemicals, genetic gene and biotechnology, there arises the needs to precisely control the micro flow of various components. Even higher demands are set for researches in micro fluids controlling technology [1, 2].

In order to realize a precise control on the flow, a type of GMM precise flow valve is introduced in this paper. A hydraulic proportional amplifying structure is adopted to the valve to improve the displacement output of the actuator. The valve discussed in the paper is featured by high precision, fast responding speed and large flow controlled.

2 Design of Precise Flow Valve Structure

In order to transfer larger GMM output displacement which is small due to the limitation of magnetic drive size, an amplifying mechanism must be put inside the valve. This hydraulic amplifier, as it shows in Fig. 1, simple in structure, owns a large displacement amplifying coefficient. When the GMM rod pushes the piston to move

© Springer Nature Switzerland AG 2019
Y. Tang et al. (Eds.): HCC 2018, LNCS 11354, pp. 666–671, 2019.
https://doi.org/10.1007/978-3-030-15127-0_66

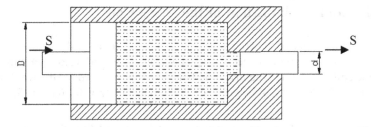

Fig. 1. Hydraulic displacement amplifying mechanism

forward with a displacement S_1, piston 2 will move with displacement S_2. Suppose hydraulic oil is incompressible, from continuity principle is resulted:

$$\pi d^2 S_1/4 = \pi D^2 S_2/4 \tag{1}$$

The structure of GMM valve, shown in Fig. 2, is designed by adding the hydraulic displacement amplifying mechanism. With GMM rod as the actuator, when the current DC is input to the drive coil, GMM rod pushes the big piston (8) forward S1 while hydraulic oil drives small piston (6) to move S2. Constants D and d represent respectively the diameter of the big piston and the small one. According to Formula 1, the displacement of spool (4) can be effectively amplified. In this paper, the diameter of the large piston and the small piston is 50 mm and 10 mm respectively, so the displacement amplification coefficient is 25.

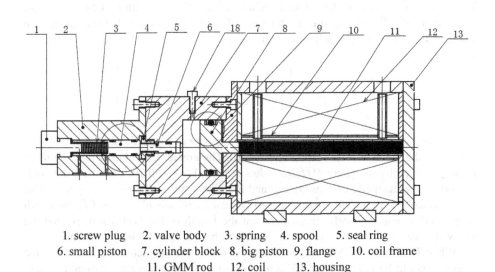

1. screw plug   2. valve body   3. spring   4. spool   5. seal ring
6. small piston   7. cylinder block   8. big piston   9. flange   10. coil frame
11. GMM rod   12. coil   13. housing

Fig. 2. Structure of GMM valve

3   Design and Optimization of the Drive Coil

The giant magnetostrictive material works under the magnetic field excitation provided by drive coil. The design of the geometric parameters of the drive coil not only matters whether the coil can provide the magnetic field intensity required for the work, but also has important effects on the efficiency of electromagnetic transformation [3, 4].

3.1   Determination of Internal Diameter and Length of the Dive Coil

The main objective of the design of the drive coil is to load the generated magnetic field to the GMM rod to the maximum extent. The axis of the hollow spiral coil gains the largest magnetic field intensity. Therefore, the inner diameter of the drive coil should be put near the GMM rod to the utmost [5].

The diameter of GMM rod is set to 10 mm. From the point of reducing magnetic flow leakage and improving the utilization of excitation magnetic field, the diameter and thickness of the central hole of the coil frame are reduced as far as possible in the design. Under the premise of ensuring machining precision, the inner diameter of the drive coil is set to 19 mm. The length-diameter ratio of the drive coil β is closely related to the distribution of the excitation magnetic field on the axis of the hollow coil. Length of the more optimum drive coil is determined through analyzing the influence of length-diameter ratio on magnetic field distribution.

The value of H/H_{max} ranges from 0 to 1, which can be calculated from *Formula* 2, indicating the ratio of magnetic field intensity at a certain point to maximum magnetic field intensity, also representing the relative intensity of magnetic field at that point. The value of δ also ranges from 0 to 1, indicating the relative position of a certain point of the coil axis. β, the length-diameter ratio of the coil, indicates that the greater the value is, the more elongated the coil will be.

$$\frac{H}{H_{max}} = \sqrt{\frac{1}{\beta^2} + 1} \left[\frac{\delta}{\sqrt{\frac{1}{\beta^2} + 4\delta^2}} + \frac{(1 - \delta)}{\sqrt{\frac{1}{\beta^2} + 4(1 - \delta)^2}} \right] \tag{2}$$

Figure 3 shows the distribution of relative intensity of magnetic field along the coil axis when the value of β is set to 1, 5, 8 and 10. The greater the value of β is, the more areas of the coil axis where H/H_{max} value exceeds 90% will be, and the better uniformity of the magnetic field as well as and the higher utilization of the magnetic field will be achieved. Therefore, on the basis of choosing a reasonable value of β, the length of the coil should generally be greater than the length of the GMM rod, so that the GMM rod can work in the area of the coil with high magnetic field intensity.

When the length of GMM rod is 150 mm and the inner diameter of coil 19 mm, the radius r should be greater than 9.5 mm. If $\beta = l_c/2r$, the coil value β should be less than 8.95 in the paper. According to Fig. 3, the greater the value of β is, the more uniform the magnetic field distribution on the coil axis will be. Referring to the $\beta = 8$ curve, it can be seen that when the coil length l_c is 1.05 times of the GMM rod

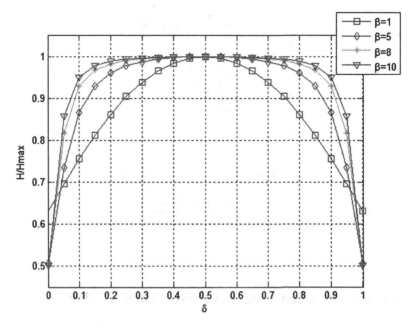

Fig. 3. Distribution of magnetic field intensity on the coil axis

length l, the GMM rod can be located in the region with high and uniform magnetic field intensity. The value of the length of the coil l_c is therefore set as 158 mm.

3.2 Determination and Optimization of the Outer Diameter of the Drive Coil

Enamelled wires which are arranged closely on the coil frame, shown in Fig. 4, form a solenoid coil, with r_1 as its inner radius, r_2 the outer radius and r_W the cross-section diameter of the enamelled wires.

Suppose $\alpha = r_2/r_1$, $\beta = l_c/2r_1$, where α and β are the form factors of the coil. The determination and optimization of the outer diameter of the drive coil is ultimately attributed to the selection of a better α value, which could make the magnetic field intensity H the maximum while making the coil loss power the minimum. It should also meet current density requirements for long-term operation and the volume and weight limit requirements for GMM valve. In the paper, the outer diameter of drive coil is ultimately decided as 69 mm and the value of α 3.63. When magnetic field intensity reaches the maximum, the maximum current density inside the coil is 2.86 A/mm^2, which meets the requirements for long-term operation.

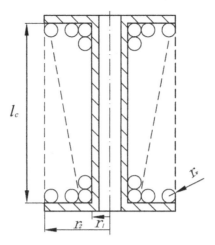

Fig. 4. Diagram of coil frame structure

4 Prototype Manufacturing and Performance Testing

The testing prototype, shown in Fig. 5, is produced by machining and assembling.

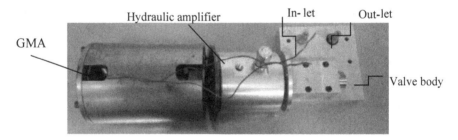

Fig. 5. The fabricated GMM flow valve

The prototype is used to produce a test device which can test the performance of the valve. Test and device are shown in Fig. 6. A 10 ml-scale measuring cup and a high precision timer are used in the measuring to record the time used for the fluid reaching the scale of 10 ml on the measuring cup under each loading current, and then the corresponding flow against each current is figured out. During the measuring, because the coil will be heated, the initial flow should not be too slow or too fast, best being controlled within 1–2 min. Meanwhile, the current cannot be loaded too high and the highest current in this experiment is set as 3 A.

In the experiment, when the hydraulic oil flows out of the orifice of the pipe under different current inputs, the volume of the oil in the measuring cup is recorded after 10 min waiting, so as to calculate the flow velocity indirectly. The experiment shows that the maximum flow rate is 8.30 L/min and the current is controlled between 0–3 A.

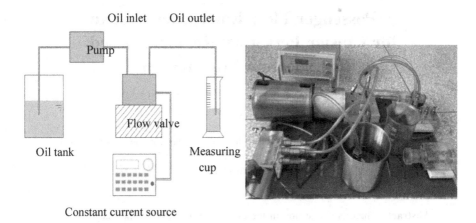

Fig. 6. Diagram of performance testing device for precision valve & picture of the device

The flow valve model has good linearity and is easy to operate. When the current variation is 1 mA, the flow variation is 18 mL/min.

Acknowledgements. The author would like to acknowledge the financial support for this research from Natural Science Foundation of Jiangxi province (No. GJJ11628, No. GJJ170986 and No. 20151BBE50044).

References

1. Brant, M.D., Wang, N.: Audio rang dynamic models and controllability of linear motion of Terfenol-D actuators. J. Intell. Mater. Syst. Struct. 345–351 (1944)
2. Lu, Q.: Research and Application of Micro Actuation Based on GMM. Wuhan University of Technology, Wuhan (2007)
3. Shrout, T.R., Zhang, S., Eitel, R.: High performance, high temperature perovskite piezoelectrics. In: IEEE International UFFC Joint Conference, Montreal (2004)
4. Chen, H., Wang, X., Zhang, X., Tan, C., Zhang, Y.: Design and modeling of precise flux valve by piezoelectric ceramics. J. Zhejiang Univ. **11**, 016 (2008)
5. Chen, P.: Design and Study of Flow Control Valve Based on GMA. Wuhan University of Technology, Wuhan (2011)
6. Zhao, Y.: Design Theory and Experiments Study of Giant Magnetostrictive Pumps. Wuhan University of Technology, Wuhan (2013)

A Passenger Flow Emergency System for Cluster Personnel Regional Based on Location Information

Binjie Xiao[✉]

Urban Construction Design Group of Shanghai Tunnel Engineering Co. LTD,
Shanghai, China
Binjiexiao@163.com

Abstract. There is congestion in the evacuation process for the places where people cluster, which will delay the evacuation time. The aggregation of personnel during the process cannot be known, nor can the remaining personnel be rescued timely. For the needs of emergency evacuation of the passenger in this site, a passenger flow emergency system based on location information is proposed, and the personnel monitoring in evacuation scenes is realized by detecting the information equipment (including wireless communication module). The functions of real-time evacuation guidance information system, information feedback and evacuation guidance information updating based on location information, artificial interpretation are based on information source as interactive tool. With the help of the information equipment, the release of evacuation information and the monitoring of evacuees can be realized, which is helpful to improve the efficiency and flexibility of evacuation and avoid evacuation congestion.

Keywords: Passenger flow cluster place · Information source detection · Emergency evacuation

1 Requirements for Emergency Evacuation System in a Crowded Field

There are more and more occasions for people clustering in cities, and there is a need for security risk prevention and control and emergency passenger evacuation in urban public places. Large passenger flow are unfamiliar to environment of urban rail transit and super-large underground space. In case of emergency, timely and effective evacuation is the key to space safety management. Intelligent perception, dynamic identification, timely warning are main evacuation rescue technology of information [1–3]. How to make use of the information perceived by the individual or the request sent at the emergency evacuation site is ignored to realize accurate early warning and improve the efficiency of emergency response.

The movement and dispersion of people is detected based on manual inspection or viewing video images [4]. The normal method is to guide the personnel to leave the scene by the evacuation sign and the evacuation light, which is more effective for

© Springer Nature Switzerland AG 2019
Y. Tang et al. (Eds.): HCC 2018, LNCS 11354, pp. 672–678, 2019.
https://doi.org/10.1007/978-3-030-15127-0_67

occasions of small space. But for indoor large space or outdoor area, evacuation exit and path in emergency have a variety of options. To choose optimal path in this situation cannot be solved by static evacuation indicator signs or light alone.

Congestion may be formed during evacuation of the cluster sites [5]. The system cannot gain the aggregation of personnel during the process, or the remaining personnel be rescued in a timely manner, which delay the evacuation time. Emergency evacuation needs to obtain emergency information immediately. At the same time, release emergency evacuation information is released in time according to the site emergency plan and evacuation path. Evacuation information is monitored and updated to improve the efficiency and speed of evacuation.

2 Guidance of Emergency Evacuation Information Based on Location Information

For the above emergency evacuation requirements, several information source detectors are set up in the emergency evacuation information guidance system based on location information.

The information source equipment (such as mobile network, Bluetooth, Zigbee and other modules) is detected by information source detectors. In the event of an emergency, evacuation information is sent to the information equipment based on the characteristics of evacuation in the area where the personnel cluster is located (such as subway and tunnel). For the evacuation process in the above personnel cluster area, the data collected can be used for monitoring, and the remaining personnel can be rescued.

The information source detectors are uniquely encoded by the cluster personnel themselves. In case of an emergency in a cluster area, the cluster personnel within the monitoring area shall be issued with the emergency information according to the collected detection information containing the location information to be encoded. In the process of evacuation, the position information of the information source is monitored. If it is found that the evacuation channel is crowded, it shall be adjusted and sent again. The evacuation site shall be checked for inspection, and the crowding condition of evacuation key channels is monitored, and induction information is issued to guide to other convenient exits and clear the crowded area. By setting up the information source detector, the personnel monitoring in evacuation scenes, the real-time evacuation guidance information push, information feedback and evacuation guidance information update based on location information can be realized.

3 Passenger Flow Emergency Evacuation System Based on Location Information

3.1 System Composition

As shown in Fig. 1, the emergency evacuation system based on location information includes: information source, information source detector, wireless communication unit, emergency evacuation management host, information access unit, camera, memory unit, evacuation guidance information sending server, broadcasting system, GIS server, clock system, upper level access unit, etc. Among them, each component equipment of the system includes necessary communication interfaces, which can realize data transmission and interaction by accessing the established communication network system in the monitoring area.

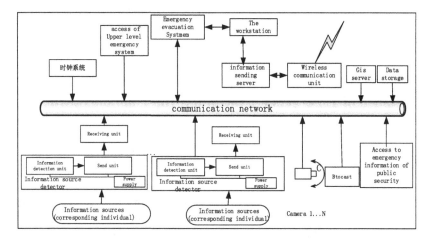

Fig. 1. Passenger flow emergency system

The information source detector is connected to the emergency evacuation management host through a detection receiving unit. Each source detector is distributed in different regions and uniquely coded. The collected detection information contains the location information that should be coded. An Omni-directional camera is set at the position of each information source detector to collect and transmit video information to the emergency evacuation management host, which serves as the source and reference information of manual intervention evacuation information.

The emergency evacuation system for cluster personnel also includes the clock system used to calculate the monitoring period and the communication interface of the upper level emergency management system to realize real-time evacuation and guidance of the surrounding area of the monitoring area. The emergency evacuation system also includes a GIS server which is connected to the emergency evacuation management host through the trunk network system. The broadcasting equipment or LED display is arranged near the information detector, and the code used corresponds to the information source detector.

Through the detection of different information sources and evacuation information push based on location information, efficient information sharing can be realized in the emergency of personnel cluster region. The evacuation information guidance and transmission based on location information and evacuation path planning is constructed, which effectively improves the evacuation effect and evacuation ability in emergency event handling through real-time interaction of information.

3.2 System Solutions

When an emergency occurs in a cluster area, the system's emergency evacuation management host and workstation determine the release mechanism of evacuation guidance information according to the emergency plan. The emergency evacuation management host realizes the functions of information processing, analysis and emergency management, and can monitor the crowding situation of key evacuation channels according to the feedback information. With GIS information, the location-based guidance information is timely sent to the information source carried by the evacuees, and send updated targeted evacuation guidance information, as well as timely and rapid guidance to other convenient exits, so as to efficiently clear crowded areas.

Process of Emergency Evacuation Method
As shown in Fig. 2, the method of emergency evacuation based on location information guidance information push includes the following steps:

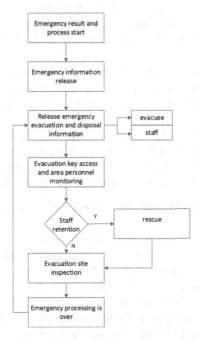

Fig. 2. System diagram of the system for emergency evacuation

S1. Several information source detectors are set in each monitoring area, and each information source detector is uniquely coded. According to the collected detection information, emergency information is issued to cluster personnel in the monitoring area in the following ways:

S11. The evacuation information sending server pushes the emergency information to the information source of the cluster personnel;

S12. Disseminate emergency information to the cluster personnel through broadcast equipment and/or LED displays;

S2. In the following ways, the group personnel in different monitoring areas are sent real-time information, including the location of the exit, the distance from the exit and the recommended evacuation path. The evacuation route is determined according to the route information of emergency plan, or the field information and GIS information collected;

S21. During the evacuation process, the information source detector feedback information to the emergency evacuation management host to confirm the change of the location information of the information source subject, and get the information of the density distribution of human flow and walking speed in the monitored area;

S22. The evacuation guidance information sending server pushes the information to the information source;

S23. Direct transmission information to the cluster personnel via the broadcast device and/or LED display;

S3. If it is found that the evacuation channels in the monitoring area and the places prone to congestion are crowded, the corresponding evacuation guidance information shall be adjusted again, and the adjusted evacuation guidance information shall be sent in step S2;

S4. For the long time stagnant personnel monitored in the monitoring area, they will continue to pay attention and send the information as per steps S21 ~ S23;

S5. The evacuation site is needed for inspection and rescue personnel are needed in the evacuation site.

The evacuation guidance information sending server is used to combine the location information distributed in different regions and send guidance information based on the location information. The event information, evacuation path and evacuation requirements can be sent to information source and information screen.

Emergency Evacuation System of Passenger Flow Based on Location Information

The method of emergency evacuation of cluster personnel based on location information is described in the exhibition center of a cluster area. There are more than 6 evacuation exits in the complex interior of the exhibition center, and the evacuation routes vary with the division of the exhibition.

The emergency evacuation in the exhibition center is divided into the first quadrant to the fourth quadrant along the main route of evacuation. When the emergency event occurs in the third quadrant, the adjacent personnel of the area should be evacuated first. The specific evacuation path can be determined according to the conventional venue emergency plan, information source collected by information source detector and GIS information collected by GIS server.

If a fire breaks out in an area, it is necessary to evacuate all people in an emergency. The methods of emergency evacuation for cluster personnel are as follows:

In the first step, the information source detector, broadcasting equipment and LED display screen shown in Fig. 1 are arranged in advance in the cluster area and key entrances. The system information processing unit is set up in the monitoring center. In case of emergency, emergency evacuation of the site is required. Issue event information and evacuation instructions according to the main information system in Fig. 1.

Each source detector is distributed in different monitoring areas and coded uniquely. Therefore, the detection information collected by the information source detector contains the position information that should be coded. The information source information detected by each information source detector includes: The MAC address information corresponding to the information source device is used to confirm the change of the location information of the information source, such as the density distribution of human flow and walking speed in the monitored area. In this way, the emergency evacuation management host will push the targeted evacuation information to the information source equipment of people in different regions according to the density and number of personnel clusters in different regions, combined with the basic venue information (such as the location of entrance and exit, distance, recommended evacuation path, etc.). By means of this way, rapid evacuation to reduce the evacuation conflict.

Step 2: when an emergency occurs, the emergency host will release the emergency information and evacuation guidance information. Among them, emergency information and evacuation guidance information are pushed to corresponding information source devices (such as mobile phone, pad and other information terminals).

According to the number of personnel in different areas monitored by the monitoring equipment in Fig. 1, the specific evacuation information is pushed to the information equipment of personnel in different areas based on the basic information of this place(such as the location of entrance and exit, distance, recommended evacuation path, etc.)

During the whole process each equipment in Fig. 1 to obtain the information of evacuees, and evacuation instructions or induction information occurs. Rescue is carried out according to the monitored legacy evacuees, and complete the whole evacuation in accordance with the evacuation procedure in Fig. 2.

Step 3: The information of personnel position and movement speed is detected by the information source detector in Fig. 1. If there is a congestion tendency in the evacuation passage, adjust the corresponding path guidance information; evacuation guidance information is pushed and updated based on information source detection information and location information for personnel who stay in a certain area for a long time, and promote or help rescue;

Step 4: Repeat the above evacuation guidance information push. The whole process all the equipment of the system in Fig. 2 to obtain the evacuation personnel's information, and to send evacuation instructions or guides information based on the location information and GIS information of the information source detector.

4 Conclusion

Personnel monitoring in evacuation scenarios is realized by taking the information source carried by personnel themselves as the information source. For event information and evacuation information, real-time evacuation guidance information, information feedback and evacuation guidance information updating based on location information, evacuation of legacy or lingering personnel, artificial interpretation and participation, etc. In the process of evacuation, the evacuation route and state of the evacuees can be monitored. In case of congestion in evacuation routes, timely adjustment and resolution can be made by information interaction. With the help of the information equipment carried by the evacuees and the attached communication module, the release of evacuation information and the monitoring of evacuees can be realized, which is helpful to improve the efficiency and flexibility of evacuation and avoid the occurrence of evacuation congestion. The evacuation effect and evacuation ability in emergency event handling are effectively improved by real-time information interaction.

References

1. Hoogendoorn, S.P., Wageningen-Kessels, F.V., Daamen, W., Duives, D.C., Sarvi, M.: Continuum theory for pedestrian traffic flow: local route choice modelling and its implications. Transp. Res. Part C **59**, 183–197 (2015)
2. Zhao, W.: Research on emergency evacuation guidance for personnel in public places. Saf. Prod. Sci. Technol. China **12**(9), 164–170 (2016)
3. Hughes, R.L.: A continuum theory for the flow of pedestrians. Transp. Res. Part B **36**(6), 507–535 (2002)
4. Li, J., Liu, M., Ran, L.: Study on individual risk of crowd gathering in public places. J. Saf. Environ. **6**(5), 112–115 (2006)
5. Fang, H.: Study on spatial evacuation vulnerability under urban emergency. J. People's Publ. Secur. Univ. China (Nat. Sci. Ed.) **17**(3), 65–70 (2011)

Research on Cluster Personnel Monitoring and Early Warning Method Based on Information Source Detection

Binjie Xiao$^{(\boxtimes)}$

Urban Construction Design Group of Shanghai Tunnel Engineering Co. LTD,
Shanghai, China
Binjiexiao@163.com

Abstract. For cluster of people monitoring in the city and early warning requirements, a method of personnel monitoring in passenger cluster area is proposed. The confirmation of emergencies reduces damage caused by delay event enlargement of cluster based on information detection personnel monitoring and early warning method and system. This method based on information source monitoring and early warning is adopted to collect the data information of cluster personnel. It mainly relies on the information source with unique identification code carried by the personnel to obtain the information such as personnel aggregation, evacuation and activity path in the region. The evacuation effect and emergency response capability can be improved through information interaction.

Keywords: Information source monitoring · Cluster personnel occasions ·
Personnel monitoring and early warning

1 Risk Analysis for Personnel Clustering

With the advancement of urbanization, there are more gathering places and large-scale activities in the city, including sports events, music festivals, art exhibitions. Large parks and exhibition centers are also built in the city, such as Shanghai Disney resort and Shanghai national exhibition center. People are crowded in these regional, which are of all ages, and the behavior and psychology for emergencies. It is of great significance to know the emergency event at the first time and inform people in the region of the optimal evacuation path, so as to alleviate the consequences of the emergency, avoid or reduce casualties, especially avoid secondary injuries caused by human factors.

For the detection method of emergency, the conventional method is to detect the abnormality through the video monitoring system by manual inspection or video analysis method. These methods rely on a lot of manpower; the analysis results are not accurate. In addition, it is difficult to set up cameras in large public areas. Large exhibition venues and other facilities after the exhibition caused obstruction, the complex path also affected evacuation [1–5].

© Springer Nature Switzerland AG 2019
Y. Tang et al. (Eds.): HCC 2018, LNCS 11354, pp. 679–685, 2019.
https://doi.org/10.1007/978-3-030-15127-0_68

In view of the shortcomings of the existing technology, a method to monitor the passenger flow in the cluster region by information detection is proposed. The emergency should be confirmed for the first time which reduces incident hazards caused by delay, and warning of the situation should be sent out based on cluster personnel monitoring.

2 Cluster Personnel Monitoring Based on Information Source Detection

In the personnel cluster region, a detector is set up to detect the information source (such as mobile network, Bluetooth, ZigBee and other modules), and evacuation information is sent to the information equipment in the event of an emergency. The evacuation effect and emergency handling capacity are improved through information interaction. The cluster personnel monitoring and risk warning system based on information source detection collects the information of cluster personnel, and obtains the information of personnel aggregation, evacuation and activity path in the region according to the information source with unique identification code carried by the personnel. The collected information is analyzed and compared with normal operation data (such as passenger flow density, number of personnel per square meter, number of passenger flow in main evacuation channels and travel speed, etc.) to make specific analysis and detection of abnormal events to realize passenger flow monitoring and warning. Based on the detection results of personnel distribution, the sudden change or aggregation of personnel in the area of the cluster can realize the detection of emergencies, and the emergency detection results can be issued to facilitate the interaction with the information terminal.

The cluster personnel monitoring system based on information source detection is shown in Fig. 1, which is composed by the information source, the source detector

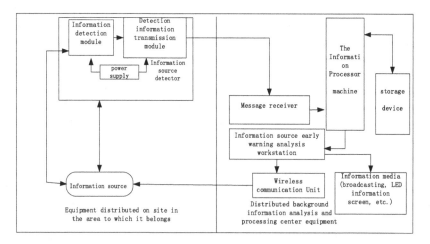

Fig. 1. Information source monitoring system composition diagram

host, wireless communication unit, emergency management, information access unit, the video camera, communication network, storage unit, induced evacuation information server, clock system, emergency management system access unit, broadcasting system, GIS server etc.

The system detects the information source of the adjacent area through the information source detector arranged in the key area. Each information source corresponds to the information source device of the unique identification code of the information source carried by a single individual in the region. The collected information is upload, summarized and analyzed. The collected detection indexes were analyzed and compared with the regular data of the information source in the region. If the result exceeds the normal allowable value, it will send out the warning information of the emergency in the cluster area and carry out emergency measures. If the result is less than or equal to the normal allowable value, it is considered normal, follow the above steps to collect data within the cycle. Finally, the passenger flow in the personnel cluster area is monitored and the emergency is confirmed by real-time and accurate monitoring.

Information source analysis workstation is a local area network server, local area network server connection broadcast and LED display. Warning information output devices, such as broadcasting and LED display, can be used as auxiliary warning information release system to release corresponding information based on location.

Information source detector is composed by information detection module and detection information transmission module. The wireless communication module adopts the integrated communication scheme. It can use WI-FI, Bluetooth, 4G LTE and other wireless communication modes by using mobile communication or wireless broadband environment. The output end of the information receiver is connected to the input end of the information processing host, the output end of the information processing host is connected to the input end of the information source risk warning analysis workstation, and the output end of the information source warning analysis workstation is connected to the wireless communication module. The regional location information is given to the collected information source after the source collector is coded without location, and the information obtained by multiple sources is combined, the regional location information of the information source, identifying information (such as MAC address), time, moving speed, trajectory, density value of the information source in the region can be obtained. The acquired information source moving speed, information source density in key areas can be gained based on these acquired information source monitoring information. In combination with the alternative route of evacuation, cluster personnel monitoring and early warning analysis were conducted.

3 Warning System of Cluster Personnel Evacuation Based on Information Source Detection

3.1 System Overview

When there is an emergency, the information equipment carried by the evacuee and the attached communication module can be used to release the evacuation information and monitor the evacuee, which is helpful to improve the efficiency and flexibility of evacuation and avoid leaving people or causing congestion.

The release of event information is helpful to emergency evacuation. Monitoring the evacuation process to ensure that evacuation is rapid, orderly, and no personnel left behind. In the case of emergency, the ability and effectiveness of individuals in the crowd to obtain evacuation information will be improved by the system. The evacuation was monitored according to the updated feedback information, and the real-time and dynamic evacuation induction information was sent based on the location information, which gains effect of real-time and targeted evacuation.

3.2 System Solutions

The information collection of the personnel distribution, activity route and aggregation state in the area of the cluster is realized by detecting the information source. It also contains local area location information, so that the cluster area can make early warning analysis according to evacuation path and evacuation ability. Based on the conventional state contrast analysis and the real-time analysis of the collected information, the emergency event detection of the personnel cluster region is realized.

In the case of an emergency in a cluster area, the system's emergency evacuation management hosts determine the evacuation induction information release mechanism according to the emergency plan. By combining the location information distributed in different regions, the Guide information based on location information is released. The corresponding information is pushed to the information source equipped with mobile network, Bluetooth, ZigBee and other communication devices through the wireless communication unit, and the corresponding induction information of each key area is sent through the broadcasting system (also available with the induction screen). Event information, evacuation path and evacuation requirements are also sent to the information terminal.

In the process of evacuation, the information source detector feedback the information source information. The change of the location information of the information source subject is confirmed by the unique MAC address of the information terminal detected by the detection equipment. The details such as the density distribution of human flow and walking speed can also be obtained. According to the feedback information, the emergency evacuation host monitors the crowding situation of key evacuation channels. With the manual interpretation information of the workstation, updating targeted evacuation induction information and location-based induction information with GIS information is timely send to the information source of the evacuation, in order to timely and quickly guide to other convenient exits to efficiently clear crowded areas.

The information detection and transmission module and information receiver of information source can adopt 4G LTE for data signal transmission to ensure flexibility and convenience. In view of small data amount of detection information, GPRS or 3G or WI-FI, Nb-iot can also be adopted. The information receiver connects the detectors of each information source and is used to receive the information collected by the information source. Information warning analysis workstation has the functions of information processing, man-machine interactive program, graphics and text display, multi-aspect data display and so on.

When the event judgment is confirmed, the system can also conveniently release the type, location, event and disposal plan information of emergency event through cable or wireless communication network. The most convenient release medium is the information source in the hands of each individual. The information is released by wireless Wi-Fi network or mobile network, and relevant information can also be pushed in combination with surrounding broadcast and LED display.

For the evacuation process in the above personnel cluster area, the data collected can be used for monitoring, and the remaining personnel can be rescued. Establish communication links with the remaining personnel, and bring relief to those who are in difficulty. By detecting the information sources of different communication modes and pushing evacuation information based on location information, efficient information sharing can be realized in the emergency of personnel cluster region. An evacuation information induction and transmission method based on location information and evacuation path planning is constructed.

As shown in Fig. 2, the information of the information source within each region is collected through the following steps. Each information source corresponds to the information source device of the unique identification code of the information source carried by a single individual within the region. Information source devices include those with messaging functions: smartphones, e-readers, tablets, automotive electronics, mobile video/entertainment devices, any or any of the wireless information terminals.

(1) The information source is detected with the detection frequency every 1–3 min
(2) Upload the MAC address of the detected information source device, and encrypt the data during the upload process
(3) The corresponding coding is carried out according to the position of the information source detector, which gives the position information of the monitoring area to the information source collected, and the area and path of the device movement of the information source within the inspection and detection cycle.
(4) The collected information is uploaded, summarized and analyzed
(5) The collected including: gather the number per unit area, personnel, travel speed, the overall number of staff, general staff and the detection index of distribution of data analysis. The results and the regularity of information sources in the area of data which belong to the density of traffic congestion degree are analyzed. If the comparison result is beyond the allowable values of conventional, into step (4), if the comparison result is less than or equal to the normal value, enter the step (5)
(6) Risk warning information of emergency events in the personnel cluster region is sent, and emergency measures from the following aspects are carried out.

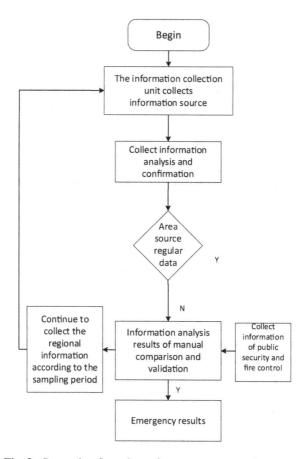

Fig. 2. Processing flow chart of emergency evacuation method

(7) The types of emergency events, places, events and information disposal plan, issued via cable or wireless communications network, distribution medium for each separate individual carries information source, broadcasting equipment can also be combined with the respective area or LED display information push;

(8) Push the fire alarm information detected by the auxiliary alarm equipment within the region synchronously;

(9) As normal, continue to follow the above steps (1–4) for data collection within a period.

4   Conclusion

A cluster personnel monitoring and early warning method is proposed. The real-time guide information of evacuation is pushed in evacuation scenes based on location information, which update induced information of evacuation by using personnel monitoring information feedback. The left or lingering persons are able to be rescued by artificial interpretation and participation. In the process of evacuation, the evacuation route and state of the evacuees can be monitored by information detectors. In case of congestion in evacuation routes, timely adjustment and resolution can be made through information interaction. By using this method, it is beneficial to obtain emergency evacuation information and improve service level in the case of personnel clustering, and accelerate the emergency evacuation of large passenger flow.

References

1. Zheng, X., Xiang, L., Jin, L.: The modified model and simulation of evacuation speed in metro station considering human-environment factors. J. Chin. Saf. Sci. **27**(8), 19–24 (2017)
2. Li, J., Liu, M., Ran, L.: Study on individual risk of crowd gathering in public places. J. Saf. Environ. **6**(5), 112–115 (2006)
3. Hoogendoorn, S.P., Wageningen-Kessels, F.V., Daamen, W., Duives, D.C., Sarvi, M.: Continuum theory for pedestrian traffic flow: Local route choice modelling and its implications. Transp. Res. Part C **59**, 183–197 (2015)
4. Zhao, W.: Research on emergency evacuation guidance for personnel in public places. Saf. Prod. Sci. Technol. China **12**(9), 164–170 (2016)
5. Fang, H.: Study on spatial evacuation vulnerability under urban emergency. J. People's Publ. Secur. Univ. China (Nat. Sci. Ed.) **17**(3), 65–70 (2011)

Author Index

Printed in the United States
By Bookmasters